Advances in Industrial Control

Advances in Industrial Control is a series of monographs and contributed titles focusing on the applications of advanced and novel control methods within applied settings. This series has worldwide distribution to engineers, researchers and libraries.

The series promotes the exchange of information between academia and industry, to which end the books all demonstrate some theoretical aspect of an advanced or new control method and show how it can be applied either in a pilot plant or in some real industrial situation. The books are distinguished by the combination of the type of theory used and the type of application exemplified. Note that "industrial" here has a very broad interpretation; it applies not merely to the processes employed in industrial plants but to systems such as avionics and automotive brakes and drivetrain. This series complements the theoretical and more mathematical approach of Communications and Control Engineering.

Indexed by SCOPUS and Engineering Index.

Proposals for this series, composed of a proposal form (please ask the in-house editor below), a draft Contents, at least two sample chapters and an author cv (with a synopsis of the whole project, if possible) can be submitted to either of the:

Series Editor

Professor **Michael J. Grimble**
Department of Electronic and Electrical Engineering, Royal College Building, 204 George Street, Glasgow G1 1XW, United Kingdom
e-mail: m.j.grimble@strath.ac.uk

or the

In-house Editor

Mr. **Oliver Jackson**
Springer London, 4 Crinan Street, London, N1 9XW, United Kingdom
e-mail: oliver.jackson@springer.com

Proposals are peer-reviewed.

Publishing Ethics

Researchers should conduct their research from research proposal to publication in line with best practices and codes of conduct of relevant professional bodies and/or national and international regulatory bodies. For more details on individual ethics matters please see: https://www.springer.com/gp/authors-editors/journal-author/journal-author-helpdesk/publishing-ethics/14214

More information about this series at https://link.springer.com/bookseries/1412

Steve S. Niu · Deyun Xiao

Process Control

Engineering Analyses and Best Practices

Springer

Steve S. Niu
Houston, TX, USA

Deyun Xiao
Department of Automation
Tsinghua University
Beijing, China

ISSN 1430-9491 ISSN 2193-1577 (electronic)
Advances in Industrial Control
ISBN 978-3-030-97066-6 ISBN 978-3-030-97067-3 (eBook)
https://doi.org/10.1007/978-3-030-97067-3

This Springer imprint is published by the registered company Springer Nature Switzerland AG
The registered company address is: Gewerbestrasse 11, 6330 Cham, Switzerland

To my wife, Min, for her continuous support;
Also to my former colleagues, who inspired
me to write this book.

—Steve S. Niu

Series Editor's Foreword

Control engineering is viewed rather differently by researchers and those that must implement and maintain control systems. The former develop general algorithms with a strong underlying mathematical basis, whilst the latter have more immediate concerns over the limits of equipment, quality of control, safety and security, and plant downtime. The series *Advances in Industrial Control* attempts to bridge this divide and hopes to encourage the adoption of advanced control techniques when they are likely to be beneficial.

The rapid development of new control theory and technology has an impact on all areas of engineering and applications. This monograph series has a focus on applications, which are now so important as the rate of new technological development continues to increase as rapidly as it does. It is the challenges from industry that often stimulate the development of new control paradigms. A focus on applications is also desirable if the different aspects of the "control design" problem are to be explored with the same dedication that "control synthesis" problems already receive. The series provides an opportunity for researchers to present new work on industrial control and applications. It raises awareness of the substantial benefits that advanced control can provide whilst tempering its enthusiasm with a discussion of the challenges that can arise.

This monograph is concerned with engineering analyses and best practice in process control systems. The authors' applications and industrial experience results in a text that is written in an entertaining style that is rather different from the more usual modelling-and-design-methods-focused texts. The questions raised and answered in Chap. 1, for example, will have been thought about by many engineers, but clear answers are not often apparent. The authors' industrial experience is evident from even this first chapter.

Chapter 2 covers some standard ideas in control, but all the problems are considered from a process control viewpoint. Moreover, some of the things—like ratio control and split-range control—that are common in process control but not used so much elsewhere are described. Some of the main ideas in advanced control are also mentioned briefly. The text has a useful role since topics like training simulators, alarm handling and life cycles may not be covered in more theoretical texts.

Chapter 3 on PID control, classed as "Essential Knowledge", has some familiar material but the explanations and diagrams make it very accessible. Moreover, the practical aspects of the use of PID are explored including the uncertainties that affect performance. An example of the practical utility of the text lies in the sections on crippled operations and shut-down and start-up.

More sophisticated PID control is considered in Chap. 4, on complex PID control loops. The multivariable nature of some control problems is explored and tools like decoupling are discussed. The section on the handling of constraints using PID-based control, for a rather special situation, is unusual and interesting. The applications from process control illustrate the special nature of process problems, although some problems like the need for bumpless transfer occur frequently in other application sectors as well.

Chapter 5 is on advanced process control and not surprisingly Model Predictive Control (MPC) is discussed. Before introducing the basic concept, the essential areas of modelling, identification and estimation are covered from a rather practical and insightful viewpoint. The need to explore the topic is partly the multivariable nature of many process systems and the constraints that are often important for quality control. Other topics, like testing and commissioning, that are valuable for engineers working on process plants are covered.

Chapter 6 is the first chapter in the section concerned with analytic skills and problem solving. Much of the material is very useful for plant process control and instrumentation engineers and is not very accessible elsewhere. The methodology of process analysis discussed in Chap. 6 involves a lot of common sense rather than theory, but it is useful to have a formal presentation and summary. The chapter is a reminder that working on a process plant involves a lot more than control loop design, and engineers bear a heavy responsibility for safe and effective operation.

Chapter 7, on the methodology of process control design, covers topics like the process hierarchy and plant-wide control. The discussion is very suitable for process engineers, who are often faced with a new system designed by a major process control supplier or automation vendor, and where it is not the detailed control design questions but more how to understand and use the system provided that is of prime importance.

Troubleshooting in Chap. 8 is one of the fun things to do if accompanied by an experienced engineer with good instincts, but for the novice it is daunting when the source of the problem is elusive and the cost of the loss of production is high. Sound advice from the text is therefore valuable. For example, asking the process plant operators for their thoughts can often be revealing. Even if their knowledge of Nyquist stability conditions is non-existent, they will know if particular sensors frequently malfunction.

Chapter 9, on control loop tuning and improvement, will be one of the most popular, dealing with the ubiquitous PID controllers. The explanations cover some standard tuning procedures. It is also evident from the examples that understanding the physical process is important if good results are to be obtained.

Chapter 10, on common calculations, deals with some typical calculations that are needed, covering for example the inferential nature of many process control problems, or the need to determine the relationships between variables and to establish the quality of data. The companion Chap. 11 is on equipment control and is more concerned with the control problems for common plant equipment, such as compressors. Both chapters have very relevant process plant control examples.

The final Chapter, 12, involves one of the most important topics, namely, plant-wide control and unit control. It is useful that both a bad design and a good design are described. In the days of cheap energy and few concerns about the environment any design that was safe and reliable might have been good enough. This is no longer the case, and it is good to be reminded that simple changes can often make a big difference if the process and control problem is well understood.

The authors have industrial experience and this text covers topics in process control very much from a plant-control-engineering perspective. The monograph should therefore be a help to engineers faced with real problems in the process industries and is a welcome addition to the series.

Glasgow, UK
September 2021

Michael J. Grimble

Foreword by Sirish L. Shah

Since the advent of real-time computing, the field of process control has advanced significantly over the last several decades. Methods and techniques with much theoretical rigor are available to solve some of the most difficult problems. However, a good working knowledge of the application of these methods requires good mathematical background, typically taught only at post-graduate levels at most universities. A plethora of books are available for researchers and engineers with advanced degrees to learn about such methods.

An equally important aspect of a successful application of these advanced techniques is process or domain knowledge that is only learned in the field in the industry. Very few books are available that complement discussions of advanced control algorithms with unit process operations to guide a control engineer to a successful application. This book is an attempt to fill this synergistic gap between control theory and process applications.

The book is divided into four parts, starting with parts I and II that discuss the essentials of control theory in a very tutorial manner. Part III focuses on the methodology of process analysis and process control design. These sections are uniquely articulated in a problem-solving setting. The last section (part IV) follows these methodologies and discusses some typical control applications that require temperature, pressure, and level regulation, including the control of rotating machinery such as centrifugal pumps and compressors. This section ends with a discussion of the important topic of plant-wide control to achieve good operational performance.

The book is an outcome of two very skilled control theoreticians and practitioners with many years of advanced industrial process control applications to their credit. The book fills that infamous gap between the theory and practice of process control.

August, 2021

Sirish L. Shah, Ph.D., FCAE, FCIC, FIEEE
Emeritus Professor, University of Alberta
Toronto, ON, Canada

Foreword by Michael W. Brown

I have been very fortunate in my career to have received a graduate degree in process control. I was even more fortunate to have had the opportunity to benefit from this education in practicing process control in many industries and companies around the world. Although the academic side of process control provides a rich set of techniques and tools to understand and optimize system performance, it is the application of process control theory to the real-world applications that has helped me build the analytical skills that I consider the greatest learning over my 35-year career. As we apply these technologies to industry successfully, the skills we have built through this practice also have a very broad impact on our understanding and analysis of data and systems in general.

For those of us lucky enough to have enjoyed a long and exciting career in the field of process control, I would suggest that there have been three critical factors in this achievement: a solid education, a good mentoring and development program, and an opportunity to practice and develop the required skills. This book contributes to advancing the effort of good mentoring and development program. I was lucky to have had some of the best mentors in the industry as the author had, and this book captures most of what I also learned over the years in the field. Parts I and II of the book present the core process control material in a simple and intuitive way to help all practitioners with their development journey. The structured analytical methodologies and best practices in Part III, which are seldom found in other textbooks, are crucial to building the necessary competence and skills in practical control. The typical control examples in Part IV demonstrate the structured thinking and work process for working with real-world applications. These knowledge and skills require many years of practical experience to capture and articulate, and are beneficial to both the practitioners and researchers pursuing excellence in the area of process control.

I had the pleasure of working with Steve Niu early in his career, and his dedication and commitment to process control are always inspiring. Now capturing these many years of knowledge and experience and sharing it with others is a testament to their continued dedication to the process control field, and to the young practitioners who also aspire to develop an exciting career in process control.

Process control has blessed me with a sort of industrial passport for over 35 years, working in a dozen countries, 50+ companies, and several industrial sectors. It is a career path I would not go back and change. My mentors and their knowledge sharing were vital to this journey. We are fortunate that these authors have now captured their knowledge to share in this book. I believe this book will help many others in their exciting journey in the area of process control.

August 2021

Michael W. Brown, M.A.Sc., P.Eng.
Executive Vice-President
Technology & Innovation
SMS Equipment Inc.
Edmonton, AB, Canada

Preface

Process control is an engineering specialty vital to process safety and operational efficiency in any modern operating plant. Process control relies heavily on technological knowledge and practical experience to excel, evident from the fact that the process control team usually has the highest concentration of higher education degrees. The process control achievements are primarily through improvements in designs and innovations in operation and typically need little to no additional hardware investment or extra energy consumption.

Process control is also full of challenges. First of all, process control is a branch of automatic control. Most of the automatic control theories and techniques originated from the aerospace and electromechanical engineering fields. There is an extensive framework of control theories and a massive collection of control technologies through its long track record of success. Unfortunately, only a (small) subset is directly applicable to the process industry due to industrial processes' unique and distinct characteristics. The process control practitioners can easily be overwhelmed by the rich selection of technologies and get lost on where to spend their effort. Therefore, it is necessary to delineate the relevant part of the theory and technology applicable to process control from the vast scope of automatic control. At the same time, it is also essential to summarize the unique challenges and particular needs of the process industry and highlight the limitations of the current technologies. The purpose is to help the practitioners make the best use of the currently available technologies and set the expectations for the new technologies.

Secondly, process control is about controlling a process. There is a well-known saying: we cannot control something that we can not measure. More importantly, we can not control something that we do not understand. It is crucial to understand the underlying processes in order to control it. In practice, due to the vast number and extreme diversity of industrial processes, a process control practitioner cannot become intimately familiar with all the processes, nor is it possible to cover all of them in any process control books. The more practical approach is developing analytical skills and effective methodologies for conducting structured process analysis, extracting the necessary information for control, and developing the most appropriate control strategy. The development and standardization of these skills and methodologies are thus crucial. Many of the techniques and methodologies can be extracted and abstracted into *best practices* to be replicated.

Thirdly, there is a substantial gap between theory and application in addressing process control's unique needs. There has also been a big disconnect between what is taught in school and what is needed in practice. "The gulf between the theory taught in academia and the practice of process control is astounding. Academia is correct to teach theory and basic principles, but in process control, the theory being taught is largely irrelevant to the practice of process control." (Smith 2009) In academic research and teaching, people are more interested in how things normally work and how to advance the theory; In practice, the real challenge is knowing when things may fail and how to prevent the failure. Students are rarely taught when and how things will fail, what to do to prevent them from going wrong, and how to respond when things do go wrong. For example, to design a PID controller and make it work is absurdly simple and straightforward. To make the PID controller reliably work in a real-world environment, the designer needs to address many "what if ..." types of practical considerations. The practical challenges are rarely taught or researched in academic research, although many have excellent research values. In an operating environment, this shift from "what it does ..." to "what if it does not ..." is the primary difference between the thinking process of an inexperienced junior engineer and that of an experienced senior engineer. This shift is also a major hurdle for most practitioners that academia could have provided more help for them to pass.

Lastly, there are often confusions about the roles and responsibilities of process control. Process control is often treated as a branch of automatic control in academic teaching without giving adequate attention to the distinct requirements of process control. In the field, process control's roles and responsibilities are often assigned to process engineers during the project design phase and then to control systems engineers for implementation and maintenance. Unfortunately, most process engineers do not know the control systems to produce a proper control design. On the other hand, control systems engineers do not have enough process knowledge or operational experience to understand the design intent and operating requirements. This *disconnect* has led to sub-optimal control designs and incorrect DCS implementation. It is the primary reason that as high as half of PID controllers can not stay in the intended control mode in many operations, especially upstream. A process engineer or a control system engineer must know enough about process control to perform process control duties.

Most textbooks or reference books in process control focus on the theoretical depth and completeness, with the practical applications as a complement to enhance the understanding of the theories. This book takes an entirely different approach. It first summarizes the essential knowledge and core technologies currently in use (and in need) in the process industry. It then focuses on developing structured problem-solving skills and best-practice methodologies. The goal is to guide the practitioners and researchers on the thinking process and general procedures for working on a new and unfamiliar process flow. The skills and methodologies should be helpful to both process control engineers working in the field and academic researchers interested in the practical side of process control. For this purpose, the presentation is deliberately made simple and intuitive by using illustrations and tables as much as possible. Control theories are reduced to the minimum and are introduced only when needed. As a result, we will limit the explanation to time-domain and avoid the advanced topics such as stability analysis and state-space model since they have minimal practical applicability due to the complexities and uncertainties of process dynamics.

The book has a practical-oriented theme, and the presentation follows the general path: overview ⇒ essential knowledge ⇒ analytical skills and best-practice methodologies ⇒ unit and plant-level control applications. This sequence coincides with the learning and progression path from awareness, knowledge, skills to mastery. Along these lines, the book's content is organized into four parts.

Part I, The Big Picture, provides a general introduction to process control. This part explains process control's unique and irreplaceable role in an operating plant and answers the general *what/why/who/where/when* types of questions about this extraordinary cross-discipline engineering specialty (Chap. 1). It then summarizes what core technologies are available and what they can do, thus defining the essential knowledge framework for process control practitioners (Chap. 2).

Part II, Essential Knowledge, provides more detailed discussions on the key concepts and core technologies of process control outlined in Chap. 2. It explains the minimum and essential knowledge that a process control engineer should acquire, from basic PID-based control loop (Chap. 3), complex PID control scheme (Chap. 4), to advanced control solutions (Chap. 5). The focus is not only on how they normally function but also on making them not fail in all practical scenarios.

A common challenge facing process control design is where to start with control design or troubleshooting when faced with a new and unfamiliar process. Part III, Analytical Skills and Problem-Solving Methodologies, introduces the best practice work processes to deal with the common process control tasks. These skills and methodologies include a structured approach for conducting process analysis (Chap. 6), general guidelines and procedures for process control design (Chap. 7), best-practice methodologies for control problem troubleshooting (Chap. 8), and control performance tuning and improvements (Chap. 9).

Some typical control applications are presented in Part IV, including common control-related calculations (Chap. 10), control of typical equipment (Chap. 11), and control of typical processes (Chap. 12). These practical examples demonstrate the whole thinking process and complete analytical procedure for solving real-world problems based on the skills and methodologies presented in Part III.

This book's targeted audiences are the process control practitioners in the field and students and academic researchers interested in the practical side of process control.

General audiences such as non-engineering personnel or operation management can use Chap. 1 as an overview to understand what process control is all about. Junior control engineers and undergraduate students can start with Part I to build their knowledge framework in process control. Part I can also serve as an introduction to process control technologies for other engineering teams to raise their awareness. For example, process engineers, control system engineers, operators, and automation managers often need to work together with process control for their decision-making; thus, it is beneficial to understand what process control can offer.

From Part I to Part IV, process control engineers and academic researchers can appreciate the unique challenges that process control technologies face in a real-world environment. It progressively builds the knowledge, skills, and eventually, the process control "sense" or "instinct" to perform their duties more effectively. Experienced process control engineers or academic researchers can benefit from the process analysis and control design methodology in Part III and enhance them through the complete real-world examples in Part IV.

Process control is cross-disciplinary and interacts with process engineering, rotating equipment, operations, control systems (DCS), instrumentation, and information technology. A process control engineer cannot be skillful in all these fields in his short career. For example, there are countless processes in different industries. It is impossible (and unnecessary) to be an expert on all the processes. That is the job of process engineers and equipment engineers, who are the best source for in-depth process knowledge. For a process control engineer, the skill and methodology to efficiently extract and abstract the specific information required by process control are more valuable and practical. Similarly, sensors, transmitters, valves, PID controllers, DCS/PLC platforms, communication networks are essential tools and platforms for process control. A process control practitioner needs to have a working knowledge of these tools and infrastructure; however, it is unnecessary to become an expert on all the fast-changing tools. That is the job of DCS and instrumentation engineers or technicians, who are the go-to persons for support.

The roles and responsibilities of a process control engineer are to utilize the latest available tools to achieve higher goals, i.e., overall operational excellence. Process control engineers must have a clear picture of their knowledge structure, and know what must know, what is good to know, and what should not be wasting time to know.

"To know what you know and what you do not know, that is true knowledge." (Confucius)

Houston, USA
Beijing, China
September, 2021

Steve S. Niu
Deyun Xiao

This book was triggered and motivated by my experience transitioning from academia to industry. Personally, when I left academia to join the process industry as an advanced process control engineer, I had a B.Sc., M.Sc., and Ph.D. degree under my belt, all in process control, plus several years of research experience in top universities in the area of advanced process control. So I felt confident and well-prepared to take up the new challenges, but only to find that I just stepped into an unfamiliar world as a complete rookie. The advanced control theories gained in school found little relevance to the jobs to be performed. State-space models, Routh stability criteria, Nyquist plot, Bode Plots, LQG control, H_∞ could hardly see any use. On the other hand, practical knowledge such as the impact of control mode changes, practical PID tuning methods, and process analysis was desperately lacking.

However, I was fortunate to have a very competent and caring mentor who guided me into the real world and helped fill up the initial gap between theory and practice (of course, the academic background did help me pick up speed quickly). Later, when I started training and mentoring other young engineers, I found they were all faced with the same challenges as I did before. There was so much missing between what was taught in school and what was needed in the field, most young engineers had to learn it the hard way on their own. I felt a strong need for a book that summarized and explained the essential knowledge and core skills in one place. Unfortunately, very few people from the industry have the time and interest to write and share their practical knowledge.

It is easy to bring a junior engineer up to speed to perform regular and routine tasks; however, it takes many years of real-world experience to become a skilled engineer. The difference is in handling new challenges, abnormal conditions, and emergencies; These capabilities can only be learned from experiencing the abnormal conditions and emergencies. This learning process can be greatly accelerated through an experienced mentor. Without a mentor, the learning is by trial and error on the actual plants, and the cost of learning can be very stiff. In this increasingly cost-sensitive operating environment, a practical-oriented book like this one could be the next best thing to having a great mentor or training program.

The content design of this book is based on my experience with academic research and training/mentoring of countless young engineers. My 10-year academic research and 25-year industrial experience covered almost all aspects of process control, from research, development, to applications; from project management, product management, to project execution; Much experience was accumulated from the design and implementation of hundreds of control solutions covering base-layer control, integrated compressor control, advanced process control, real-time optimization. These projects range from small control improvement projects to multi-billion-dollar mega-projects spanning oil & gas upstream production, refining, petrochemical, chemical, and pulp & paper processes.

I hope this book can help make the typical process control tasks a little more technical and less artistic.

Houston, USA
September 2021

Steve S. Niu

Contents

Abbreviations

AI	Analog In; Artificial Intelligence
ALARP	As Low As Reasonably Practical
AO	Analog Out
APC	Advanced Process Control
ARM	Alarm Rationalization and Management
ARWU	Anti-Rest Windup
BEP	Best Efficiency Point
BFW	Boiler Feedwater
BHP	Brake Horse Power
BPD	Barrels per Day
BS&W	Basic Sediments and Water
BTU	British Thermal Unit
CAPEX	Capital Expenditure
CCR	Central Control Room
CET	Cause and Effect Table
CG	Chromatographs
COD	Control Overview Diagram
CPI	Controller Performance Index
CPM	Control Performance Monitoring
CSTR	Continuous Stirred Tank Reactor
CV	Controlled Variable
DCS	Distributed Control System
DMC	Dynamic Matrix Control
DOF	Degree of Freedom
DP	Differential Pressure
DV	Disturbance Variable
E&P	Exploration and Production
ERP	Enterprise Resource Planning
ESD	Emergency Shutdown
FAT	Factory Acceptance Test
FCE	Final Control Elements

FCS	Fieldbus Control System
FEED	Front End Engineering Design
FGS	Fire and Gas System
FIR	Finite Impulse Response
FMEA	Failure Mode and Effect Analysis
FOPTD	First Order plus Time Delay
FSR	Finite Step Response
FWKO	Freewater Knockout
GLR	Gas Liquid Ratio
GOR	Gas Oil Ratio
GPC	Generalized Predictive Control
HART	Highway Addressable Remote Transducer
HAZOP	Hazard and Operability
HMI	Human Machine Interface
HRSG	Heat-Recovery Steam Generators
HSE	Health, Safety and Environmental
I/O	Input/Output
I/P	Current to Pressure Converter
IMC	Internal Model Control
IPE	Inferential Property Estimation/Estimator
ISA	Instrument Society of America
ISO	International Standard Organization
KPI	Key Performance Index; Key Performance Indicator
KPV	Key Process Variables
LDPE	Low Density Polyethylene
LOPC	Loss of Primary Containment
LP	Linear Programming
LTI	Linear Time-Invariant
MCSF	Minimum Continuous Stable Flow (of a pump)
MCTF	Minimum Continuous Thermal Flow
MIMO	Multi-Input Multi-Output
MISO	Multi-Input Single-Output
MMSCMD	Million Standard Cubic Meters Per Day
MOC	Management of Change
MOP	Maximum Operating Pressure
MOV	Motor Operated Valve
MPC	Model Predictive Control
MPFM	Multi-Phase Flow Meter
MSV	Multi-port Selection Valve
MV	Manipulated Variable
NN	Neural Netwrok
NPSH	Net Positive Suction Head (of a pump)
NPSHA	Net Positive Suction Head Available (of a pump)
NPSHR	Net Positive Suction Head Required (of a pump)
NRV	Non-Return Valve

OOS	Out of Service
OPC	Open Platform Commication; OLE for Process Control
OPEX	Operating Expenditure
OTS	Operator Training Simulator
P&ID	Piping and Instrumentation Diagram
PAS	Process Automation System
PCN	Process Control Narrative
PFD	Process Flow Diagram
PI	Propotional-Integral algorithm
PID	Proportional-Integral-Derivative
PIR	Percentage In Range
PLC	Programming Logic Control
POD	Process Overview Diagram
PPE	Personal Protection Equipment
PSD	Plant Shutdown
PV	Process Variable
QP	Quadratic Programming
RGA	Relative Gain Array; Relative Gain Analysis
RO	Restriction Orifice
ROI	Return On Investment
RTO	Real Time Optimization
RTU	Remote Terminal Unit
RVP	Reid Vapor Pressure
SAT	Site Acceptance Test
SCADA	Supervisory Control and Data Acquisition
SIA	Secure Instrument Air
SIL	Safety Integrity Level
SIS	Safety Instrument System
SISO	Single Input Single Output
SOP	Standard Operating Procedure
TTSS	Time To Steady State
TVP	True Vapor Pressure
VRU	Vapor Recovery Unit

Symbols

English Letters

A	Cross-section area
C_p	Heat capacity at constant pressure
C_v	Heat capacity at constant volume
F_m	Mass flow rate
F_v	Volumetric flow rate
K_c	PID controller gain (min)
K_p	Process gain
$\overline{K}_p$	Normalized Process gain
H_p	Polytropic head (of a compressor)
M	Gas molecular weight
N	Speed (of a pump or compressor
P_s	Suction pressure (pump, compressor)
P_d	Discharge pressure (pump, compressor)
T_i	PID controller integral time (min)
T_d	PID controller derivative time; Compressor discharge temperature
T_s	Compressor suction temperature; Sampling time in process model
W	Power (of a motor, pump, compressor)
Z	Gas compressibility
d	Time Delay (discrete-time)
n	Compressor polytropic exponent
u	Process Input
v	Gas velocity
y	Process Output
z	Process Output with Noise

Greek Letters

κ Ratio of specific heat = C_p/C_v
λ First order filter constant in PV filtering; Closed loop performance indicator in λ tuning
ρ Density of gas, vapor, or liquid
σ Compressor polytropic index
τ Time constant (lag)
θ Process time delay (dead time)

Constants

g Gravitational Acceleration, $g = 9.80665\ \mathrm{m/s^2}$
P_0 Atmospheric pressure, -101.35 kPa
R Gas constant, $R = 8.31446\ J \cdot \mathrm{K}^{-1} \cdot \mathrm{mol}^{-1}$
T_0 Absolute zero temperature, -273.15 °C

Functions

ln Base-e logarithm (natural)
log Base-10 logarithm

Fonts

italics Used for emphasis or for special terminologies
Typewriter Used for tag names, DCS variables, or programming codes

List of Figures

List of Tables

Part I
The Big Picture

In a modern operating plant, once a plant is built and running, a core facility remaining in the plant/field is the control room, which hosts the process control system and personnel. Process control as an engineering specialty is indispensable to process safety and production efficiency. Process control has its unique roles and responsibilities, supported by distinct know-how and skills. It is thus irreplaceable by other engineering specialties.

Process control works with ideas, algorithms, and software and is thus less visible or transparent than other teams or specialties. As a result, although the process control achievements are evident and substantial, how process control works is usually mysterious to most people and thus leads to many misunderstandings or under-valuing. On the other hand, by the cross-disciplinary nature, process control must collaborate closely with many other teams and specialties such as process engineering, rotating equipment, control systems engineering, operations, instrumentation, and measurement. It is crucial to raise the general awareness of process control among all the related teams and specialties to unleash the full potential.

For this purpose, Chap. 1 of this book starts with a high-level overview of process control and explains to the general audience *what* process control is, *why* it is needed, and *what* it needed. More importantly, it clarifies the roles and responsibilities of process control engineers and explains how they are different from others and why close collaboration is necessary to achieve the best control design and optimal control performance.

Process control evolved from general automatic control. Over the past century, automatic control has developed a rich collection of theories and technologies. However, these theories and technologies are mainly targeted at aerospace or electromechanical applications. Only a small subset is applicable to process control due to industrial processes' distinct characteristics and unique challenges. A process control practitioner, especially those new graduates joining the industry, may find it discouraging: on the one hand, they are armed with sophisticated control theories and techniques, from classical to modern; but on the other hand, they found little relevance between what they have learned and what is actually needed. Lacking the necessary knowledge and skills to perform their assigned duties, they have to re-start a long learning journey, and often need to learn the hard way.

To help junior engineers and people from other engineering backgrounds pick up speed quickly, Chap. 2 in this part of the book summarizes the process control technologies prevalent in practical use. This summary defines a general knowledge framework based on practical needs, which may significantly differ from the knowledge framework built from academic leanings. This summary can also serve as an awareness education to non-process control people such as process and equipment engineers, control system engineers and technicians, operators, and operation management. It helps them understand what process control is capable of and the irreplaceable roles in process safety and production efficiency.

Chapter 1
Process Control Overview

Process control is a vital part of process operation. On the one hand, it is indispensable by function and irreplaceable by other engineering specialties. On the other hand, process control is such a unique engineering specialty that it is not widely understood or appreciated in an operating plant. Raising general awareness of process control is a great way to help operation and production. This chapter provides a high-level introduction to process control by starting in Sect. 1.1 with a brief explanation on *why* and *when* process control is needed, *where* and *how* process control fits in. This chapter also tries to clarify some confusion and misunderstandings on process control. Section 1.2 describes the plant operation requirements and shows how operating objectives and control objectives are linked and achieved. Finally, Sect. 1.3 puts process control into the context of overall plant automation and explains the integral role that process control plays in safe and efficient operation.

1.1 Introduction to Process Control

Process control is a unique engineering specialty often confused with other engineering specialties such as process engineering and control system engineering (instrumentation, control systems, etc.). In fact, there is a large gap between process engineering and control systems engineering, and process control bridges this gap and plays a critical role in a modern operating plant.

1.1.1 What Is Process Control?

A *process* stands for a series of unit operations to produce a material in large quantities, either continuously or in batch. Typical industry processes include

S. S. Niu and D. Xiao, *Process Control*, Advances in Industrial Control,
https://doi.org/10.1007/978-3-030-97067-3_1

- oil and gas production and refining;
- petrochemical and chemical processing;
- biochemical and pharmaceutical;
- pulp and paper manufacturing;
- power generation plants;
- food and beverage.

Process control is a branch of automatic control that focuses on the theory and technology to operate the above industrial processes *safely* and *efficiently*. Process control continuously monitors a process for deviations and abnormalities, takes immediate action to correct those abnormalities, and brings the process back to its optimal state.

In a typical work process, process control starts with basic process engineering, translates the engineering design and operating requirements into control objectives and requirements, designs the proper process control strategy, specifies the functional requirements, and eventually transforms it into the final control scheme on the targeted control system infrastructure. Thus, process control's primary role and responsibility are designing, implementing, and maintaining a sound process control solution to ensure safe and efficient operation.

To achieve this goal, process control works closely with process engineering during process design to ensure that the engineering design delivers the required controllability and stability. During the commissioning and operation phase, process control works with control systems personnel to ensure the process design is implemented and operated as designed. Process control is a very cost-effective solution for improving process safety and production efficiency, typically with low investment and high return (Liptak 2006).

Process control bridges the gap between process engineering, operation and maintenance, and control systems (Fig. 1.1). It overlaps and thus collaborates with several

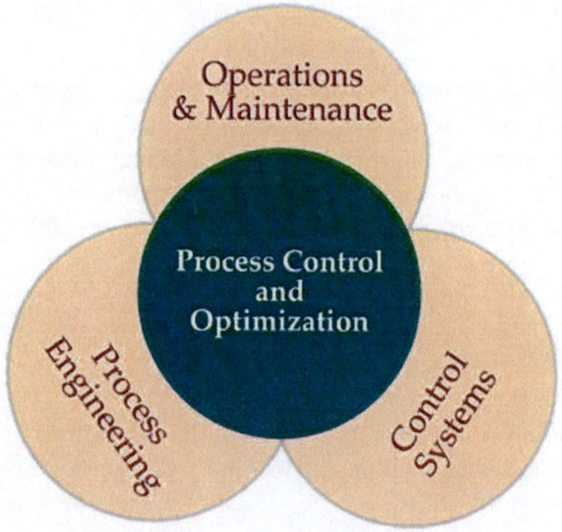

Fig. 1.1 Process control is cross-disciplinary

related engineering specialties. For this reason, process control engineers often have a more coherent view of the entire process than other engineers.

The critical deliverable from process control is the control scheme in the control systems and the up-to-date process control narrative (PCN) document.

1.1.2 Why Do We Need Process Control?

All commercial productions are for-profit, and thus maximizing profit is the ultimate goal of the operation. In general, maximum profit is achieved through the highest production rate of the right products at the best selling price and with the minimum capital expenditure (CAPEX) and lowest operating expenditure (OPEX) (Fig. 1.2). Full-capacity operation with the highest up-time ensures maximum production while keeping the products consistently on-spec and meeting the high demand leads to the best selling price.

In this highly volatile market, and with increasingly stringent requirements on health, safety, and environmental (HSE), the operation must stay safe and efficient to maximize profit. The key process variables must be maintained at their desired values during normal operation and within a safe range during plant upsets. For example, the vapor pressure of the crude oil from a stabilizer must be maintained close to the design value to meet environmental regulations even when the feedrate or the temperature/pressure condition fluctuates. Likewise, the temperature inside a packed-bed reactor must be maintained at its design value to ensure the desired conversion and prevent reaction run-away.

Automating operation is the proven way to improve safety and efficiency, and process control is a crucial enabler of automated operation (Marlin 2015). Take an upstream exploration and production (E&P) process as an example. As shown with a process overview diagram (POD) in Fig. 1.3, the multi-phase fluid from the production wells is collected and sent to the processing facility for separation into gas, oil, and water, which are further processed and conditioned for sale, injection,

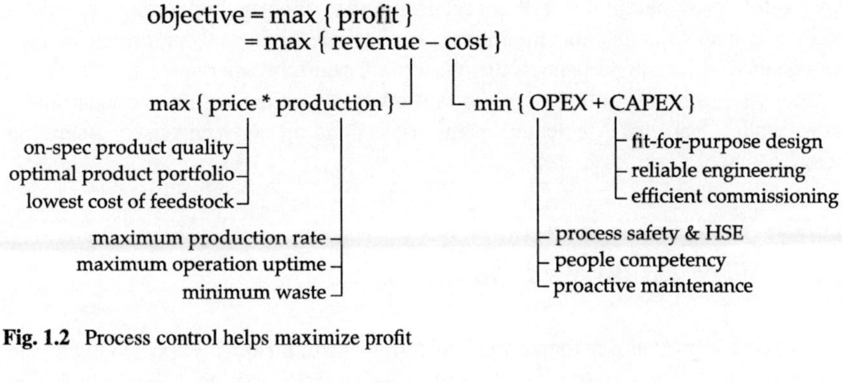

Fig. 1.2 Process control helps maximize profit

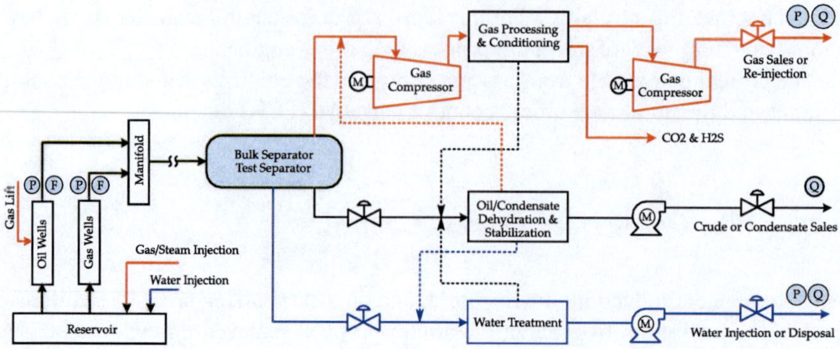

Fig. 1.3 Process overview diagram for an E&P plant

or disposal (Campbell 2004). The operating objective is to safely and efficiently separate the well fluid, meeting the requirements on both *quality* and *quantity* for the produced oil, gas, and water by controlling the process variables such as pressure (P), flow (F), and quality (Q).

A sound process design (piping, equipment, line-up, material, and energy balance) makes it *possible* to achieve this objective. However, in actual operation, constant attention and intervention are needed to maintain the desired *quality* and *quantity* due to inevitable disturbances such as continuous fluctuations in supply and demand, frequent process upsets, and occasional failures. The attention and intervention are provided by process control. Plant operators and the process control system work together to ensure that the plant runs as desired. Therefore, process control consists of manual process control (operator control) *and* automatic process control. Better automatic control typically results in less operator control and vice versa.

Fully automated operation ("un-manned" operation) has long been an aspiration. The automated operation is necessary for some process operations under harsh conditions or with less stringent control requirements. However, fully automated operation relies on *credible data*, *reliable information*, and *optimal decisions*, which are not yet up to the expectation in complex operations. For mission-critical processes, the fully automated operation is still not realistic due to abnormal operating conditions that are beyond what machine intelligence can handle. Abnormal conditions include, for instance, severe disturbances, the onset of a hazard, or emergencies.

Section 1.2 continues the discussion on operating objectives and control objectives, while Chap. 6 is dedicated to the understanding and analysis of operating needs.

1.1.3 When Do We Need Process Control?

A plant typically takes one to five years to design and build and is expected to operate for 20 or more years. Experience shows that many operating difficulties can be traced

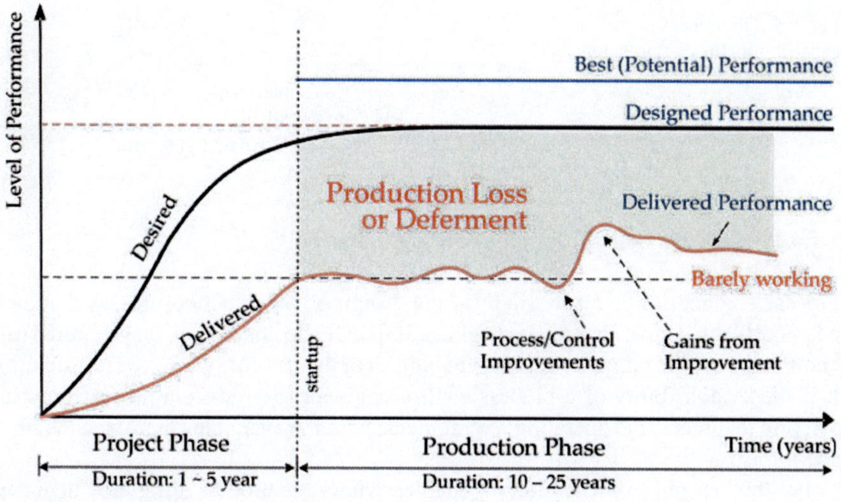

Fig. 1.4 Project execution: desired versus delivered

back to the process control design flaws. Operations have to live with or work around these flaws for the plant's entire life cycle. Even a tiny flaw in the design may result in a significant accumulated loss over the many years of plant operation. Figure 1.4 illustrates the project life cycle and the impact of a bad design.

There has been a growing awareness of the criticality in integrating the process design and the control design at the early stage in a new project (Sakislis et al. 2004). Process design should take process operability and controllability as critical considerations. For this reason, process control should be involved as early as possible in the project life cycle, preferably from the early conceptual design phase, to ensure a sound overall control strategy. Early process control involvement can eliminate many operating issues from the source and is thus much more cost-effective. Otherwise, the design flaws may be carried over to operation phase to be fixed. Once the plant is in operation, the troubleshooting effort is typically limited to local areas and *ad hoc* improvements. As a result, there are limited opportunities to re-consider and improve the control design's overall integrity and optimality.

1.1.4 What Is Process Control Not?

Process control has a significant overlap with process engineering and control system engineering (Fig. 1.1). However, process control is a very different specialty (King 2016; Lee and Weekman 1976), and the difference is not widely recognized. There is a significant disconnect between process and control (Fig. 1.5) in many practical operations, contributing to many day-to-day operational challenges.

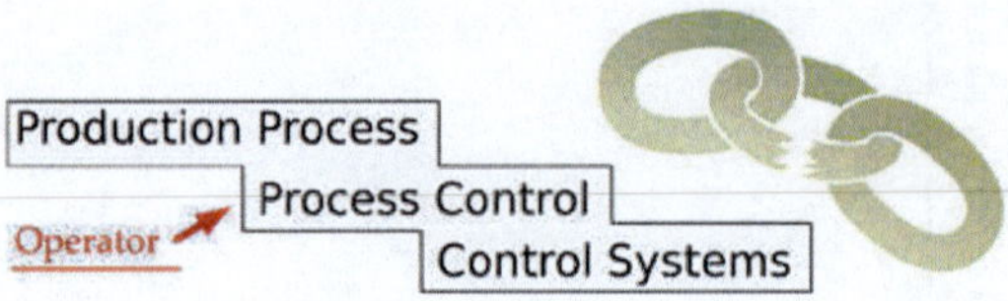

Fig. 1.5 Disconnect between process engineering and control systems

1.1.4.1 Process Control Is Not Process Engineering

Process engineering is responsible for the design of a production process that can potentially run. However, process engineers typically do not (nor are they required to) know the available process control technologies or the control systems infrastructure. It is the responsibility of a process control engineer to produce a process control solution that meets the operating requirements with the available control technology and infrastructure.

For this reason, a process control engineer views the process differently than the process engineering people:

1. Static versus dynamic: Process engineers are more focused on the static behavior of the process, such as steady-state material and energy balance. However, for process control engineers, above and beyond the steady-state relationship, their primary interest is in the dynamic behavior of the process (process dynamics), including the dynamic material balance, dynamics cause and effect relationships, and dynamic response characteristics of the process variables of interest.
2. Feedforward versus feedback: Process engineers are more customized to rely on first-principle calculations to meet the material or energy balance requirements, while process control engineers count more on feedback control to achieve the same. For example, to control a temperature, a process engineer is more inclined to propose elaborate calculations based on heat duty, heat transfer, and energy balance to determine the required energy input. On the other hand, a process control engineer is more likely to request an online temperature measurement and use a simple feedback controller to adjust the fuel flow to achieve the temperature target.
3. Absolute values versus incremental changes: Another difference between the two professions is that process control engineers are more interested in the "delta changes" than the absolute steady-state values that the process engineers are interested in. For example, a temperature controller compares the temperature measurement (process value) with the temperature setpoint (desired value) and calculates the amount of heat duty to be removed or added (the incremental changes in control action) to bring the temperature back to its setpoint. As long as the temperature measurement is not the same as the desired temperature, the controller will keep adding or removing heat! After all, the controller does not care about the absolute value of the heat duty. The downside of this powerful mechanism is that it can often drive the heat duty to saturation if it is not informed of the heat duty limit.

This difference in thinking significantly influences the decision on how the plant should be designed and controlled.

1.1.4.2 Process Control Is Not Control System Engineering

Control systems engineering is responsible primarily for the control system infrastructure, including the sensors, transmitters, final control elements, the distributed control system infrastructure, safeguarding system, and the controllers in isolation. As a result, they rarely (nor are they required to) care about the process design philosophy or overall control strategy. Process control engineers develop the functional specification based on the control strategy and pass them on to the control systems engineers for implementation. With better knowledge of the available control system infrastructure and capability, the control system engineers implement the control solution in the targeted control platform. The focus of a process control engineer and a control system engineer is thus quite different:

1. Software versus hardware: While a process control engineer is responsible for providing the control strategy and the functional specification, a control system engineer's mission is to implement the functional specification with the available hardware/software in the targeted control systems infrastructure.
2. Holistic view versus localized view: The process control engineer must have a holistic view of the control solutions that connect with the process design, the control systems, and the operation; while a control systems engineer is charged to precisely achieve each specified control function on the given platform. They do not necessarily care about the purposes of the control functions and if/how they all work together.

Control systems engineers talk about *devices*, while process control engineers work on *ideas* (Åström and Kumar 2014). Ideas are less visible than devices, and therefore, there is a tendency of under-valuing process control.

1.2 Operating Objectives and Control Objectives

The ultimate goal of production is to maximize profitability, which demands safe and efficient operation. Therefore, the control objectives must serve the operating objectives to make the operation safe and efficient.

1.2.1 Process Flow and Operating Objectives

A process is the continuous or batch processing of certain materials. Feeds go into the process, and products come out of the process. Process engineering design puts

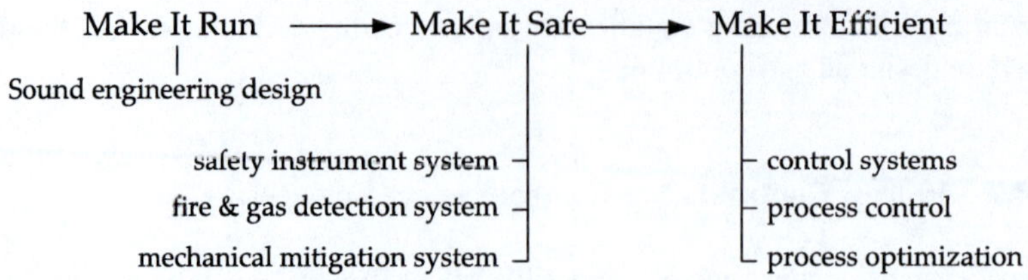

Fig. 1.6 Operating objectives: from design to operation

together the proper piping and equipment to ensure the material's movement and quality meet the business requirements (Fig. 1.3). The ultimate objectives are to make the plant run, and run it safely and efficiently (Fig. 1.6). A sound process design is a prerequisite for a solid process control solution. By design, a process must be inherently *operable*. In other words, *a plant must be able to run manually* (start, run, and shutdown) before it can be operated automatically.[1] The process must be able to continuously operate to produce the desired material up to specifications if the designed conditions are provided.

Chapter 6 provides a detailed discussion on how to dissect and understand a given process in order to control it.

1.2.2 Control Objectives and Control Strategy

Process control makes an *operable* process *operative*, i.e., run as per desired performance requirements. Process control, consisting of *operator control* and *automatic control*, provides the required operating condition for the process to continuously operate to produce the on-spec product in the desired *quantity* and *quality*.

To serve the operating objectives, a process control solution must meet the following operating requirements:

1. During normal operation, deliver the right product with the right quality and quantity.
2. During process upsets, temporarily deviate from the operating target but stay inside the operating envelope.
3. During operation change, ensure a safe and quick transition from one condition to another.

[1] Operators always have the privilege to override automatic control in case of an emergency. One of the design flaws in the Maneuvering Characteristics Augmentation System (MCAS) in Boeing 737 Max that caused the two catastrophic crashes in 2019 was that the MCAS system is given the privilege of overriding the pilot's control. Consequently, one of the improvements is "... MCAS will never override the pilot's ability to control the airplane using the control column alone." (from `https://www.boeing.com/737-max-updates/mcas`).

The general control strategy typically consists of the following three levels of control actions:

1. Regulatory control: Maintain the operating point at (or move it to) the desired *quantity* and *quality* target.
2. Protective control: Keep the operating point within the operating envelope during process upsets.
3. Operator control: When automatic control fails to react correctly or timely, the operator intervenes and makes manual corrections.

The basis for control design is the process flow, supply and demand model, cause and effect relationship, and dynamic response behavior of the concerned process variables. Therefore, it is essential to follow a systematic approach to understand the underlying process and ensure the control strategy is adequate at all levels of the plant operation (Stephanopoulos 1984). See Chap. 7 for the in-depth discussion on control design.

The control strategy is typically designed by process control engineers and implemented by control system engineers. The implementation involves the following activities:

1. Place the right sensors and transmitter at the right location to provide and necessary measurements. Process measurements provide information on the current status of the process conditions.
2. Position the proper final control elements such as valves or motors at the right location. The final control elements are the means to influence the operation.
3. Perform the required calculations of the corrective moves based on the measurements and adjust the final control elements to maintain the operation at the desired targets, specified either by the operators or higher-level application in the control hierarchy.

Figure 1.7 shows the locations of some process measurements such as pressure, flow, and quality, along with the locations of the final control elements such as control valves and electric motors for pumps and compressors.

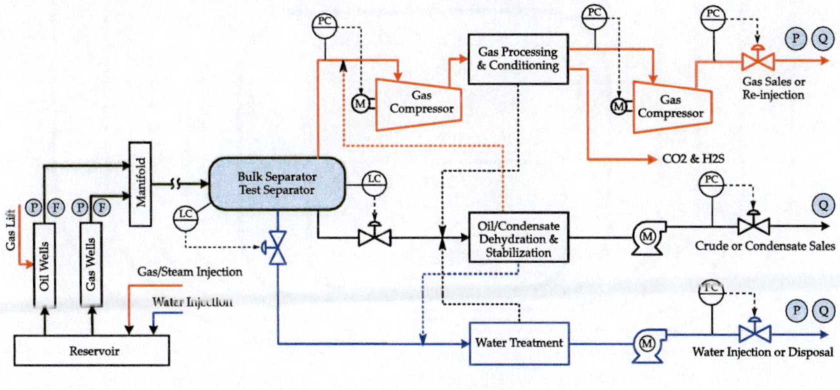

Fig. 1.7 Overall control strategy for a general E&P process

1.2.3 *Process Control for Process Safety*

Operating safety is of paramount importance in the design and operation of a plant. Unsafe operation is the most significant risk to profitability since one fatal accident can easily wipe out a whole year's profit or even force the plant to close down permanently.

Operating safety is achieved through multiple layers of safety measures, as shown in Fig. 1.8, where process control is the first layer of protection against unsafe operations! This layered setup is sometimes nicknamed the "safety onion."

Operating safety typically consists of two parts:

1. Process safety: Process safety concerns the prevention of fires, explosions, and accidental releases of hydrocarbons and chemicals in an operating facility dealing with hazardous materials. Process safety is related to design and engineering and has not received adequate attention compared to human safety. From the safety onion in Fig. 1.8, inherently safe process design is the fundamental requirement by process safety, while process control is the first line of defense against unsafe operations.

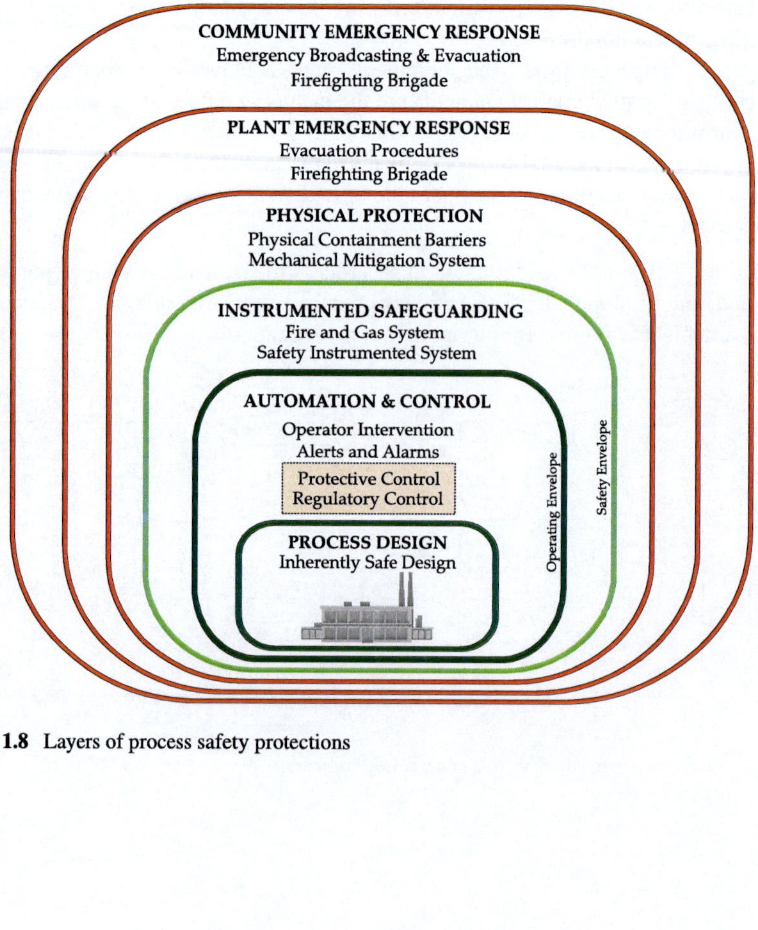

Fig. 1.8 Layers of process safety protections

2. Human safety: If both the process control and safeguarding layers fail, incidents may happen, and human safety may be endangered. Proper personal protection equipment (PPE) and standard operating procedures (SOP) are the primary tools to protect human safety. However, as shown in Fig. 1.8, improving the control and safeguarding effectiveness is more efficient than passive protection for human safety.

In this layered protection design, each layer is a vast topic by itself. Only a very brief description is provided here:

1. Process design: The first and utmost assurance for safe and profitable plant operation is through an engineering design that is *inherently* safe and stable. That means, during normal operation, the safety requirements such as material balance, energy balance, equipment sizing, the material of construction, and pressure rating are all adequately fulfilled. When operation conditions change, the plant can typically settle by itself to a new steady-state, which may not be the desired condition but is still safe.
   Most processes are inherently stable. For example, even without control, a heat exchanger's outlet temperature cannot increase without bound due to the limit in heat duty. However, there are also many unstable processes such as vessel levels and most exothermic chemical reactions. These unstable processes require much more stringent control and safeguards to keep the operation within the safe limits.
2. Process control: All plants are designed to operate within a particular operating region, defined by constraints and limits based on safety, operability, and operating efficiency. This operating region is called the *operating envelope*. There are usually one or several optimal *operating points* within the operating envelope that yield the best profitability for operation.
   Process control ensures that the plant operates at or near the optimal operating points during normal operation and stays inside the operating envelope during process upsets. Operators perform process control manually in the old days, while automatic control has been the norm in a modern operating plant. However, operator intervention is required during abnormal operating conditions such as severe disturbance, extreme turndown, emergency, or startup/shutdown. Alarms and alerts are the means to notify the operator of abnormal conditions.
   In a modern control system, alarms are so easy to add to the extent of being abused. Operators are often distracted or even overwhelmed by *nuisance alarms*. Alarms are meant for the operator to take action. Matters that are not worthy of operator attention should not be configured as an alarm. To ensure the effectiveness of alarms, many operating plants have taken *alarm rationalization* as a mandatory exercise.
3. Safeguarding system: The purpose of the safeguarding system is to protect against human injuries, loss of assets, and pollution of the environment by proactively and quickly shutting down the operation area at risk once an out-of-control situation occurs. A shutdown of the operation incurs costs and losses by itself; however, it is a proactive action to prevent more severe consequences.

A safeguarding system typically consists of a *safety instrument system* (SIS) and a *fire and gas mitigation system* (FGS). SIS typically triggers a unit shutdown, while a fire and gas incident usually forces a plant-wide shutdown. The standard safeguarding functionalities include trips, interlocks, and emergency shutdowns:

a. Safety instrument system (SIS): In the case of a major process upset, the process control layer may be too slow or insufficient in response to prevent the operation from going outside the operating envelope. The "out of control" operation can result in off-spec products or even unsafe operations.
The SIS layer takes automatic, independent, and fast action to prevent a hazardous incident from occurring and protect personnel and plant equipment against potentially serious harm. Once a severe failure is detected, the safeguarding layer proactively shuts down (trip) the operation to prevent the condition from further deterioration.
b. Fire and gas system (FGS): If the SIS system also fails, a major incident may follow, typically in the form of *loss of primary containment* (LOPC). A gas leak or liquid spill is a direct result, leading to potential fires and explosions. For certain operations with toxic gases (e.g., with high H_2S concentration), loss of containment can be a direct fatal threat to human life in the surroundings.
The F&G system takes proactive action to reduce the hazardous event's consequences *after* it has occurred. Its purpose is to mitigate and minimize the damages. Upon detecting a fire or gas incident, the FGS proactively shut down the plant operation (emergency shutdown).

4. Mitigation by physical or mechanical protection: The safeguarding system typically relies on power and instrument air to stop the plant operation and bring it to safety. In case of loss of power or instrument air, mechanical protection provides another layer of protection. The goal is to prevent the damage from further escalation. Examples include pressure relief valves or containment barriers. For example, in a high-pressure low-density polyethylene (LDPE) plant, the operating pressure can be as high as over 2,000 atm. The hyper-compressor and the pipings must be surrounded by a concrete protection wall ("blast wall") up to 10 meters high and 2 meters thick against a potential explosion.
5. Emergency response: The plant and community emergency responses, including firefighting brigade and site/community evacuation procedures, are the last resort to respond to a major incident.

In summary, sound process engineering provides the basis for safe and efficient operation; process control is responsible for running the plant at the designed operating point and protects it from running out of the operating envelope during production upsets.

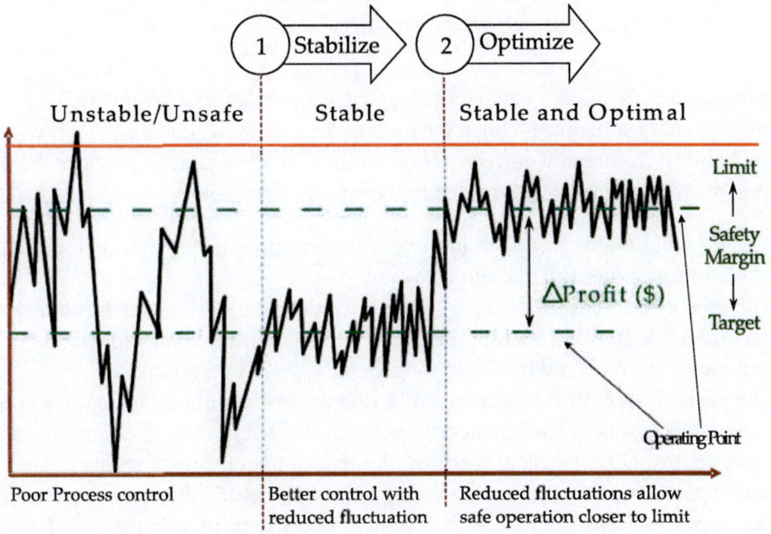

Fig. 1.9 More profitable production through stabilization and optimization

1.2.4 Process Control for Production Efficiency

Process control stabilizes operation and reduces uncertainty; therefore, it allows the plant to operate closer to the operating limits for better profitability. See Fig. 1.9 for an illustration of this *stabilize* and *optimize* philosophy.

Most *advanced process control* solutions such as *model predictive control* (MPC, Chap. 5) have optimizing engines built-in to drive the operating point closer to the most profitable operating region. More advanced production optimization can be performed by dedicated applications such as *real-time optimization* (RTO, Sect. 2.3.7). Chapters 6 and 7 provide in-depth discussions and general guidelines on analyzing the process requirements and producing the proper process control solutions.

1.3 Process Control in Context

From the operating objectives (Fig. 1.6) and the layers of process safety protection (Fig. 1.8), it is evident that process control is an essential and integral part of the overall operation and safety establishment. They work together to achieve the desired safe and efficient production.

1.3.1 Process Automation Hierarchy

Process control is implemented in a layered fashion, as shown in Fig. 1.10. This layered layout is sometimes called the *automation pyramid*. Each layer relies on the layers below it to function correctly. At the same time, each layer provides the basis for the layers above it to perform the higher-level functions:

1. The physical plant: The safe and efficient operation of the plant is the ultimate goal of process control and optimization.
2. The sensor and transmitter layer provides the necessary view into the plant operation variables (process measurements). The final control elements such as valves and motors provide the means to influence the plant operation.
3. The baselayer controls, predominantly PID controllers implemented in a control system environment such as distributed control system (DCS) or programmable logic control (PLC) system, perform the real-time control functions to keep the plant operation at the optimal operating points and within the operating envelope.
4. Advanced process control (APC) typically consists of complex PID control schemes and model predictive control (MPC) applications, focusing on the stable operation of a more extensive process area such as an entire distillation column or a compressor unit.
5. Real-time optimization (RTO) uses a combination of detailed mathematical models, explicit economic objective functions, and real-time access to the control systems to maximize the profitability of an operating unit or even the entire plant.

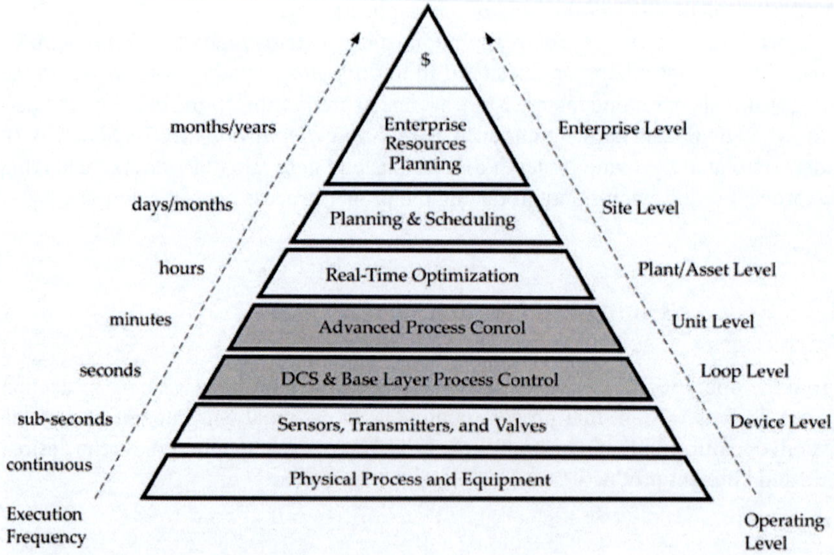

Fig. 1.10 Plant automation hierarchy

6. Planning and scheduling: Management information and decision system for making mid-term planning and scheduling decisions on production, inventory, shipping, and purchasing.
7. Enterprise resource planning (ERP): ERP is tied to corporate supply-chain management and sets the long-term operating targets for the control and optimization applications.

The higher-level applications have a broader scope of concern and a lower frequency of execution. They generate the operating directives for the next layer below.

1.3.2 Process Automation System (PAS)

A *plant automation system* (PAS) is typically a combination of *distributed control systems* (DCS) and *programmable logic control* (PLC) based platforms. A typical PAS infrastructure consists of the following three subsystems, as illustrated in Fig. 1.11:

1. Distributed control system (DCS): A DCS is the core platform and operator interface for monitoring and control. The DCS system typically also acts as the *human–machine interface* for the operators for all other subsystems.
2. Safety instrumented system (SIS): SIS is typically a PLC-based system running condition-based logic and sequences. In case of emergency, it performs shutdown of process and equipment, isolation of hydrocarbon inventories, switch off of electric systems, depressurization, blow-down, and emergency ventilation control.
3. Fire and gas system (FGS): The fire and gas system continuously monitors all plant areas for abnormal conditions such as a fire or combustible/toxic gas release via various fire and gas detection instruments. In the event of a hazardous situation, the system alerts the operator of the hazardous situations. Depending upon the hazard level, the fire and gas system will also proactively shut down the process equipment and activate automatic fire fighting systems to prevent escalation of the incident.

In addition to the DCS and PLC, there may be vendor-provided, dedicated control and protection packages. Ideally, all vendor packages are interfaced with the DCS to provide the operator with a single standard interface. Through PAS, operators and engineers can remotely monitor and control the plant from a central control room (CCR).

For maximum reliability and integrity and to minimize the chance of common failures affecting both control and safeguarding, field instruments for each subsystem are functionally and physically segregated from each other. However, all the displays and alarms are typically routed to the DCS to be viewed and operated in the *central control room* (CCR). Sufficient measurements and associated display functions such as trends and recording functions, indications, and alarms are provided to help detect

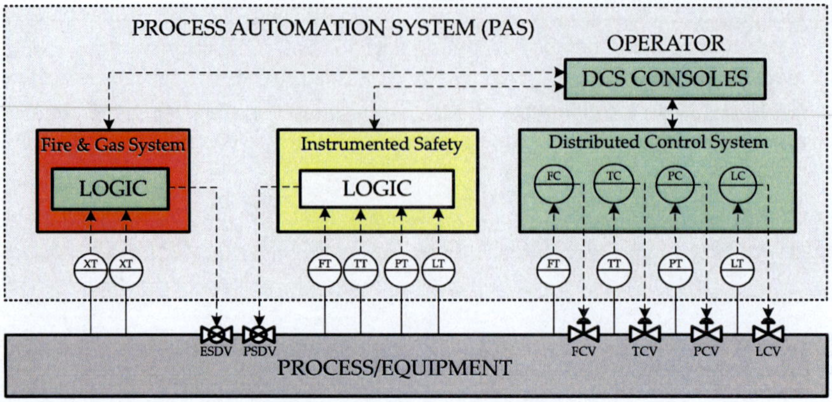

Fig. 1.11 Process automation system

abnormal conditions in the plant and give the operator the possibility to observe the development of imminent disturbances.

For legacy or practical reasons, full PAS integration may not be realistic, and variable levels of integration exist from plant to plant.

1.3.3 Symbology and Identification

A process variable may need to be measured, monitored, controlled, and safeguarded. A standard representation, including naming and symbols, facilitates communication, documentation, and operation. The process control schematics typically follows the symbol and identification standards of ISO 3511, or ISA S5.1 and S5.3, which include the standard naming convention as follows:

`nXY-nZ` Example: `32FICA-101A, 12AI-201.`

The first part (`n-`) is a two to three-digit number indicating the process unit. The two or more letters following the unit number designate the type of measurement (`X`) and its function (`Y`); The dash "-" is optional but preferred for clarity. The number after the letters is a two to four-digit serial number, and the optional suffix (`Z`) can make the tag name uniquely identifiable.

For example, tag name `32FICA-101A` indicates that this is a process variable in unit #32. It is a flow measurement (`F-`) with an indication in DCS (`-I-`). It performs a control function (`-C-`) with alarm(s) (`-A-`) configured. The tag serial number is `101` with a suffix (`-A`), indicating that there may be other tags with a similar tag name, e.g., `32FICA-101B`. See Table 1.1 for the commonly encountered identification letters. For a complete list, please consult the ISO-3511 or ISA-S5 standards.

Table 1.1 Instrument identification letters

Letter	As measurement code	As function code	Examples
A	Analysis	Alarm	21AI-102 analyzer indicator
E	Voltage	Sensor	21FE-101 flow sensor
D		Differential	21PDI-101 differential pressure indicator
F	Flow	Ratio	12FICA-101 flow control 32FFC-211 flow ratio control
H	Hand		21HIC-101 hand control
I	Current	Indicator	21AI-102 analyzer indicator
J	Power		21JI-101 power indicator
K	Time, schedule		21KS-121 on/off switching control
L	Level		21LIC-101 level control
P	Pressure		21PC-101 pressure control
Q	Quantity	Totalize	21FQI-101 flow totalization
S	Speed, frequency	Switch	21FS-101 flow switching control
T	Temperature	Transmitter	21PT-101 pressure transmitter
U	Multivariable	Multi-function	21UC-102 MPC control or complex control
			21UZ-102 a SIL trip function
V		Valve	21FCV-101 flow control valve 21FZV-101 flow shutdown valve 21KSV-101 on/off switching valve
X	Unclassified		23XC-101 load controller
Y	Event, State	Calculation	23FYI-101 flow compensation calculation
Z		Safeguarding	23FZA-101 flow tripping with alarm

The full tag names are unique within the same operating plant, while the serial number is unique in each group of instruments sharing the same measurement and function codes. On the other hand, the different functions for the same process variable share the same serial number. For instance, the sensor, transmitter, indicator, control, and safeguarding for a flow variable can be named 12FE-101A, 12FT-101A, 12FI-101A, 12FC-101A, 12FZ-101A, respectively, all with the same serial number. The indication and alarm function may be added to the control function to become 12FICA-101A. A redundant flow measurement is named 12FICA-101B.

Although it is necessary to specify the full tag name on P&ID for proper construction and configuration, the tag names are often simplified for clarity where space is of concern. For example, on a DCS schematic, the flow tag 23FICA-101A may be reduced to 13FC101A without sacrificing clarity. In addition, the instrument symbol 23FT-101A is often omitted without loss of clarity (see Fig. 1.12).

Some of the typical instrument symbols for indicators, controllers, or safeguarding units used in this book are illustrated in Fig. 1.13, and more details can consult the ISO and ISA standards. For clarity, the unit number of all the tag names is omitted for the clarity of the graphics.

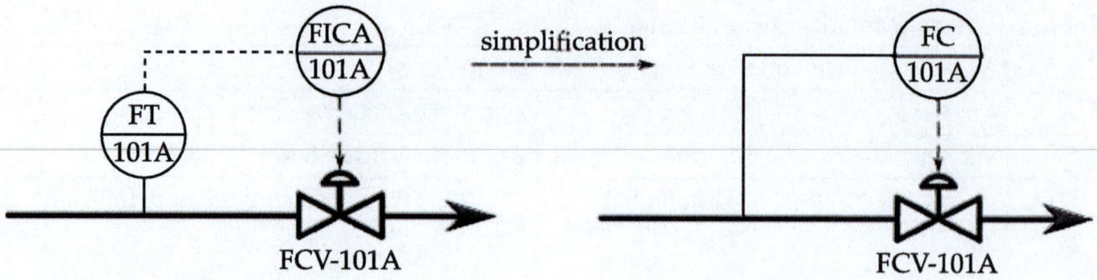

Fig. 1.12 Simplification of symbols

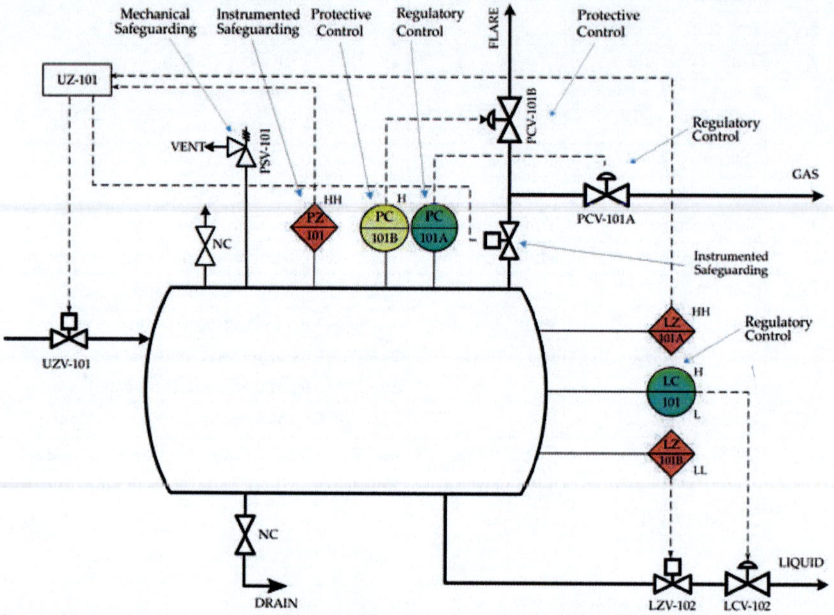

Fig. 1.13 Example of control and safeguarding

1.3.4 Control and Safeguarding

The process safety protection setup in Fig. 1.8 shows that the safety instrument system (SIS) is the next layer of defense against unsafe operation after process control. Actions from the safeguarding layer are typically unit shutdown or plant-wide shutdown and are thus very disruptive. For instance, the emergency shutdown of large equipment such as a gas compressor can potentially be damaging. Close integration between process control and safeguarding is thus crucial in an integrated design.

Example 1.1 Control and safeguarding of a gas/liquid separator. Figure 1.13 is an example of a typical two-phase separator in an upstream oil and gas production facility. A separator receives the multi-phase fluid from wells and separates it into gas and liquid under the proper pressure and interface level.

There are four layers of control and protection in this process configuration:

1. Normal regulatory control: The level controller `LC-101` and the pressure controller `PC-101A` provide continuous regulatory control during normal operation. The stable level and pressure ensure satisfactory separation of gas and liquid in the separator.
2. Abnormal protective control: If the regulatory pressure control `PC-101A` fails to maintain the desired pressure, the pressure may rise above the high limit set by the protective controller `PC-101B`, which will open the valve to flare. Temporary flaring will stop the pressure from going dangerously high while waiting for the operation to come back to normal.
3. Instrumented safeguarding: If the pressure continues to go up even with the protective controller active (or failing), then the safeguarding pressure trip `PZ-101` proactively triggers a process shutdown through `UZ-101` to close all the inlet/outlet isolation values to ensure positive containment of the hydrocarbons. Similarly, `PZ-101A` and `PZ-101B` safeguard the vessel against high or low levels by proactively triggering a shutdown and close the inlet or outlet isolation valves.
4. Pressure relief valves: In the rare case of power or instrument air failure, the safeguarding layer fails to act, and the pressure goes up to a dangerous level; the relief valve `PSV-101` will pop open to stop the pressure from bursting the vessel.

1.3.5 Operation by Setpoints and Limits

The different layers of control and protection are achieved by operating at different setpoints and limits in a staggered fashion. Figure 1.14 provides a list of all the potential setpoints and limits, with SP being the desired setpoint, SPH/SPL the setpoint high/low limit, H/L the high/low alarm limits, and HH/LL the high/low trip limits.

The setpoints and limits are carefully specified based on process design and operating requirements. Most of the time, only a subset of the limits and setpoints is required. For example, an operation may be high constrained (Fig. 1.14 left), low constrained, or both (Fig. 1.14 right).

In Fig. 1.13, the normal operating pressure inside the vessel is controlled by the pressure controller `PC-101A`. The protective controller `PC-101B` is assigned with a setpoint higher than that of `PC-101A`. During normal operation, the protective controller `PC-101B` is inactive, and the flare valve `PCV-101B` remains closed. Only when the normal regulatory pressure controller `PC-101A` fails to maintain the pressure at the normal setpoint and the vessel pressure rises above the `PC-101B` setpoint will this protective controller `PC-101B` becomes active. The protective controller opens the flare valve to let go of the vessel's excessive gas to flare (a waste!).

If both the normal controller `PC-101A` and protective controller `PC-101B` fail to maintain the vessel pressure and the pressure surpasses the high alarm limit, an alarm

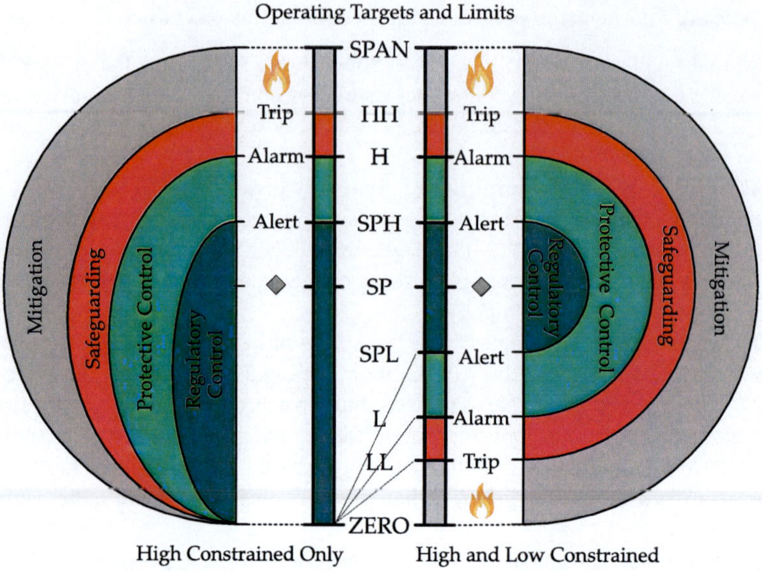

Fig. 1.14 Operating setpoints and limits

will sound to call for operator attention and intervention. If the operator intervention is unavailable or unsuccessful, and the pressure continues to rise and reaches the trip limit, the safeguarding element `PZ-101` will kick in and proactively shut down the separator operation by closing all the inlet and outlet isolation valves.

In summary, a process is designed to operate under different setpoints and limits, including normal control setpoint, protective control setpoint, alarm limits, and tripping limit. If automatic control fails, an alarm alerts the operator for manual intervention. If operator control fails, instrumented safeguarding such as SIS kicks in to shut down the operation. Therefore, the setpoints and alarm limits must be properly aligned and spaced to allow ample time for the next defense line to react.

1.4 Summary

The key messages conveyed by this chapter include the following:

1. Process control is essential for process safety and production efficiency.
2. Process control has its unique roles and responsibilities, supported by distinct know-how and skills. As a result, it is indispensable in a modern operating plant and is irreplaceable by other engineering specialties.
3. Process control achieves its objectives by improving control strategy and modifying software implementation. Therefore it is cost-effective with low investment and high return.

4. Process control is cross-disciplinary and very much application-oriented. The collaboration and interactions with other teams, especially process engineering, control systems, and operations, are instrumental.
5. Process control has been one of those most undervalued or misunderstood engineering specialties, and there are tremendous practical benefits in raising awareness of the roles and contributions of process control.

References

Åström KJ, Kumar PR (2014) Control: a perspective. Automatica 50(1):3–43

Campbell J (2004) Gas conditioning and processing—the equipment modules, vol 2, 8th edn. John M, Campbell and Company

King M (2016) Process control: a practical approach, 2nd edn. Wiley

Lee W, Weekman VW (1976) Advanced control practice in the chemical industry. AICHE J 22(27)

Liptak BG (2006) Process control and optimization, instrument engineers' handbook, vol II, 4th edn. Taylor and Francis

Marlin TE (2015) Process control—designing processes and control systems for dynamic performance, 2nd edn. McGraw-Hill

Sakislis V, Perkins JD, Pistikopoulos N (2004) Recent advances in optimization-based simultaneous process and control design. Comput Chem Eng 28(10):2069–2086

Stephanopoulos G (1984) Chemical process control—an introduction to theory and practice. Prentice-Hall, New York

Chapter 2
Process Control Knowledge Framework

Process control evolves from automatic control. For historical reasons, most of the automatic control theories and techniques originated from the aerospace and electrical engineering fields, of which only a (small) subset is directly applicable to the process industry due to the unique characteristics and distinct requirements. On the other hand, some new technologies have been developed over the years to address the particular requirements of the process industries. The scope has become so large that it is necessary to delineate the relevant part of the knowledge for process control from the general scope of automatic control and complement it with the relevant new technologies developed over the recent years, such as those related to process engineering and process modeling. This chapter presents a knowledge framework that summarizes the core process control technologies currently in use (or in need) in the process industries, from simple PID controllers (Sect. 2.1), complex PID control schemes (Sect. 2.2), to advanced process control (Sect. 2.3) solutions. The control design philosophy (Sect. 2.4) dictates how these technologies are utilized.

2.1 Simple Control Loops

Most industrial control loops are *simple PID control loops* (Åström and Kumar 2014; Lee and Weekman 1976; Shinskey 2001; Smith 2010). A simple PID control loop is also called a *standalone PID control loop*. It typically consists of only one variable to control (*control target*) and one variable to manipulate (*control handle*).

2.1.1 Open-Loop Versus Closed-Loop Control

Control can be divided into the four broad categories as shown in Table 2.1:

S. S. Niu and D. Xiao, *Process Control*, Advances in Industrial Control,
https://doi.org/10.1007/978-3-030-97067-3_2

Table 2.1 Type of controls

	Manual operation	Automated operation
Open-loop control	No control	Sequential control
Closed-loop control	Operator control	Automatic control

1. Open-loop control: A block diagram of open-loop control is given in Fig. 2.1. In an open-loop control system, the process (①) to be controlled is usually well understood and well behaved with no significant uncertainties such as noises and disturbances, or the effect of the uncertainties on the controlled parameter is within the acceptable range. The control action (③) is generated by a fixed logic or sequence (⑤) following a pre-defined control target (④), "start it and trust it"! There is no need to check whether the controlled parameter responds as desired (②), and thus the next control action is independent of the control result of the current step. Some open-loop control examples include firing a bullet, switching traffic lights, and toasting bread with a toaster. Sequential control such as automated plant startup and shutdown, the transition from one operating mode to another, and batch operation are primarily open-loop control with little or no feedback.
2. Closed-loop control: Figure 2.2 is the block diagram of a typical closed-loop control. Most industrial processes have significant noises and disturbances (①) with unpredictable impacts on the controlled variables. The manipulated variable (③) must be adjusted continuously to correct the deviations caused by noises and disturbances and maintain the controlled variable at their target (④). The action is based on the feedback (②) from the control result of the previous step.
   Feedback control is one of the most commonly used mechanisms for closed-loop control. The basic principle of feedback control is

   `Measure → Compare → Correct.`

   Both operator control and automatic control are closed-loop controls. They complement each other to achieve the operating requirements. The controller (⑤) in Figs. 2.1 and 2.2 can be the operator, the control software, or both:

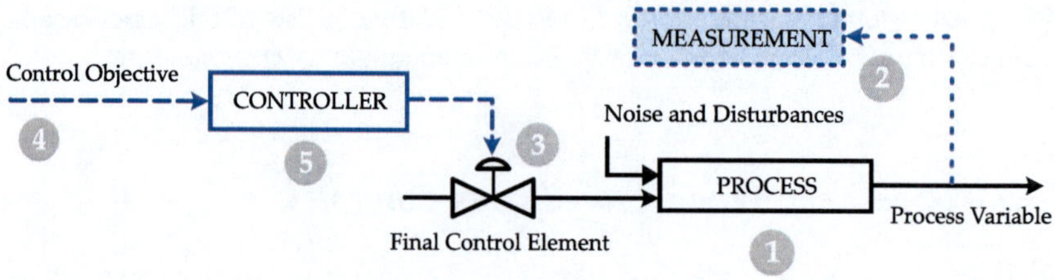

Fig. 2.1 Open-loop control

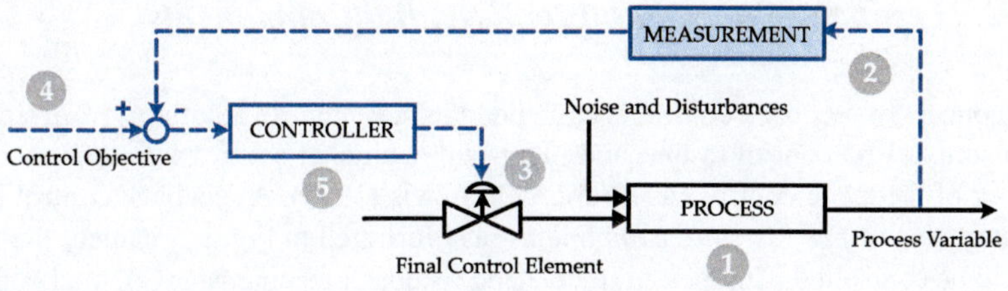

Fig. 2.2 Closed-loop control (feedback control)

a. Operator control: Manual control by the operator is the primitive form of closed-loop control. The operator monitors the value of the controlled variable, compares it with the desired value, and adjusts the manipulated variable (e.g., control valves) to keep the controlled variable at or around its target value. This checking and adjusting keep on as frequently as necessary.
   Driving a car is an example of manual closed-loop control. The driver (operator) periodically checks the speedometer for the driving speed, compares it with the posted speed limit, and makes the necessary adjustment to the gas pedal to stay below (but close to) the speed limit.
b. Automatic control: The `measure-compare-correct` process is highly repetitive and tedious for an operator (e.g., the driver). For continuous operation, this type of manual operation is very prone to human errors. An automatic control system replaces the operator to perform this repetitive work more efficiently and reliably, becoming automatic closed-loop control. For example, cruise control in a modern car provides drivers with relief from fatigue on a long-haul trip.
   Although automatic control is intended for normal operation, operator manual control often needs to take over when the situation is too complicated for automatic control to handle.

The variables in a control loop may have different names in a different context, see Fig. 2.3 for some examples. These names are sometimes used interchangeably in the book.

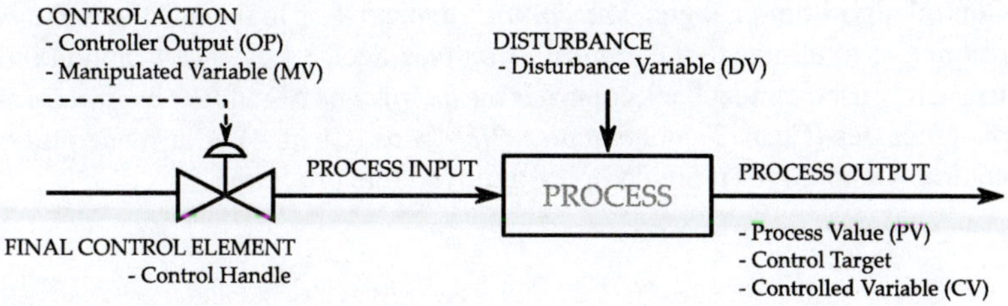

Fig. 2.3 Naming of variables in a control loop

2.1.2 *Feedback Control and Its Essential Components*

The concept of feedback control is embedded in most control solutions, from single-loop simple PID control to unit and plant-wide complex control solutions. A brief history of feedback control can be found in Lewis (1992). A feedback control loop is comprised of *five essential components*, as illustrated in Fig. 2.2, namely the process to be controlled (①), measurements and feedback mechanism (②), final control element (③), the control objective (④), and the control engine (⑤):

1. Process: A *process* refers to the physical plant or equipment to be controlled. The critical information required on the process is the dynamic cause and effect relationships among all the variables and the dynamic response characteristic of each relationship, including the noises and disturbances that affect the process operation (see Sects. 2.1.3 and 6.4 for details).
2. The measurement, monitoring, and feedback mechanism: The *measurement* is the window into the process and must provide an undistorted view of changes in the process variables. We cannot control what we cannot measure.
   A process measurement typically includes a sensor and a transmitter, with the former providing the physical measurement and the latter transmitting the measurement value to the controller as feedback (see Sect. 2.1.4). The availability and credibility of process measurements are essential requirements for process control. A wrong measurement is sometimes worse than no measurement due to the false sense of security.
3. Final control elements (FCEs): The *final control element* is a manipulable device that can influence the process and cause the process output to change in a predictable fashion. The most common final control element is a *control valve*. A control valve is sometimes called a *throttling valve* to distinguish it from an on/off valve. Other control elements such as electric current, motor speed, or switches are also commonly encountered (see Sect. 2.1.5 for more discussions).
4. Control objectives: The *control objectives* serve the operating objectives and dictate the control strategy and functionality of the control solution. For a simple loop, the control objective is to keep the controlled variable at or around its setpoint. In the context of control design, the control objectives determine what variable to control and how to set the target values (see Sect. 2.1.6 and Chap. 7 for detailed discussions).
5. Control algorithm or logic: The *control algorithm* or logic calculates the *control action* to eliminate the *control error* (see Sect. 2.1.7). The technique most frequently used for feedback control is the *on/off control* and *PID control* for simple processes (Chap. 3) and *complex PID-loops* (Chap. 4) or advanced process control (Chap. 5) for more complex multivariable processes.

2.1.3 Control Loop Component: Process Dynamics

Process dynamics is the transient behavior of a process variable responding to a change in another variable. Response to a change is typically not instantaneous but takes some time to re-settle. Some processes (e.g., level) may never settle, and some processes may even diverge. In addition, there are always uncertainties in the process value, such as the effect of measurement noises or process disturbances. For example, a load change[1] is a common disturbance that affects the process response behavior. Figure 2.4 shows a block diagram for an abstract process. The input $u(t)$ is typically a control handle such as a valve; $y(t)$ is the noise-free process output. The actual available process output is the measured variable $z(t)$, which includes the artifacts of noise and disturbances.

Figure 2.5 is an example of an output response $z(t)$ to a step-change in input $u(t)$ and is thus often called *step response*.

With a step change in input $u(t)$, the process dynamics are about the following transient behaviors in the process output $y(t)$:

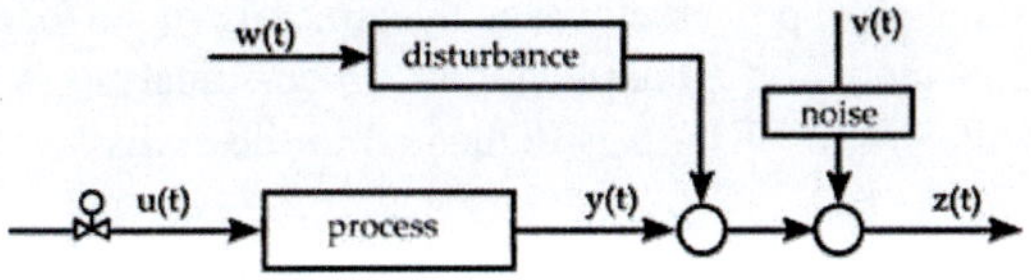

Fig. 2.4 A process with noise and disturbance

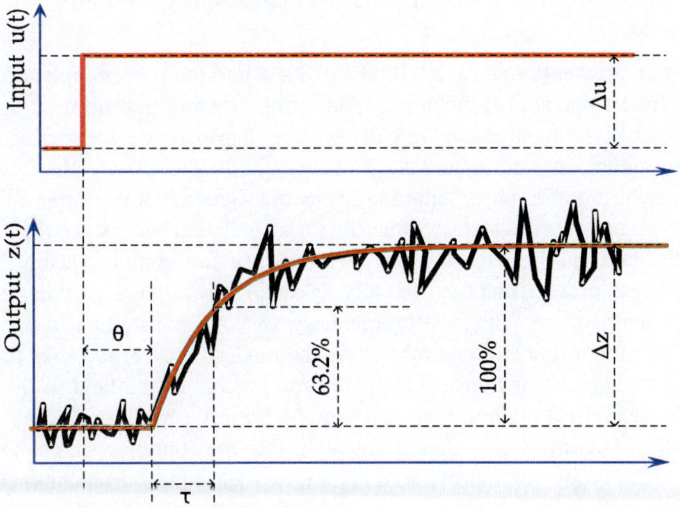

Fig. 2.5 Process dynamics from a step response test

[1] Many times load is used interchangeably with disturbance.

1. Gain (K_p): How much is the eventual change in output $y(t)$? For a stable process, the process output will eventually settle to a new steady state. While for integrating or unstable process, the new steady state may not exist, and the definition of gain is different.
2. Lag (τ): How fast is the change? The process lag is usually related to the material or energy capacitance. The different types of processes differ from each other, mainly in the lag behavior or pattern.
3. Delay (θ): When does the output $y(t)$ start to change? Delay is also called *time delay* or *dead time* and is typically caused by transport delay.

Note that the true output response $y(t)$ is not known due to noises and disturbances. Instead, only the noise-contaminated process value $z(t)$ is available. The true process dynamics $u(t) \mapsto y(t)$ must be deduced from the apparent dynamics $u(t) \mapsto z(t)$. The impact of noises and disturbances is one of the main challenges facing process modeling in model-based applications (Chap. 5).

Most industrial processes can be described with the three parameters above; however, some can have more complex dynamics and require more sophisticated mathematical representations. Sections 6.6 and 9.1 provide more discussions on this topic.

A good understanding of process flow and dynamics is crucial for advanced control applications such as model-based applications. Process analysis is the core content of this book and the entire Chap. 6 is dedicated to the skills and methodology of process analysis.

2.1.4 Control Loop Component: Process Measurements

Process measurement provides a window of view into the process operation and is essential for all operations, including monitoring, control, optimization, and safeguarding. Process measurement typically includes a primary device such as a sensor and a transmitter. The sensor senses the changes in the process variables, while the transmitter converts the sensor output to an electric signal suitable for transmission to other devices or control systems. Sometimes a secondary device is needed to further process the data, such as a multi-phase flow meter (a flow computer) (Fig. 2.6).

The process measurement is typically an analog signal such as pressure, flow, level, and temperature. The measurement needs to be transmitted to the control systems for monitoring and control. Inside the modern control systems, the signal is all digital. Signals for transmission from the process in the field to the control system in the control room are typically in a 4–20 mA current or a 1–5 V voltage signal, which is converted to digital signals inside the control systems. The latest Fieldbus technology, which provides a digital, two-way, multi-drop communication link among intelligent field devices and automation systems, allows the digital signal to be extended to the field.

The control action is calculated based on *credible* and *reliable* process measurements. In an operating plant, metering and measurement is an engineering specialty

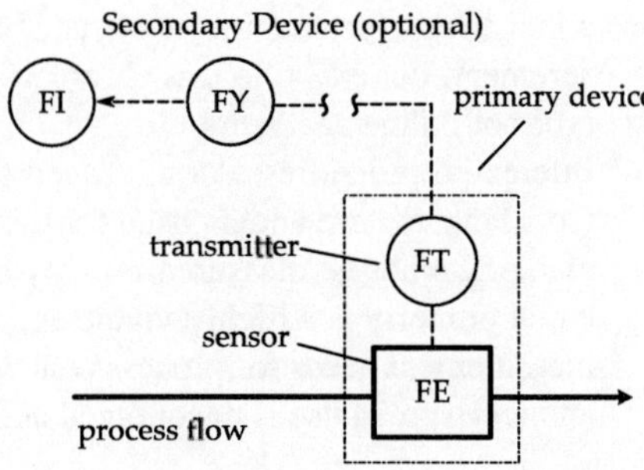

Fig. 2.6 Process measurement components

by itself. For example, custody meters and online analyzers are big topics and require dedicated teams to take care. However, process control practitioners must know what is available and how to use them.

The typical process measurements and the major types of technologies for each measurement type are summarized in Table 2.2:

1. The four conventional process variables are flow, pressure, level, and temperature. A flow can be a liquid, gas, or even solid flow (e.g., polymer granules). A pressure can be gas pressure, steam pressure, or liquid pressure.
2. Multi-phase flow meter: In upstream oil and gas production, there is a need to measure the flow rate at the wellhead. However, the challenge is that the flow is multi-phase (oil, gas, water, and other impurities), and accurate phase flow measurement is not practical with currently available technology. Much progress is being made on this front, including multi-phase flow meters (MPFM). MPFM is a small computer installed at the wellhead to analyze the fluid and infer the individual phase flow rate.[2]
3. Quality: With more stringent requirements on operation and thanks to the technological advances, quality measurement via online analyzers is becoming more

Table 2.2 Typical process measurements

Measurement	Instrument type
Flow	DP-based, vortex, ultrasonic, Coriolis, electromagnetic
Level	DP-based, RF induction, capacitance, radiation
Pressure	Capacitance, piezoelectric, differential pressure
Temperature	RTD, thermal couples, infra-red, optical
Multi-phase flow	Multi-phase flow meter
Quality	Online analyzer, lab sampling
Inferential property	Soft sensor

[2] A related application is the well production measurement discussed in Sect. 12.3, where a process control solution must be provided to work around the limitation of multi-phase flow measurement.

widely available and affordable and has become another common process measurement. For example, gas chromatographs (CG) are one of the most common types of online analyzers.

4. Inferential properties: For advanced control applications, inferential properties (soft sensors) are widely used for those properties that are not directly measurable or cannot be measured reliably or fast enough by online analyzers. Inferential property is a highly viable and cost-effective way of complementing the measurement needs for process control and real-time monitoring. More creative and widespread use is encouraged and expected.

The fundamental requirements for a good measurement are *repeatability* and *accuracy*. For process control, measurement repeatability is typically more important than accuracy except for custody transfer and constrained control since feedback control can automatically compensate for the inaccuracies as long as the measurement is consistently repeatable.

Flow measurement is fundamental because the physical movement of material and energy is a crucial operating requirement. The different flow meters, measurement principles, and measurement results are summarized in Table 2.3 for a quick look-up (Mulholland 2016).

The essential properties of a flow meter that is important to process control include the following:

1. Service type: Certain flow meters are best for gas flow measurement, while others are more suitable for liquid. Steam measurement poses some unique challenges and may have different requirements on flow meters. There are also flow meters such as DP-based that can work equally well on both gas and liquid.
2. Mass flow versus volumetric flow: A flow meter can produce either a mass flow or volumetric flow. Certain flow meters such as DP-based can be configured to produce either.
3. Measurement range: Measurement instruments all have an effective range. The normal operating values should be typically within 10–80% of the measurement range. Oversized or undersized meters may impact the measurement accuracy.

Table 2.3 Type of flow meters and their measuring principles

Flow meter type	Measurement principle	Indicated flow
Orifice plate	Differential pressure $\mapsto$	Raw mass flow
Orifice plate	Differential pressure $\mapsto$	Raw volume flow
Orifice plate	Differential pressure $\mapsto$	Standard volume flow
Vortex	Vortex frequency $\mapsto$ Velocity $\mapsto$	Actual volume flow
Ultrasonic	Radar signal transit-time $\mapsto$ velocity $\mapsto$	Actual volume flow
Ultrasonic	Doppler frequency shift $\mapsto$ velocity $\mapsto$	Actual volume flow
Electromagnetic	Induction voltage $\mapsto$ velocity $\mapsto$	Actual volume flow
Coriolis	Motion $\mapsto$ density $\mapsto$	Actual mass flow

4. Turndown ratio: The measurement typically loses its accuracy when the value falls below the lower range. For a DP-based flow meter such as an orifice, the effective range is typically 10% of the full range in DP measurement. Since the flow is proportional to the square-root of the pressure drop, the effective flow measurement has a turn-ratio of about 30% ($\sqrt{10} : 1$).
5. Cutoff threshold: While all measurement accuracy decreases when the measurement value approaches zero, certain types of flow meters, notably vortex meters, will completely lose measurement readings when the flow rate is below a certain threshold. This threshold is called the *low cutoff* flow. For critical variables, special measures should be taken to take care of the non-availability measurement below the cutoff limit.
6. Pressure loss: Some flow meters such as an orifice and a Venturi tube are based on pressure drop across a measurement device. The pressure loss is usually relatively small compared with the operating pressure. However, if the nominal operating pressure is already low, the pressure loss across the flow meter may become a concern.
7. Response time: All measurements have delays caused by factors such as the location or response time of the measurement devices. For specific applications such as compressor control (see Sect. 11.3), the response time of the measurement becomes vital.

Certain flow meters require square-rooting on the measured value to arrive at the correct flow, while others do not. Those requiring square-rooting are also called the *square-root types*, while those that do not are called the *linear types*. Table 2.4 provides a summary of the typical applications of common flow meters:

1. Differential pressure-based (DP) meters are significant due to their low cost and high reliability. However, the accuracy of DP-based measurement is related to the operating condition. Most of the time, compensation of the measurement results is needed to get the correct measurement. Flow compensation and conversion are discussed in detail in Sect. 10.3.

Table 2.4 Comparison of flow measurement technologies (G—gas, L—liquid, S—steam, M—mass flow, V—volume flow)

Flow meter	Service Type	Mass or Volume	Square Root?	Accuracy	Turndown Ratio	Pressure Loss	Cost
Orifice plate	G/L/S	M/V	Y	M	~3:1	M	L
Venturi tube	G/L/S	M/V	Y	L	~3:1	L	L
Ultrasonic	G/L	V	–	H	>10:1	L	H
Vortex	G/L/S	V	–	L	~10:1	M	M
Electromagnetic	L	V	–	H	>10:1	L	H
Coriolis mass	G/L	M	–	H	>10:1	M	H
Multi-phase	MF	V	–	L			H

2. Coriolis flow meter: A Coriolis flow meter (CM) has a high turndown ratio and accuracy and is typically used for smaller lines where a high-pressure drop is available. Coriolis mass flow meters can work in uni-directional and bi-directional mass flow measurements with large viscosity and density variations.
3. Ultrasonic flow meters: Ultrasonic (US) flow meters are used for gas measurement for larger line sizes. It usually has a high turndown ratio (20:1), with almost no pressure loss.
4. Electromagnetic flow meter: Electromagnetic (EM) flow meters are typically used for the services with a minimum conductivity above the manufacturer's requirement.
5. Vortex meter: Vortex meter (VM) is typically used for gas measurement. A significant drawback is the value cutoff. When the flow is lower than a threshold, the vortex meter will simply stop producing a value.

2.1.5 Control Loop Component: Final Control Elements

Final control elements (FCEs) can be control valves, motors, and switches, with control valves being the most common (Fig. 2.7).

The final control element receives a control signal in the range of 4–20 mA. The opening or closing of the control valve is usually done automatically by electrical, hydraulic, or pneumatic actuators. Most control valves are still pneumatic, driven by a pressure signal of 3–15 psi (0.2–1.0 bar). So the control signal has to be converted from the 4–20 mA signal to 3–15 psi by a current to pressure (I/P) converter. A positioner is usually used to control the opening or closing of the valve more accurately.

The final control element is another overlapping area that process control needs to know *just enough* about what is available and where to go for more information. For this reason, detailed discussions on valves are not provided here. Instead, a list of common valve types, along with their advantages and disadvantages for process control, is provided in Table 2.5.

The most commonly used control valves are globe valves and butterfly valves. Ball valves are typically used for on/off control or process isolation. A check valve, also called a non-return-valve (NRV), is used to prevent reverse flow. A solenoid

Fig. 2.7 Control valves

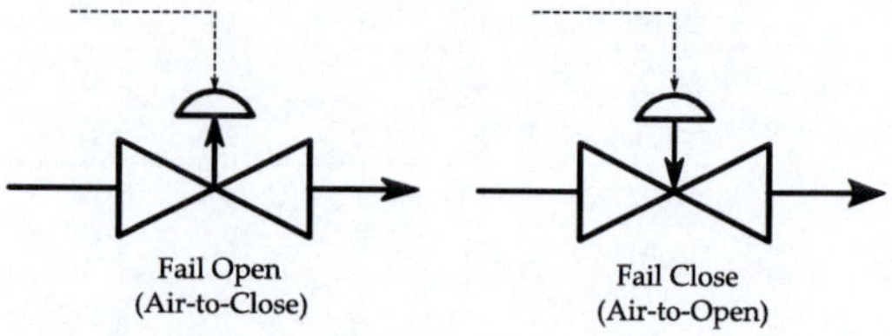

Table 2.5 Types of valves and their pros and cons for control

Valve type	Throttling	Isolation	Tight shutoff	Pressure loss
Globe valve	* * *	*	*	H
Butterfly valve	* *	*	*	L
Choke valve	* *	* * *	* * *	H
Needle valve	* *		* * *	H
Gate valve	*	* *	* * *	L
Ball valve		* * *	* * *	L
Plug valve		*	* *	L
Check valve		*	*	L
Solenoid valve		* * *		

valve is an electromechanically operated valve typically used for responding to a trip signal (or loss of signal). When the solenoid valve is de-energized, it vents the instrument air in the valve, thus causes the valve to go to the fail-safe position.

Key valve properties of concern to process control include the following:

1. Valve characteristic: Valve characteristics refer to the dynamic response behavior. It is crucial to process control because the valve dynamics are an integral part of the loop dynamics, affecting the control performance. Commonly available characteristics include *linear*, *equal-percentage*, and *fast-open*. See Fig. 2.8. The valve *flow coefficient* C_v is defined as the volume of water at 60 °F in US gallons that will flow through a valve per minute with a pressure drop of 1.0 psi across the valve at a *fully open* position. Therefore, the actual flow rate is a function of the valve opening *and* the differential pressure. Valve characteristics provided by vendors are the flow rate versus valve opening under *constant* pressure across the valve. However, the pressure across the valve typically does

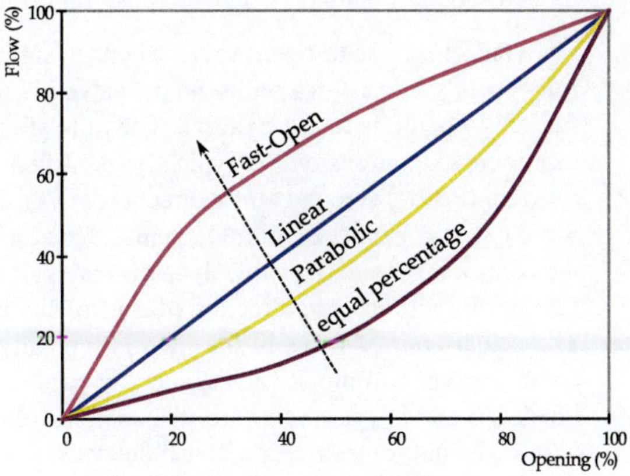

Fig. 2.8 Valve characteristics

not remain constant in practice but decreases as the valve is opened up. The actual valve characteristic is thus different than in the manufacturer's specification. The actual behavior is called the *installed characteristics* as opposed to the manufacturer's characteristics or *inherent characteristics*. For example, an equal percentage valve by specification behaves more like a linear valve when installed. For this reason, control valves in practical applications are often of equal percentage characteristics.

2. Valve sizing: The size of the valve determines the rangeability, which affects the controllability of the control valve. Valve is inherently nonlinear in characteristics (see Fig. 2.8), especially when approaching the two ends of the range. The typical effective range for a valve is between 10 and 80%. The valve sizing should match the operating range so that the valve operates at the mid-range (more linear) during normal operation. It is a more common mistake to oversize a control valve than undersizing it because of the false sense of additional safety. Another consideration is valve leaking (also called *passing*). Due to wear and tear over time, a valve may no longer be able to close tightly. The leaked flow may become a problem for measurement or control. See Sect. 12.3 for an example. An oversized valve may aggravate the leaking problem.
3. Fail-safe position: The position of a failed valve can significantly impact the associated process, equipment, and control design. "Fail" refers to the loss of "power," the medium that moves the valve actuator. The most common "power" medium is the instrument air, which is compressed air at a pressure of approximately 7 to 8 bar(g). As per ISO-5208 standard , control valves may fail in various positions, including the following:

    a. Fail-open (FO): Upon loss of instrument air, the valve jumps to a fully open position under the force of the spring.
    b. Fail-closed (FC): In the case of instrument air failure, the valve quickly goes to a fully closed position.
    c. Fail-last/fail-locked (FL): If instrument air is lost, the valve position stays where it was. A motor-operated valve (MOV) is a typical example.
    d. Fail-indeterminate (FI): The fail position is uncertain or indifferent.

    For a fail-close (air-to-open) valve, a control signal of 20 mA drives the valve to fully open, while a 4 mA will send the valve to fully close. The opposite is valid for a fail-open (air-to-close) valve. The fail-safe position is dictated by process safety considerations and is typically identified on the piping and instrument diagram (P&ID) (see the arrow direction in Fig. 2.7).
    Since a stable feedback control loop must have an overall negative gain to reduce the control error and the valve dynamics are part of the loop dynamics, the fail-safe position affects the direction of control action of the controller.
    The actual control valve in the field is typically an assembly consisting of a control valve, two block valves, and a bypass valve, as shown in Fig. 2.9. The block valve and bypass valve are all manual valves. When the control valve needs to be taken out for service, the two block valves are closed, and the control valve is removed. The flow is then manually adjusted by the bypass valve.

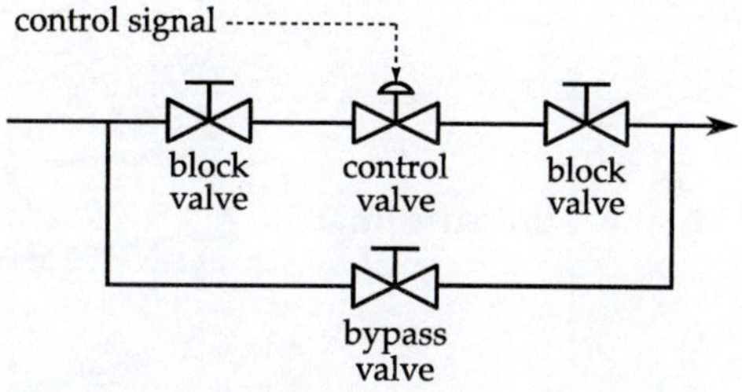

Fig. 2.9 Control valve arrangement

4. Pressure loss: A control valve is a constriction for flow and will always result in some permanent pressure loss. In some applications, the pressure loss may be too significant to be acceptable. The flow-balancing control in Sect. 4.1.4 is an example of unacceptable pressure loss caused by the control valves, and a unique flow-balancing control scheme is designed to minimize the loss.
5. Positioning accuracy: Out of the five components in a feedback control loop (Fig. 2.2), the control valve causes the most control problems. The control valve problems are typically associated with the valve being unable to move to the position as requested. Common issues include *stiction*, *slip*, *hysteresis*, *passing* (leaking), and *deadband*. The critical requirements on valve positioning include the following:

    a. Valve range: For an air-to-open valve, a 100% output from the controller, converted to a 20 mA electric current, should put the valve in a fully open position, while 0% output should produce a 4 mA current to close the valve fully.
    b. Dead angle: All air-to-open valves need to have more than 4 mA current to lift the valve from its seat (for a globe valve) or start turning (for a butterfly valve) from the fully closed position. However, if the valve requires a considerably higher current than 4 mA to start moving, the control performance may be degraded. Similarly, an air-to-close valve should not need too much less than 20 mA to cause the valve to start closing.
    c. Transient response: Upon signal change, the valve should move quickly and smoothly to its new position without much delay or overshoot.

    See Sect. 8.4 for more details on valve problems and their troubleshooting.

2.1.6 Control Loop Component: Control Objectives

Control objectives serve the operating objectives. Generally speaking, from the control point of view, the control objective is to drive the control error to zero. From an operation point of view, the control objectives are to maintain the desired operating point at the target value during normal operation and keep the operating point within

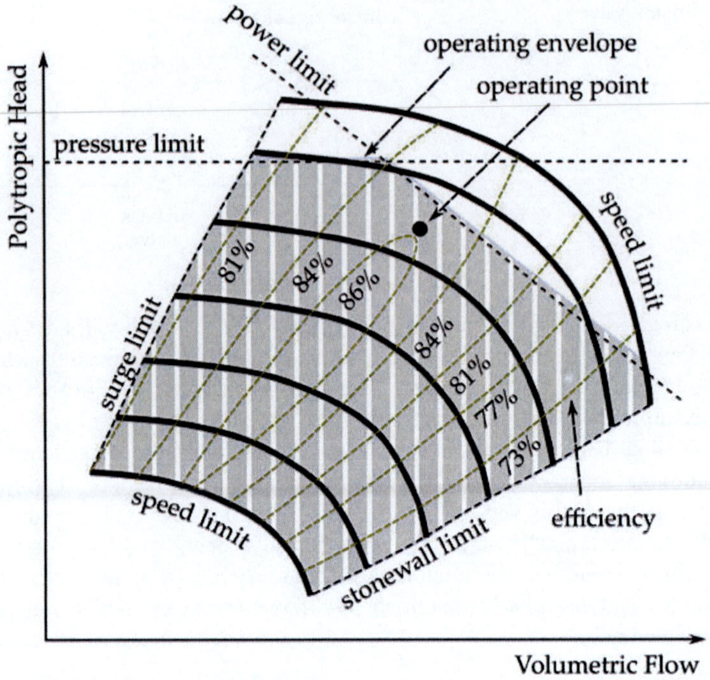

Fig. 2.10 Control objectives and constraints

the operating envelope during a process upset (Fig. 2.10). The control objectives typically include one or more of the following:

1. Quality control: Ensure that production meets all the product specifications.
2. Capacity control: Maintain the dynamic supply and demand balance (in both material and energy).
3. Protective control: Keep the operating point within the operating envelope.
4. Sequential control: Ensure quick and stable transition between different operating conditions.

The control objective dictates what process variable is to control and at what value it should be controlled. For a simple control loop such as a flow controller, this is usually a trivial decision. For a complex control scheme consisting of multiple interacting control loops, however, determining the control objective for each control loop can be the most important and most challenging part of the control design. It requires a good understanding of the process design and operating requirement and a holistic view of all the control loops. Unfortunately, the importance of control objective analysis is often underestimated when multi-input multi-output (MIMO) variables are involved with multiple levels of control. Many controllability and operability issues are related to poor objective setting, which results in a flawed control strategy.

For example, it is unrealistic to expect the home A/C system to maintain a temperature below 10 °C on a hot summer day even if the targeted temperature at the thermostat is set to 0 °C. Another typical example is surge tank level control. The purpose of the surge volume in the tank is to absorb the flow fluctuations. The level should be allowed to fluctuate within a range. Keeping a tight control on the level simply passes the fluctuations downstream and defines the design purpose of the surge tank. A more subtle example is given below.

Example 2.1 "Goal-zero" vision. With the increased awareness of the health safety and environmental (HSE) impact, many companies run HSE "goal-zero" campaigns on HSE incidents.

Is "goal zero" achievable? The answer is no, at least not economically! From a feedback control perspective, once goal zero is reached, the measurement is lost since the number of incidents cannot be negative. See illustration in Fig. 2.11. There is no more indication of the difference (the *control error*) between the actual number of incidents (*measurement*) and the targeted number of incidents (*setpoint*) since both are zeros now. We do not know how safe we are from the next incident until the next incident actually occurs. However, it is precisely the next incident that we try to prevent. A catch-22 situation!

The same or more effort needs to be continuously invested in maintaining the zero incident status, but with no confidence in the result. Therefore, "goal zero" is a good vision but not a realistic control objective or setpoint.

A *near miss* is a precursor to an incident. Many companies implement the number of near misses as another performance indicator with the belief that the number of near misses can predict incidents. Reducing the number of near misses can effectively

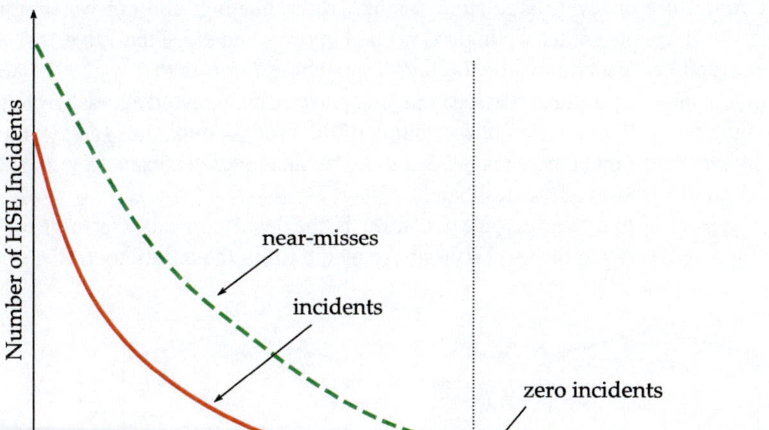

Fig. 2.11 "Goal Zero" for HSE incidents

reduce the probability of incidents. Since the number of near misses is a non-zero target, it can serve as a viable control objective.

Chapter 7 provides in-depth discussions on setting the control objectives and achieving the objectives with the proper control strategy.

2.1.7 Control Loop Component: Control Algorithms

At the center of the control loop is the controller. The controller compares the process value with the setpoint and calculates the next control move.

The controller can be a simple standalone PID controller with one control handle and one control target or a complex control scheme consisting of multiple PID controllers, supporting function blocks, and other elements. The selection of control algorithms depends on many factors, and the critical considerations include performance, reliability, and the life cycle ownership cost:

1. On/off control: On/off control is the simplest controller and thus is also the least expensive in life cycle ownership. The manipulated variable can assume just two values, e.g., open/close for a valve, on/off for a switch, or run/stop for a pump. For this reason, on/off control is sometimes called *snap-acting* or *switching control*. In some level control applications, on/off control is also called *dump control*. Figure 2.12a shows an example of a simple on/off switching control, represented by `LS`, where `S` stands for switching, with an on/off valve `LSV-101`.
   On/off control is a very reliable and cost-effective option for control problems where sloppy control performance is acceptable, such as with large capacity temperature or level, large storage tanks, room heating, and hot-water supply tanks. A simple switching logic is needed to open and close the valve.
   A crucial consideration for the on/off switching logic is that a gap (or deadband) must be provided between opening and closing to avoid excessive chattering/hunting. Some basic understanding of the process dynamics (delay and lag on overshoot/undershoot) is needed to set an adequate deadband. See Fig. 2.13 for an illustration of the deadband.
   A typical application of on/off control is the level control of a large vessel (Fig. 2.12a). When the level rises above a high limit, the on/off controller opens

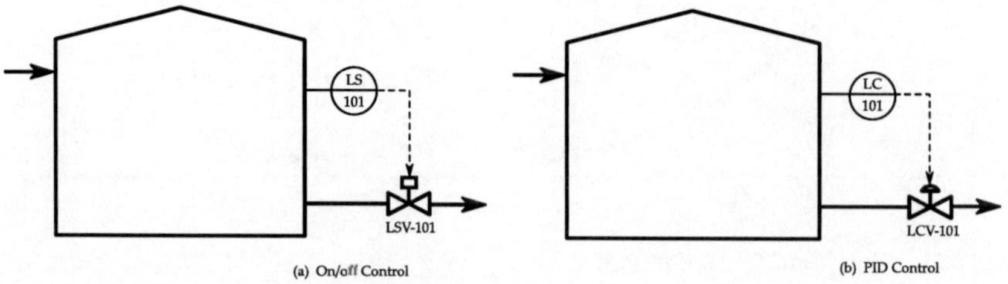

Fig. 2.12 On/off control and PID control

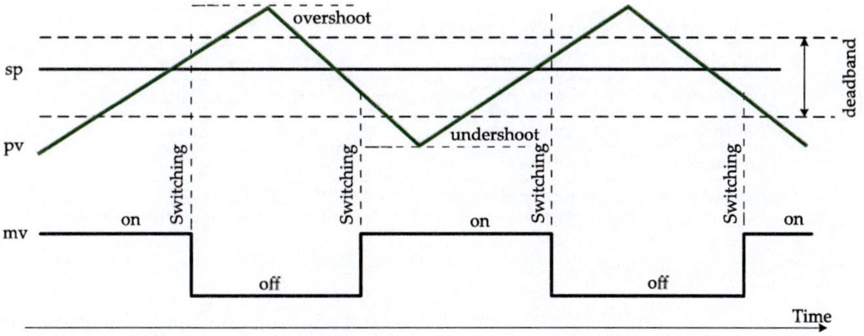

Fig. 2.13 Deadband in on/off switching control

the valve to "dump" the liquid. Conversely, when the level falls below a specific low limit, the valve is closed, and the level will start to rebuild. A ball valve is typically used for on/off control.
Home air-conditioning control is another example where on/off control is used for simplicity and reliability. In some utility plants, the supply of nitrogen or instrument air is controlled by simply starting and stopping the compressor.
The major downside of an on/off control scheme is that it cannot provide the required granularity in control moves to maintain a tight setpoint.

2. PID control: The majority of industrial process control is by PID control loops. A PID controller offers three types of control actions, as shown in Fig. 2.14:

   a. Proportional action is a linear function of the control error and provides a quick initial response to a change: For example, for temperature control, the control action for a 20 °C error is twice that for a 10 °C error. However, since proportional action is proportional to the control error, it requires a non-zero control error to sustain the control action. This non-zero control error results in a *steady-state offset*, which is a significant drawback of proportional control.
   b. Integral action eliminates steady-state errors: The control action is proportional to the time integral of the control error. For example, for the temperature control, as long as there is a difference between the desired temperature and

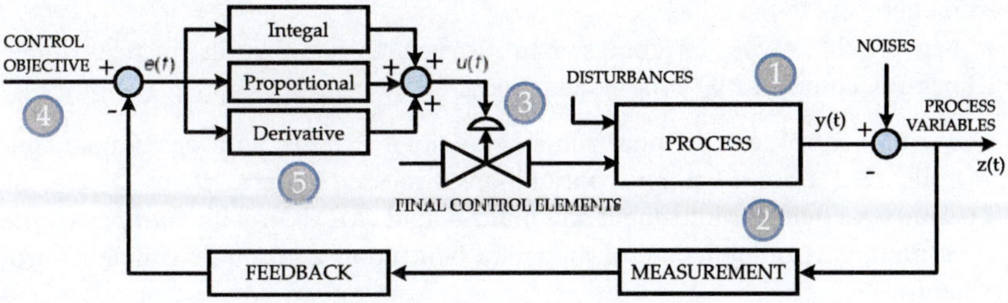

Fig. 2.14 Structure of a PID control loop

Table 2.6 Comparison of on/off control and PID control

Control algorithm	Key advantages	Key disadvantages
On/off control	Simple, reliable, inexpensive	Inaccurate, constant cycling
Proportional control	Simple, fast	Steady-state offset
Integral control	Eliminates offset	Extra time lag
Derivative control	More responsive	Sensitive to noise

actual temperature, the integral action will continue to change the heat input in the same direction.

c. Derivative action provides anticipatory action for better disturbance rejection by sensing the rate of change in the controlled variable. The control action is proportional to the rate of change of the control error. However, noises and disturbances are prevalent in the process variables. The derivative action amplifies noises and may result in too aggressive control actions. For this reason, most controllers in the process industry are P or PI controllers. D action is rarely used.

A PID controller `LC-101` (Fig. 2.12b) can replace the on/off switching control `LS-101` (Fig. 2.12a) for better control performance. Table 2.6 lists the pros and cons of on/off control and PID control schemes. More detailed discussions on PID control are provided in Chap. 3.

2.2 Complex Control Loops

The simple control loops in Sect. 2.1 are comprised of one controlled variable and one manipulated variable. Although simple controllers are dominant in practice, many applications involve multiple controlled variables and manipulated variables and have to be considered together due to the interactions among the variables. One approach is to have multiple simple controllers work together to address the complexities. Table 2.7 provides a list of process control loops by the input/output structure, with detailed explanations in the following subsections, as indicated by the section numbers in the table.

There are three types of process control structures prevalent in practice: simple PID control, complex PID control, and model-based control.

1. Simple control: Single-input and single-output (SISO), with one simple controller. These are the simple control loops discussed in Sect. 2.1.
2. Complex control: Multi-input and multi-output (MIMO), with *multiple simple controllers*. Complex control loops are built upon a group of simple control loops.

Table 2.7 Common complex control loops

	Single output	Multiple outputs
Single input	On/off control (Sect. 2.1.7)	Cascade control (Sect. 2.2.4)
	Simple PID control (Sect. 2.1.7)	Ratio control (Sect. 2.2.6)
	Protective control (Sect. 4.1.3)	Override control (Sect. 2.2.7)
Multiple inputs	Split-range control (Sect. 2.2.1)	Selective control (Sect. 2.2.8)
	Dual controller control (Sect. 2.2.2)	Decoupling control (Sect. 4.1.2)
	Fan-out control (Sect. 2.2.3)	Model predictive control (Sect. 2.3.4)
	Feedforward control (Sect. 2.2.5)	Advanced process control (Sect. 5.2)

3. Model-based control: Multi-input and multi-output, with *one complex controller* performing sophisticated calculation and control logic. Model predictive control is a successful example of model-based control.

PID-based control is sometimes called *baselayer control* as they are implemented in the baselayer DCS/PLC platform; model predictive control is called *advanced process control* and is typically implemented in dedicated software and hardware separate from DCS/PLC. Adding to the confusion, complex PID control and model predictive control are sometimes collectively called advanced process control as opposed to simple PID control.

This section briefly introduces the common complex PID control schemes in widespread use and explains *what* they are and *when/where* they should be used. Discussions of advanced features and practical considerations associated with each complex loop are deferred to Chap. 4.

2.2.1 Split-Range Control

Split-range control is a popular complex control scheme for the following application scenarios:

1. Extending the control range by manipulating more than one control handles.
2. Providing preferential control if there is a difference in the priority of meeting the control targets.

Figure 2.15 is an example of split-range control, where the process flow valve `PCV-101A` is used to maintain the vessel pressure during normal operation. In case the pressure goes abnormally high, the flare valve `PCV-101B` will start to open. With split-range controller `PC-101`, the controller output between 0 and 70% will cause pressure valve `PCV-101A` to open from 0 to 100%. When the controller output increases to more than 70%, the main control valve `PCV-101A` will remain at 100%, and the flare valve `PCV-101B` will open from 0 to 100%, corresponding to 70–100%

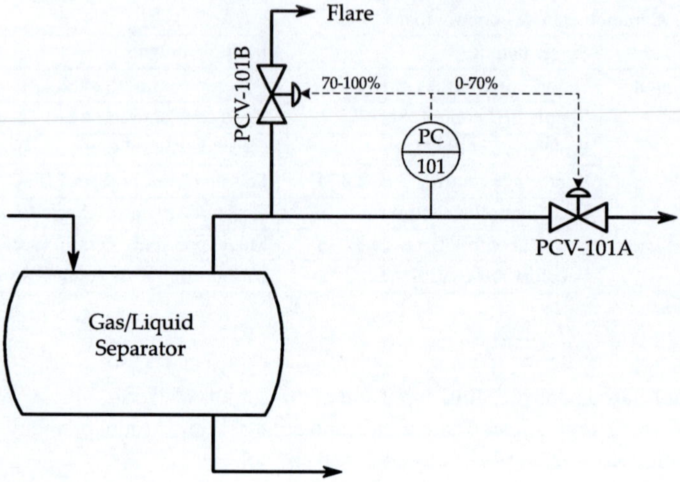

Fig. 2.15 A split-range control loop

of the controller output. The value at which the signal is split is called the *split point*, which is 70% in this example.

This split-range control arrangement provides both preferential control and extended range. During normal operation, the preferred flow direction is through `PCV-101A` to downstream. In case of excessive gas causing high back pressure, the capacity of the main valve `PCV-101A` is not sufficient, and thus the control is extended to the flare valve `PCV-101B` for additional capacity.

Some essential considerations or implementation challenges of split-range control include the following:

1. Proper configuration of the split point: Many split-range control applications use the default 50% split point for convenience or out of negligence. The default 50% is rarely the optimal split point. The correct split point should be calculated based on the dynamics of the two ranges. In the above example, the split point is 70%, and the two split ranges are 0–70 and 70–100% since the dynamics for flaring are much faster.
2. Initialization during mode change: When the controller has switched from automatic mode to manual mode and later back to automatic mode again, the controller output may no longer be in the same range. Proper re-initialization of the controller is required; otherwise, the transition may potentially be bumpy.

See Sects. 4.1.5 and 4.3.4 for discussions on how to address the challenges.

2.2.2 *Dual-Controller Control*

Split-range control is widely used in practice; however, it is often misused or abused due to the negligence of the unique challenges associated with it. In many applications, dual-controller control is a cleaner and more reliable scheme than split-range control.

With dual-controller control, the split-range controller `PC-101` is replaced with two independent controllers (`PC-101A` and `PC-101B`, respectively). They share the same measurement, but each has its own (but different) setpoint and manipulates a separate valve (see Fig. 2.16).

Compared with split-range control, the configuration, tuning, and troubleshooting are overall much simpler. The only downside is that the two controllers must have sufficiently different setpoints to meet the degree-of-freedom requirement (see Sect. 4.1.2 for more discussions).

2.2.3 *Fan-Out Control*

Fan-out control is a special case of split-range control where the same output is sent to two or more control elements simultaneously. That is, the manipulated variables are "linked in parallel". Figure 2.17 is an example of a fan-out control, where the controller output is "fanned out" to more than two valves. This configuration is common in upstream oil and gas operations, where it is not unusual to see a flow controller output fanned out to dozens of injection wells.

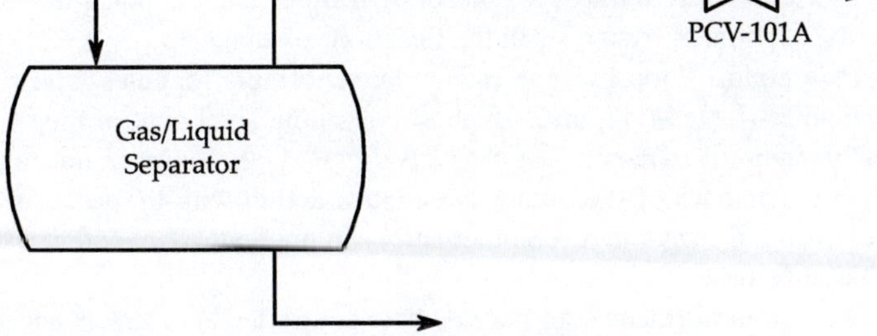

Fig. 2.16 Dual-pressure control in place of split-range control

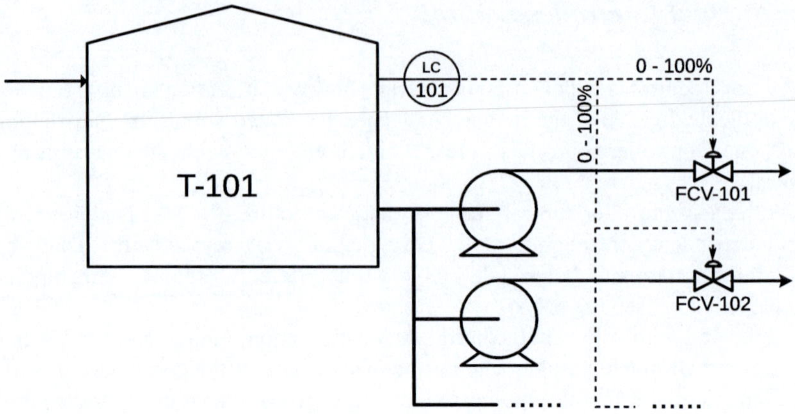

Fig. 2.17 A fan-out control loop

Although fan-out control is straightforward in concept, there are some practical challenges with the implementation and operation:

1. Controller tuning: The number of manipulated variables that the controller output is fanned out to frequently change in response to operating needs. Consequently, the same tuning may not be adequate for all the scenarios. Some kind of gain-scheduling mechanism may be needed.
2. Back initialization and bumpless transfer: When some manipulated variables are taken offline and put back online later, proper initialization is essential but may not be automatically guaranteed.

Section 4.3.3 provides discussions on how to handle these challenges.

2.2.4 *Cascade Control*

Cascade control is probably the most popular complex PID control loop (Marlin 2015). It is widely used to improve control performance by reducing the impact of nonlinearity, stiction, or hysteresis in the final control element.

A cascade control loop has one manipulated variable (handles) and two controlled variables (targets). Figure 2.18 shows a simple level control loop `LC-101` that directly manipulates the flow control valve `FCV-101`. However, due to the slow and complex dynamics of the pump, the control action with the valve is not fast enough to maintain a smooth level. Fluctuations in the flow affect both the flow and level at the same time.

For improved performance, an intermediate controller `FC-101` is added to the control loop to "reject" the fluctuations locally and minimize their impact on the level. The two PID controllers are linked in series, with one controlling the setpoint

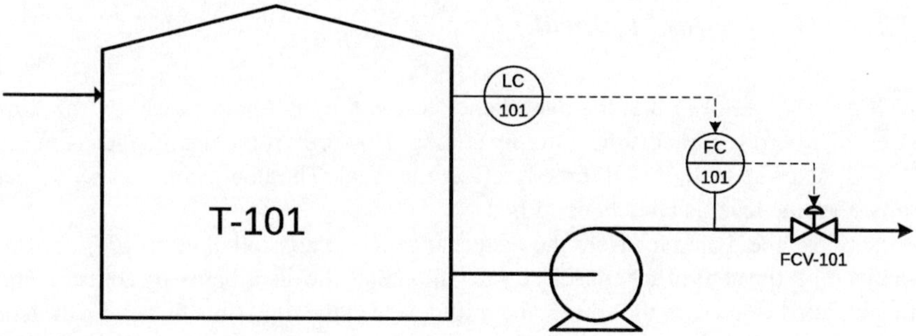

Fig. 2.18 A cascade control loop

of the other. The former is called the *master controller*, the *primary controller*, or the *outer control loop*, while the latter is called the *slave controller*, the *secondary controller*, or the *inner control loop*.

A valve positioner is an example of cascade control. A valve positioner is a simple feedback control loop (P-only controller) to ensure that the actual valve position tracks the desired setpoint. The position setpoint is received from the upstream control actions, and the valve positioner serves as the secondary control loop. The valve positioner suppresses the uncertainties such as stiction or hysteresis imperfection and "hides" them from the PID controller, and ensures that the actual valve position is what the controller requests.

Cascade control is very popular in practice. However, some requirements must be met, and some practical challenges must be addressed to use cascade control effectively:

1. There must be a definite cause and effect relationship from the final control element to the secondary variable and from the secondary controlled variable to the primary controlled variable. In this example, the cause and effect relationships between the valve, flow, and level are self-evident.
2. Cascade control only adds value to the overall control performance if the secondary variable responds to the disturbance sooner and quicker than the primary variable. The secondary loop should typically be five times faster than the primary loop.
3. Tuning: The tuning of cascade control loops should start from the secondary loop with the primary loop in manual mode. When tuning the primary loop, the secondary loop should be in cascade mode.

To cascade is to delegate. Cascade control is like running a business between headquarter and branch offices. The local branch office is charged with making timely decisions on local matters without escalating to the main office. The headquarter only sets the long-term goals and address high-level issues. See Sect. 4.1 for more discussions on the application scenarios and practical considerations.

2.2.5 *Feedforward Control*

In Fig. 2.18, it is seen that the disturbance to the flow is "suppressed" by the flow controller before it affects the outer level loop. However, if the disturbance is in the feed flow to the tank, it will directly affect the level. The flow controller will react only after the level is already disturbed.

Suppose the dynamics from the disturbance to the controlled variable is known. In that case, the control scheme can try to "intercept" the disturbance by adjusting the manipulated value with the correct magnitude and at the right time, in anticipation of "canceling out" the effect of the disturbance before it reaches the controlled variable. This "intercept" and "cancel" approach is the principle of *feedforward control*.

Figure 2.19 shows a boiler heater process where the fuel oil is adjusted following the changes in feedwater flow to maintain the desired temperature (or steam quality) in the outlet flow. Here the feedwater flow rate `FI-101` is the primary flow, and its fluctuations directly affect the temperature `TC-102`. Due to the pure feedback nature, the cascade feedback control loop `TC-102/FC-102/FCV-102` does not take any corrective action before the controlled variable `TC-102` is already affected.

The feedwater flow `FI-101` can be added as a feedforward variable to the existing temperature feedback control loop for improved control performance. The water flow rate is measurable, and the dynamics from the flow `FI-101` to the temperature `TC-102` are easily known and relatively stable.

Feedforward control is an open-loop control loop added to a feedback control loop. The feedforward control action is calculated based on a good understanding of the dynamics between the disturbance and controlled variable and must meet the following conditions to function correctly:

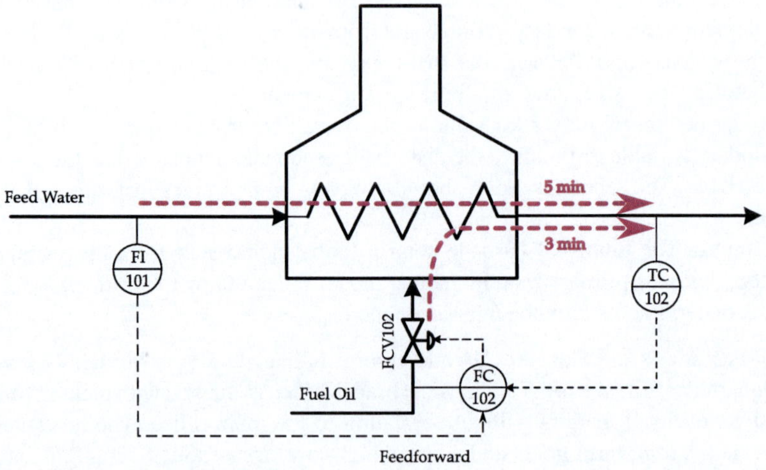

Fig. 2.19 A feedforward control loop

1. The dynamic relationship between the feedforward and controlled variable (mainly the gain and delay) must be known and remain relatively constant over time.
2. The time to reach the controlled variable `TC-102` from the feedforward variable `FI-101` must be longer than that from the manipulated variable `FC-102` for best performance ("complete cancellation").

The feedforward compensation to feedback control can be either dynamic or static. A static compensation only requires the gain compensation for the dynamics, while a proper dynamic compensation requires knowledge on the full dynamics, including gain, delay, and lag. Section 4.1.2 provides discussions on some practical considerations for the implementation and operation of feedforward control.

2.2.6 Ratio Control

Ratio control is a special case of feedforward control. A ratio control scheme has two inputs, the primary flow and the secondary flow. The primary flow is also called the *wild stream*. The ratio controller manipulates the secondary flow stream to achieve a specified ratio with a primary flow stream. The ratio controller's setpoint is thus the ratio between the secondary flow and primary flow:

$$\text{ratio} = \frac{[\text{secondary flow rate}]}{[\text{primary flow rate}]}.$$

This output of the ratio controller is sent down to a PID flow controller to manipulate the secondary flow.

Figure 2.20 shows an example of a mixer with a primary stream `FI-101` and a secondary stream `FC-102`. The control objective is to maintain the desired ratio between the primary and secondary flows to meet the quality requirement `AI-101` on the mixed flow.

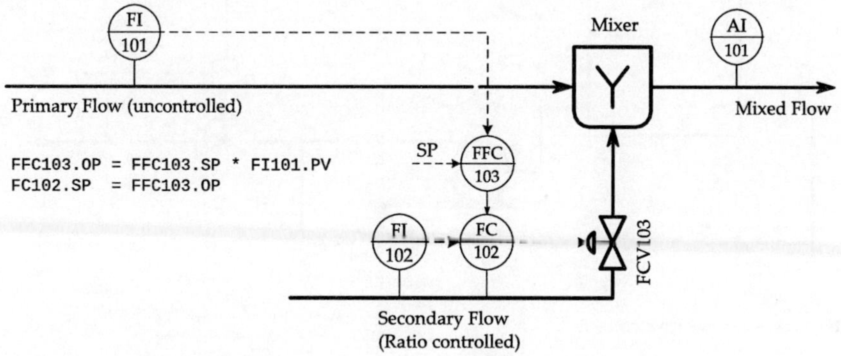

Fig. 2.20 A ratio control loop

Ratio control is similar to cascade control, except that the ratio control loop does not need to be five times slower than the flow control loop. Besides, the ratio typically is not the ultimate goal of control. In this example, the quality variable after the mixer `AI-101` is the true objective.

A special consideration for ratio control is that ratio calculation is subject to noise and failures, explicit division calculation should be avoided, and the DCS built-in ratio control blocks should be used instead.

2.2.7 *Override Control*

Override control is used to maintain one controlled variable on target without violating constraints on other controlled variables. In an override control, there are two or more controlled variables (targets) sharing one manipulated variable (handle), typically through a high selector (for high override) or low selector (for low override). At any time, only one controller is active, i.e., only one output is selected by the selector and sent down to the next level of control elements such as a valve. The control action requested by the most demanding controller is selected and accepted.

An example in our daily life is the so-called adaptive cruise control in newer cars. The regular cruise control maintains a speed target by manipulating the fuel. Adaptive cruise adds another control target to the requirement: the distance with the car in front. When following too close, the adaptive cruise control overrides the speed control and automatically slows down to ensure a safe distance.

An override control can be an overriding protective control such as the pressure controller `PC-103` in Fig. 2.21, where the primary controller `LC-101` is the regulatory control during normal operations, and the protective controller `PC-103` only kicks in during abnormal operating condition that cause the pressure to go higher than its limits.

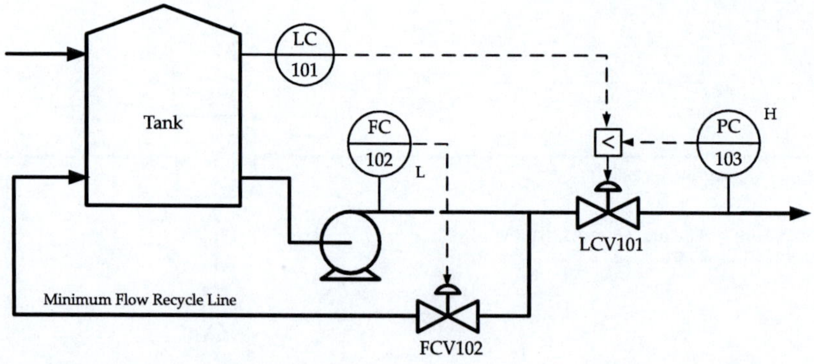

Fig. 2.21 A protective control

An override control can also be an overriding selective control where all the controllers have an equal chance to be selected by the selector. There is no distinction between normal operating conditions and abnormal conditions in their roles. For example, two tanks share one pump for level control. A selector can be used to select which level controllers should be in control. Whether a high selector or low selector should be used depends on whether the two tanks' maximum or minimum level is maintained.

The principle stays the same for both protective and selective override control: each controller in the override control scheme is an independent PID control loop. The output of the most demanding controller is automatically selected as the control action, while the outputs of the other controllers are simply ignored.

There are more control targets than control handles. How to handle the inactive controllers whose outputs are ignored is the challenge for override control. For those controllers, which are effectively open-loop, the integrating action of the PID controller may drive the output to saturation. When the controller is required to become active, the output needs time to come out of saturation before being selected. This delay is typically unacceptable for a protective control.

This saturated condition is called *reset windup* or *integral windup*. Prevention of reset windup is crucial when implementing override control. Section 4.3.3 provides detailed discussions on the prevention of reset windup.

2.2.8 *Selective Control*

When multiple handles are available to control multiple targets, and at any given time, only one pair of target and handle needs to be connected, then a switch, either manual or automatic, can be used.

The practical consideration for selective control is protecting those control handles or control targets that are not currently selected. In the case of multiple targets, it is similar to the scenario of overriding control (Sect. 2.2.7) and requires anti-reset windup protection, while in the case of multiple control handles, the same bumpless transfer behavior as in fan-out control (Sect. 2.2.3) needs to be considered.

2.3 Advanced Process Control

PID is powerful but has its limitations! Advanced process control (APC) technologies are developed to overcome the limitations of PID control. Model predictive control (MPC), adaptive control, and nonlinear control are examples of advanced process control.

2.3.1 Limitations of PID Control

While PID controllers can produce an acceptable performance for most applications, they can perform poorly in some. The fundamental limitation of PID control is that it is based on a feedback mechanism with a simple structure. Besides, the tuning parameters are not based on direct knowledge of the underlying process. They remain constant even when process dynamics have changed. As a result, many industrial processes are beyond the capability of PID controls due to the following challenges:

1. Multiple inputs and multiple outputs with strong interactions.
2. Long time delay and very slow dynamics.
3. Nonlinear time-varying dynamics.
4. Design limits and operating constraints.

Advanced process control solutions are developed to overcome the limitations. It is built on top of baselayer controls, deals with a more extensive scope of problems, and aims at higher level performance targets such as quality, efficiency, and even profitability. The relationship between baselayer control and advanced process control is illustrated in Fig. 2.22.

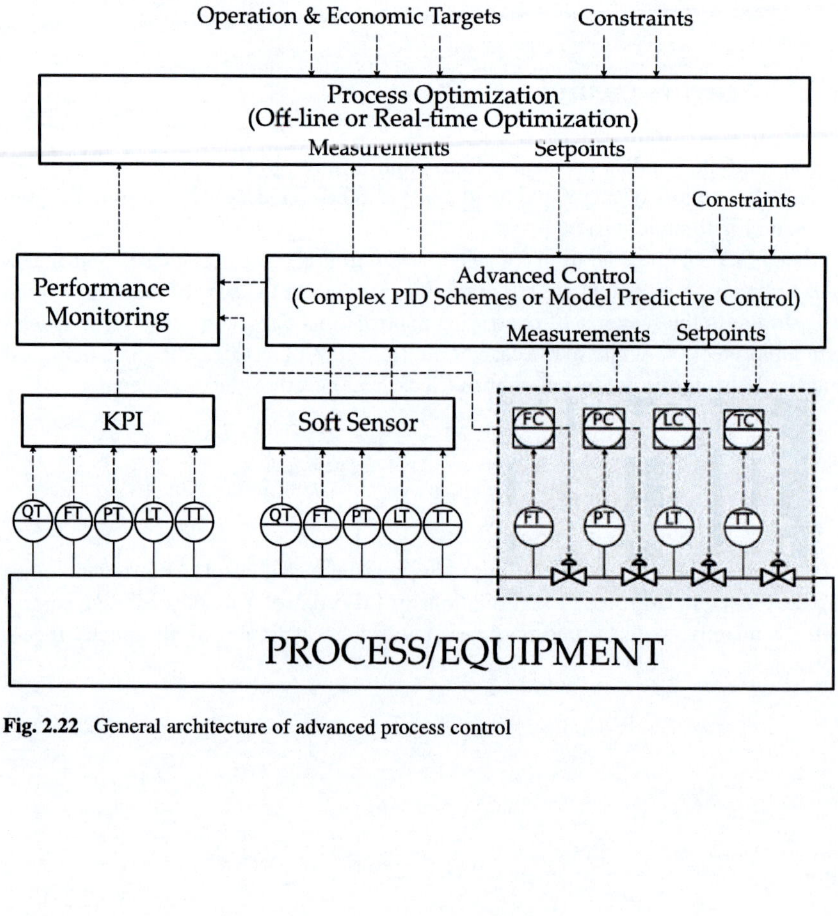

Fig. 2.22 General architecture of advanced process control

2.3.2 Process Models: Abstraction and Generalization

There are countless industrial processes of various types, and each requires dedicated expertise. It is unrealistic for a process control person to have intimate knowledge of all processes. Instead, process control relies on a generalization and abstraction of the processes to do control. This generalization and abstraction are via *process models*.

A process model is a mathematical representation of the external cause and effect relationship between the process inputs and outputs. The process model is derived from process input/output data, which are typically obtained through deliberately disturbing the process variable to reveal the causal relationship. This activity is called *plant test*, *step test*, or *bump test*.

Model is at the center of most advanced process applications. Once the process model is available, it can be used for many purposes, including simulation, control, and fault detection. See Fig. 2.23 for an illustration:

1. Simulation (Fig. 2.23b): With a known process model, we can observe the process output response to the specified changes in the process input.
2. Control (Fig. 2.23c): With the process model known, we can back-calculate the process inputs to achieve the desired process outputs.
3. Fault detection (Fig. 2.23d): With the process model known, and with both process input and output data available, trying to determine if the process has deviated from what it was when the model was built, i.e., to detect if a model/plant mismatch has developed.

There are many types of process models, from heuristic, empirical, to first-principle-based. Advanced process control applications are all based on a particular type of process model. The more advanced the control is, the more reliance is on process understanding and process models. The technique for obtaining process

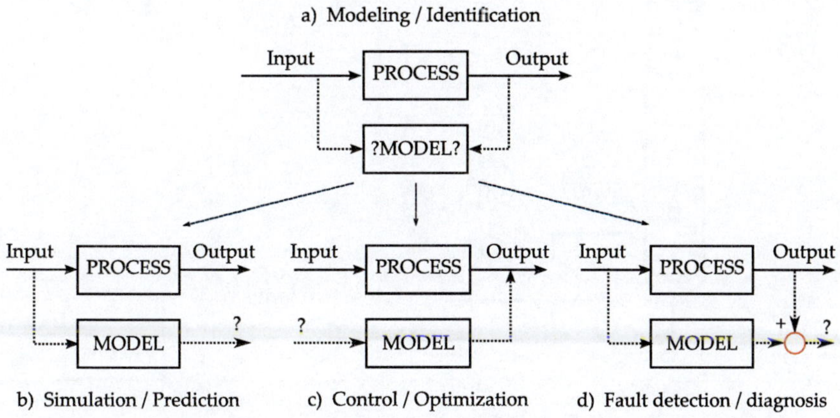

Fig. 2.23 Process model and its applications

models from input/output data is called *process identification* and is a core competence for advanced process control (APC) practitioners. See Sect. 5.2.1 for more detailed discussions on process models and process modeling.

2.3.3 *Inferential Property (Soft Sensor)*

Reliable and accurate measurement is a prerequisite for control. As process control moves from controlling a specific process variable to controlling product quality, online analyzers and lab sampling become increasingly important. However, despite rapid technological advances, most online quality analyzers are still not sufficiently *reliable*, *fast*, or *accurate* for closed-loop control.

Inferential property estimator, commonly known as *a soft sensor*, is a *predictive analytic* that uses faster and more reliable process measurements to infer or predict product quality. Online analyzer measurement or lab result provides the feedback to correct the prediction periodically.

Figure 2.24 shows an example of a crude oil stabilizer, where the *true vapor pressure* (TVP) is measured downstream of the stabilizer. The TVP value is a quality target for control. However, the TVP online analyzer is neither reliable nor fast enough for a PID controller to operate in closed-loop confidently.

An inferential property estimator (IPE) is built based on a process model between the pressure/temperature in the stabilizer and the TVP value. This model runs online to predict the TVP value using the fast and reliable pressure and temperature measurements. A PID controller can then be used to control the predicted TVP value instead of the unreliable measurement from the analyzer.

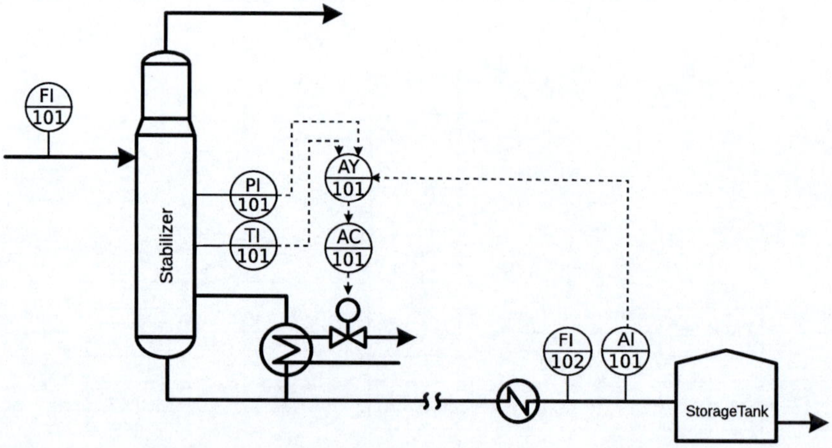

Fig. 2.24 An example of inferential property application

The predicted TVP value is periodically corrected with feedback from the online analyzer `AI-101` when it deems that the analyzer reading can be trusted. If the analyzer is unavailable (e.g., offline for calibration) or is believed to be untrustable, then the TVP controller `AC-101` can continue to run on the predicted value (within a limited time).

There are many practical challenges for successful IPE implementations, such as the following:

1. Availability of accurate predictive models, including both the model structure and model parameters.
2. Quality assessment of the online analyzers for deciding if the analyzer value can be trusted to correct the IPE value.
3. Timestamp matching between the predicted values and the analyzer values.

See Sect. 5.3 for more detailed discussions on inferential property estimators.

2.3.4 Model Predictive Control

Model predictive control (MPC) is an advanced process control paradigm that has gained widespread applications in the process industry. Figure 2.25 illustrates the control architecture. It is evident from a comparison with Fig. 2.2 that MPC is a natural extension of the basic feedback control.

At the center of the MPC technology is the process model. The process model captures the multivariable interactions in the process dynamics and provides the convenience for more sophisticated calculations such as constraint handling and optimization. Compared to PID-based baselayer control, MPC offers the following capabilities beyond the reach of a PID controller:

1. Multivariable: Multiple controlled variables (CV) are controlled with multiple manipulated variables (MV). The cause and effect relationship and the dynamic response characteristics are elegantly captured in the multivariable process models.

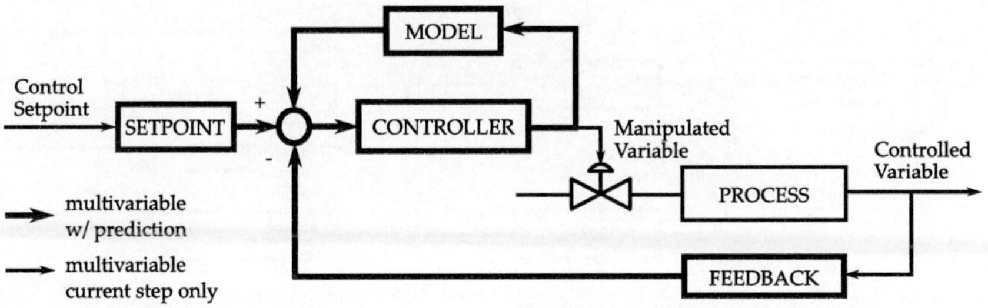

Fig. 2.25 Model predictive control

2. Predictive: The process model provides the capability of predicting the process outputs into the future. Therefore, MPC controls the process based not only on where the unit is currently running but also on where it is predicted to run in the next few minutes or even hours.
3. Constrained: MPC monitors and honors the engineering limits and operating constraints when computing the control actions. The sophisticated MPC calculation allows constrained optimization to produce the next moves that will not drive the operating points outside the various constraints.
4. Optimizing: MPC pushes the operating points toward the most profitable operating areas, typically near the constraints.

The primary challenge for MPC is in obtaining high-quality process models.

2.3.5 *Self-Tuning PID Control and Adaptive Control*

Another important but very challenging direction for process control is adaptive control. Due to the nonlinear and time-varying nature of the process dynamics, the performance of a controller with a fixed structure and constant parameters will typically deteriorate over time. Adaptive control modifies the controller (structure and parameter) online following the changes in process dynamics or operating conditions. Figure 2.26 illustrates the concept and the adaptation mechanism for general adaptive control.

A fully adaptive MPC is highly desirable, but the current technology is still not sufficiently robust to handle abnormal conditions. The challenge is in the credibility of the process model identified online, dictated by the data quality.

A small step in the direction of adaptive control is self-tuning PID, which performs automated PID tuning by introducing step changes, model identification, tuning improvement, all online. However, because of the same concern on data quality, self-tuning is yet to gain more trust and acceptance by field engineers.

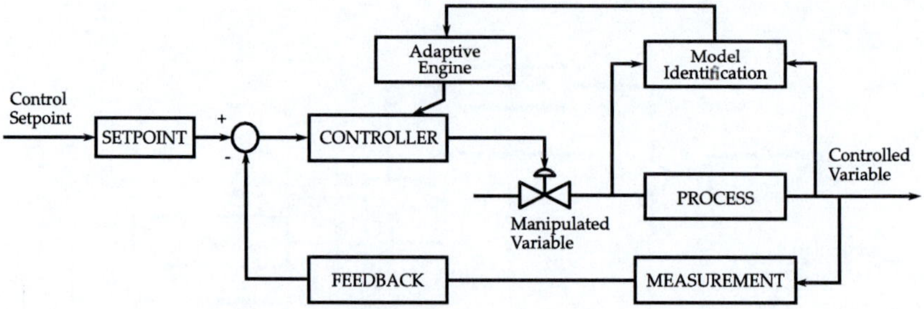

Fig. 2.26 Generic architecture of adaptive control

2.3.6 Control Performance Monitoring

Performance monitoring can involve all levels and aspects of the operation and production, including data quality, analyzer health, control loop performance, equipment status, unit-level summary, and plant-level overview. Monitoring the performance of the control loops is an essential component of performance monitoring.

Control performance monitoring relies on well-defined and well-understood KPIs. *Big data* and *predictive analytic* are playing an active role in performance monitoring in recent years. *Artificial intelligence* (AI) is a hot topic for achieving operational excellence. However, machine intelligence has not (yet) reached the level of human intelligence. The most effective monitoring and control are still based on sound engineering insight into the physical working of the process. Process models, from mathematical to first-principle, are still preferred whenever available. Reliability is the biggest hurdle from passive monitoring to real-time closed-loop control.

2.3.7 Real-Time Optimization

Real-time optimization (RTO) is an extension of closed-loop process control. RTO optimizes the process performance on a larger scale (Fig. 1.10). Unlike traditional process controllers or model predictive control, the scope of RTO typically covers an entire unit or even a plant. RTO is built upon large-scale steady-state first-principle models and relies on a sophisticated optimization engine to optimize the process performance, usually measured in terms of profitability, by taking into account the real-time operating data, market data, inventory data, operating constraints, etc. The results of the optimized operation condition are sent down to model predictive control (MPC) or baselayer control as control setpoint or operating limits.

The fundamental nature of the models allows physically meaningful parameters to be adjusted to reflect changes in plant equipment performance, such as catalyst deactivation or heat exchanger fouling, or compressor blade erosion. The model parameters are adjusted to make the calculated values match the actual measurements (data reconciliation) before each optimization calculation. RTO continuously checks and validates the measurement data to ensure that the process is at steady state before using the measurements to fit the parameters.

The main challenge of RTO is in the quality and quantity of the equations. A modestly complex process unit may require hundreds or even thousands of equations and is usually a very time-consuming activity to build. Besides, the solving of the equations is computationally intensive for online minute-to-minute execution.

2.3.8 Dynamic Simulation and Operator Training Simulator

Dynamic simulation and *operator training simulator* (OTS) system are valuable tools for many purposes:

1. Operator training for plant startup: An OTS system has been most successful in training operators to start a new plant well before the plant is ready to start. Experience shows that a well-utilized OTS can reduce the startup time by more than one-third.
2. Operator training for knowledge retainment: With the successful implementation of MPC, the operators gradually lose their operating skills, especially in handling plant emergencies. So far, OTS is the only viable training method to help the operators retain and refresh their operating skills.
3. MPC modeling with dynamic simulation data: Developing MPC models with dynamic simulation or OTS data even before a new plant is built is a promising and cost-efficient way of process modeling. The high-fidelity dynamic models built for dynamic simulation have sufficient accuracy to generate the input/output data required by MPC modeling.
4. "What if" scenarios analysis: Dynamic simulation during the design phase of the plant can often identify many operability issues. The author had the experience in a recent large project where a dynamic simulation study before the start of detailed design revealed that the designed operating points are too close to the pressure trip limits and affect the operating capacity. A large part of the process design was subsequently modified to correct the problem in time.

Process control is usually the advocate for OTS and typically leads the effort in the initial system development. The main problem, however, is with the ownership of the system after delivery. Dynamic simulation and OTS need model updates following process changes. Process control rarely has the staffing and budget to update the models. It is more appropriate for the ultimate user, operations, to take over and assume the responsibility, but they rarely do. Most OTS systems end up losing sync with the actual plants and eventually are abandoned.

2.3.9 Alarm Rationalization

Modern control systems have made alarm configuration so easy that alarm flooding has become a common problem across the industry. A large percentage of the alarms are *nuisance alarms* that distract the operators instead of helping them. *Alarm rationalization and management* (ARM) are becoming mandatory in many operating plants. Industrial standards such as EEMUA-191 and ANSI/ISA88 are in place to guide the alarm management effort.

After a new plant is built, many alarm configurations need to be reviewed, adjusted, and a large part of them need to be disabled or removed based on the initial operation experience. Process control engineers are typically invited to be part of the alarm rationalization effort.

Alarm rationalization is typically based on a careful evaluation of the HSE impact and urgency on each alarm. The HSE impact is measured by the potential loss in terms of health, safety, or environment impact if the alarm event was not addressed. The urgency is the estimated response time for the operator to address the alarm event. Based on the HSE impact and urgency, the alarms are classified into different types and priorities and assigned appropriate limits.

The alarm configuration is dependent on the operating mode. The alarm limits are typically different for startup mode, normal operating mode, scheduled shutdown mode, emergency shutdown mode, and turndown mode. In a sophisticated alarm management system, the new alarm limits and priority can be automatically downloaded to the control systems during mode change. For example, during startup and shutdown, the flow rate and pressure can go down to zero, and thus many low flow and pressure alarms may go off. For a particular mode of operation, many alarms may have to be disabled to avoid nuisance alarms. This mode-based alarm suppression is called *dynamic alarm suppression*. In addition, many process variables have a cause and effect relationship between them. An alarm on one variable may trigger the alarm on another. In this case, only the alarm on the first variable is of concern.

Alarm configuration is included in the control design as a critical step. When designing a complex PID control scheme, it is crucial to validate the control setpoints against alarm limits, protective control limits, and SIS trip limits to ensure sufficient spacing (thus response time) between the limits (Fig. 1.14).

2.4 Process Control Design

Process control is required during the design of a new plant or troubleshooting of an existing plant. Process control design requires solid process control knowledge and a high level of problem-solving skills, which can be possessed only by senior process control staff.

2.4.1 Process Control Philosophy

Process design dictates what kind of process control strategy is needed, and the process control design validates the process design for feasibility and operability. This interaction ensures that process design has process controllability in mind in order to achieve high operability, while process control design truly understands the intent of process design to address the operating needs effectively. Below are a few questions at the philosophy level that need to be answered early in a process design:

1. The big picture versus the details: It is critically important to start with a "full-view" picture in mind when designing new (or troubleshooting existing) control solutions. The interaction and inter-operation between processing units and even plants dictate the high-level control objectives and overall control strategy, which in turn determine how the next level of details should be laid out. A top-down approach for process control design is recommended following the different phases of project execution.
2. Level of automation: The degree of sophistication of the control solution can vary drastically from totally manual control to fully automated control. The level of automation to achieve is always a trade-off between desire and practicality. It is a compromise of many factors, including HSE impact and economics. In general, the same ALARP (as low as reasonably practical) principle (NOPSEMA 2015) should be followed.

   The car driving is a good example of this compromise between desire and practicality and the choice is a compromise of many factors including financial situation and risk tolerance:

   a. Normal driving: Manual closed-loop control. The driver

      - `measure` speed and direction;
      - `compare` with posted speed and lane position;
      - `correct` with fuel, steering wheel, and brake.

   b. Cruise control: Automatic closed-loop control on speed. The cruise control system

      - `measure` speed;
      - `compare` with the posted speed;
      - `correct` with fuel.

      The driver takes care of direction and brake.

   c. Adaptive cruise control: Automatic closed-loop control on speed and brake. The cruise control system

      - `measure` speed and following distance;
      - `compare` with the posted speed and the pre-set following distance;
      - `correct` with fuel and brake.

      The driver takes care of direction only.

   d. Self-driving car: Fully automated closed-loop control. The self-driving system

      - `measure` speed and following distance;
      - `compare` with the posted speed, the pre-set following distance, and lane position;
      - `correct` with fuel, brake, and steering wheel.

      The driver is hands-free.

3. Simplicity versus sophistication: The simplest is often the most reliable. For process control, reliability is often more important than sophistication. The most advanced may not necessarily be the most appropriate! Most of the time, the best solution is the "fit-for-purpose" solution, which has the best value on return in terms of "total cost of ownership."
   Various techniques may be used to improve the control of a process. However, as the complexity of the control system increases, so does the cost for operator training and maintenance. The reliability of a complex control system also tends to decrease. It is common to see sophisticated control loops taken offline by operators because they are too "sophisticated!" Therefore, it is compulsory to strike the right balance between manual and automated process control and between the complexity of the control system and the benefits provided.
4. First-principle versus feedback: For process control design, the governing principle is the material and energy balance and process dynamics; therefore, it is imperative to always go with the physical laws rather than fighting them. However, process engineers rely heavily on first-principle-based calculations to enforce the material and energy balance, which a reliable and straightforward feedback control can readily achieve. In a project team comprising both process engineers and process control engineers, this difference in thinking is often a clash point in design philosophy.
5. Role of the operators: The level of operator intervention is another crucial consideration in process control design. Some sites rely heavily on panel operators to make most operation decisions, while others pose many restrictions on what the operator can do. For example, some operations may forbid the operator to make setpoint changes to the protective controllers, while others may allow the operator even to change PID tuning parameters at will.
   This operation decision can have a significant impact on the process control strategy and performance. For instance, a protective controller is often locked by the operator, and no changes are allowed even to the controller setpoint and mode. However, a bad value in the measurement may shed the controller to manual mode. Since the controller has been locked from the operator, including mode change, back-end logic needs to be added to unlock the controller to allow the operator to switch the controller back to *normal mode* (but not any other mode).
6. Maintainability and sustainability: The control performance will inevitably deteriorate over time. A poorly performed control scheme not only reduces production profits but may also become a liability. Sustainable performance and easy maintainability should be critical considerations in process control design.

2.4.2 Process Control Life Cycle

Many operating challenges originate from poor overall control design or incorrect implementation. Flaws in the overall control strategy are very challenging to diagnose

Table 2.8 Process control deliverables

Project phase	Process control deliverables	Check point
	Initial process control narratives (PCN)	
Conceptual design	+ Process flow analysis + Overall control strategy + Control overview diagrams	Design review
Front-end design (FEED)	+ Key process control loops + Process control schematics	Design review + HAZOP review
Detailed design	+ Detailed function specifications + Loop configuration details	Design review + HAZOP review
Implementation	+ DCS control modules and sequences + Logic flow diagrams	Factory acceptance test (FAT)
Commissioning	+ "As-Built" updates	Site acceptance test (SAT)
Maintenance	+ "As-Built" updates	MOC approval
Improvement	+ "As-Built" updates	MOC approval

and costly to correct after the plant is built and operating. Therefore, process control inputs at the early stage of process design are critical to eliminating many operational problems from their source.

Large projects are typically carried out in phases such as *appraise*, *select*, *define*, *execute*, and *operate*. The project design goes through *conceptual design*, *front-end engineering design*, and *detailed design*. Table 2.8 is a simplified list of the project phases and the process control deliverables at each phase. The process control narrative (PCN) document, the key project deliverable from process control, is progressively developed through the different phases of a project to become the authoritative document on process control implementation and operation. Even after the plant starts, the PCN document remains a live document and is updated by incorporating all the ongoing changes, following proper *management of change* (MOC) procedures.

2.5 Summary

Process control plays an indispensable and irreplaceable role in process design and operation. This chapter puts together a knowledge framework summarizing the core process control technologies currently in use (or in need) in the process industry and how they have helped production and operation:

1. Feedback control is ubiquitous in process operations. On/off controllers and standalone PID controllers account for most controllers in operation. PID control is the most successful feedback control and consists of five essential components, which should always be viewed together when discussing a PID control loop.

2. Complex PID control: Most practical problems consist of multiple control targets and control handles that interact with each other. Several PID controllers and supporting function blocks work together as a complex control scheme to address the process complexity and become the complex PID control scheme.
3. Advanced process control: Anything that is above and beyond complex PID control is considered advanced process control. The most popular advanced process control is model predictive control (MPC), while other advanced control technologies in use (or in need) include nonlinear control and adaptive control. As PID control is standard for simple control problems, MPC has become the *de facto* standard for complex control problems.
4. Process control design: Process control plays a critical role in the entire project life cycle. Process control narrative is the crucial deliverable by process control and is a progressively developed document valuable for many purposes.

References

Åström KJ, Kumar PR (2014) Control: a perspective. Automatica 50(1):3–43

Lee W, Weekman VW (1976) Advanced control practice in the chemical industry. AICHE J 22(27)

Lewis FL (1992) Applied optimal control and estimation. Prentic-Hall, Chapter 1. A Brief History of Feedback Control

Marlin TE (2015) Process control—designing processes and control systems for dynamic performance, 2nd edn. McGraw-Hill

Mulholland M (2016) Applied process control—essential methods. Wiley-VCH, Weinheim, Germany

NOPSEMA (2015) ALARP Guidance Note. https://en.wikipedia.org/wiki/ALARP

Shinskey F (2001) Process control: as taught versus as practiced. Ind Eng Chem Res 41

Smith CL (2010) Advanced process control–beyond single-loop control. Wiley

Part II
Essential Knowledge

Process control is a branch of automatic control. On one hand, automatic control has been a vibrant field for over a century with a rich collection of theories and technologies, from classical control to modern control, from time-domain techniques to frequency-domain approaches, from differential equations to state-space representation, from continuous-time techniques to discrete-time, and from PID control to H_∞ control. On the other hand, industrial processes have many unique and distinct characteristics, such as multivariable, nonlinear, time-varying process dynamics. The complexities in the control structure, unpredictable load disturbances, unreliable measurements, deteriorating valve characteristics, are among the critical challenges in control design and operation, resulting in the most significant uncertainties to control performance. As a result, only a (small) subset of the automatic control theories is directly applicable to the process industry. The control technologies that are in prevalent use in the process industry include the following:

1. Simple feedback control (Chap. 3): Single variable on/off control and PID control are the dominant control solutions for the process industry, thanks to their simplicity and reliability. Although building a PID control loop is absurdly easy, making a PID controller work reliably under all operating scenarios is far from trivial. Many practical enhancements and protections are required.
2. Complex PID control schemes (Chap. 4): A complex PID scheme is a natural extension of the basic feedback control loop to handle complex control problems, especially multivariable processes. A complex PID scheme consists of multiple PID controllers and other supporting function blocks. In a complex control scheme, the control signal can travel a complex route to reach the final control elements, and the route may change automatically as the operating conditions change. The change in signal path implies a change in the control structure that can have a significant impact on the control performance. Therefore, protecting the control loop integrity and ensuring a smooth transition during control structure change is critical.
3. Advanced process control (Chap. 5): Complex PID control solutions can address some complicated control problems but are not effective for many others. Besides,

complex PID control solutions can become too complex for implementation. Loop integrity is a primary concern. Advanced process control offers more elegant and robust solutions for complex control problems and has caused a mindset change in control design. The power of advanced control technologies comes from the process models, which are the basis for almost all advanced control technologies and applications, including real-time optimization, operator training simulator, and equipment health monitoring. Model predictive control and inferential property estimator are among the most successful model-based technologies for the process industry.

Ironically, but not surprisingly, most of the above technologies prevalent in the process industries originated from practice rather than academia. Their vast success in practice slowly caught the attention of academia, which provided theoretical proof and improvements to establish their soundness and helped proliferate to other industries.

PID control was invented over a century ago, while model predictive control (MPC) has been around for four decades. They have been very successful so far, but there is an urgent need for more tools and technologies to address the process industry's increasingly stringent demands. Chapter 5 provides a summary of the process industry's unique challenges and particular needs and highlights the limitations of the current technologies (Sect. 5.1) to help the practitioners make the best use of the currently available technologies and set the expectation for the new technologies.

Process control is a practical-oriented engineering specialty. The technologies that are suitable are those that are useful.

Chapter 3
Basic PID Control

PID is a simple three-parameter control algorithm. It is deceptively simple yet highly competent to meet the needs of most applications. The majority of the control loops in the process industry are of PID type. Although the success of PID control primarily attributes to the simplicity and robustness of the algorithm, the many functional enhancements in practical implementations play a critical role in its versatility and reliability. The different forms of the PID algorithm (Sect. 3.1), the various operating modes (Sects. 3.2 and 3.6), the loop integrity protection (Sects. 3.3 and 3.4), and the rich configuration options (Sect. 3.5) are all crucial in making the PID control not only work but also not fail. A good understanding of the operating requirements and the control design is the key to the proper operation of the PID control algorithm.

3.1 PID Control Algorithms

The first practical application and theoretical analysis of the PID algorithm can be traced back to the early 1920s (Minorsky 1922; Samad et al. 2020). Over the last century, the PID algorithm has seen many generations of evolution in practical implementations, from the early mechanical, pneumatic, electronic, digital, to the latest Fieldbus (Åström and Hägglund 2006; Marlin 2015; Visioli 2001). Nevertheless, the PID algorithm remains a simple three-parameter equation.

A brief history of PID can be found in Åström and Hägglund (1995) and Åström and Kumar (2014).

3.1.1 Introduction to Laplace Transform

Classical control analysis and design are based on linear control system theory and limited to single-input single-output (SISO) processes. The underlying processes are typically described with (high-order) linear differential equations that are often

S. S. Niu and D. Xiao, *Process Control*, Advances in Industrial Control,
https://doi.org/10.1007/978-3-030-97067-3_3

cumbersome to handle. A *transfer function* representation allows algebraic manipulations of the linear systems rather than working directly with the differential equations. *Laplace transform* is a critical notion for creating transfer functions.

As a practical-oriented book, we strive to limit the mathematics and theoretical derivations to the minimum possible. However, Laplace transform (and the Z-transform mentioned later) are concise and convenient representations that we find hard to avoid. Therefore, we will assume our readers have a basic familiarity with the Laplace transform and Z-transform for representing transfer functions. On the other hand, even though state-space models as an important development in modern control theory are more elegant in describing multivariable systems and are being explored by process control applications such as model predictive control, we choose to avoid them so that we can focus our discussion to the core and essential concepts only.

The Laplace transform of a time-domain function $f(t)$ is given by $\mathcal{L}(f(t))$ as

$$\mathcal{L}(f(t)) = F(s) = \int_0^{\infty} f(\tau)e^{-s\tau}d\tau \tag{3.1}$$

which is a linear transformation,

$$\begin{aligned}\mathcal{L}[C(f_1(t) + f_2(t))] &= C\,F_1(s) + C\,F_2(s)\\ &= C\int_0^{\infty} f_1(\tau)e^{s\tau}d\tau + C\int_0^{\infty} f_2(\tau)e^{s\tau}d\tau.\end{aligned} \tag{3.2}$$

For example, a second-order differential equation of the following form

$$\frac{d^2y(t)}{dt^2} + 2\,\frac{dy(t)}{dt} + 10\,y(t) = \frac{du(t)}{dt} + 0.5\,u(t) \tag{3.3}$$

leads to a Laplace transform of

$$s^2\,Y(s) + 2\,s\,Y(s) + 10\,Y(s) = s\,U(s) + 0.5\,U(s). \tag{3.4}$$

The transfer function is thus given by

$$G(s) = \frac{Y(s)}{U(s)} = \frac{s+0.5}{s^2+2s+10} = \frac{s+0.5}{(s+1)^2+3^2}. \tag{3.5}$$

There are numerous references and resources on Laplace transformation, and the reader is advised to review the basic concept and manipulation as needed.

3.1.2 PID Control Algorithms, Ideal Form

The basic PID controller provides a combination of three types of control actions

$$[\text{Output}] = [\text{Proportional Action}] + [\text{Integral Action}] + [\text{Derivative Action}] \quad (3.6)$$

or in equation format

$$u(t) = K_c \left(e(t) + \frac{1}{T_i} \int_0^t e(\tau)\, d\tau + T_d \frac{d\, e(t)}{dt} \right) \quad (3.7)$$

where

- $u(t)$ control signal or control action (controller output OP),
- $e(t)$ control error $e(t) = y_{sp}(t) - y(t)$,
- $y(t)$ controlled variable (the process value or PV),
- $y_{sp}(t)$ setpoint (SP) of the controlled variable,
- K_c controller gain (proportional action),
- T_i integral time, the parameter that scales the integral action, and
- T_d derivative time, the parameter that scales the derivative action.

The three terms are illustrated in Fig. 3.1.

Assume a constant control error after a step change; the three types of actions are illustrated in Fig. 3.2. The proportional action produces an initial "kick" proportional to the control error and remains constant as long as the control error remains constant. The integral term acts on the control error and accumulates over time at a fixed rate if the control error remains constant. The derivative action produces another initial "kick" proportional to the rate of change in the control error and then vanishes. As a linear controller, the three actions are additive.

The Laplace form of the PID control algorithm is given by

$$U(s) = K_c \left(1 + \frac{1}{T_i\, s} + T_d\, s \right) E(s). \quad (3.8)$$

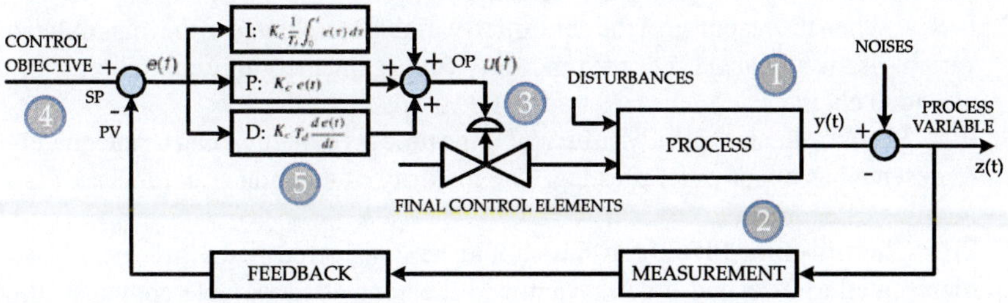

Fig. 3.1 A PID control loop

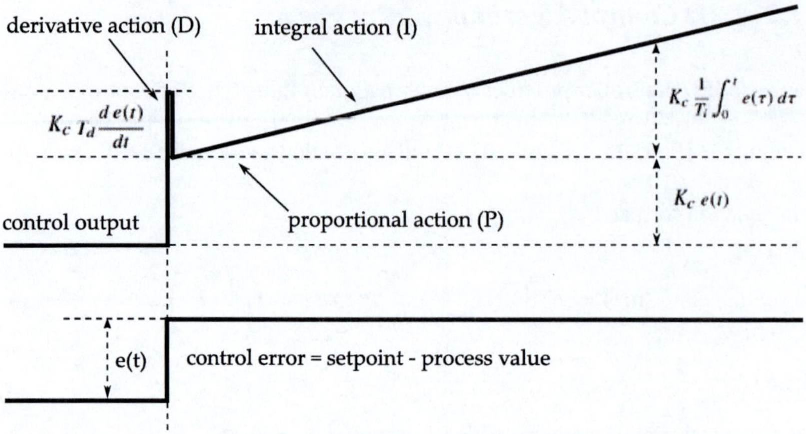

Fig. 3.2 The P, I, and D response of a PID controller following a setpoint change

It is well known that the derivative action amplifies process noises and can cause instability. Therefore, in actual implementation, a variation of Eq. 3.8 is sometimes used to smoothen the derivative action:

$$U(s) = K_c \left(1 + \frac{1}{T_i\, s} + \frac{T_d\, s}{\frac{T_d}{N} s + 1}\right) E(s) \tag{3.9}$$

where N is a constant and generally assumes a value between 1 and 33 (Visioli 2006).

In practice, the derivative term ($T_d\, s$) is rarely used, except in some special applications. The selection of PID structure is based on the operating requirements and process dynamics, i.e., how the controlled variable reacts to a change in the manipulated variable. Although the three-term PID algorithm in Eq. 3.8 is useful for some special temperature or pressure control loops, the vast majority of PID applications are running the PI algorithm and occasionally I-only or P-only algorithms. For example,

1. P-only: If the process is best represented at an integrator level, then a P-only controller provides simplicity and reliability.
2. I-only: When the response of the controlled variable to a change in the manipulated variable is instantaneous, i.e., the process is a pure gain, the I-only control typically is a good choice.
3. PI: A PI controller is typically sufficient if the process dynamics can be adequately represented as a first-order plus lag. The majority of the industrial processes fall into this category.
4. PID: The full-form PID algorithm should be considered if the process is best represented as a second-order system and the controlled variable contains little noise.

The full-form of PID has a second-order transfer function, the PI controller has a first-order transfer function, while the P-only algorithm has a scalar gain only:

$$U(s) = \begin{cases} K_c\, E(s), & \text{P-only} \\ K_c \dfrac{T_i\, s + 1}{T_i\, s}\, E(s), & \text{PI} \\ K_c \dfrac{T_d\, T_i\, s^2 + T_i\, s + 1}{T_i\, s}\, E(s), & \text{PID.} \end{cases} \tag{3.10}$$

Some general rules of thumb for choosing control algorithms based on process type are provided below (Brannan 2018; Smith 2009):

1. Flow loops: Flow response is typically fast and the flow measurement is usually noisy due to turbulent flow. A PI control algorithm is generally used with a moderate gain and strong integral action.
2. Pressure loops: Depending on the process flow, the dynamics of a pressure loop can be fast or slow. The speed of response is also different for steam, vapor, or liquid. In some processes, such as a distillation column, the pressure response behaves more like a ramp process within the operating range of interest. PI controller is typically used for pressure control with a significant gain value.
3. Temperature loops: The temperature response is moderately slow because of heat transfer lags and sensor response delay. Both PI and PID controllers can be used. A significant gain is usually needed for tight control, and the integral time is in the same order of magnitude as the process time constant. If the derivative term is used, its value should generally be no larger than one-fifth of the integral time. A PV filter should be considered if PV noise is significant.
4. Level loops: Most level controls are for inventory control with good surge capacity available, and the level is only required to be maintained within a safe range. There are also scenarios where tight level control is desired. For example, a continuous stirred tank reactor (CSTR) level must be tightly controlled to maintain the expected residence time. The level in an oil/gas separator requires the level to be stable for phase separation. A PI controller is typically used for level control, although in theory P-only controller can be used for level control without worrying about steady-state offset. The gain and integral time are correlated for pure level control, so a fixed control gain of about 1.0 is typically used, and the integral time is the only tuning parameter to adjust.
5. Quality loops: Quality control has become essential as plant operation shifts from process variable control to product quality control. Quality measurements from online analyzers or lab sampling are typically slow and unreliable. For example, a gas chromatograph (CG) may take a few minutes to an hour to produce the required results. Quality control is thus the most important but also most challenging control. A direct PID control is typically not feasible except for some quality measurements that are fast and reliable such as oxygen analyzers. Advanced control supplemented with inferential property estimation (Chap. 5) is generally recommended. If a PID control is acceptable, a PI controller with a small gain and considerable integral value is recommended.

3.1.3 PID Control Algorithms, Other Forms

Although the latest PID implementations have converged to a standard format, many other forms of the PID algorithm are still around, either on legacy systems or for reasons of back compatibility. The three popular PID equation forms are shown below:

$$G_c(s) = \begin{cases} K_c\left(1 + \dfrac{1}{T_i\, s} + T_d\, s\right), & \text{Ideal (Non-Interacting) Form} \\ K_c\left(1 + \dfrac{1}{T_i\, s}\right)(1 + T_d\, s)\,, & \text{Serial (Interacting) Form} \\ K_c + \dfrac{K_i}{s} + K_d\, s, & \text{Parallel Form.} \end{cases} \tag{3.11}$$

If the derivative value T_d is set to zero, the interactive and non-interactive forms become identical. In practical implementation, there are many more variations based on these three forms.

The difference in the PID forms impacts the tuning of the controllers. The popular tuning rules such as Ziegler–Nichols (Sect. 9.2.3), Cohen–Coon, and IMC-based methods (Sect. 9.3.4) all assume the controller has the ideal non-interactive form. It is not difficult to convert the PID parameters from one form to another. For example, the parallel form is related to the standard form as follows:

$$G_c(s) = K_c + \frac{K_i}{s} + K_d\, s = K_c\left(1 + \frac{K_i}{K_c\, s} + \frac{K_d}{K_c}\, s\right) = K_c\left(1 + \frac{1}{T_i\, s} + T_d\, s\right) \tag{3.12}$$

$$\mapsto \quad T_i = \frac{K_c}{K_i}, \quad T_d = \frac{K_d}{K_c}. \tag{3.13}$$

In addition, a particular form may be more suitable for a specific application. For example, the parallel form has separate gains for P, I, and D terms and is better suited for online gain adjustments (e.g., gain scheduling in Sect. 4.1.5).

We will assume the standard PID form in all our discussions in this book for clarity and convenience, unless otherwise mentioned.

3.1.4 PID Equation Types ("PID Structure")

PID control aims to bring the *process value* (PV) to its *setpoint* (SP) by eliminating the *control error*. See Fig. 3.3 for an illustration of the three variables in the PID calculation. The control error is defined as

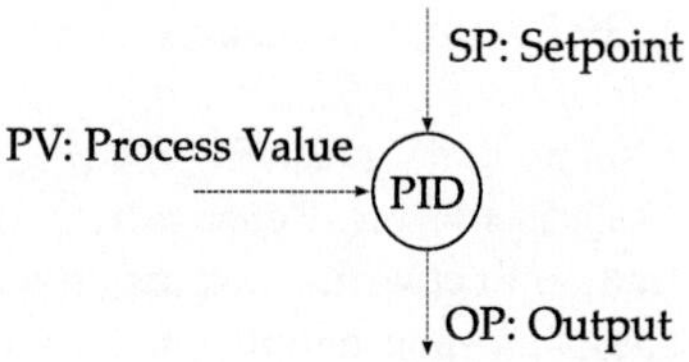

Fig. 3.3 PID process value (PV), setpoint (SP), and output (OP)

$$[\text{Control Error}] = [\text{Setpoint}] - [\text{Process Value}] = \text{SP} - \text{PV}. \quad (3.14)$$

The control error can be caused by changes in either the SP or the PV. Setpoint changes are typically initiated by the operator or another controller, while PV changes are caused by load disturbances, measurement noises, and other changes to the PID control loop. Based on the error source, the PID control serves two purposes:

1. Setpoint tracking: When an `SP` change occurs, the PID controller moves the `PV` value to follow the SP.
2. Disturbance rejection: When the `PV` moves away from the `SP` due to process noise or load disturbance, the PID controller brings the process value back to the SP.

In aerospace and robotic control, setpoint tracking is the primary objective, while quick and smooth rejection of disturbance is critical in the process industry.

Due to the simplicity of the PID algorithm, it is often difficult to achieve optimal performance for both setpoint tracking and disturbance rejection at the same time. As the proportional (P), integral (I), and derivative (D) actions have a different response on the control error (Fig. 3.2), the PID control implementations by commercial control systems all provide configuration options to allow the prioritization or trade-off on the two objectives. The choice is typically between

1. balanced performance on both setpoint tracking and disturbance rejection and
2. prioritized performance on disturbance rejection only.

Table 3.1 provides a list of commonly available configuration options for a commercial PID controller.

Table 3.1 PID equation types

Equation type	Description
`PID`	P, I, and D act on control errors (both SP and PV changes)
`PI-D`	P and I act on control errors, D acts only on PV changes
`I-PD`	I acts on control errors, P and D act on PV changes only
P-Only	P acts on control errors
I-Only	I acts on control errors
`PI`	PI acts on control errors
`PD`	PD acts on control errors

For example, equation type `I-PD` indicates that the integral action is on control error (i.e., for both SP and PV changes), while the proportional and derivative terms respond to PV changes only (i.e., disturbance rejection). This option eliminates the "kick" action on SP change by the `P` and `D` terms (see Fig. 3.2) and provides the desired trade-off between quick action on PV changes (because of proportional action) and smooth response to SP changes (integral action only).

Commercial control systems all have their variation in the implementation of the PID equations. For example, in the Yokogawa Centum system, the following five equation types are available:

1. Basic type (`PID`): This is the standard PID above, and all three terms act on control error.
2. PV proportional and derivative type (`I-PD`), where the P and D actions only apply to the PV changes, not to SP changes.
3. PV derivative type (`PI-D`): The derivative action applies to the PV changes but not to SP changes.
4. Automatic determination (default): The equation is `I-PD` in `AUT` mode, but automatically changes to `PI-D` if the controller mode is switched to `CAS` or `RCAS` mode (see next section for control modes).
5. Automatic determination type 2: Equation `I-PD` is used in `AUT` and `RCAS` mode. The equation automatically switches to `PI-D` if the controller is switched to `CAS` mode.

One interesting detail to note is that in Yokogawa Centum, the equation type may change automatically following a controller mode change. For example, the default control equation type is `Automatic Determination Type 2`, with which the equation type is `I-PD` in `AUTO` and `RCAS` mode and automatically changes to `PI-D` when mode changes to `CAS`. This default configuration is inappropriate if the setpoint is provided by an external program such as model predictive control since proportional action is required to track the setpoint change. As part of MPC commissioning, the PID equation type needs to be changed to `automatic determination type I`, where the PID controller takes `PI-D` when the mode changes to `RCAS`.

For comparison, in the Honeywell Experion system, the following four equation types are usually provided:

1. Equation A (`PID`): All three terms act on the control error.
2. Equation B (`PI-D`): The proportional and integral terms act on the control error, while the derivative term only acts on the PV change.
3. Equation C (`I-PD`): The integral term acts on the control error, while the proportional and derivative terms only act on the PV change.
4. Equation D (`I-Only`): This equation provides integral action only.

The differences between PID, I-PD, and PI-D equations are as follows:

$$\text{PID:} \quad U(s) = K_c\left(1 + \frac{1}{T_i\,s} + T_d\,s\right)\text{E}(s) \tag{3.15}$$

$$\text{I-PD:} \quad U(s) = K_c\left(\text{PV}(s) + \frac{1}{T_i\,s}\text{E}(s) + T_d\,s\,\text{PV}(s)\right) \tag{3.16}$$

$$\text{PI-D:} \quad U(s) = K_c\left(\text{E}(s) + \frac{1}{T_i\,s}\text{E}(s) + T_d\,s\,\text{PV}(s)\right) \tag{3.17}$$

with $E(s) = \text{SP}(s) - \text{PV}(s)$ being the control error. Note that the integral action, if present, always acts on the control error since PID relies on the integral action to eliminate the steady-state offset between setpoint and process value. On the contrary, the derivative term D should never apply to the setpoint changes because a setpoint change is typically a step change, and the derivative of the step change may be an unacceptably large value.[1]

The derivative term produces anticipatory action and is sensitive to noisy data, i.e., "amplifying" the process noise. It may produce a "kick" action (Fig. 3.2) that is rarely desirable for setpoint changes. For this reason, the full-form PID equation is usually not recommended. If the derivative action is needed, the I-PD or PI-D form is preferred, and PV filtering should be considered.

Because disturbance rejection (on PV changes) is typically more critical than setpoint tracking (error changes) in the process industry, a general recommendation is to configure all PID loops as I-PD by default and use PI-D or PID on an exception basis. For example, all standalone and protective control should be configured as I-PD, while the secondary controller in a cascade loop or MPC scheme shall be PI-D since fast setpoint tracking is desired.

3.1.5 PID Dimension Analysis

One crucial detail to notice is that in the PID calculation, all variables are dimensionless. The control error as the input to PID calculation is 0–1 or 0–100%, so is the controller output. The internal PID algorithm does not know or care about the engineering units.

The basic PID calculation in Eqs. 3.7 or 3.8 can be rewritten as

$$[\text{Output}] = [\text{Gain}] \cdot [\text{Lead/Lag Compensation}] \cdot [\text{Control Error}]. \tag{3.18}$$

All the variables must be normalized to dimensionless by dividing their value with their measurement range before entering the equation:

---

[1] An alternative to disabling derivative action is to use a setpoint filter, where a step change in setpoint is dampened via a first-order filter.

1. The controller gain K_c is dimensionless (or, with unit %/%) in the PID calculation. The control gain K_c is calculated based on the process gain K_p. The process gain is typically provided in the engineering unit and must be normalized as follows:

$$K_p = \frac{\text{PV}_2 - \text{PV}_1}{\text{OP}_2 - \text{OP}_1}, \qquad \overline{K}_p = \frac{\dfrac{\text{PV}_2 - \text{PV}_1}{[\text{PV Range}]}}{\dfrac{\text{OP}_2 - \text{OP}_1}{[\text{OP Range}]}} \cdot 100\% \tag{3.19}$$

   where K_p is the process gain in the engineering unit, and $\overline{K}_p$ is the normalized gain with no dimension.
   From Eq. 3.19, it is also evident that the process gain is a dynamic gain that is the change of process output (= controller `PV`) over the change of process input (= controller `OP`).
   Since the gain is normalized against the instrument range, *re-ranging* the instrument also changes the process gain and inadvertently affects the controller gain.

2. The control error must be in %, even though the `PV` and `SP` are all in the engineering units. That is

$$[\text{Control Error}] = \frac{\text{SP} - \text{PV}}{[\text{PV Range}]} \cdot 100\%. \tag{3.20}$$

   The `PV` and `SP` must have the same engineer unit and the same measurement range.
3. Controller output: The result of the internal PID calculation is always in percentage (0–100%). However, the controller output `OP`, which is the control action to be sent to the final control element downstream, may be provided as 0–100% or in the engineering unit, depending on the type of control system. For example, Honeywell TDC or Experion system allows the output to be in percentage 0–100% only, Schneider Foxboro allows engineering unit only, but Yokogawa Centum and Emerson DeltaV allow both. Nevertheless, the 0–100% signal must be converted to engineering unit to drive a secondary control element (see Sect. 10.1 for signal transformation).

Figure 3.4 is an example of a simple level-to-flow cascade control loop. Assume that the implementation is in a Foxboro DCS, and the values, units, and measurement ranges are given in Fig. 3.4.

The level controller `LC-101` and the pressure controller `PC-101` are standard overriding control through a low selector. The level controller is the active primary controller, with an output value of 750 m^3/h.

If a change of 80 m^3/h in the flow rate (`FC-101`) causes the pressure (`PC-101`) to change by 2 bar, then the dynamic process gain and the normalized process gain are given, respectively, by

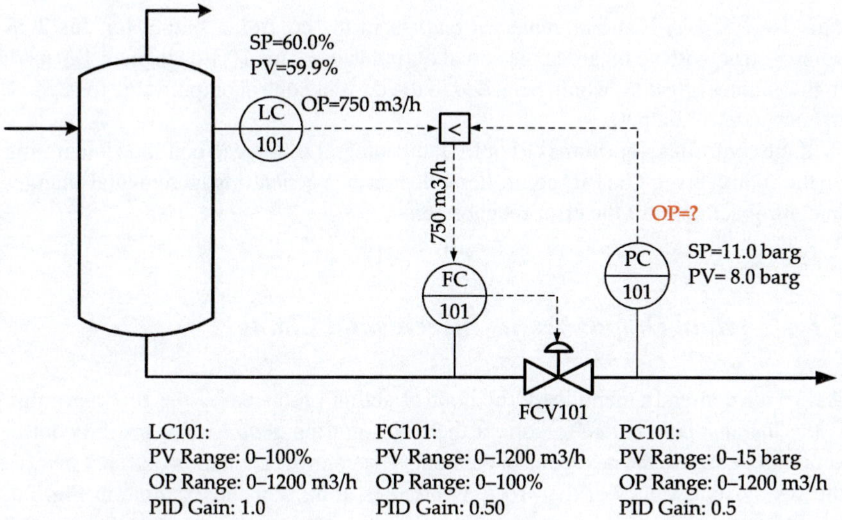

Fig. 3.4 Dimensions of PID controller

$$K_p = \frac{2}{80} = 0.025 \text{ (bar/m}^3\text{/h)}, \quad \overline{K}_p = \frac{\dfrac{2}{15-0}}{\dfrac{80}{1200-0}} = 2 \text{ (\%/\%)}. \tag{3.21}$$

The flow FC-101 has a measurement range of 0–1200 m³/h, while the pressure PC-101 is 0–15 bar. The implication is that a flow change between 0 and 1200 m³/h corresponds to a pressure change in the range of 0–15 bar, leading to a steady-state gain from flow to pressure as $(15 - 0)/(1200 - 0) = 0.0125$ (bar/m³/h). This steady-state gain is not far off from the dynamic gain of 0.025 (bar/m³/h), provided that the selection of the measurement ranges is in line with the physical laws. However, these two gains are different. In addition, the gain K_c of the PID controller is based on the normalized dynamic gain of 2%/% in Eq. 3.21, which has contributions from both the dynamic gain and steady-state gain.

The pressure controller PC-101 is an override controller, and its output is not selected by the low selector during normal operation. As shown in Fig. 3.4, the pressure control error is $11 - 8.0 = 3.0$ bar. If the control algorithm is proportional only (P-only), the controller output would be given by

$$[\text{Control Error}] = \frac{\text{SP} - \text{PV}}{[\text{PV Range}]} = \frac{11.00 - 8.0}{15.0 - 0.0} = 20\% \tag{3.22}$$

$$\begin{aligned} u(t) &= [\text{Controller Gain}] \cdot [\text{Control Error}] \\ &= 0.5\,(\%/\%) \cdot 20\% = 10\% \\ &= 10\% \cdot (1,200 - 0)\ \text{m}^3\text{/h/\%} = 120\ \text{m}^3\text{/h}. \end{aligned} \tag{3.23}$$

This 10% of $u(t)$ is the *incremental change* in the controller output for this 20% control error, with no regard to the actual output value. This 10% change, or 120 m^3/h in the engineering unit, would be added to the current control output value to become the new control output.

If the controller algorithm is PI or PID, the integral action will continue integrating on the control error. The PID controller will continue generating incremental changes in control action until the error reaches zero.

3.1.6 Actual Output Versus Incremental Changes

As we have already mentioned, the control action produced by the PID algorithm is the *incremental change* to correct the error, not the *actual value* for the control action to be sent to the next control element. The control calculation does not involve the steady-state values of the process variables at all. For the example in Fig. 3.4, the PID algorithm produces a 10% change in the `OP` in response to a 20% control error in the `PV`, whether the 20% control error is built on a 5 bar pressure or a 10 bar pressure is of no concern to the controller. Similarly, the 10% change in the `OP` is purely driven by this 20% control error. The PID calculation does not know or care whether the current `OP` is 40% or is already saturated at 100%.

Similarly, all the variables in Fig. 3.1, including the process input $u(t)$ and process output $y(t)$ and $z(t)$, are incremental changes rather than absolute values, to be technically correct. This concept makes perfect sense in explaining the PID control theory. However, in reality, the controller output will have to be converted to an absolute value to drive the final control element, such as a valve. The valve expects to receive an absolute value, e.g., a specific valve opening like 40%. Besides, the measurement of the process variable is in absolute values, e.g., 8 bars. Even the controller setpoint must be supplied in absolute values.

In other words, the process inputs and outputs are the actual values, while the internal PID control calculation is with *changes*. The absolute values from the process measurements must be converted to incremental values to perform PID calculation. The PID calculation results, which are incremental changes, must then be converted back to absolute values to drive the final control elements, e.g., valves. For this reason, the full-form PID control algorithm in Eq. 3.7 should have been given as

$$\mathrm{OP} = K_c\left(e(t) + \frac{1}{T_i}\int_0^t e(\tau)\,d\tau + T_d\frac{d\,e(t)}{dt}\right) + \texttt{bias} \qquad (3.24)$$

where OP is the absolute value of the controller output, and `bias` is an offset term to account for the steady-state values of the control output signal. This bias term is tricky and is a source of significant confusion in interpretation and implementation.

The conversion of the incremental change in controller output to the absolute valve opening is typically done by adding the incremental change to the current value of the final control element, such as the current valve opening or motor speed, to become the full value for the next move. In Fig. 3.4, the next control output from the pressure controller `PC-101` is given by

$$\begin{aligned} \mathrm{OP}_{\mathrm{new}} &= \mathrm{OP}_{\mathrm{prev}} + u(t) \cdot [\mathrm{OP\ Range}] \\ &= 750\,(\mathrm{m}^3/\mathrm{h}) + 10\% \cdot 1200\,(\mathrm{m}^3/\mathrm{h}) = 870\,\mathrm{m}^3/\mathrm{h}. \end{aligned}$$

The accumulation of incremental changes is a powerful feature of PID control and is critical for eliminating the control error. On the other hand, the conversion between absolute and incremental values introduces many challenges for practical implementations. For example, if the control algorithm for `PC-101` is `PI` or `PID`, the integral action will keep producing incremental control actions added to the controller output. Since the output of the pressure control `PC-101` is not selected by the selector, its control action would not affect the pressure, and the control error will not go away. The continuous accumulation of the incremental changes will eventually cause the control output to wind up to its high limit (100% or $1200\,\mathrm{m}^3/\mathrm{h}$). This output saturation is called *reset windup* and is discussed in Sect. 4.3.3 in more detail.

Thinking in dynamic changes is a crucial difference between the mindset of a process control engineer and engineers in most other engineering specialties.

3.1.7 PID Discrete Implementation

The PID equations in the above discussions are all in continuous time (e.g., Eq. 3.7) or Laplace form (e.g., Eq. 3.8). Because the practical implementation of PID control is almost all in digital form nowadays, the PID equations must be converted from continuous form (differential equations or Laplace form) to discrete form (differential equation or Z-transform).

The discretization is through digital sampling, where a crucial parameter is the *sampling time* T_s. In a digital implementation, the sampling time is determined by the DCS *scan-time* or *scan rate*, which in turn is determined by the execution frequency:

$$u(k) = \begin{cases} K_c\left(e(k) + \frac{T_s}{T_i}\sum_i e(i) + \frac{T_d}{T_s}\Delta e(k)\right) & , \text{ Ideal Form} \\ K_c\left(e(k) + \frac{T_s}{T_i}\sum_i e(i)\right)\left(e(k) + \frac{T_d}{T_s}\,\Delta e(k)\right) & , \text{ Serial Form} \\ K_c\,e(k) + K_i\,T_s\sum_i e(i) + \frac{K_d}{T_s}\,\Delta e(k) & , \text{ Parallel Form} \end{cases} \tag{3.25}$$

where $\Delta e(k) = e(k) - e(k-1)$ is the change in control error between the two scans.

3.1.8 Positional Versus Velocity Forms

The PID algorithm in Eq. 3.7 is also called the *positional form* as it directly calculates the control action (the incremental change). In some practical applications, the so-called *velocity form* is sometimes preferred. The velocity form calculates the difference between the last control action and the current one and is given by differentiating both sides of Eq. 3.7:

$$\frac{d\,u(t)}{dt} = K_c \left(\frac{d\,e(t)}{dt} + \frac{1}{T_i} e(t) + T_d \frac{d^2\,e(t)}{dt^2} \right). \tag{3.26}$$

The two forms of the PID equation are mathematically equivalent. However, the effect of the three types of control actions is more amenable to the understanding of the velocity form. For example, the rate of change in control action is directly proportional to the control error (the proportional action). As long as the control error exists, there is always a change in control action (due to the integral action). The accumulated integrated value in the position form is contained in the previous output in the velocity form.

The discrete-time equations for the velocity form of the three PID equations in Eq. 3.25 are given by

$$\Delta u(k) = \begin{cases} K_c \left(\Delta e(k) + \frac{T_s}{T_i} e(k) + \frac{T_d}{T_s} \Delta^2 e(k) \right) & \text{Ideal Form} \\ K_c \left(\Delta e(k) + \frac{T_s}{T_i} e(k) \right) \left(\Delta e(k) + \frac{T_d}{T_s} \Delta^2 e(k) \right) & \text{Serial Form} \\ K_c \, \Delta e(k) + \frac{T_s}{T_i} \, e(k) + \frac{T_d}{T_s} \, \Delta^2 e(k). & \text{Parallel Form} \end{cases} \tag{3.27}$$

where

$$\Delta e(k) = e(k) - e(k-1)$$
$$\Delta^2 e(k) = \Delta e(k) - \Delta e(k-1) = e(k) - 2\,e(k-1) + e(k-2).$$

The control output is then given by

$$u(k) = u(k-1) + \Delta u(k). \tag{3.28}$$

Note that the control action $u(k)$ is the incremental change added to the previous control output (see Eq. 3.24). $\Delta u(k)$ is, in a sense, an incremental change to the incremental change $u(k)$, and Eq. 3.28 is only a variation of the equation in the internal computation. In this velocity form, $\Delta u(k)$ is proportional to the control error via the integral terms. If the error is constant for a particular reason, the incremental control

Table 3.2 PID discretization

	Positional form		Velocity form	
PID term	Continuous-time	Discrete-time	Continuous-time	Discrete-time
Output	$u(t)$	$u(k)$	$\frac{du(t)}{dt}$	$\Delta u(k)$
P-action	$K_c\, e(t)$	$K_c\, e(k)$	$K_c \frac{de(t)}{dt}$	$K_c \Delta e(k)$
I-action	$\frac{K_c}{T_i} \int_0^t e(\tau) d\tau$	$\frac{K_c T_s}{T_i} \sum_{i=1}^{k} e(i) T_s$	$\frac{K_c}{T_i} e(t)$	$\frac{K_c T_s}{T_i} e(k)$
D-action	$K_c T_d \frac{de(t)}{dt}$	$\frac{K_c T_d}{T_s} \Delta e(k)$	$K_c T_d \frac{d^2 e(t)}{d^2 t}$	$\frac{K_c T_d}{T_s} \Delta^2 e(k)$

action becomes constant as well, and there is no integration on control error. The discrete form of the three terms in the PID algorithm (Eq. 3.8) is given in Table 3.2.

The following explicit equation, obtained by expanding Eqs. 3.27 and 3.28, is suitable for computer implementation:

$$\begin{aligned} \mathrm{OP_{new}} &= \mathrm{OP_{prev}} + u(k) \\ &= \mathrm{OP_{prev}} + u(k-1) + \Delta u(k) \\ &= \mathrm{OP_{prev}} + u(k-1) \\ &\quad + K_c \left[\left(1 + \frac{T_s}{T_i} + \frac{T_d}{T_s}\right) e(k) - \left(1 + \frac{2T_d}{T_s}\right) e(k-1) + \frac{T_d}{T_s} e(k-2) \right]. \end{aligned} \tag{3.29}$$

One advantage of using the velocity form is that the implementation of *gain scheduling* (Sect. 4.1.5) can be more straightforward, combined with the parallel PID equation (third equation in Eq. 3.27). Gain scheduling refers to the technique of changing the PID gain online in a running controller. This online gain adjustment is often desirable since a process may have very different dynamics under different operating conditions. With a positional form PID algorithm, swapping in the new proportional gain K_c may incur a bump in the control action since the proportional action works directly on the control error. The proportional "kick" does not exist with the velocity form since the rate of change in control error is zero before and after the proportional gain swapping.

The pseudo-codes for a bare-bone implementation are given in Table 3.3. The control error is calculated from the controller setpoint and process value, and the PID algorithm calculates the control action as the controller output. The calculation waits for T_s seconds until the next sampling time.

The two variables in the pseudo-codes in Table 3.3, `setpoint` and `process_value`, must first be normalized to 0–100% before entering the calculation. The calculation result, `output`, is also in percentage. Following the discussion in Sect 3.1.6, the calculated PID output is the incremental change required to eliminate the control `error`. It needs to add to the actual valve position to produce the new valve position.

A practical implementation of the same PID algorithm is substantially more complicated than what the pseudo-codes have shown because various operating modes

Table 3.3 Pseudo-codes for PID implementation

```
initialization:
    output_k1  = 0
    error_k1   = 0
    error_k2 = 0

loop:
    error   = setpoint - process_value
    output = output_k1 + Kc*(1+Ts/Ti+Td/Ts) *error
                       - Kc*(1+2*Td/Ts)*error_k1
                       + Kc*Td/Ts*error_k2
    output_k1 = output
    error_k2  = error_k1
    error_k1  = error
    wait(Ts)
    goto loop
```

and failure scenarios must be taken into consideration. The practical considerations are discussed in the loop integrity section later in this chapter.

3.1.9 PID with Primitive Function Blocks

PID controllers can be built with primitive function blocks (see Sect. 4.2.1) if a PID control block is unavailable or the available PID block does not have the desired functionality. For example, the PID functionality in most remote terminal units (RTU) is very limited. It typically does not have built-in support for anti-reset windup in an override control configuration. In this case, building the PID with primitive function blocks is a last resort to work around the limitation.

Figure 3.5 shows a PI controller built with the gain and lead/lag function blocks (in a Foxboro DCS system). It is easy to verify that the transfer function is given by

$$G_c(s) = \frac{\mathrm{OP}(s)}{\mathrm{E}(s)} = K_c \frac{1}{1 - \dfrac{1}{T_i\, s + 1}} = K_c \left(\frac{\dfrac{1}{T_i\, s}}{\dfrac{1}{T_i\, s + 1}} \right) = K_c \left(1 + \frac{1}{T_i\, s}\right) \tag{3.30}$$

where OP(s) is the output and E(s) = SP(s) − PV(s) is the control error.

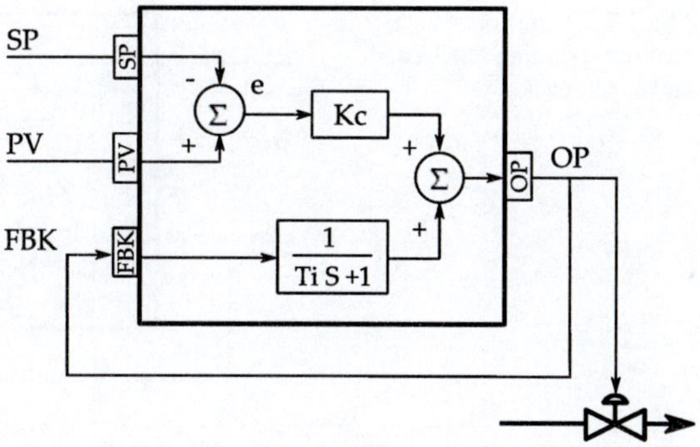

Fig. 3.5 PI controller with primitive building blocks

Similarly, an I-only controller can be constructed with primitive blocks as in Fig. 3.6, and the transfer function is easily verified to be

$$G_c(s) = K_c \frac{\dfrac{1}{T_i\, s + 1}}{1 - \dfrac{1}{T_i + 1}} = \frac{K_c}{T_i\, s}. \tag{3.31}$$

From Figs. 3.5 and 3.6, it is seen that the controller feedback comes from the output that it produces, and the controller does not know if the output reaches the final control element. As long as there is a control error, the integrating action would carry on, and the controller output would continue to change and may eventually wind up or down to saturation.

Consider a scenario where a *smart valve* is used, and the actual valve position is available as a *read-back*. The above controllers can take the valve read-back as its feedback (`FBK`) instead of the output it generates, then the windup problem caused by valve malfunction is conveniently avoided. See Fig. 3.7.

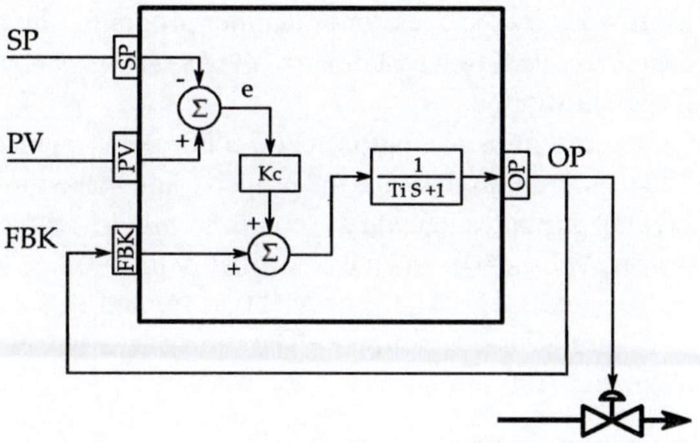

Fig. 3.6 I-only controller with primitive building blocks

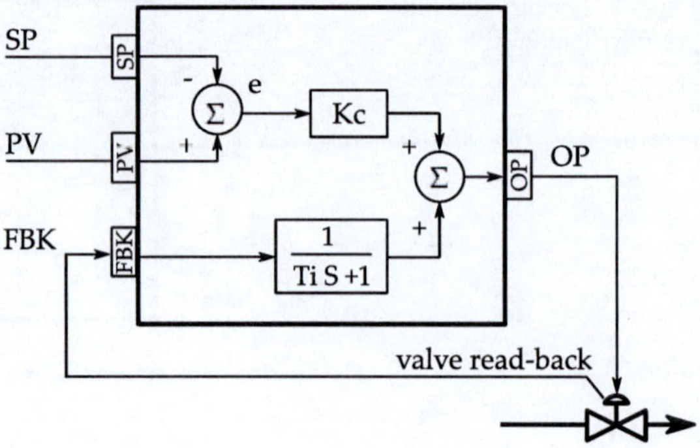

Fig. 3.7 PI controller with primitive building blocks and valve read-back

3.2 PID Mode of Operation

Although straightforward in theory, the real-world implementation of the PID controller is quite complicated since it has to accommodate all types of application scenarios. One of the requirements is the different operating modes. For example, the process may need to be controlled manually by the operator, automatically by the PID controller, or programmatically by external software/logic through the PID controller. For this reason, a practical implementation of the PID controller always provides the various operating modes to meet the different application requirements. The transition between the controller modes poses a significant challenge in implementation, and many practical considerations are required to ensure a reliable and smooth transition.

3.2.1 *PID Control Mode*

A basic PID function block has two inputs and one output. The *process value* (`PV` or `CV`[2]) is typically pulled from the process measurement, and the *setpoint* (`SP` or `SV`) is provided by the operator or another program. The controller output (`OP` or `MV`) is the corrective action sent down to the *final control elements*, e.g., valves. See Fig. 3.8 for an illustration.

In actual implementation, a PID controller is represented as a tag such as `FC-101`. The tag attributes describe the property and behavior of the controller. The essential PID tag attributes include the controller *mode* (`MODE`), the *setpoint* (`SP` or `SV`), the *process value* (`PV`), and the controller *output* (`OP` or `MV`). They can be referenced as `PC101.MODE`, `PC101.SP`, `PC101.PV` and `PC101.OP`.

Figure 3.9 is a simplified illustration of the basic PID controller implementation in Yokogawa Centum.

---

[2] The nomenclature in different control system implementation is mentioned in Sect. 3.2.2, and for now, we focus on the common concept and generic implementation only.

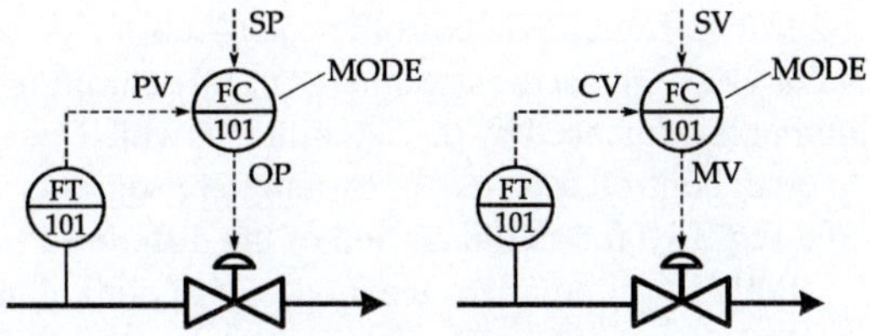

Fig. 3.8 Variables in a basic PID controller (Left: Honeywell, Right: Yokogawa)

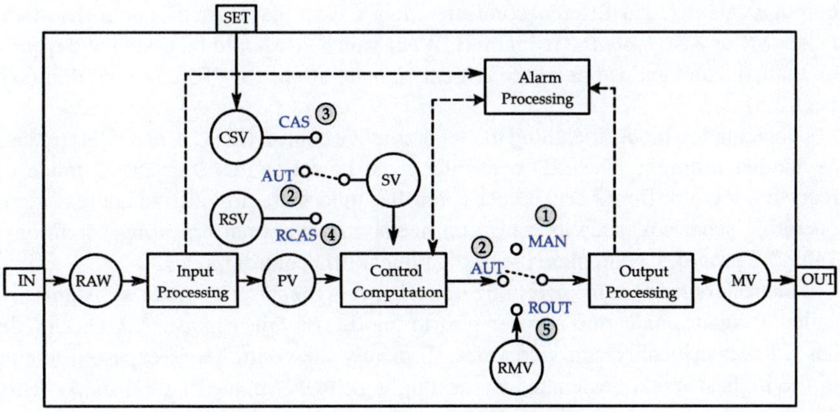

Fig. 3.9 PID control function block in Yokogawa Centum (simplified)

The `MODE` attribute determines how the controller produces the control action. The most basic controller modes include manual (`MAN` ①), automatic (`AUT` ②), and cascade (`CAS` ③), as shown in Fig. 3.9. The remote cascade mode (`RCAS` ④) and remote output mode (`ROUT` ⑤) are typically provided for advanced control applications. These values are assigned to the `MODE` attribute to change the controller behavior, e.g.,

$$\texttt{FC101.MODE} = \texttt{CAS}.$$

The operating requirement determines the mode of operation. When the operating requirement changes with the operating condition, the control mode may need to be changed. For example,

1. Manual control: The operator directly sets the valve position through the so-called *face-plate* (the *human–machine interface* HMI) of the PID controller.
2. Automatic control: The valve position is automatically calculated and set by a PID controller, whose setpoint is provided by the operator through the PID face-plate.
3. Supervisory control or custom control: The valve position is calculated and set directly or indirectly through a PID controller by a non-PID calculation program such as a calculation block or a piece of control logic.

In other words, the controller output can be supplied directly by the operator (`MAN` mode), by the PID controller, or by another program (`ROUT` mode) through the PID controller. If the PID controller calculates the output, it compares the active process value (PV) against the setpoint (SP) and determines the control action. The controller output is influenced by the setpoint SP, which can be set by the operator (AUTO), by another controller (CAS), or by another program (RCAS) through the PID controller. See Fig. 3.10 for an illustration of the different PID control modes.

Table 3.4 summarizes the modes of operation in Yokogawa Centum and Fieldbus control system (FCS), showing where the setpoint is obtained and how the controller output is calculated in different controller modes. Note that when the controller mode is in MAN or ROUT, its SP is ignored. What value SP should be assumed depends on many factors and is one of the critical considerations for PID configuration (see Sect. 3.5).

A calculation block, including model predictive control (MPC, Chap. 5), is treated as another program. The PID controller must be set to RCAS or ROUT mode to receive a setpoint from MPC. Other controller modes are provided to address more operating scenarios, such as transition needs and abnormal operating conditions. Table 3.5 provides a complete list of the typical PID control modes.

Controller mode has *priority* and *criticality*. A higher priority mode is customarily called a *higher mode*, and a lower priority mode is a *lower mode*. A higher mode has a lower criticality and vice versa. Typically the controller is expected to run in the highest mode designated by design, e.g., RCAS mode or CAS mode. This design mode is also called the NORMAL mode in some control systems. In abnormal operating conditions, the controller mode can or should "shed" to a lower mode for

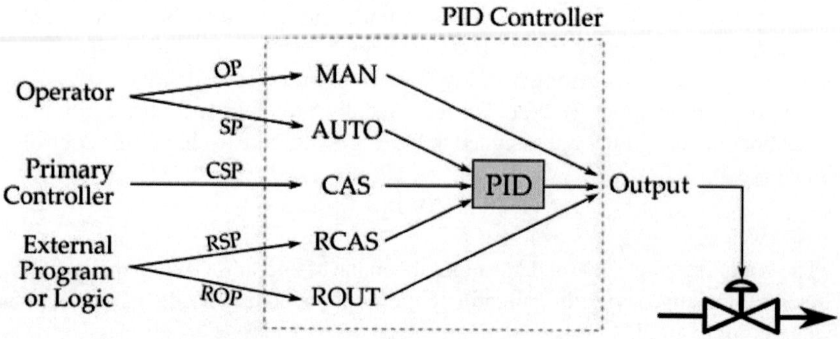

Fig. 3.10 Common modes of PID controllers

Table 3.4 Mode of operation for PID controller in Yokogawa Centum

PID mode	Description	Setpoint, set by	Output, provided by
MAN	Manual mode	–	Operator
AUTO	Automatic mode	Operator	PID
CAS	Cascade mode	Upstream PID	PID
RCAS	Remote cascade mode	Upstream program	PID
ROUT	Remote output mode	–	Upstream program

Table 3.5 Common operating modes of PID controllers

MODE	Controller mode	Source of setpoint (SP)	Source of output (OP)
OOS	Out-of-service	N/A	System
LO	Local override	Master controller	Controller
IMAN	Initialization	System	System
MAN	Manual	Operator or tracking	Operator
AUTO	Automatic	Operator	Controller
CAS	Cascade	Master controller	Controller
ROUT	Remote output	External program	Controller
RCAS	Remote cascade	External program	Controller

safety, e.g., from CAS to AUTO or even to MAN. This mode change can be performed by the operator or initiated by a back-end logic. For example, in a cascade control loop (Sect. 2.2.4), the setpoint for the secondary controller is automatically calculated by the primary controller. If the operator does not trust the primary controller, he/she can switch the control mode of the secondary control from CAS mode to AUTO and manually assign the setpoint.

Typically, a PID controller running in lower mode than design would result in less satisfactory control performance. A key performance indicator (KPI) for the PID controller is the percentage of time the controller operates in its NORMAL mode. This KPI is also called controller *uptime*.

3.2.2 *Variations in PID Implementation*

Although the PID concept is the same, PID implementation in different control systems can be pretty different for legacy reasons, not only in the functionality but also in the naming convention.

First of all, the three parameters in the PID formula have different names in different control systems. Some examples are listed in Table 3.6.

The proportional band is defined as the percentage change at the controller input required to produce 100% controller output, and thus

Table 3.6 Different names for the PID parameters

DCS vendor	DCS system	Control gain	Integral action	Derivative action
Honeywell	TDC/experion	Gain	Integral	Derivative
Yokogawa	Centum	Proportional band	Integral	Derivative
Emerson	DeltaV	Gain	Reset	Rate
Schneider	Foxboro	Proportional band	Integral	Derivatives
Foundation fieldbus	FCS	Gain	Reset	Rate

$$[\text{Proportional Band}] = \frac{100}{\text{Gain}}, \qquad \text{Gain} = \frac{100}{[\text{Proportional Band}]}. \tag{3.32}$$

A larger controller gain, or smaller proportional band, results in stronger proportional action.

The `reset` is also called the *integral* or *I-gain*. It determines how much to change the output over time due to the error. Reset or I-gain implies that a larger number will have more effect. Integral implies the opposite:

$$\text{Reset} = \frac{1}{\text{Integral}}, \qquad \text{Integral} = \frac{1}{\text{Reset}}. \tag{3.33}$$

The reset is typically in reset/min, and the integral is in minutes. In specific control systems such as DeltaV, the unit for integral time is in seconds, and correspondingly, the reset should be in reset/sec. Similarly, the unit for the derivative parameter may be in minutes or seconds, depending on the control system.

The most critical PID attributes, such as `MODE`, `SP`, `PV`, and `OP`, also have different names in different control systems. See Table 3.7 for some examples.

Note that the Foxboro system uses multiple bit flags, including `.RL` for remote/local and `.MA` for manual/auto, to define the controller mode. This difference can be a special challenge for some applications such as control performance monitoring.

Although all major control systems support the common controller modes such as `MAN`, `AUTO`, `CAS`, support for additional modes of operations varies from one control system to another. However, the control modes provided by Foundation Fieldbus and Emerson DeltaV have offered some uniformity that all can appreciate. The typical control modes include the following:

1. `OOS` for *out of service*: The controller freezes its output at the last calculated value, and the controller status changes to "`Bad`."
2. `LO` for *local override*: The controller output is fixed at its last calculated value due to a detected fault within the device. This mode is typically used by controller self-tuning. It is also known as *output tracking*. The `TRK` mode in Centum provides a similar function to `LO`.
3. `PRD`: In Yokogawa Centum, there is a *primary direct* (`PRD`) mode. When the PID controller is put in `PRD` mode, the PID calculation is bypassed, and the upstream control block directly sets the controller outputs. The output of the primary con-

Table 3.7 Common PID attributes

DCS vendor	DCS system	Mode	Setpoint	Process value	Output
Honeywell	TDC/experion	`.MOD`	`.SP`	`.PV`	`.OP`
Yokogawa	Centum	`.MD`	`.SV`	`.PV`	`.MV`
Emerson	DeltaV	`.MODE`	`.SP`	`.PV`	`.OP`
Schneider	Foxboro	`.RL + .MA`	`.SP`	`.MEAS`	`.OUT`
Foundation fieldbus	FCS	`.MODE`	`.SP`	`.PV`	`.OP`

troller goes directly to the next control element downstream of the failed controller. The necessary unit conversion and range scaling are automatically performed as needed.

4. `IMAN` or `INIT` for *initialization manual* or simply *initialization*: `IMAN` or `INIT` mode indicates that the forward path to a physical output is broken, and the output is tracking the downstream block. The PID output is typically fixed at its last calculated value because the output signal path is broken. The PID output is back-calculated to provide bumpless transfer when the loop is re-connected.
5. `MAN` for *manual*: The controller output is fixed at a value determined by the operator. The controller status is "`Good`."
6. `AUTO` for *automatic*: The PID controller calculates the output with the selected PID equation. The operator sets the setpoint to the controller.
7. `CAS` for *cascade*: The PID controller calculates the output with the selected PID equation. Another controller sets the setpoint to the PID controller.
8. `RCAS` for *remote cascade*: This mode is also called *supervisory* mode in older control systems. The controller calculates the output with the selected PID equation, while a calculation block or external program produces the setpoint.
9. `ROUT` for *remote output*: This mode is also called *direct digital control* in some older control systems. A sequence or an external program sets the controller output and the PID calculation is bypassed.

See Table 3.8 for the different modes of operation supplied by the four major DCS vendors and the Foundation Fieldbus controller standards.

Some control systems may provide additional modes of operations in their PID implementation. For example, in the Fieldbus control system or Emerson DeltaV system, the controller modes can have different statuses, such as the `target mode`, `actual mode`, `permitted mode`, and `normal mode`. Honeywell DCS has additional modes called `NORMAL` and `NONE`. These modes are mainly for convenience or for facilitating the handshake during mode change.

There are many other differences in the actual implementation, and it is outside the process control scope to discuss in further detail. A process control engineer should

Table 3.8 Operating modes of PID controllers by DCS vendors

Controller mode	Fieldbus control	DeltaV	Centum	TDC/experion	Foxboro
Out of service	OOS	OOS	O/S		
Local override	LO	LO	TRK		
Primary direct			PRD		
Initialization	IMAN	IMAN	IMAN	INIT	
Manual	MAN	MAN	MAN	MAN	LOCAL+MAN
Automatic	AUTO	AUTO	AUT	AUTO	LOCAL+AUTO
Cascade	CAS	CAS	CAS	CAS	REMOTE+AUTO
Remote cascade	RCAS	RCAS	RCAS	BCAS	
Remote manual	ROUT	ROUT	ROUT		

be aware of these differences and know when to seek help from control system engineers if more in-depth information is needed.

3.3 PID Loop Integrity

The working principle of a feedback control loop, including PID control, is that the controller sees a control error between the controlled variable and its setpoint. It then calculates the corrective action (as the controller output) to send down to the manipulated variable, anticipating that the process output, which is the controlled variable, would change in the direction, magnitude, and speed as desired.

The controller is designed and configured based on a good understanding of the loop dynamics. If the loop dynamics significantly deviate from the baseline condition for any reason, the control performance would be affected, from minor degradation in performance to complete breakdown. For this reason, monitoring and safeguarding the loop integrity is the most important practical consideration in a real-world PID implementation and is the primary difference between a PID control in theory and PID control in practice.

Control loop stability is a significant interest of research in the academic world. Routh stability criterion (Routh 2016) was established over a hundred years ago and has been the foundation for stability analysis. However, as shown in this section, the many uncertainties in the control loop, especially in the loop data flow, have reduced the value of stability analysis.

3.3.1 Process Flow and Data Flow

Figure 3.11 illustrates how the feedback control in Fig. 2.2 is implemented in a real-world environment. It shows the five essential components in the control loop and the communication between the field and the control system.

Figure 2.2 is the purest feedback control loop that we are familiar with in text-books, where all the implementation details are swept into two or three abstract blocks (the process, the measurement, and the controller), not even considering sampling and signal conversion that are essential for digital implementation. Figure 3.11 provides a more realistic but still highly simplified view of the feedback control loop. These two drawings illustrate the fundamental difference between a laboratory simulation and real-world implementation of the PID control loop.

Each block in Fig. 3.11 is called a *function block* or *control element*. When discussing a specific function block, we call it the *primary control element* or *primary element*. The block downstream of it is called the *secondary control element*, *secondary element*, or *secondary block*. A primary control element can manipulate a secondary control element. A valve is a *final control element* since it is the ultimate control element to drive the process.

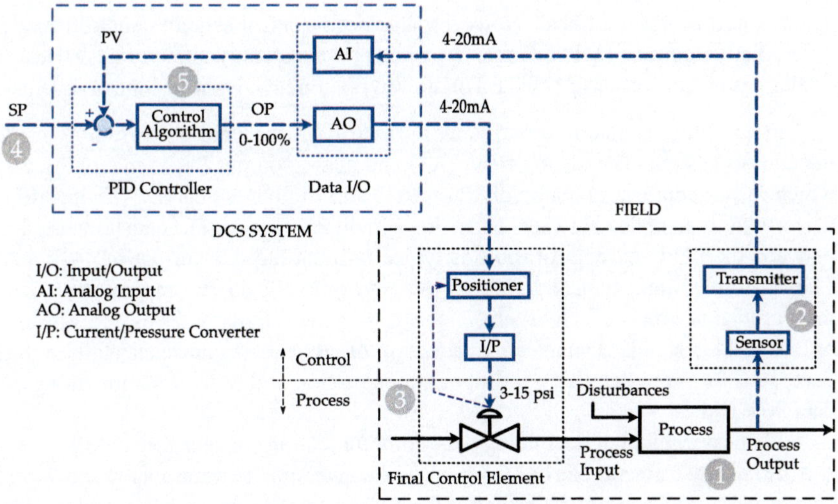

Fig. 3.11 Closed-loop control

For engineering analysis, two types of flows are essential for process control: the *process flow* (the black/solid lines in Fig. 3.11) and the *data flow* (blue/dashed lines in Fig. 3.11).

1. Process flow: Process operation is the continuous movement of material and energy facilitated by piping, vessels, and equipment. A process can refer to a simple device such as the flow through a control valve or a complex unit operation such as a distillation column or an entire plant. We will call this physical flow of material and energy the *process flow*.
   As illustrated in Fig. 3.11, the process flow starts from the *process input*, through the process, to the *process output*. Figure 1.3 is a process overview diagram (POD) that shows the process flow of a typical oil and gas processing facility.
   The process flow is the result of process design and dictates the operability of the process. The process flow design is documented with *process flow diagram* (PFD), which contains the critical information of all the equipment, the process flow streams, the utility streams, and the critical control loops (Turton et al. 2018).
2. Data flow: The purpose of process control is to regulate the material and energy flow as per operating requirements. Process control is based on process measurements and control signals. We will call the movement of measurement data and control signals as the *data flow*.
   For a feedback control loop, the data flow is a closed loop that starts from the controller as the controller output `OP`, travels to the actuator, works on the process, and returns to the controller as the process value `PV`. The data flow represents the control strategy and is part of process engineering and process control design.
   The design and implementation of the data flow are reflected in the *piping and instrumentation diagrams* (P&ID, see P&ID standards ISA-S5 and ISO-3511)

as dotted or dashed lines.[3] However, advanced control schemes such as model predictive control (MPC, Chap. 5) cannot be adequately shown in P&ID, and there is no consensus or standard on how to document advanced control schemes.

The data flow is analog (continuous) primarily in the field but is digital inside the control system. The analog input (`AI`) and analog output (`AO`) function blocks, which at the center are analog/digital (A/D) and digital/analog (D/A) converters, perform the conversion between the analog and digital signals. The data transmission between the field and control room is typically in an electric current of 4–20 mA. The 4–20mA signal typically follows the NAMUR NE 43 recommendation. The most crucial notion is the "live zero," i.e., a zero value is transmitted as 4 mA instead of 0 mA. This standard enables the detection of signal loss and sensor failure. See Sect. 10.1 for more details on the signal transmission and transformation along the data flow path.

With the new Fieldbus technology, the communication between the control system and the field instruments can be all digital. The conversion between analog and digital occurs inside field equipment instead of the `AI` and `AO` modules in the control system.

The actual design and working of control data flow are complicated and involve substantially more details than shown in Fig. 3.11, and are not part of core knowledge for process control engineers. Nevertheless, process control engineers should understand how the process flow and data flow affect process control design and functioning.

The primary concern is the integrity of the data flow with the following three common abnormal scenarios:

1. Bad data: The data along the data flow path may become unavailable or unusable. For example, communication between control elements may be lost; the controller may receive a `Bad PV` value that it cannot act upon; a secondary element on the data flow path such as the `AO` block may be out of service (`OOS`) and does not propagate the control signal further.
   A common cause of `Bad PV` is the loss of signal, or the PV value is out of the measurement range. In some control systems, there is an option to clamp out-of-range values. A clamped value is not reported as a bad value, but it stays unchanged not tracking the actual value.
2. Disconnected data flow: The data flow involves multiple components and can become disconnected for many reasons, most commonly by operating needs such as mode changes. A broken connection causes the loop to open and thus puts the loop out of control. For example,
   - A secondary element such as a controller or `AO` block may no longer be in cascade mode or becomes inactive and thus breaks the loop.
   - A high-selector may have selected the control output from another controller.
   - If the secondary element is in the initialization mode, the loop is also broken.

[3] Some process engineers tease the process control contribution as "only the dotted lines on P&ID," which grossly undervalues process control engineers' role in a project.

3. Constrained data flow: When the control output is sent down to the final control elements, it may travel a complicated path (see complex PID control loops in Chap. 4). The control signal may exceed the high or low limits set by the secondary elements and becomes constrained. For example,
   - A valve has a limit of 0–100% in its opening, and a motor has a speed or power limit.
   - A control block has a high and low limit on its setpoint when in cascade mode.

   When the control signal is limited, the effectiveness of the control action will be compromised. The primary controller must be aware of the problem and take mitigating actions as needed.

3.3.2 Process Dynamics and Loop Dynamics

Process control aims to control the process, and the *process dynamics* are the basis for control. In Fig. 3.11, the *process dynamics* is the time response characteristics from the process input to the process output, i.e., the dynamics of the process flow. For example, in Fig. 3.4, the process dynamics for the level controller `LC-101` is from the flow `FI-101` to the level `LI-101`. In Fig. 1.7, the process dynamics for the compressor suction pressure controller is from the compressor speed to the pressure.

However, from Fig. 3.11, it is evident that the process is only a part of the control loop. A complete control loop comprises many essential components, including the sensor/transmitter, the valve, the valve positioner, analog/digital converters, and the current/pressure converter. They all have dynamics of their own, and the *loop dynamics* are the combined dynamics of all the components along the control loop, i.e., the dynamics of the *data flow*. That is,

$$G_{\text{loop}}(s) = G_{\text{AI}}(s) \cdot G_v(s) \cdot G_p(s) \cdot G_{\text{Tx}}(s) \cdot G_{\text{AO}}(s) \cdot G_f(s) \cdots \tag{3.34}$$

For example, the dynamics of the `AI` and `AO` function blocks, $G_{\text{AI}}(s)$, $G_{\text{AO}}(s)$, the sensor and transmitter $G_{\text{Tx}}(s)$, the control valve $G_v(s)$, and the PV filter $G_f(s)$ are all part of the loop dynamics. As a result, the *loop gain* and the *process gain* can be significantly different.

Controller design and operation are based on the loop dynamics rather than the process dynamics alone. Although the process dynamics is the most critical component in the loop dynamics, there are applications where the dynamics of the other components may be significant. For example, the compressor surge phenomenon (see Sect. 11.3) is extremely fast. Therefore, the compressor anti-surge control is required to execute every 100 ms or faster. The speed of transmitter response is at the same order of magnitude as the controller, but the valve dynamics are two order-of-magnitude slower. In this case, the valve dynamics become dominant in the loop dynamics. Besides, some discrete events inside the control loop, such as the disconnection of the data flow or limiting of the signal, may have a more detrimental effect on the overall loop dynamics. The detection and mitigation of these discrete events is a critical consideration in a real-world PID implementation.

PID control loop design is based on an assumed cause and effect dynamic relationship between the manipulated variable (controller output `OP`) and the controlled variable (the controller `PV`), as shown in Fig. 3.11. It also assumes that this relationship remains (relatively) unchanged during operation. In practice, however, these assumptions often become invalid due to changes to the data flow. Therefore, the integrity of the data flow is the primary challenge to a control loop's reliable operation. A mismatch between the PID tuning parameters and the changed loop dynamics can deteriorate control performance, while broken loop dynamics can result in the PID loop going completely off control.

Below are the potential consequences from loop dynamics issues:

1. Bad data: Data is not available due to `Bad Data` or control elements out-of-service. In this case, the cause and effect relationship established by the loop dynamics ceases to exist (G_{loop} in Eq. 3.34 becomes meaningless), and the controller is effectively off control.
2. Data flow is broken: In this case, the controller may still receive `PV` updates, but the change in process output is not the result of the controller action. The established cause and effect relationship is no longer valid since the control action will not reach the final control elements and thus would not affect the controlled variable.
3. Limited data: If data value becomes limited (e.g., saturated) somewhere on the data flow, the cause and effect relationship exists but is no more accurate. A change in the control action will not cause the expected change in the PV.
4. Drifting in loop dynamics: The gradual deviation of the loop dynamics from the design condition may be caused by various reasons such as a fouled heat exchanger, a worn-out valve, or a flow meter out-of-calibration. The changed loop dynamics result in a gradual deterioration in the control performance. These abnormal conditions are more difficult to detect reliably, and remains a significant challenge in both theory and practice.

The first three scenarios mentioned above are partially addressed by proper control design and implementation. The fourth problem is related to performance monitoring, and there is no standard solution yet. See Sect. 10.2 for more detailed discussions.

3.4 Tracking and Initialization

In a real-world environment, as shown in Fig. 3.11, the control action produced by a PID controller may not always reach the final control element that it intends to due to problems in the data flow. Therefore, it is critical to track the data integrity, detect abnormal conditions, and mitigate. Tracking and initialization is a common approach taken by most control system implementations.

3.4.1 Tracking

It is evident from the internal architecture of a PID controller (Fig. 3.9) that the control modes and data processing options can potentially alter the data flow in a control loop. For instance, switching the PID controller from cascade mode to automatic mode interrupts the data flow in a cascade control loop.

The first step in protecting the loop integrity is tracking the data flow for disconnections or constraints. In most control systems, a control element can inform the primary control element of its operating mode, error condition, and constraint status. If all the control elements in a control scheme support status monitoring and broadcasting, then the control loop's overall integrity can be tracked. In case of any issues as listed in Sect. 3.3.1, the controller can take appropriate action to mitigate.

Both the process and the controller have many operating modes other than the normal mode; some are legitimate normal operating modes, while others may be crippled modes due to failures somewhere else in the loop. Mode changes, either intentional or unintentional, are frequent events in operation. A mode change typically involves the following three steps:

1. Transition from the normal mode to a different mode: For a mode change, a crucial requirement is that the transition must be bumpless. One example is the transition from the normal operating mode to a crippled mode after a failure is detected. The transition may require changes to the controller mode, the setpoint, or the output following pre-defined operating procedures. It is critical to minimize the disturbance to the process operation.
2. Abnormal mode of operation: In case of abnormal mode operation such as crippled process operation or shutdown, the controller mode, setpoint, and output settings require careful considerations. For example, should the controller be switched to manual mode and the output be set to fully closed?
3. Transition back to the normal mode: When the abnormal condition is resolved, and the controller needs to resume normal control, the condition may have significantly changed. Simply re-connecting the data flow may cause a bumpy transition. For example, if the controller output is 50% and the `AO` has been in manual mode with a 30% output, closing the loop would bump the `AO` setpoint (thus the `AO` output) to 50% in one shot. Since the `AO` output is typically connected to a control valve, this sudden change in `AO` output may be an unacceptable disturbance to the process operation.

For a smooth transition, it is critical to track and monitor the loop dynamics for changes continuously. The current practice includes the following:

1. Tracking the integrity of the data flow: The control systems tracks and monitors the data flow for status and value changes such as disconnection or limitation in the data flow. If the operating mode or data status poses a danger to the data flow integrity, the PID controllers are informed, and appropriate mitigating actions would be taken.

Due to technological limitations, the data flow tracking is usually only implemented from the primary controller down to the AO block. Currently, the tracking is limited to discrete events such as a change in the controller's operating mode or a change of selection in the control selector (see Fig. 3.11). If smart devices are utilized, the tracking may extend to the final control elements but cannot go further with the analog process data.

2. Monitoring data health: Control implementation continuously monitors and processes the I/O data for anomalies. The data may be categorized into GOOD (normal), BAD, or UNCERTAIN to help the controller make appropriate decisions. For example, an out-of-range data value is labeled BAD by default. However, if the clamping option is enabled, the out-of-range value will be clamped to the at-range value and be marked as UNCERTAIN.
   The data health monitoring is part of the I/O processing in most control function blocks but is still primitive. The health checking is typically limited to out-of-range verification only. More practical and reliable monitoring technologies are needed to protect the loop integrity and improve the control loop reliability.

For reliable PID control, tracking and monitoring must cover the data flow of the entire control loop. However, many events that significantly impact the loop dynamics do not have standard or mature theories or techniques to detect reliably due to technological and practical reasons. For example,

1. Gradual changes in process dynamics that are caused by operating condition changes, equipment fouling, or seasonal temperature changes.
2. Changes in valve characteristics, such as increased stiction, deadband, or hysteresis are caused by wear and tear
3. Process measurements have drifted away, e.g., due to lack of calibration.

3.4.2 Back-Calculation

When the normal data flow is re-established after disconnection, all the control elements on the data flow path are expected to quickly and smoothly restore the normal operating conditions without introducing significant disturbances to the controller output. The resumption of control should not cause the final control elements to change if the operating conditions have not changed. Even if the operating condition has changed, the controller should not generate abrupt changes in the final control elements. This smooth transition is called a *bumpless transfer* and is typically achieved through proper initialization (or re-initialization) of the control elements. Because the goal is to eliminate or minimize the transition bumps in the *final* control elements, the initialization starts from the final control elements and goes backward along the data flow. This type of initialization is thus called *back-initialization*.

Back-initialization is typically facilitated by an additional information flow called *back-calculation* data flow, also called *secondary data flow*, that complements the primary data flow. The primary data flow propagates forward, carrying the control

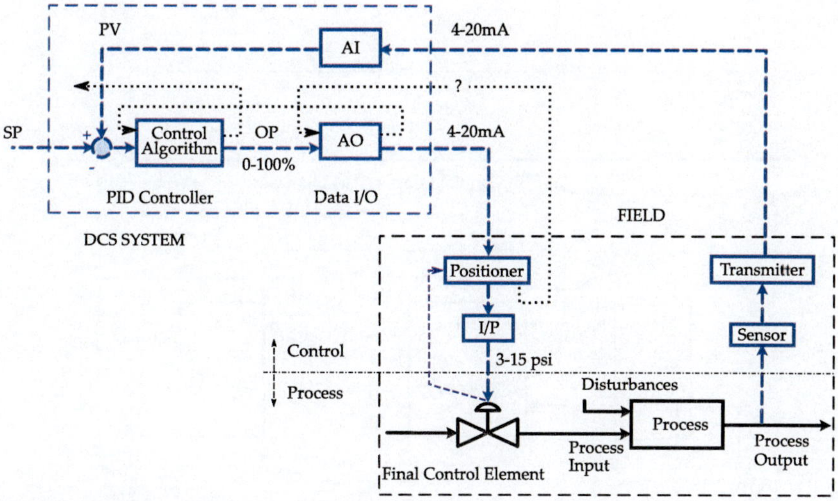

Fig. 3.12 Process flow, data flow, and back-calculation

data (the calculated control action) from the controller to the final control elements, while the back-calculation flow propagates backward, carrying the control elements' values and statuses back to the controllers. See the dashed lines in Fig. 3.12 for the data flow and the dotted lines for the back-calculation information flow.

The back-calculation flow of a simple flow control loop is shown in Fig. 3.13, where the normal control loop is shown on the top, and the connection of the function blocks is shown at the bottom. If supported, the back-calculation flow starts from the control valve, propagates back to the `AO` block, and then to the `PID` block.

Back-calculation flow is provided to support back-initialization. It complements the data flow and is part of the control information flow. The relationship between the process flow, the data flow, and the back-calculation is illustrated by Eq. 3.35:

$$\begin{array}{c}\text{Closed-Loop}\\ \text{Control}\end{array}\left\{\begin{array}{l}\text{Process Flow}\left\{\begin{array}{l}\text{Physical Material Flow}\\ \text{Physical Energy Flow}\end{array}\right.\\ \text{Data Flow}\left\{\begin{array}{l}\text{Process Data + Monitoring}\\ \text{Control Data + Back-Calculation.}\end{array}\right.\end{array}\right. \tag{3.35}$$

The back-calculation information flow is primarily system-related, so it is beyond this book's scope to provide in-depth coverage. Only the critical information essential to the control loop's functional integrity is briefly discussed here.

Every control function block has an input connection and an output connection for back-calculation. Different control systems may have different names for the back-calculation signals. For example, in the DeltaV system, the back-calculation output is called `BKCALO`, and the back-calculation input is called `BKCALI`. In the Foxboro

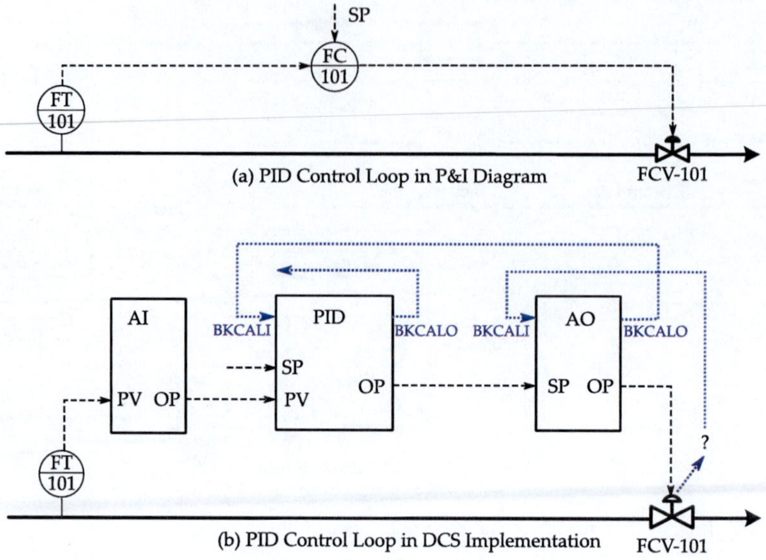

(a) PID Control Loop in P&I Diagram

(b) PID Control Loop in DCS Implementation

Fig. 3.13 PID back-calculation in a flow control loop

system, the back-calculation input and output are called `BCALCI` and `BCALCO`, respectively. They need to be manually and explicitly connected in the control loop configuration. In Honeywell Experion and Yokogawa Centum systems, the connections between the primary function block and the secondary function block are automatic and implicit, hidden to the user.

Without loss of generality, let us call the two connections as `BKCALI` and `BKCALO`. Combined with other general-purpose connections that carry the control element's status, they provide the *value* and *status* of all the control elements for communication between the control elements. The value and status typically include, among others:

1. Control modes: These are primarily the control modes of the function blocks. For example, a function block in manual mode indicates that the data flow is disconnected at this function block.
2. Connection status: If selectors and switches are used, the primary controllers must know which controller's output is selected.
3. Constraint status: The constraint status refers to whether the data flow at this function block is high-limited, low-limited, or both.
4. Actual value: The process value (PV) of the function block. The purpose is to inform the upstream function block of the current function block's process value (PV) or output value (OP).
5. Handshake information: The handshake information is communicated between function blocks during mode changes.

The BKCALO connection is for the current controller to report its value to the upstream control blocks. The current control element also receives similar back-calculation information as BKCALI from the control elements downstream. The controller responds to the secondary function block's status changes and performs appropriate initialization accordingly.

In the simple flow control loop in Fig. 3.13, the AO block sends its mode status and output value to the PID controller via the connection from BKCALO to BKCALI. The PID controller propagates its status and value to the upstream function block if exists.[4] This daisy chain connection of BKCALO to BKCALI constitutes the back-calculation information flow. This back-propagation is possible only if all the function blocks on the data flow path fully support back-calculation, which is not always true in practice. For example, in Fig. 3.13, the AO block may not receive the on/off status or position from the downstream valve FCV-101. Besides, many arithmetic calculation blocks do not support back-calculation by design.

The information flow, including the data flow and back-calculation, can become overwhelmingly complicated for a complex control scheme. Its impact on control performance is often underestimated.

Example 3.1 Back-calculation in a cascade control loop. The data flow of the simple cascade control loop in Fig. 2.18 is shown as Fig. 3.14.

The control action from level controller LC-101 is sent down to the secondary controller FC-101 and further down to the control valve FCV-101. The change in the valve opening causes the flow FT-101 and level LT-101 to respond to maintain the flow and level at their respective targets. However, this data flow can be disrupted by many events, including the following:

1. The flow controller FC-101 is switched to manual (MAN) mode by the operator, and thus the controller is disconnected from the primary controller LC-101. The data flow path from LC-101 to the valve is broken.

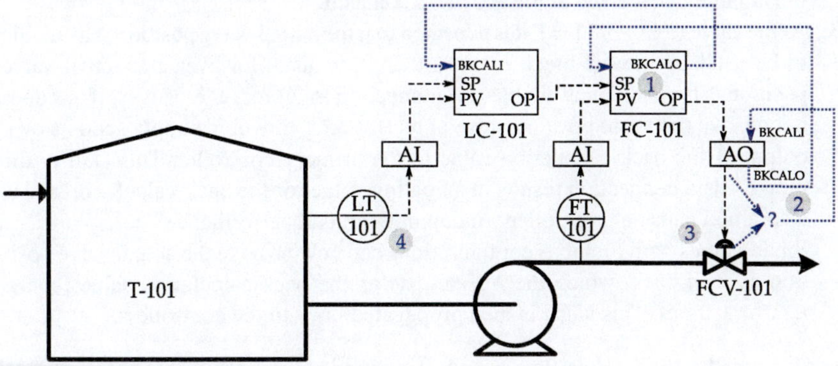

Fig. 3.14 Information flow in a cascade control loop

4 There may be a limit on how many levels of upstream blocks the back-calculation and can propagate in one scan cycle.

2. The flow controller automatically sheds to manual mode upon detecting bad flow measurements (BAD PV) in FT-101.
3. The valve FCV-101 is stuck and does not follow the control output.
4. The analog output (AO) module is switched to manual mode and no longer propagates the control output to the controller valve FCV-101.
5. The level measurement LT-101 is frozen or saturated. It still produces a value, but the value remains constant.
6. Finally, even the control valve FCV-101 can be switched to manual and bypass the control action in some control systems and with Fieldbus devices.

The feedback control loop that we have been familiar with refers to the data flow. However, the data flow has no information on whether the change in the process value is a result of the control action that it sends down to the valve. The back-calculation flow (dotted blue line in Fig. 3.14) tracks the integrity of the data flow and "informs" the controllers of the status and value of the downstream control elements. In this sense, the back-calculation flow serves as the feedback for the data flow. In other words, the data flow makes the control loop work, and the back-calculation flow tries to make the control loop not fail:

1. At the secondary controller (FC-101, ①), the mode of the flow controller is part of the information fed back to the primary controller. Once the level controller is informed that the flow control is no longer in cascade mode, the level controller automatically switches to initialization mode (INIT or IMAN, depending on control system type). The level controller's output (LC101.OP) would typically track the flow controller PV or setpoint (FC101.PV) provided via the BKCALO to BKCALI connection and remain ready for bumpless transfer when the flow controller is switched back cascade.
2. At the analog output block(AO), the AO output is fed back to the flow controller as BKCALO through the back-calculation flow. The downstream valve information, if available, is received at the BKCALI connection.
3. At the valve (FCV-101 ③), it is expected that the actual valve position is available to be used by the AO block as the BKCALO value. However, the actual valve position is typically unavailable (disconnection at ②) for older valves. If AO does not receive the value position valve at its BKCALI, it would simply send its own output as the back-calculation value to the primary controller. This "tail in the mouth" data connection results in a floating value for the back-calculation and is a common cause of controller windup or degraded performance.
   Smart valves with Fieldbus configurations can now provide the actual valve position as read-back, which the AO can use as the back-calculation value for the BKCALI input. This value is then propagated back to the controllers.

Compared with the data flow, which is a closed-loop from the controller to the process and back to the controller, the back-calculation information flow is only a partial loop, from the final control element back to the controller at best. The status and health of the rest of the data flow are missing. Tracking and monitoring the entire data flow is necessary but is still a challenge.

3.4.3 Initialization

Bumpless transfer during transition is a crucial requirement for process control. When the data flow is disconnected, the controller output will not reach the final control element. The output of the primary element becomes out of synchronization with the setpoint of the secondary element. The values must be back in-sync again before closing the loop to avoid the so-called *transfer bump*, which is the purpose of controller initialization. Several common scenarios may trigger an initialization, including the following:

1. The control element is started or activated the first time.
2. A request from an external user program or logic.
3. The control output was floating (indisposable), and now one or more output connections have become disposable.

Many practical considerations in PID configuration options are to ensure bumpless transfer. The typical approach is back-initialization, which requires several pieces of information to make the right decisions:

1. When to perform the back-initialization? When the loop is re-connected via mode change or selection change, there is a complex procedure of status tracking and broadcasting within the control system to inform the relevant control elements of the status change. Most of the time, this internal tracking and communication are sufficient to initiate the initialization procedure.
2. Is back-calculation possible? A simple controller shown in Fig. 3.12 has one secondary control element, so the back-initialization is straightforward. For a more complex control scheme where the control output is shared by multiple secondary control elements such as a split-range control, fan-out control, or override control, back-initialization may not be possible as the controller output is already connected to some secondary elements. In this case, alternative actions must be taken by the controller to achieve bumpless transfer. In extreme cases, extra function blocks may need to be provided in the control design to ensure bumpless transfer. See Sect. 4.3 for discussions.
3. What value should the controller output be back-initialized to? In the simplest case shown in Fig. 3.11, the PID controller is expected to align its output to match the `AO` block's setpoint before re-connection. However, how would the PID controller be informed about the setpoint value of the `AO` function block?

Initialization of the function blocks starts from the final control elements and goes backward along the data flow. When a disconnected data flow is re-connected (e.g., a secondary element is switched back to cascade mode), the secondary element will make an initialization request to its primary element. For example, if the `AO` output in Fig. 3.14 is at 30%, and the flow controller `FC-101` output is at 50%, the controller would set (re-initialize) its output to 30% before re-connecting with the `AO` block. The PID controller `FC-101` initializes itself to match the `AO` output, not the other way around. The back-initialization propagates backward until it reaches the topmost controller `LC-101` or reaches a control element that cannot be back-initialized.

In most cases, the primary element can simply adjust its output accordingly to match the secondary element. However, there are many scenarios that the back-initialization may not be possible. For example, for a fan-out control (see Sect. 2.2.3), the controller output may already be connected to one or more control elements. If one additional control element (e.g., a valve) comes back online, the controller cannot back-initialize to setpoint value of the newly connected element. As a result, an initial gap exists between the primary element's output and the newly connected element's setpoint, and a transfer bump may ensue at the time of re-connection. In this case, it becomes the secondary element's responsibility to initialize itself to the primary controller. In a fan-out control scheme, assume the controller output is at 50% and is connected with two valves; if a third control valve having a valve opening of 30% is switched back to cascade, the controller cannot back-initialize its output to 30% since it is already connected to the two valves at 50%. This 20% gap will cause a bump in the valve opening upon re-connection.

A built-in bumpless transfer feature exists in many control function blocks, including the control splitter used for fan-out control. The bumpless transfer function gradually decays this initial gap to achieve a smooth transition. See Sect. 4.3.4 for a detailed discussion. Process control engineers must be aware of the potential transfer bump in the control scheme and ensure a bumpless transfer feature is available in the design. If not, a dedicated function block may need to be inserted to achieve bumpless transfer. The manual loader (or AUTOMAN block) is the simplest function block in DCS that can be used for this purpose.

Another challenging scenario is with bad PV. If a PID controller's PV goes bad and subsequently comes back to normal, a large gap between the setpoint and PV may already have developed. Restoring the PV may lead to a bump in the control error, thus the controller output. If the bumpy behavior is unacceptable, then operator intervention is expected to align the setpoint with the process value first and gradually bring the setpoint to the desired value.

In summary, tracking and initialization are automatically achieved for most of the commonly encountered control schemes consisting of standard control elements. However, some control schemes with complex data flow do not have the full support for back-calculation. These complex scenarios would require the process control engineer to either change the control design or use back-end logic to ensure the continuity of the back-calculation data flow. Therefore, it is crucial to know when the control system can automatically take care of the back-calculation and when extra help should be provided in the control design.

3.5 PID Configuration

PID configuration is an overlapping area between process control and control systems. Although primarily the responsibility of control systems engineers/technicians, some configuration options are closely related to the process design and control strategy and only the process control engineers are qualified to specify.

Table 3.9 Typical configuration parameters of a PID controller

Configuration	Description, options, or examples
Algorithm type	P, PI, PID etc.
Equation type (structure)	`I-PD`, `PI-D`, `I`-only, etc.
Normal mode	`AUTO`, `CAS`, `RCAS`, etc.
Tuning parameters	System default values typically are NOT acceptable!
Direction of action	`DIRECT/REVERSE`
Setpoint high/low limit	The valid value range for setpoint
Fastest scan rate/sequence	50 ms, 100 ms, ..., 1 s, ...
Output format	Positional/Velocity
Output scaling	0% = Fully closed, 100% = Fully open
Fail-safe position	Fail-open, Fail-close, Fail-stay
PV filtering	Noisy signal can cause poor performance
PV tracking	Setpoint tracks PV when `MODE` in `MAN`
Anti-reset windup	Integral windup can cause delayed response
Bumpless transfer	No "bump" during `MODE` change

As discussed in previous sections, implementing a PID controller is much more complex than just solving a three-parameter equation. Many things can go wrong and probably will go wrong if not correctly configured or prevented. Table 3.9 provides a list of the crucial PID configurations that are common in all control systems. Their incorrect configurations may have a significant impact on the control performance.

These configurations require a good understanding of the control objectives and control strategy and are thus best determined by process control engineers as part of process control design. Note that the control systems have default values for all the configurations, but many of the default values are not acceptable and must be supplied with the correct values!

3.5.1 Output Scaling

Controller output, typically in the range of 0–100%, is converted to a 4–20 mA analog signal by the analog output (`AO`) module in the control system and then sent to the valve's actuator in the field. See Fig. 3.11 for an illustration. For *fail-close* (*air-to-open*) valves, a 4 mA signal drives the valve to the fully closed position (*fail-safe* position), and a 20 mA signal drives the valve to the fully-open position. However, for a *fail-open* (*air-to-close*) valve, a 4 mA signal corresponds to a fully open position (fail-safe position), and a 20 mA signal would drive the valve to a fully closed position.

Table 3.10 Control signal and valve fail-safe position

Controller output (%)	Intended valve position	For fail-close valve (mA)	For fail-open valve (mA)
0	Fully closed	4	20
100	Fully open	20	4

With fail-open valves, a signal inversion has to be performed along the way from the controller to the valve to ensure *negative feedback*. This inversion can be done either in the controller, where 100% is defined as a closed valve, or in the analog output (AO) module, where a 100% signal is mapped to a 4 mA instead of 20 mA. Both configurations exist in practice, and which one to choose has been mostly a personal preference. This decision, however, does affect the configuration of PID direction of action and is also a source of confusion.

In modern process control design, to avoid confusion, it is customary to follow a convention that an output of 0% from the PID controller always represents a fully closed valve and 100% a fully open valve. The signal inversion, when required, is performed in the AO module. See Table 3.10 for an illustration.

3.5.2 *Direction of Action*

The PID controller can be configured as either `DIRECT` or `REVERSE` acting. Along the closed data flow path of the control loop, the process can have a positive or negative gain, the control valve can be air-to-open or air-to-close, and the AO signal can be direct or inverted. Therefore, the loop gain can be either positive or negative, out of a combination of 16 possibilities (see Table 3.11). A stable control must be a negative feedback control, which requires that the overall loop dynamics have a negative gain. If the direction of action is wrong, nothing else will matter because the controller output will quickly run off-scale in the wrong direction regardless.

The convention mentioned above in Sect. 3.5.1 requires that 0% of controller output cause the valve to close fully, and 100% of controller output fully opens the valve. That means the AO is configured so that the combined dynamics of the AO block and the control valve always have a positive gain. The analog output (AO) needs to perform a signal inversion to produce a positive gain for an air-to-close valve, as illustrated in Table 3.12.

With this simplification, the AO and controller valve is taken out of the picture. The sign of the controller gain (direction of action) is determined by the process gain alone and should always be opposite to the sign of the process gain to ensure negative feedback. See Table 3.13.

Table 3.11 Loop components and the sign of loop gain

PID controller	Analog output (AO)	Control valve	Process
Direct acting	Normal scale	Air-to-open	Positive gain
or	or	or	or
Reverse acting	Inverted scale	Air-to-close	Negative gain

Table 3.12 AO and controller valve should produce a positive gain

Control valve	AO block	Combined gain
Air-to-open	× Normal signal	↦ Positive gain
Air-to-close	× Signal inversion	↦ Positive gain

Table 3.13 Determining PID direction of action

Process gain	PID direction of action	Example
Positive gain	→ Reverse acting	Flow control FC-101 in Fig. 3.14
Negative gain	→ Direct acting	Level control LC-101 in Fig. 3.14

In practice, the following rule of thumb can be followed to determine whether a controller should be configured as DIRECT or REVERSE acting: when the process value (PV) to be controlled is higher than the controller setpoint (SP), if the desired action is to increase the controller output (OP), the controller action is defined as DIRECT action, and vice versa.

For example, in Fig. 3.15, the level controller LC-101 should be DIRECT acting because if the level is above the desired setpoint, the outlet flow should be increased. Similarly, the flow controller FC-101 should be configured as REVERSE acting so that if the flow were to increase suddenly, the valve opening would be decreased to reduce the flow. This decision is completely decoupled from the decision on the direction of action of the AO block and the fail-safe position of the valve.

3.5.3 Setpoint High/Low Limit

When the PID control is in AUTO mode, the setpoint is entered by the operator. The setpoint high and low limits (see SPH and SPL in Sect. 1.3.5) should be configured to protect against human entry errors. The decision on the range between SPL and SPH is a joint decision between process engineering and operations.

In a cascade control loop, the setpoint high and low limits of a secondary PID controller are part of the status tracking. When the limits are violated, the data flow

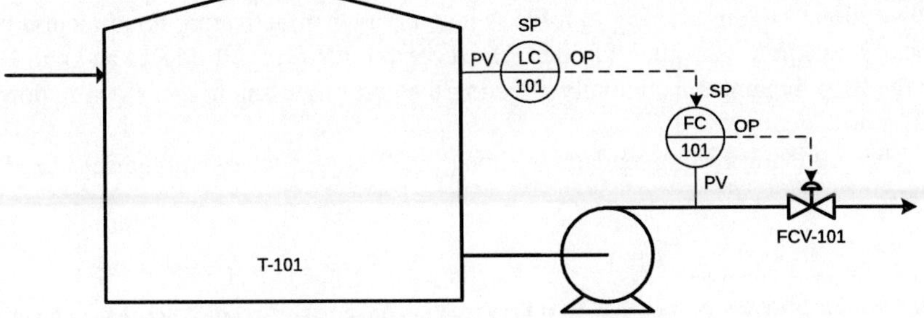

Fig. 3.15 A basic PID cascade control loop

is reported as limited, and the primary control automatically activates the anti-reset windup action.

3.5.4 Execution Frequency and Execution Order

The controllers have a default execution frequency (*scan time* or *scan rate*), typically every second. Any controller operating faster or slower than the standard execution frequency shall be mentioned explicitly in the control design. For example, compressor anti-surge controllers typically need to execute at a much higher frequency, such as 100 ms per execution.

Another important consideration during implementation is the *order of execution* of the functional blocks. In a control system, each control block along the control path is assigned a number to specify the order of execution.[5] For a complex control scheme, the calculation may be delayed by several scan cycles if the order of execution is incorrectly specified. For example, for the cascade control loop in Fig. 3.15, the execution should start from the primary loop `LC-101`; the calculated output is sent down to the secondary loop `FC-101`, where the control action is calculated and the output sent down to the valve `FCV-101`. Following this execution order, all the calculations in this complex scheme are completed with the latest measurements and within the same execution cycle. If the order is reversed and the secondary loop `FC-101` is calculated first, its setpoint would be the result from the last execution of controller `LC-101`, based on the previous level measurement. This cascade control will take at least two scan cycles to complete, with level and flow measurements from two different cycles.

For a complex control that includes a selector block (such as in override control) in its data flow path, the order of execution is crucial. All the primary blocks must run before the selector block so that the feedback is correct.

3.5.5 PV Filtering

On a modern control system, filtering is typically provided as a standard configuration option. Adding filtering to a PV is just a matter of activating this function by specifying a non-zero filter constant. However, the PV filter introduces an extra lag to the loop dynamics, potentially affecting the controller tuning and slowing down the control action.

The following recommendations should be followed when configuring the PV filter:

---

[5] In Honeywell Experion, for example, if the execution order is not specified, the default order is assigned which is based on the order that the function blocks are added to the system.

1. The PV filter constant should be no larger than 1/5 of the controller integral time.
2. When the PV filter time is added or changed, its impact on the controller tuning should be assessed to decide if the controller needs to be re-tuned.

3.5.6 Initial Tuning Values

Different control systems have different implementations of the PID control algorithms. They all have a set of system default values for the PID tuning parameters. For example, on the DeltaV system, the system default tuning values for a PID loop has a gain of 0.25 and an integral of 10 s. Unfortunately, these default values are typically not acceptable for most control loops. As an extreme example, all the level control valves of the oil/gas separators in one new plant wore out (severe leaking or passing) in just two years. Investigation revealed that all the PID tunings were left at the system default at gain = 0.25 and integral = 10 s. The level controllers are expected to have an integral time of about 20–30 min instead of 10 s!

Although the PID control parameters cannot be finalized until the plant is in operation, it is essential to specify the initial PID tunings during the design phase of a new project since the system default values are rarely appropriate. These initial values are meant to set the controller on its foot for the startup. Some recommended initial tuning parameters can be found in Table 9.1. During commissioning, the parameters should be fine-tuned.

Tuning parameters for critical control loops, particularly those with long dynamics, can sometimes be calculated based on dynamic simulation results if available. As *dynamic simulation* becomes more popular, controller tunings can be calculated based on simulation results even before the construction of the actual plant starts. The high-fidelity process models from dynamic simulation can provide sufficiently accurate data for PID tuning.

3.5.7 PV Tracking

When `PV tracking` is enabled, the PID controller setpoint will track the PV value if the controller has switched away from cascade (e.g., to `MAN`), and the original setpoint value is discarded. Enable `PV tracking` will force the control error to zero so that when the PID controller is switched back to cascade mode again, there are no transition bumps in the control action. However, some control applications require that the controller setpoint retain its value during controller mode change, such as protective and level controllers. It is thus a trade-off between retaining the setpoint and ensuring bumpless transfer. The decision is thus controller-specific but should be documented in the control narrative. Generally speaking, standalone controllers and protective controllers should have PV tracking disabled, so the operator does not need to remember and reset the setpoint values.

3.5.8 *Anti-Reset Windup*

Reset windup is a common concern with PID control. In modern control systems, with the help of a back-calculation information flow, the anti-reset windup function can be automatically achieved through proper connection and configuration of the control blocks (controller, selectors, splitters). However, there are many cases in complex control loops where back-calculation is not fully supported. In this case, external feedback may need to be manually configured. Reset windup is an essential consideration for a process control engineer during process control design. Detailed discussion is provided in Sect. 4.3.3.

3.5.9 *Bumpless Transfer*

The data flow in a feedback control loop is round-trip. The controller's output causes the process to have the desired change, reflected by the process value fed back to the controller. The path from the controller output to the final control elements must remain connected. If the connection is interrupted anywhere along this path, e.g., the secondary controller in a cascade loop is switched away from the cascade mode (`CAS`), and the output of the upstream controller (the master controller in the cascade loop) is then "unconsumed" and left floating.

When the secondary controller is switched back to `CAS` mode, the output of the master controller is expected to be the same as the value of the controller output just before the switch. For this purpose, the primary controller should automatically initialize itself to align with the secondary controller. Most controller initialization requirements have been built into the control modules (via back-calculation) on a modern control system; no special attention is required. However, under some special circumstances (e.g., with a `CALC` function block), the controller initialization may be more complex than the default functionality can support, and extra help is needed from the process control engineer to complete the controller configuration during control design. See Sect. 4.3.4 for detailed discussions.

3.6 PID Mode Changes

The different modes of the PID controller provide the flexibility of handling different application requirements. On the other hand, the transition between different modes also creates many challenges that require careful considerations during process control design and implementation.

3.6.1 PID Mode Change in Normal Operations

A detailed discussion of the sequence of events during the mode change can be quite involved and is outside the scope of a process control book. However, it is essential to understand how the mode change affects the controller setpoint and output. Some features are automatically taken care of by the built-in PID function modules, and some features must be explicitly configured during implementation. Incorrect configurations can lead to degraded performance, such as bumpy transition and reset windup. Let us take the simple cascade control loop in Fig. 3.15 as an example.

The primary controller `LC-101` can be in either `AUTO` or `MAN` mode, while the secondary controller `FC-101` can assume `MAN`, `AUTO`, or `CAS` mode during normal operation. The `NORMAL` operating mode for `LC-101` is `AUTO`, while for `FC-101` is `CAS`. The transition between the different modes triggers a chain of events:

1. `MAN` to `AUTO`: When the PID controller `LC-101` is switched from `MAN` mode to `AUTO`, the following occurs:
   a. If `PV Tracking` (see Sect. 3.5) is enabled, the `SP` equals the `PV` value when the controller is in `MAN` mode; thus, the control error is zero. The mode switch from `MAN` to `AUTO` will not create a control error, and thus there will be no change in output following this mode transition. The `OP` will remain unchanged until a non-zero control error is developed. The switch is thus bumpless. However, the original setpoint value is lost, and the operator needs to bring the setpoint to the desired value.
   b. If `PV Tracking` is not enabled, the SP retains the previous value when the controller was in `AUTO` mode. This setpoint valve is most likely different from the current process value (most `PV` values start from zero during startup), and a significant control error may exist. The PID controller can produce a sudden change in the output when the controller is switched to `AUTO` due to the proportional kick, resulting in a bumpy action. On the other hand, with `PV Tracking` disabled, the `SP` value is preserved during the mode switch, and the operator does not need to remember and re-set the setpoint after the mode change.
2. `AUTO` to `MAN`: When the PID controller is switched from `AUTO` mode to `MAN`, the controller output (`OP`) will retain the last good value until the operator manually adjusts it.

The mode change is more complex for the cascade control loop as two controllers are involved:

1. `AUTO` to `CAS`: When the secondary controller (`FC-101`) is switched from `AUTO` to `CAS`, the controller setpoint (the cascade setpoint, or `CSV`, in Fig. 3.9) will take the value of the current setpoint, and the output of the primary controller (`LC-101`) is automatically initialized to the setpoint value of the secondary controller.

After completing a series of handshakes, the mode of the primary controller changes from INIT (or IMAN on some other control systems) mode to AUTO mode. Depending on whether PV Tracking is enabled, the primary controller may have an existing control error between SP and PV, and the mode change may immediately generate a significant change in the output and send it down to the secondary controller as a setpoint change.

2. CAS to AUTO: When the secondary controller mode is switched ("shed") from CAS to AUTO, the secondary controller is bumpless because the SP will take the value of the cascade setpoint (CSV); therefore, no transient control error.
The primary PID controller will change to INIT (or IMAN) mode, and the output is left floating, and its value is dependent on the specific configuration.

The handshake process can be pretty complicated and is not guaranteed to succeed if the mode change is not allowed or other permissions are not met. Figure 3.16 is a simplified illustration of the handshake process for mode change from AUTO to CAS in a DeltaV DCS that involves coordinating the current mode, requested mode, target mode, and actual mode.

Although this handshake process is the scope of control systems instead of process control, it is beneficial to have a basic appreciation of the complexities in the actual mode switching. It is essential to understand how the mode change affects the control setpoint and output and what initial values are assigned when the new control mode is in effect.

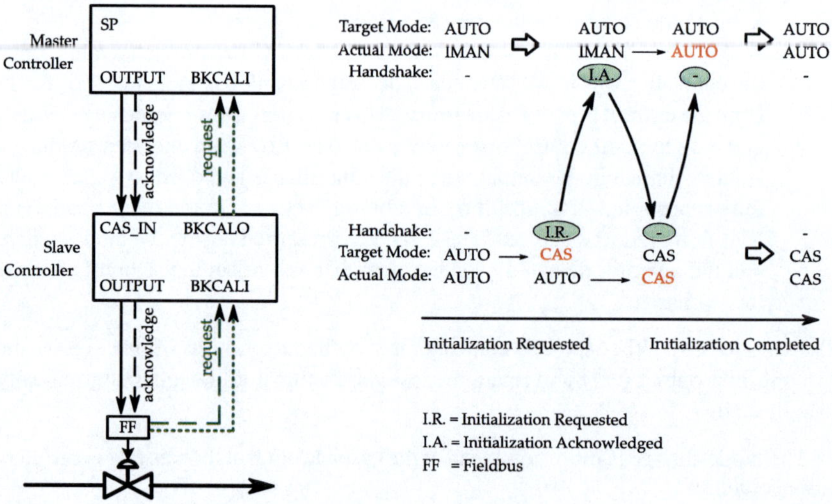

Fig. 3.16 Handshake process for a PID mode change

3.6.2 PID Modes During Crippled Operations

The crippled mode operation concerns the controller's reaction to the sudden loss of the process values (PV) due to, for example, measurement failures. Crippled mode is a practical scenario that is often neglected in implementation. In case a measurement failure is detected, the following decisions should be considered regarding PID control response:

1. Controller mode: In the case of BAD PV, should the controller remain in control (e.g., in AUTO or CAS) or be switched to manual operation (MAN mode)? Usually, the controller should not be allowed to be in control if the measurement is lost or cannot be trusted. The controller should be configured to automatically "shed" to MAN mode when a BAD PV status is received.
2. After the controller sheds to MAN mode, what value should the controller output assume? There are typically three options, depending on the criticality of the control loop and the availability of the operator:

   a. Retain the *last good value*: Keeping the last good value is the most common option and is recommended for operations where operator attention is available reasonably quickly after alarm notification. It is a safety concern if the valve position remains fixed for an extended duration.
   b. Move to the fail-safe position: Set the valve to its fail-safe position in one shot (fully close or fully open). This option may be the preferred action for unmanned or safety-critical operations, although this could potentially introduce a significant disturbance to operation.
   c. Ramp to a safe position: In theory, a better solution is to slowly ramp the output to its fail-safe position at a pre-defined rate while waiting for operator intervention. However, most control systems do not support this option out-of-box, and back-end logic needs to be implemented to handle the ramping of the output. Therefore this option is not practical or worthwhile for most operations due to the added complexity in the configuration, except for very critical operations or equipment. Compressor anti-surge control (see Sect. 11.3) is an example of this use case.

   After the controller sheds to MAN mode, should the controller retain its setpoints or make the setpoint track the process value? Enabling PV Tracking will help achieve bumpless transfer when the controller is switched back to AUTO; however, some control applications such as level control and protective control typically prefer to retain the setpoint during mode changes. See PV Tracking in Sect. 3.5 for details of PV Tracking configuration.
3. Once the failure is cleared, should the controller be automatically put back to NORMAL mode by back-end logic or manually by the operator? Automated operation is undoubtedly desirable, but a necessary condition is that the failure condition is genuinely resolved. For critical operations, it is still safer to count on the operator to verify the health of the measurement and make the call.

3.6.3 PID Modes During Shutdown and Startup

During emergency shutdowns (ESD), the controller mode, output, and setpoint should be set to the correct values or position to ensure a safe shutdown and the readiness for re-start. Typical considerations include the following:

1. Controller mode: Should the controller be switched to MAN mode or remain in AUTO?
2. Controller output: Should the controller output be set to the valve's fail-safe position, or just retain the last good value?
3. Controller setpoint: Should the controller setpoint be retained or be set to different values, such as tracking the process value?

Conventionally it is assumed that all the controllers should be put to manual (MAN) mode during shutdown and switch to normal mode by the operator during or after startup. For a large plant with an increasingly large number of control loops and fewer operators, managing the controller mode, output, and setpoint during shutdown and startup becomes a burden and distraction.

Fortunately, many control loops are safe to remain in their NORMAL mode during the shutdown and startup process. Consider the example in Fig. 3.17:

1. Flow control loops: Flow control loops (e.g., FC-101) are typically REVERSE acting. During shutdown with no process flow, the controller would drive the valve to the fully open position. If an open valve is acceptable during the startup, then the flow loop can be left in CAS or AUTO mode all the time (except during BAD PV).
2. Level control loops: Most level control loops (e.g., LC-101) are manipulating a downstream flow or valve and are thus DIRECT acting. During shutdown with a

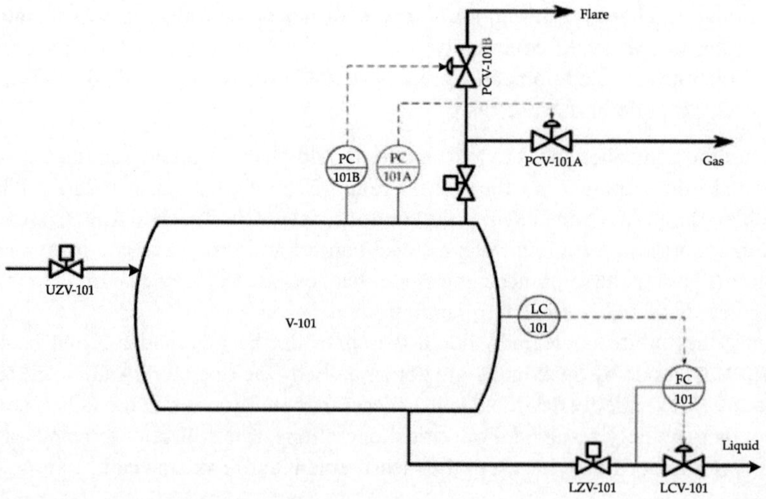

Fig. 3.17 PID controller mode during shutdown

low level, the controller automatically drives the flow setpoint or valve position to zero. So it is typically safe to keep the level control in `AUTO` mode during shutdown and startup. For instance, the level-flow-valve cascade loop (`LC-101` to `FC-101`) in Fig. 3.17 can be left in `AUTO` mode all the time (except with `BAD PV`).

3. Protective controllers: For a high-pressure protective controller such as `PC-101B` in Fig. 3.17, the SP is higher than PV during the normal operating mode. As a `DIRECT` acting controller, the controller output is zero when `PV < SP`. The flare valve is fully closed as expected. During shutdown and startup, the pressure is lower than the normal value, the control valve will remain in the fully closed position, so it is safe to have the controller in `AUTO` mode all the time.
   However, a low-pressure protective controller must start from `MAN` mode during startup and switch to `AUTO` only after the process value is above the setpoint.
4. Overriding Control: The protective controller in an override control loop follows the same principle as a standalone protective controller. For high protection, it is typically safe to remain in `AUTO` all the time, while for low protection, it is typically not.
5. Normal regulatory controller: Controller for normal operating control such as `PC-101A` in Fig. 3.17 will need to be slowly brought to normal setpoint by the operator during startup, either in `MAN` or `AUTO` mode. So these control loops will typically be put to `MAN` during a shutdown. They are the process variables that require the most attention from the operators during startup, shutdown, and other mode transitions.

During design, the process control engineer should review each control loop individually and specify how the controller should behave during startup, shutdown, and mode transition. If left unspecified, they will assume the default system behavior, and the default is often unacceptable.

3.7 Summary

This chapter discusses the three essential requirements for a practical PID implementation:

1. Versatility to meet different application requirements: Different PID equations have a direct impact on the control performance (Sect. 3.1). Various controller modes provide flexibility to perform manual control, automatic control, and supervisory control (Sect. 3.2).
2. Reliability to ensure safe operations: A vital difference between a simulation study and actual implementation is in reliability. It is easy to make a PID controller work, but it is overwhelmingly difficult to make it not fail, especially in a complex control loop (Chap. 4). Back-calculation is a common approach to monitoring the health of data flow in a control loop (Sect. 3.4).

3. Scalability to support complex control loops: A complex control loop is comprised of multiple PID controllers and supporting function blocks. The communication and inter-operation between the function blocks can become very complicated, and the proper configuration is critical to operating a control loop (Sect. 3.5). Many configuration options require in-depth knowledge of the control design, and only process control can make the correct decision.

Although PID configuration is primarily the responsibility of control systems engineers, many configuration options are determined by operating requirements and dictated by the process control strategy. They depend on process control to provide the right decisions or guidance.

References

Åström KJ, Hägglund T (1995) PID controllers: theory, design, and tunings, 2nd edn. Instrument Society of America

Åström KJ, Hägglund T (2006) Advanced PID control. ISA

Åström KJ, Kumar PR (2014) Control: a perspective. Automatica 50(1):3–43

Brannan C (2018) Rules of thumb for chemical engineers—a manual of quick, accurate solutions to everyday process engineering problems, 6th edn. Elsevier

Marlin TE (2015) Process control—designing processes and control systems for dynamic performance, 2nd edn. McGraw-Hill

Minorsky N (1922) Directional stability and automatically steered bodies. J Am Soc Navig Eng 34:280

Routh EJ (2016) Dynamics of a system of rigid bodies. Hansebooks

Samad T, Bauer M, Bortoff S, Cairano SD, Fagiano L, Odgaard PF, Rhinehart RR, Sanchez-Pena R, Serbezov A, Ankersen F, Goupil P, Grosman B, Heertjes M, Mareels I, Sosseh R (2020) Industry engagement with control research: perspective and messages. Ann Rev Control 49:1–14. Open access

Smith CL (2009) Practical process control—tuning and troubleshooting. Wiley

Turton R, Shaeiwitz JA, Bhattacharya D (2018) Analysis, synthesis, and design of chemical processes. Prentice Hall

Visioli A (2001) Tuning of PID controllers with fuzzy logic. IEE Proc—Control Theory Appl 148(1):1–8

Visioli A (2006) Practical PID control. Advances in industrial control. Springer, London

Chapter 4
Complex PID Control Loops

Many practical problems are more complicated than a simple PID controller can handle. Multivariable, constrained, or nonlinear problems require complex PID control solutions comprising multiple PID controllers and other function blocks working together in concert (Sect. 4.2). Although PID-based control schemes are surprisingly versatile in addressing many complex control problems (Sect. 4.1), the design and implementation can be surprisingly tricky. As there are many ways to achieve the same control objective with these building blocks, a good understanding of the operating requirements and adequate knowledge of these function blocks are crucial in choosing the best complex PID control design. Ensuring the loop integrity is one critical requirement that relies on in-depth knowledge of the data flow and the control design (Sect. 4.3). For a complex control solution, simplicity and reliability is always the top priority in the control design and the best solution is thus always the one that is *fit-for-purpose*.

4.1 Applications of Complex PID Control Schemes

Section 2.2 has summarized the commonly available complex control loops and explained what functionality they provide and when they should be used. These loops all consist of one or more PID controllers, supported by some auxiliary function blocks, having a fixed structure and performing a predictable task. We will call them the *standard complex loops*. These standard complex loops serve as the building blocks for more complex control solutions (Smith 2010).

4.1.1 Main Challenges for Complex Control Loops

A PID control loop is straightforward and does what it is supposed to do most of the time. In practical applications, however, the main challenge is not how to make

S. S. Niu and D. Xiao, *Process Control*, Advances in Industrial Control,
https://doi.org/10.1007/978-3-030-97067-3_4

the PID loop work but how to make the PID loop not fail, especially in a complex control scheme. From practical experience, more than 90% of the effort is spent on less than 10% of the possibility of PID loops not working as per design.

We will start with a compressor control example to demonstrate how a complex control scheme is built with the basic PID controllers and standard complex loops. This example is used here solely to help the discussions in this chapter on complex control loops. Thorough coverage of this example is in Sect. 11.3, so there is no need to delve into the control strategy for now.

A typical process flow of a gas compression unit is shown in Fig. 4.1.

A compressor is a critical piece of equipment in many plants. Its failure often results in the shutdown of the processing unit or even the entire plant and thus must be efficiently controlled and reliably protected.

Some compressor variables have very fast dynamic responses, typically in tens to hundreds of milliseconds, so computation-intensive control schemes such as model predictive control (MPC) are out of the question. The only viable choice is a complex PID control scheme. Figure 4.2 illustrates an integrated compressor control scheme designed for implementation and operation in a standard DCS platform.

As shown in the control schematic, most of the control loops are the standard complex PID loops discussed in Sect. 2.2. For example,

1. Standalone PID control: `PC-004` is a simple PID controller responsible for capacity control.
2. Cascade control: `PC-002` to `XC-101` to `SC-101` is a three-level cascade control loop.
3. Protective control: `PC-001` is a protective controller with a higher setpoint than `PC-002` to protect the suction header against excessively high pressure.
4. Overriding control: Several override controls are implemented in this example, including `PC-002/PC-003`, `UC-111/PC-112`, and `JC-101/SC-101`. Control selectors, both high selection and low selection, are extensively used to support the overriding controls.
5. Split-range control: The output of `PC-002` is sent to `XC-101` and the two control valves `ASCV-111` and `ASCV-121`, via a split-range control block `XY-101`.

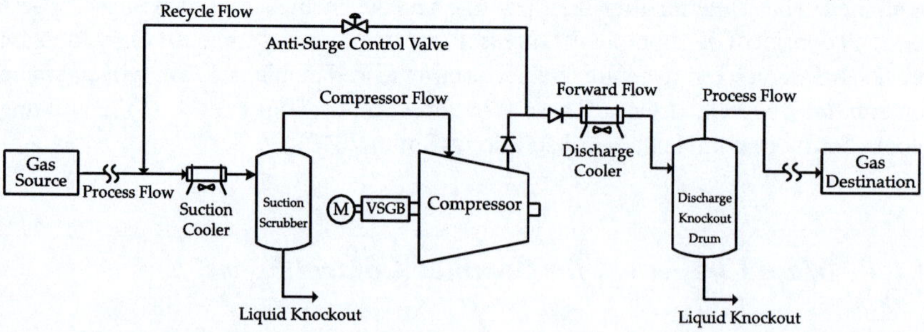

Fig. 4.1 A typical gas compression process flow

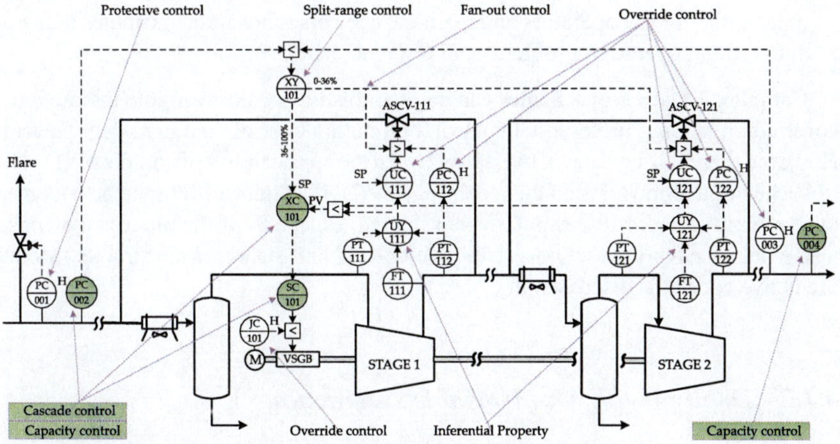

Fig. 4.2 A complex control loop: integrated compressor control

6. Fan-out control: The output of `PC-002` is sent to the two recycle valves `ASCV-111` and `ASCV-121` through a fan-out control scheme.
7. Inferential Properties: The process values (`UY111.PV` and `UY121.PV`) for the two anti-surge controllers (`UC-111` and `UC-121`) are inferential properties, calculated with `UY-111` and `UY-121`, respectively.

This example shows that a PID-based complex control solution can become very sophisticated. As the complexity increases, selecting the best control scheme and protecting the loop integrity become the primary challenges:

1. The standard complex PID control loops discussed in Sect. 2.2 are the building blocks for complex control schemes. For complex control problems, there can be more than one way to achieve the same control objective, e.g., a split-range control versus a dual-controller control, a ratio control versus or a feedforward control. Each scheme has its pros and cons and may be appropriate for one control problem but not for others. A sound control solution requires a deep understanding of the process dynamics, operating requirements, and the functionality of the control function blocks. A good design and a bad design may all work (to some extent), but the difference is in the control performance, with a last-lasting impact on the operation.
2. The data flow in a simple PID control loop is relatively simple but still has many practical complexities affecting the control performance (see Sect. 3.2). The information flow can become overwhelmingly complicated in a complex control scheme comprising multiple PID controllers and supporting function blocks. For example, with control selectors (override control) and splitters (cascade and fan-out control), the possibility of reset windup and bumpy transition is significantly higher. Protecting the integrity of the data flow in process control design requires significant knowledge and experience. The complexity is also a consider-

able hurdle for the operators and maintenance personnel and contributes to many sub-optimal operation problems and even incidents.

Complex PID control schemes can be used to address multivariable interactions, constraint handling, preferential control, optimizing control, and nonlinear control. However, there is a significant overlap between the applicability of complex PID control schemes and model predictive control (MPC). If a control PID scheme becomes overly complicated, MPC should be considered, especially if the process dynamics also exhibit heavy interactions, extensive delays, and stringent constraints. Model predictive control is discussed in Chap. 5.

4.1.2 Dealing with Multivariable Interactions

Complex control loops are designed to solve complex control problems that consist of multiple control targets with multiple control handles. The control targets and control handles may exhibit a strong correlation and interaction and must be considered altogether.

Degree of Freedom Requirement

A control handle is one *degree of freedom* (DOF) and can control one target. All the control targets and control handles must be paired up in a complex control scheme. In other words, the number of control targets must match the number of control handles to leave no DOF unconsumed. Unconsumed control handles will eventually be driven to saturation and cause degradation in control performance.

It is common to find that the number of control targets does not agree with the number of control handles in a complex control problem. In this case, the standard complex control loops in Sect. 2.2 can be used to adjust the balance of DOFs by "merging" or "splitting" the control targets or control handles. For example, a fan-out control loop has a single target manipulated by two or more control handles. An override control loop has one control handle but multiple control targets, although only one target is controlled at any given time:

1. Reduce the number of controlled variables by correlating the targets: Override control (Sect. 2.2.7) is the most popular way of "reducing" the number of control targets. For example, many override control schemes are utilized in the compressor control example in Fig. 4.2. The two controlled variables, `SC-101` and `JC-101`, share one control handle, the gearbox speed. The compressor suction pressure `PC-002` and discharge pressure `PC-003` are "merged" via a selector and consumes one DOF. Ratio control (Sect. 2.2.6) "ties" two targets together and thus manipulates only one control handle. A selective control selects one target at a time and thus is counted as only one controlled variable in terms of DOF requirement.

Another option is to use inferred variables. For example, instead of controlling both temperature and pressure, a pressure-compensated temperature reduces the number of targets from two to one (see Sect. 10.4). In Fig. 4.2, compressor surge is physically related to the suction pressure `PT-111`, discharge pressure `PT-112`, and discharge flow `FT-111`. Instead of controlling all three variables, a simple calculation is performed in `UY-111` to relate the three variables to a single variable, the *surge indicator*. That is, one controlled variable is used instead of three.

2. Reduce the number of manipulated variables: Techniques such as selective control (Sect. 2.2.8), fan-out control (Sect. 2.2.3), split-range control (Sect. 2.2.1), and feedforward control (Sect. 2.2.5) can be utilized to change the DOF provided by the manipulated variables. For example, the compressor suction pressure in Fig. 4.2 can be influenced by the compressor speed `SC-101` and the two recycle valves (`ASCV-111` and `ASCV-121`). A split-range control block `XY-111` "links" the two types of final control elements and effectively treats them as one DOF.
3. Increase the number of controlled variables by "splitting" or "duplicating" the measurements: Cascade control uses one control handle to control two targets (Sect. 2.2.4). Dual-controller control duplicates the control loop but with different setpoints (Sect. 2.2.2). For example, in Fig. 4.2, the compressor suction pressure is "split" into two variables and controlled with two controllers, `PC-002` and `PC-001`, and thus consumes two DOFs. The normal regulatory controller `PC-002` treats the suction pressure as a regulatory control target (capacity control), manipulated by the control speed. The overriding protective controller `PC-001` treats the suction pressure as a constrained variable, controlled with the flare valve. Similarly, the split-range control block `XY-111` "splits" the selector output into two signals, one goes to the load controller `XC-101`, and the other is fanned out to the two anti-surge control valves (`ASCV-111` and `ASCV-121`).
4. Increase the number of manipulated variables: The number of control handles is usually fixed once the process engineering is complete. It is typically not possible to increase the number of manipulated variables to meet the DOF requirement without process change.

Table 4.1 summarizes the options of using complex PID control schemes to adjust the number of DOFs.

Similarly, when troubleshooting an existing control scheme, it is essential to validate that all the controlled variables and manipulated variables are paired up, and thus, all the DOFs are consumed.

Interactions Among Process Variables

One of the key challenges for multivariable control problems is the interaction between the process variables. One manipulated variable may affect multiple controlled variables, and conversely, one control target may be impacted by more than

Table 4.1 Changing number of degree of freedom with complex control loops

Complex PID loop	To increase targets	To reduce targets	To reduce handles	Reference
Split-range control			✓	Section 2.2.1
Dual-controller control	✓			Section 2.2.2
Fan-out control			✓	Section 2.2.3
Cascade control	✓			Section 2.2.4
Feedforward control			✓	Section 2.2.5
Ratio control		✓		Section 2.2.6
Override control		✓		Section 2.2.7
Selective control		✓	✓	Section 2.2.8
Model predictive control	✓	✓	✓	Section 2.3.4

one control handle or disturbances. Complex control loops such as feedforward control, ratio control, and decoupling control provide some capabilities to address multivariable interactions; However, they all require a good understanding of the process dynamics:

1. Feedforward control: A controlled variable is controlled by a manipulated variable, subject to noises and disturbances that can be measurable or unmeasurable. Feedforward control is a common approach to deal with measurable disturbances. Feedforward control is an open-loop control and must work along with a feedback control loop (Stephanopoulos 1984) in practical applications. Figure 4.3 shows the data flow of both the feedback and feedforward channels. Feedforward control "intercepts" the change in the disturbance variable and adjusts the manipulated variable in anticipation to "cancel out" the effect of the disturbance.
   It is vital to know both the feedforward and feedback dynamics to achieve "perfect" cancellation. If we use $G_d(s)$ to represent the disturbance dynamics and $G_p(s)$ for the process dynamics, the feedforward control $G_{ff}(s)$ is given by

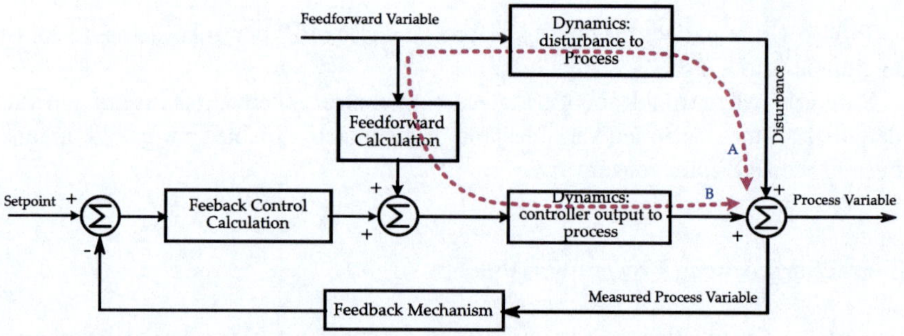

Fig. 4.3 Block diagram of feedforward control

$$G_d(s)\,U_d(s) + G_p(s)\,G_{\text{ff}}(s)\,U_d(s) = 0 \quad \mapsto \quad G_{\text{ff}}(s) = -G_d(s)/G_p(s) \tag{4.1}$$

where $U_d(s)$ is the disturbance input.
In most cases, we can simplify both the process dynamics and disturbance dynamics into gain, dead time, and time constant (i.e., a first-order process model). In that case, feedforward control can be interpreted as follows:

a Gain: The process gain and the disturbance gain are related by $K_d + K_{\text{ff}} \cdot K_p = 0$, which gives $K_{\text{ff}} = -K_d/K_p$, with K_d being the gain of disturbance channel, K_p being the gain of process channel, and K_{ff} being the gain for feedforward compensation.
b Dead Time: A necessary condition for feedforward control is that the dead time from the disturbance variable to the controlled variable must be greater than or equal to that from the manipulated variable to the controlled variable. That is, $\theta_d \geq \theta_p$, where θ_d is the delay of the disturbance channel, and θ_p the delay of the process channel. The feedforward control must have a time delay equal to $\theta_{\text{ff}} = \theta_d - \theta_p$.
c Time constant: For best performance, the feedforward control should provide lead/lag compensation for the difference in the dynamics of the two channels.

The calculation of the feedforward can be either *static* or *dynamic*. Static compensation involves only a gain adjustment to the disturbance when being added to the controller output. While a dynamic compensation needs also to consider delay and lead/lag compensation. For example, assume the process dynamics $G_p(s)$ and disturbance dynamics $G_d(s)$ are given by

$$G_p(s) = \frac{K_p}{\tau_p s + 1} e^{-\theta_p s}, \qquad G_d(s) = \frac{K_d}{\tau_d s + 1} e^{-\theta_d s} \tag{4.2}$$

then the feedforward transfer function is given by

$$G_{\text{ff}}(s) = -\frac{G_d(s)}{G_p(s)} = -\frac{K_d}{K_p}\frac{\tau_p\, s + 1}{\tau_d\, s + 1}\, e^{\theta_d - \theta_p} \tag{4.3}$$

with $\theta_d \geq \theta_p$. If the process dynamics or disturbance dynamics have more complex characteristics than first-order-plus- dead-time, the feedforward would be more complicated. Due to the time-varying nature of the dynamics, it is unrealistic to expect "perfect" dynamic compensation with a fixed lead-lag term. As a result, most feedforward control in practice only implements the gain and dead-time calculation.
Feedforward control is typically implemented in its incremental form for better stability. The delta change in the feedforward variable is used instead of the absolute value to ensure bumpless calculation in case of a bad value in the feedforward variable. That is

$$u(k) = u(k-1) + \Delta u(k) + K_{\text{FF}}\,[y_{\text{FF}}(k) - y_{\text{FF}}(k-1)] \tag{4.4}$$

where $u(k)$ is the new output, $u(k-1)$ is the previous output, $\Delta u(k)$ is the feedback control output, and $y_{FF}(k)$ is the feedforward variable.

The feedforward output is typically added to the feedback control output (*additive* feedforward), as shown in Eq. 4.4. There is also the option of multiplying the feedforward value with the feedback output (*multiplicative* feedforward), but it is rarely needed.

Feedforward control is based on the assumption that the dynamics of both the feedback and feedforward channels are known and stay relatively constant over time. This assumption is rarely valid. Inaccurate compensation can sometimes do more harm than good, so feedforward control should be used with discretion.

2. Ratio control: Ratio control is another example that two process variables can impact the controlled variable. Ratio control forces the secondary variable to change proportionally with a primary variable (the wild variable).

Continuing with the example in Fig. 2.20, the internal ratio calculation is usually given as follows:

$$\begin{aligned} \texttt{FC102.SP} &= \texttt{FFC103.OP} \\ &= \texttt{FFC103.SP} \cdot \texttt{FI101.PV}. \end{aligned} \tag{4.5}$$

The primary variable (the wild variable) is the process value for the ratio controller, and the desired ratio value is the setpoint. The output of the ratio controller is the setpoint for the secondary controller (the PID controller). Direct ratio calculation should not be used to avoid divide-by-zero errors.

Ratio control is similar to a static feedforward control with a fixed gain term to relate the two process variables. They both are open-loop control and must be used along with a feedback control loop. If there are significant differences in the dynamics between the two variables entering the ratio calculation, then lead/lag and dead-time compensation may need to be added for the primary measurement (`FI-101.PV` in this case). For example, if the primary flow `FI-101` in Fig. 2.20 has a two-minute transport delay to the mixer, while the manipulated variable `FC-103` has a one-minute delay, then flow `FI-101` should be delayed by one minute before entering the ratio controller `FFC-102`.

Ratio control may appear similar in configuration to cascade control. However, the difference is that the ratio controller does not have its dynamics and thus no tuning requirement. Internally, it is a simple scalar calculation as in Eq. 4.5.

Decoupling control

Decoupling control is an early effort to address the multivariable interactions among input and output variables (Gordon 2005; Shinskey 1994). Figure 4.4 shows an example of strong interactions between level (`LC-101`) and temperature (`TC-101`).

Adjusting temperature control valve `TCV-101` affects both the temperature (`TC-101`) and the flow `FC-101` (thus the level `LC-101`). Similarly, controlling

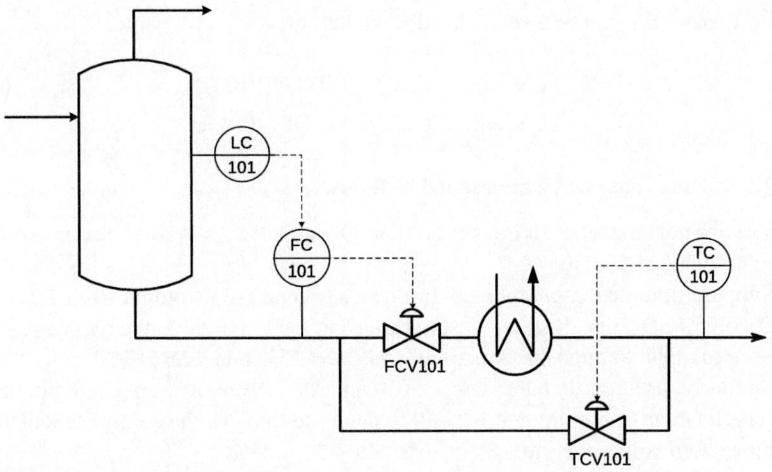

Fig. 4.4 Example of decoupling control

the level `LC-101` by manipulating the flow control valve `FCV-101` would also affect the temperature `TC-101`.

With decoupling control, the level and temperature can be controlled independently of each other. When the flow or temperature needs to be adjusted, the decoupling control scheme simultaneously moves the flow and temperature control valves. One controlled variable will change as desired, and the other controlled variable remains unaffected. Consider the following two scenarios:

1. The flow rate needs to be adjusted to change the level, but the temperature must remain constant. In this case, the two valves would move together so that the total flow change will meet the level control requirement, but the ratio of the two flows remains constant so that the temperature will remain unchanged.
2. The temperature needs to be adjusted, but the total flow rate must remain constant to maintain the same level. For this to happen, the two valves will move in the opposite direction such that the total flow remains the same, but the ratio between the two flows will drive the temperature to the new setpoint.

For a control valve, the flow through the valve is proportional to the valve coefficient C_v, the fluid density ρ, and the differential pressure across the valve ΔP. That is, FC $\propto C_v\sqrt{\Delta P\,\rho}$. If the two control valves are of identical design, i.e., $C_{v,1} = C_{v,2}$, and the pressure drop across the two valves are the same, then the two scenarios discussed above can be described as

$$\text{FC.OP} = \frac{\text{FCV} + \text{TCV}}{2} \tag{4.6}$$

$$\text{TC.OP} = \frac{\text{TCV}}{\text{FCV} + \text{TCV}} \cdot 100\%. \tag{4.7}$$

FCV and TCV can be easily solved to yield

$$\text{FCV} = 2 \cdot \text{FC.OP} \cdot (1\text{-TC.OP}/100) \tag{4.8}$$
$$\text{TCV} = 2 \cdot \text{TC.OP} \cdot (\text{FC.OP}/100). \tag{4.9}$$

The calculations can be interpreted as follows:

1. The normal control position is `FC.OP` = 50% and `TC.OP` = 50%, leading to `FCV` = 50% and `TCV` = 50%.
2. Suppose that the temperature control needs to increase its output from `TC.OP` = 50 to 75%. To keep the flow rate unchanged at `FC.OP` = 50%, the two valves will need to move in opposite directions, to `FCV` = 25% and `TCV` = 75%.
3. Similarly, assume that `FC.OP` = 50% and `TC.OP` = 50% and a temperature setpoint change require that `FC.OP` to decrease to 25%, the control scheme will move both valves, i.e., to `FCV` = 75% and `TCV` = 25%.
4. Assume the temperature is stable and a level setpoint change requires that the flow controller output `FC.OP` to change from 50 to 75%. In this case, both valves need to open at the same time and in proportion to keep the temperature unaffected. The final valve positions will be `FCV` = 75% and `TCV` = 75%.
5. Similarly, assume that `FC.OP` = 50% and `TC.OP` = 50% and the level controller requires a change of `FC.OP` = 50% → 25%. The decoupling control scheme will move both valves simultaneously and in proportion to keep the temperature unaffected. The final valve positions will be `FCV` = 25% and `TCV` = 25%.

In reality, the two valves are usually of different sizes. The above formulas need to be expanded to accommodate this general scenario. For simplicity, let us define $C = C_v\sqrt{\Delta P}$. The two scenarios described in Eqs. 4.8 and 4.9 become

$$\text{FC.OP} = \frac{C_1\,\text{FCV} + C_2\ \text{TCV}}{C_1 + C_2} \tag{4.10}$$
$$\text{TC.OP} = \frac{C_2\,\text{TCV}}{C_1\,\text{FCV} + C_2\ \text{TCV}} \cdot 100\%. \tag{4.11}$$

To control the temperature but maintain a constant flow, the total valve changes shall remain the same. Similarly, to maintain the temperature constant, the relative ratio of the temperature control valve versus the total valve should remain constant. For convenience, define

$$K = \frac{C_2}{C_1} = \frac{C_{v,2}}{C_{v,1}}\sqrt{\frac{\Delta P_2}{\Delta P_1}} \tag{4.12}$$

and solve for FCV and TCV,

$$FCV = FC.OP \cdot (1 + K) \cdot \left(1 - \frac{TC.OP}{100}\right) \tag{4.13}$$

$$TCV = TC.OP \cdot \left(1 + \frac{1}{K}\right) \cdot \frac{FC.OP}{100}. \tag{4.14}$$

Decoupling control is in essence feedforward control. The disturbance to be rejected for one controlled variable is the normal process reaction to another controller. Therefore, it is crucial to have the exact dynamic cause and effect relationships between the valves and the controlled variables to decouple the interacting variables.

The example above is static decoupling control and already appears complex to understand and implement. To dynamically decouple the multiple input and output variables, the control design would become much more complicated and is simply impractical for baselayer control. For this very reason, decoupling control has never received widespread acceptance in practice. On the other hand, model predictive control (MPC) has provided a more elegant solution.

4.1.3 Handling Constraints

Contrary to popular belief, PID-based baselayer control can also handle constraints in the form of protective control. A normal regulatory control is responsible for maintaining the operating point during normal operation. When the operating point deviates from the target value by too much, the protective controller kicks in to keep the operating point within the operating envelope (see Sect. 1.3.5). During normal operating conditions, the protective control remains inactive.

There are three types of protective control implementations: standalone, dual-controller, and overriding.

1. Standalone protective control: A standalone protective control is a standard PID controller, except that the controller is designed to be inactive during normal operation. This is achieved by setting the setpoint higher (for high protective control) or lower (for low protective control) than the normal operating value (e.g., the design value). During normal operations, the controller output is saturated to fully closed (high protective control) or fully open (low protective control) position. Only when the normal regulatory control can not maintain the setpoint and the process value deviates from the setpoint to reach the protective control setpoint (the constraint limit) will the protective controller "wake up" and take action.

   Figure 4.5 shows a control scheme for a centrifugal pump. Centrifugal pump has a minimum flow requirement, below which the pump may become unstable.

   Flow controller `FC-102` acts as a minimum flow recycle controller. Its purpose is to open up the recycle valve to re-circulate a portion of the flow back to the tank to maintain a minimum flow through the pump in case the pump flow drops

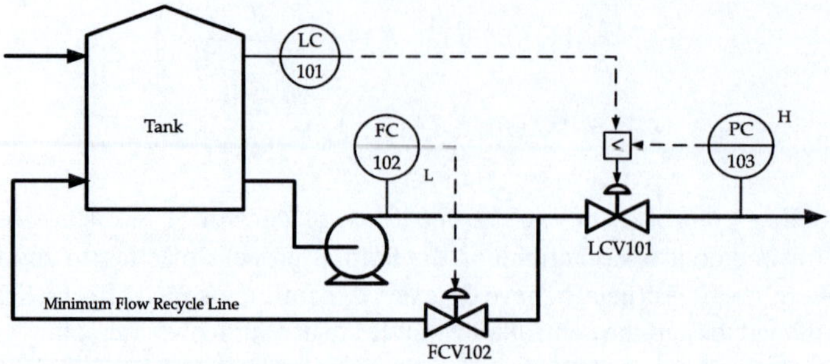

Fig. 4.5 Example of protective control

below the minimum flow threshold. See Sect. 11.2 for a detailed discussion on pump control.

2. Dual-controller protective control (Sect. 2.2.2): Each controller in the dual-controller configuration is similar to a standalone protective control in that each has its own control handle and consumes one DOF. However, unlike a standalone protective control, dual controllers work in pairs on the same process variable: one is the primary controller for normal regulatory control. The other is for constrained control to protect the process value from going over the constraint limit. The two pressure controllers, `PC-101` and `PC-102` in Fig. 4.2, or `PC-101A` and `PC-101B` in Fig. 2.16, are typical dual pressure controller schemes.
3. Overriding protective control (Sect. 2.2.7): An overriding protective controller shares the control handle with the normal regulatory control with the control selector. It has a more tolerant setpoint so that the controller remains dormant and its output is not selected during normal operation.
   The pressure controller `PC-103` in Fig. 4.5 is an example of an overriding protective controller. It protects the downstream piping or equipment by restricting the downstream pressure to a safe limit. If the pressure rises above the setpoint, the pressure control will override the flow controller and force the control valve `LCV-101` to open less.

In the integrated compressor control example in Fig. 4.2, there are several protective control loops. For example,

1. `PC-001` and `PC-102` constitute a dual-controller protective control scheme. The controlled variable is the compressor suction pressure. The normal regulatory control loop for the suction pressure is `PC-002`. The protective controller `PC-001` remains inactive if the suction pressure is within normal range. If the pressure rises above the protective setpoint, the protective controller `PC-001` opens the flare valve to release the excess gas to the flare.
2. `PC-003` at the compressor discharge is an overriding protective control. When the discharge pressure goes too high, it overrides the normal regulatory controller `PC-002` and slows down the compressor to reduce the discharge pressure.

3. `UC-111` and `UC-121` are two overriding controllers. They share the recycle valve with another two PID controllers. Whichever controller reaches its constraint limit, it would open the recycle valve through the high selector, allowing the highest demanding request to pass through.

As demonstrated by Fig. 4.2, protective control can become highly complicated and require careful design for the overall control scheme to function in concert. Some crucial considerations include the following:

1. Rationalization of the normal control setpoint, the protective control setpoint, and the safeguarding limits so that they are adequately spaced to allow sufficient time for the next level of protection or the operator to respond.
2. Controller tuning: Protective controllers are typically tuned more aggressive for quick response than the normal regulatory control loop.

4.1.4 *Optimizing Flow Control*

Baselayer PID control schemes can achieve a certain degree of profit maximization through the creative use of complex PID schemes. Preferential control, flow balancing, and limit pushing are several examples.

Preferential Flow Control

A commonly encountered control requirement is preferential flow. When a material or energy stream feeds multiple consumers, there is often a requirement to prioritize the flow in case the supply is insufficient. For example, in a water processing unit, the produced water from the wells is preferably sent to water lifting wells first. Any excess should then be sent to the injection wells. If still with excess, the remaining should go to deep water disposal (DWD) wells.

To achieve this type of preferential flow arrangement, the first approach coming to mind is a split-range control scheme (Sect. 2.2.1). The high-priority flow path is put at the lowest leg of the split range, while the least-preferred flow path should be in the upper leg. In theory, the split-range control can be 1 to n split, where n is 2 or more, but it is rare to see more than two splits due to controllability.

In the example in Fig. 2.15, the process gas is expected to be sent to the downstream compressor as the normal process flow. However, if excessive supply causes the pressure to go higher than a limit, the excess gas will be sent to flare. Dual-controller control can also achieve the same preferential control, see Fig. 2.16. The downside is that the setpoints of the two controllers must be set sufficiently apart to avoid interfering with the normal regulatory control.

Another application scenario is multiple flows differentiated with a bias term. For example, in a fan-out control, the same output is sent to multiple control elements. Although this equal split is acceptable for most applications, there are scenarios where the different control elements need to receive different flows. A common approach is to utilize a gain/bias function block to either apply a different ratio (gain) or a fixed offset (bias) to the output received by each control element. As an example, a pressure control output is sent to three flow controllers at the wellhead of three water injection wells. If one of the wells has a higher capacity than the others, an offset (bias) can be added to the output to cause a differentiated flow for this well. In other words, instead of 40, 40, 40%, the output is distributed as 35, 50, and 35% after applying a 15% bias term to the second well. The feedback mechanism automatically determines the output value received by each flow control, but the bias is maintained.

Note that the gain/bias function block typically does not support back calculation and may need a calculation block to provide the back-calculation information to the primary controller explicitly.

Flow Balancing Control

Related to preferential flow control is an application scenario of flow balancing control. Two parallel streams may need to maintain a specific ratio in their flow rates, which can generally be achieved through a simple ratio control scheme. However, an additional requirement is that the pressure loss across the control valves shall be as low as possible. Permanent pressure loss results whenever a piece of device, equipment, or pipe is added to a flow system. This pressure loss makes the pump or compressor work harder to generate the same flow in the established system. Every bit of pressure loss equals extra energy used (electricity, steam, or natural gas) to pump or compress the fluid, which translates into more money spent to maintain operations.

Example 4.1 Flow balancing control in a water treatment process. In the simplified flow diagram shown in Fig. 4.6, unprocessed water passes through two de-sanding vessels to remove the entrained sands.

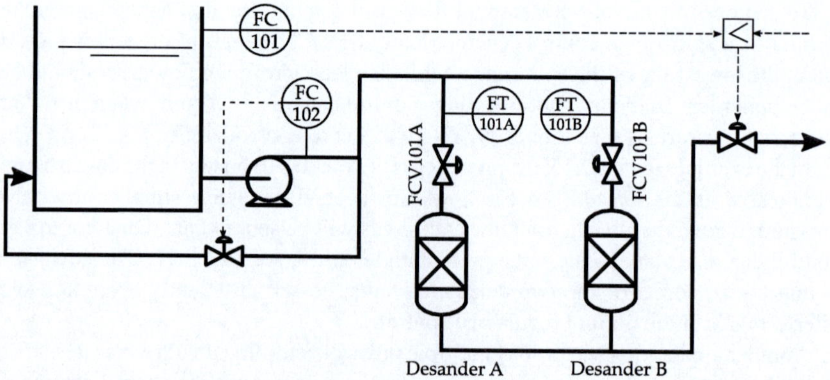

Fig. 4.6 A process with flow balancing requirements

The two de-sanders are of different capacity, and thus a flow ratio of 3:2 between `FI101A` and `FI101B` is required to achieve the best efficacy and lifespan. At the same time, the pressure losses across the flow control valves are expected to be as low as possible to meet the pressure requirements by the de-sanders and downstream units because a lower pressure loss means less demand on the pump.

A conventional design would lead to a ratio control with a split-range logic on the valve opening, as shown in Fig. 4.7a (left). As the split-ratio controller output goes from 0 to 100%, flow controller valve `FCV101A` closes from 100 to 0% while flow controller `FCV101B` opens from 0 to 100%. The normal operating area is around 50% for both valves for the desired controllable range.

One obvious flaw with this design is that at 50% of openings, significant pressure losses across the valves will incur, which is not acceptable due to the capacity limit of the pump.

A different split-range control is recommended, as shown in Fig. 4.7b. With this improved design, both valves have an opening near 100% at the design condition, thus resulting in minimum pressure losses to the process. Note that the pipe sizing is selected such that when both valves are fully open, the flow split-ratio is at the vicinity of the desired value.

A control design is provided in Fig. 4.8. The ratio control block `FFC101` calculates the flow ratio. The setpoint of the ratio controller is the desired split ratio, and the output of the ratio controller provides the setpoint for flow controller `FC101`, whose process variable is the flow rate of stream A. The flow controller `FC101` manipulates the control valve `FCV101A` to achieve the desired flow ratio. The openings of control valves `FCV101A` and `FCV101B` are calculated by `FY101` based on the split logic in Table 4.2.

Valve Position Control

Valve position control (VPC) is a simple and effective optimization scheme that can be achieved with standard PID controllers. The opening of a control valve can vary

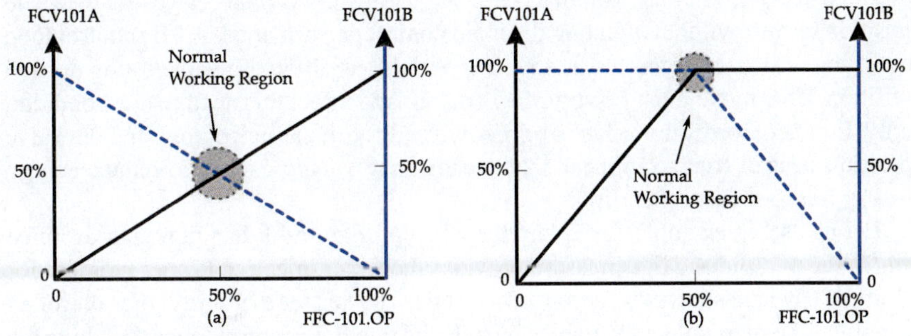

Fig. 4.7 Split point for flow balancing control

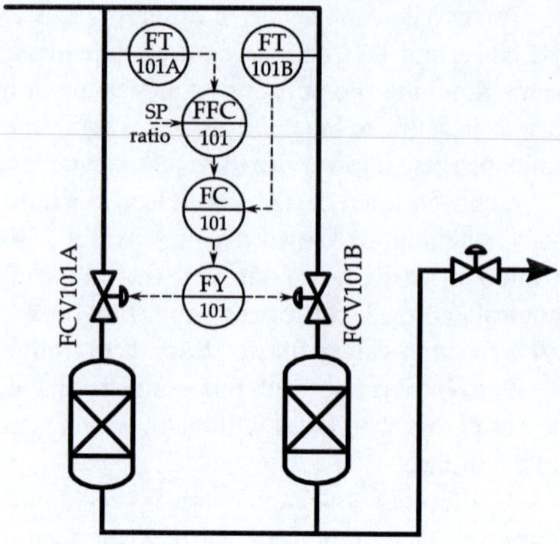

Fig. 4.8 Split-range design for flow balancing control

Table 4.2 Flow balancing control logic

Controller output	Valve openings	
FC101.OP (%)	FCV101A (%)	FCV101B (%)
0–50	100	0–100
50	100	100
50–100	100–0	100

between 0 and 100%, typically driven by a controller. Except for *full-bore* valves, all control valves will cause pressure losses even when the valve is at 100% open. Besides, all valves have a preferred range of operation, such as 10–80% for best controllability. There are thus incentives to limit the valve position to a range that can either save energy or improve controllability.

Valve position control (Luyben and Lyuben 1997; Shinskey 1978) is a smart approach to optimize the valve operation (subject to other constraints). The basic idea of valve position control is quite simple. Assume a variable somewhere else is available to vary or "float" within a range. Adjusting this variable can cause the valve position to shift without affecting the main control performance. A PID control loop can then be implemented to use this free variable to drive the valve to the desired position. This move must be very slow and aims to affect the steady-state condition only. In other words, the valve is primarily used for dynamic control; and the valve position control tries to influence the steady-state valve position to reduce energy loss or improve controllability.

Let us use an example to explain the concept. Figure 4.9 is a pumping example where the pump discharge pressure provides the water injection header pressure for multiple (two are shown) water injection wells. The header pressure is the controlled variable. The manipulated variables include the two flow control loops that drive the two flow control valves.

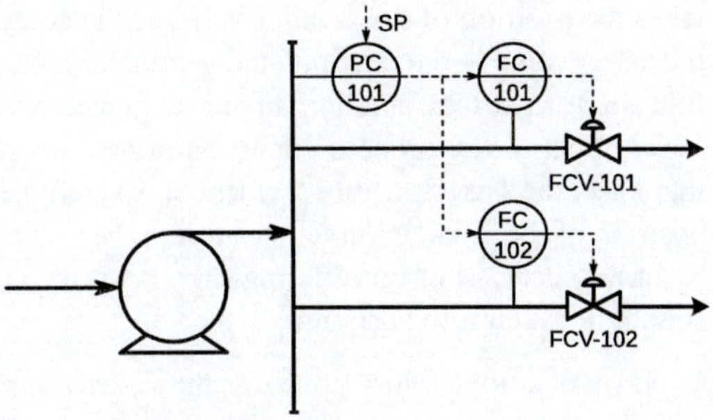

Fig. 4.9 Valve position control example

The valve flow is determined by the valve opening and differential pressure across the valve. A higher setpoint will result in a higher discharge pressure and lower valve openings, which means more pressure loss will occur across the control valves. A lower discharge pressure will cause the flow control valve to open more. In other words, some of the energy that the pump consumes to raise the discharge pressure is wasted at the control valve as pressure loss. It has some economic benefits to reduce the energy loss at the valve by shifting the steady-state valve position toward the fully open direction without affecting the controllability.

The setpoint of the pump discharge header pressure is an operator input and can be regarded as a *free variable* or an extra DOF. The setpoint can vary slowly within a range without negatively impacting the dynamic flow control performance. The valve position control is illustrated in Fig. 4.10. The control objective is to maintain the steady-state valve position (the larger of the two valves) at around 70%.

Another example is a control scheme with two parallel valves, a larger one for coarse control and a smaller valve for fine control. The smaller valve is often found in saturation due to the smaller control range. The smaller valve may remain at the saturated position for a long time until the larger valve is adjusted (often manually) or the process condition changes back. A saturated valve renders the variable out of control and is thus undesirable. In this case, a simple PID controller can be added that

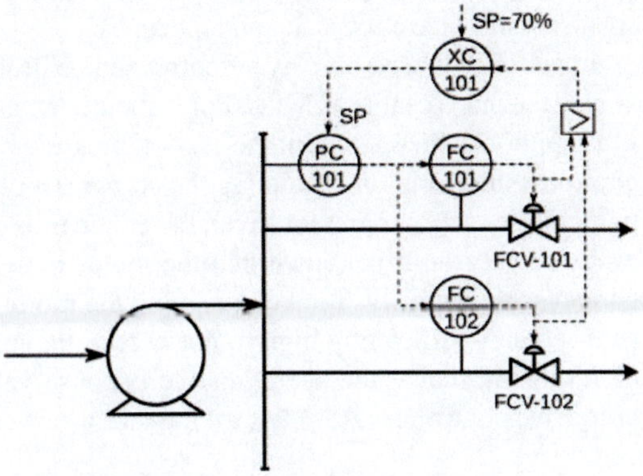

Fig. 4.10 Valve position control scheme

takes the position of the smaller valve as the controlled variable and the position of the larger valve as the manipulated variable. For example, with a setpoint of 50%, the fine control with the smaller valve is achieved by varying the valve position around 50%. Suppose the process condition drives the valve position away from 50%. In that case, the valve position control slowly adjusts the larger valve and causes the position of the smaller valve to return to the vicinity of 50% gradually.

In summary, when considering valve position control, the following characteristics should be taken into account:

1. Valve position control concerns the *steady-state* or *average* position of the valve. The primary purpose of the control valve is to control other variables, and thus the valve is meant to move freely for that purpose. Value position control intends to drive the *steady-state* or *average* valve position to the desired range. For this reason, the valve position control action must be very slow and smooth to not hamper or interfere with the normal dynamic control function of the valve. An I-only PID control algorithm or a PI algorithm with very low gain are typically used for valve position control.
2. A clear cause and effect relationship must exist. The basis for valve position control is that a *free variable* that has a deterministic cause and effect relationship with the valve position is available. This free variable is typically another DOF reserved for the operators, such as the setpoint of another controller. It serves as the manipulated variable of the valve position controller to influence the valve position.

The valve position control provides some incremental benefits beyond stable and efficient control. However, it should be used with discretion. If misapplied, it can also do much harm.

Limit Pushing

Process control improves the production efficiency by pushing the operating point closer to the operating limits, typically the more profitable operating region. Limit pushing is one of the primary reasons that model predictive control (MPC) has been so successful (see Sect. 2.3.4 and Chap. 5).

In fact, PID-based baselayer control can be designed to push limits, although not as elegant and versatile as MPC. For instance, the gas compression process in Fig. 4.2 is a supply-driven operation; the suction pressure controller `PC-002` is the capacity control responsible for regulating the compressor throughput. A lower pressure setpoint pushes the compressor to run faster and thus deliver a higher throughput. From process analysis, it is known that the motor power `JC-101` sometimes limits the compression capacity. To avoid tripping the motor, the operator often sets the pressure setpoint sufficiently high to not exceed the maximum motor power limit. This high setpoint makes the operation overly conservative. A simple remedy is to add a protective controller, `JC-101`, on the motor power. The pressure control `PC-002`, with a setpoint slightly lower than the normal value, forces the compressor to run at the highest speed possible within the motor power limit.

For limit pushing, a vital requirement is that a reliable "stopper" must exist to limit how far the pusher can push. In this compressor example, a lower setpoint makes the pressure controller `PC-002` a capacity "***pusher***". At the same time, the motor power `JC-101` is a reliable "***stopper***" to prevent the pusher from pushing too far and too hard. Limit pushing is a standard feature in MPC that is discussed in Chap. 5.

4.1.5 Tackling Nonlinearity

Nonlinear behavior can be caused by various operating conditions and can also be caused by control structure changes such as by a switch or selector. PID controller has a fixed set of tuning parameters for the control of fixed loop dynamics. In a complex control scheme, changes in the data flow may cause drastic changes in the loop dynamics and render the existing PID tunings inadequate. Online modification to the controller is one way to adapt to the changed loop dynamics. The modification can be as simple as *gain scheduling* or as sophisticated as control structure change.

Gain Scheduling

Gain scheduling is a common approach for handling some specific types of process nonlinearity. Gain scheduling applies a different controller gain to the running controller as the operation transitions from one condition to another. In this way, the controller gain more closely aligns with the changed loop dynamics.

Fan-out control (see Fig. 2.17) is a typical application where gain scheduling may drastically improve control performance. For example, a pressure controller on the injection manifold in a water injection process distributes the high-pressure water to a dozen injection wells. A fan-out control loop is used to manipulate a dozen injection flow controllers, one for each injection well. Any injection wells can be freely taken out of, or put back in, service at the operator's discretion based on operation needs. Although the *process* dynamics remain the same, the number of wells online can drastically change the *loop* dynamics and renders the tuning inadequate.

The unpredictable change in the number of secondary elements poses the following challenges:

1. How to tune the pressure controller?
2. How should the controller respond to the change in the number of flow controllers in cascade?

The recommendation is to tune the controller assuming all the flow controllers are in cascade control. Suppose the gain is K_c and the number of flow controllers is N, and the number of flow controllers in cascade is n, then the controller gain is given by $N/n \cdot K_c$. This approach also assumes that the dynamics of each injection well are identical, and there is an online mechanism to count the number of flow controllers (n) in the cascade.

The controllers that are taken out of cascade control may have a zero flow (out of service) or a fixed flow (in MAN). Either way, the loop dynamics for the pressure controller are not affected; only the steady-state output of the pressure control will be shifted.

Gain scheduling for nonlinear process behavior such as asymmetric response (different process gain in the opposite direction) is more complicated. It needs considerably more accurate modeling of the process dynamics and the benefits may not be justifiable for the incremental performance improvement.

Gap Control

The built-in control functionalities for nonlinear PID control typically include PID gap control and quadratic gain. There are many variations in different control systems. Gap control adjusts the PID gain in a pre-defined fashion based on the magnitude of the control error, as illustrated in Fig. 4.11.

A gap control or quadratic gain function produces a smaller gain when the control error is small (the process valve is near the setpoint). A larger controller gain is produced when the error is significant and needs more aggressive action. That is, for simple gap control

$$K_{\text{gap}} = \begin{cases} K_c & \text{outside the gap: err} < -\text{gap}/2 \text{ or err} > \text{gap}/2 \\ k \cdot K_c & \text{inside the gap: } \text{gap}/2 \leq \text{err} \leq \text{gap}/2. \end{cases} \tag{4.15}$$

We can assume that the quadratic gain function has the following format:

$$K_{\text{gap}} = a \cdot \text{err}^2 + b. \tag{4.16}$$

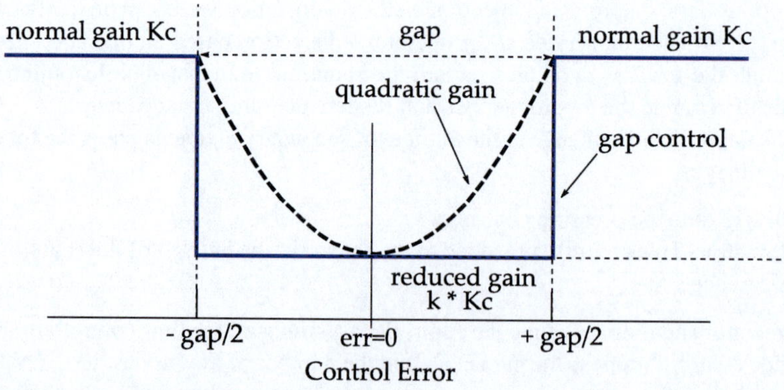

Fig. 4.11 Gain scheduling: gap control and quadratic gain

Assume that the gain is $k \cdot K_c$ when the error is zero, and the gain is the regular gain K_c outside the gap; it is easy to find that the quadratic function is related to gap control as below:

$$K_{\text{gap}} = \begin{cases} K_c & \text{outside the gap} \\ (1-k)\,K_c \dfrac{4}{\text{gap}^2} \cdot \text{err}^2 + k \cdot K_c & \text{inside the gap.} \end{cases} \tag{4.17}$$

The gain scheduling function is typically built in the PID module, with the gap width and scaling factor k as the tuning parameters. The default value for the scaling factor is typically 0.25. That is, the control action is four times less aggressive inside the gap. The selection of the gap size depends on the nonlinear characteristics and the control objective to achieve. For example, the objective of surge tank level control is to maintain a stable outlet flow at the cost of fluctuations in the tank level. The level fluctuation is not a concern as long as it stays inside a specific range. In this case, a gap control can be implemented with a gap size of, for example, 40%. Assume the level control setpoint is 50%; the controller has a smaller gain when the level is between 30 and 70% with slow control actions. The gain automatically switches to the regular value for more aggressive action if the level goes above 70% or below 30%.[1]

Gain scheduling may effectively improve the control performance, but the approach is highly empirical and typically not based on rigorous process information; and thus, the tuning parameters are always approximate.

4.1.6 Rationalizing Control Scheme

There is often more than one way to achieve the same control objective, but the control performance may be drastically different. The difference between a good and not-so-good control design is sometimes subtle. The impact may not be make-or-break but rather between optimal and sub-optimal. A sub-optimal design may deliver sub-par performance over its entire life cycle and is often a maintenance burden as well.

Nested Loop and Cascade Loop

We will use a nested control loop to demonstrate that the control design can be pretty creative once the basic principle is well understood. It needs a decent process control "sense" (mastery skills) to turn a mediocre (but acceptable) control solution into a great solution.

[1] On the other hand, the level is inherently nonlinear with the flow for a horizontal cylindrical vessel. The process gain is the smallest at 50% and becomes higher when the level moves more away from 50%. With a fixed PID gain, the control performance is equivalent to a nonlinear controller with quadratic gain.

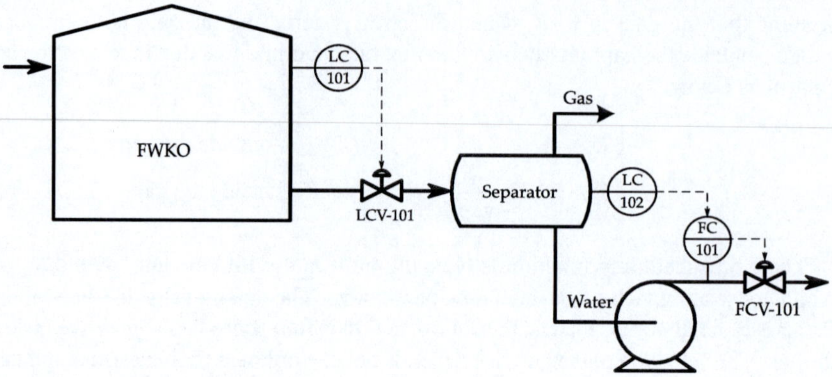

Fig. 4.12 Tank level control: simple cascade control

Figure 4.12 shows a water processing process where the produced water from wells is sent to the free water knockout (FWKO) tank and then flows to a gas/liquid separator to remove the entrained gas in the liquid. The water is pumped out and sent downstream for further processing. The production rate (the capacity) is regulated by the flow controller at the pump discharge line.

The operating requirement is to maintain the two vessel levels by manipulating the two flow control valves. Level and flow measurements are available as needed. For the FWKO tank, the level is the controlled variable, and the outlet flow valve `LCV-101` is the manipulated variable. For the separator, due to the pump's nonlinear dynamics, a cascade level (`LC-102`) to flow (`FC-101`) control loop is utilized to improve the level control. The original design is illustrated in Fig. 4.12.

However, a significant challenge is that the separator has a much smaller capacity than the FWKO does. The fluctuations in the FWKO outlet flow can easily overwhelm the separator level controller `LC-102`. During the first few years of operation, the separator was frequently tripped on a high level, resulting in multiple shutdowns. To prevent the high-level trips, a second level controller `LC-103` off the same level transmitter on the separator was implemented as an overriding protective control on FWKO tank level controller `LC-101`. See Fig. 4.13.

In case the separator level goes too high, the override controller `LC-103` would override the FKWO level control output and reduce the inlet flow to the separator. This design provides the desired protection against high levels in the separator and prevented high-level trips.

Although improved over the original scheme, this control scheme still has serious flaws. In addition to the added complexity introduced by the protective controller `LC-103` and the overriding mechanics through the selector, the scheme may potentially cause a high FWKO tank level. Once overridden by `LC-103`, the FWKO tank level controller `LC-101` is out of control. Even though the capacity of the FWKO tank is large to allow some grace time, there is no protection against the level going overly high to trip the tank in the design. As will be explained in Chap. 6, the fun-

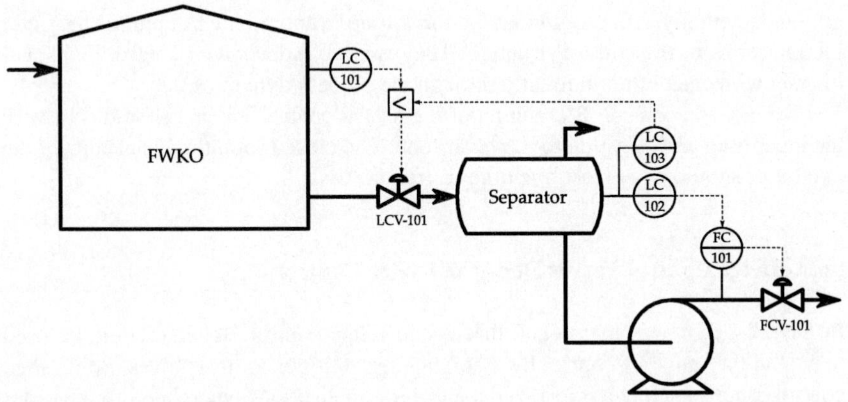

Fig. 4.13 Tank level control: with protective control

damental flaw in this design is that the process is supply-driven, and the original control solution is based on a supply-driven operating model. Introducing the protective controller LC-103 can force the operation into a demand-driven operation without a swing stream at the FWKO.

Since the separator has a much smaller capacity (and thus much faster dynamics) than the FWKO, a nested control scheme was proposed to improve the control reliability, as shown in Fig. 4.14.

The faster separator level control is the inner loop, while the slower FWKO level control acts as the outer loop. The separator level control is with the inlet flow control valve LCV-01, and the FWKO tank level is controlled by the control valve at the pump discharge (LCV-02). This nested control scheme is a much cleaner and more reliable design, delivering better overall performance.

A nested loop is similar to a cascade loop in that the inner loop is the one with a faster response and thus has tight control to suppress disturbances. However, they

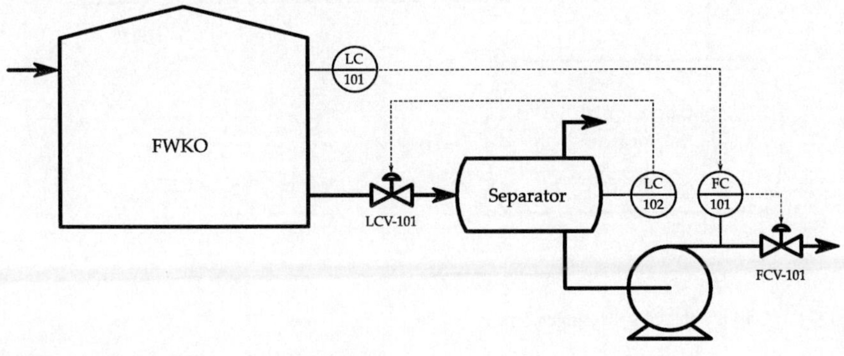

Fig. 4.14 Tank level control: nested control

are fundamentally different in that the inner loop is not an integral part of the outer loop in terms of the control structure. They are two independent control loops that interact with each other indirectly through the process dynamics.

Like cascade control, PID tuning of the nested control scheme should start with the inner loop and move on to the outer loop. The closed-loop time constants of the two loops should be at least five times different.

Split-Range Control Versus Dual-Controller Control

In Sect. 2.2.2, it was mentioned that a split-range control scheme could be used to extend the range of control by manipulating multiple control valves. Split-range control can also achieve prioritized control for applications where a preferred order of sequence exists among the multiple manipulated variables for controlling the same target.

Figure 4.15 shows a pressure control scheme for a production separator which is very common in an E&P production facility. The control scheme has a split point of 70% where the first 0∼70% of controller output is mapped to 0∼100% for pressure valve `PCV-101A`; pressure controller output from 70% to 100% is mapped to 0∼100% for control valve `PCV-101B`. This output mapping is illustrated in Table 4.3 and Fig. 4.16.

This standard split-range control simultaneously provides the following two functions:

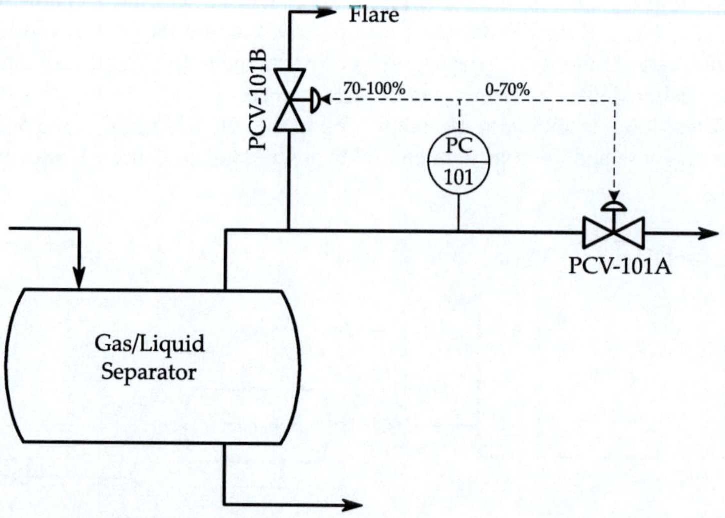

Fig. 4.15 Split point in split-range control

Table 4.3 Output mapping in a split-range control

PC-101 output (%)	PCV-101 (%)	PCV-102 (%)
0–70	0–100	0
70–100	100	0–100

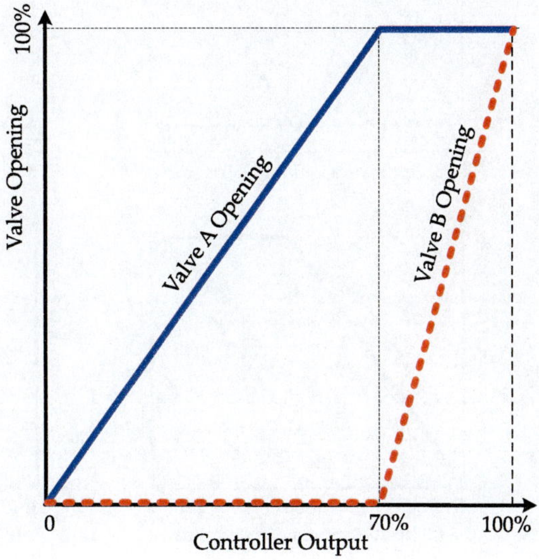

Fig. 4.16 Split point calculation in split-range control

1. Extending the control range: The pressure control during normal operating conditions is via pressure control valve PCV-101A. In case of a significant plant upset, the capacity of PCV-101A is not sufficient. The pressure controller opens up flaring valve PCV-101B to extend the control range.
2. Prioritizing flow control: With this scheme, the control valve PCV-101A is the preferred control. The process flow goes to PCV-101A first. When the capacity of PCV-101A is exceeded (PC-101.OP>70%), the controller opens PCV-101B. Conversely, when both PCV-101A and PCV-101B are open, if the pressure returns to normal and the pressure controller needs to reduce outlet flow, it will reduce and close PCV-101B first. Valve PCV-101A is throttled only after PCV-101B is fully closed.

The use of split-range control can be very flexible and creative. Figure 4.17 shows several variations of split-range control. For example,

1. The direction of action of the two valves can be the same or different, e.g., Fig. 4.17a, b.
2. There can be a gap or an overlap at the split point, e.g., Fig. 4.17c, d.
3. The output of the controller can be sent to a valve or another control element.
4. The split-range control can use more than two valves if justifiable, although it is strongly discouraged for the reasons mentioned below.

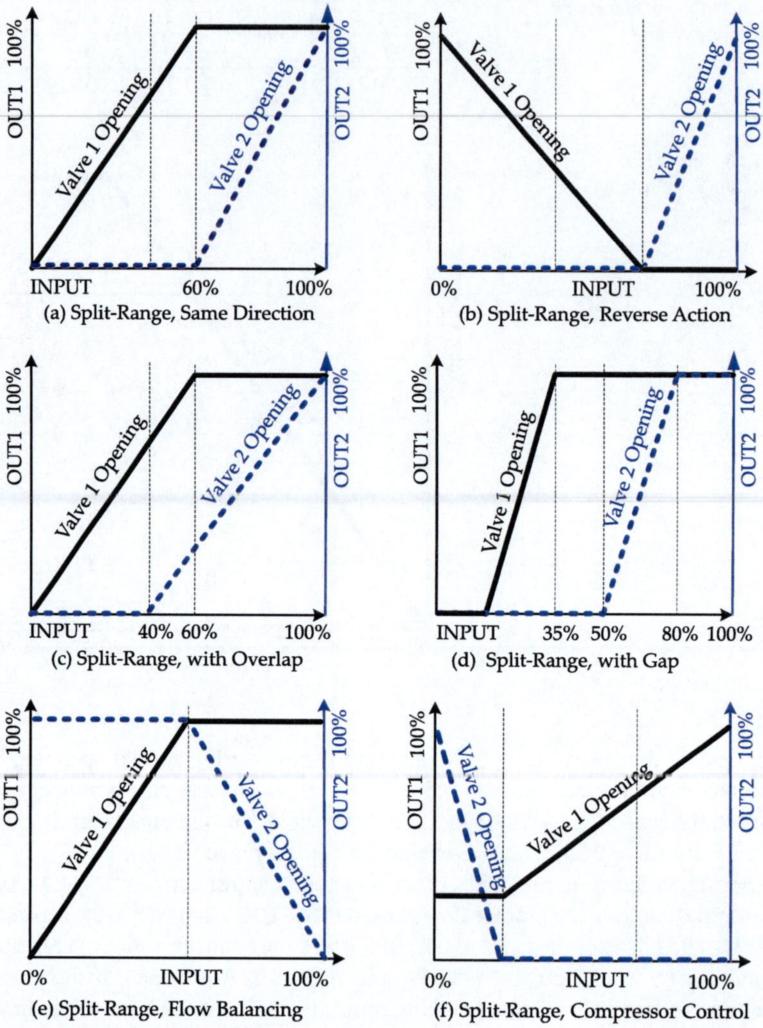

Fig. 4.17 Examples of split-range control loop

Split-range has been in widespread use to the extent that it is sometimes overused or even abused. Some inherent drawbacks require careful considerations in the control design and implementation to ensure integrity and performance. The considerations include the split-point calculation, controller tuning, and back initialization:

1. Split-point calculation: A split point at 50% is typically the default, although this default value is often inappropriate. The PID controller in a split-range control only has one set of tuning parameters. The 50% split point implicitly assumes that the dynamics of the two legs are identical. In reality, the two dynamics

can be significantly different, e.g., for the split-range control arrangement in Fig. 4.15. As a result, the same control tuning is not adequate for the full range of control. If the controller is tuned to produce an optimal performance for the suction pressure valve, the same tuning would be too slow for the flare valve to react to pressure upset. The split point, or transition point, should be calculated based on the actual dynamics of the two control ranges to ensure the controller tuning is adequate for both legs in the full range.

Example 4.2 Split-point calculation. A split-range pressure controller directly manipulates two control valves for two steam flow streams. The two valves are sized as 50 and 150 kg/h, respectively. The desired split point is then given by

$$\text{Split-Point} = \frac{50}{150 + 50} \cdot 100\% = 25\%. \tag{4.18}$$

The control range is 0–25% for the smaller valve and 25–100% for the larger valve.

Another approach to determining the split point is via step tests. Apply a 1 or 2% change to each valve, one by one. Assume that the change in pressure is 1.5 bar and 0.5 bar, respectively, then the split point can be calculated as

$$\text{Split-Point} = \frac{K_{p,1}}{K_{p,1} + K_{p,2}} = \frac{0.5}{0.5 + 1.5} \cdot 100\% = 25\%. \tag{4.19}$$

See Fig. 4.18 for an illustration of the split-point calculation based on loop gains. The practical challenge is that the calculation of the split point can be tricky and inaccurate. It is often difficult to find a fixed set of PID tuning parameters to achieve good performance for both control valves. For critical control loops, it may be worthwhile considering a more complex gain scheduling scheme.

2. Bumpless transfer: When the secondary control elements in the control loop are disconnected and subsequently re-instated, the controller output will initialize to the one that switches to `CAS` first. This behavior can potentially cause a bumpy transition. For the example in Fig. 4.15, suppose both valves, `PCV-101A` and `PCV-101B`, were taken out of service and subsequently put back to `CAS`. If both valves are at 0% opening, and now `PCV-101B` is switched back to `CAS` first, it is expected that the pressure controller `PC-101` be initialized to 70%

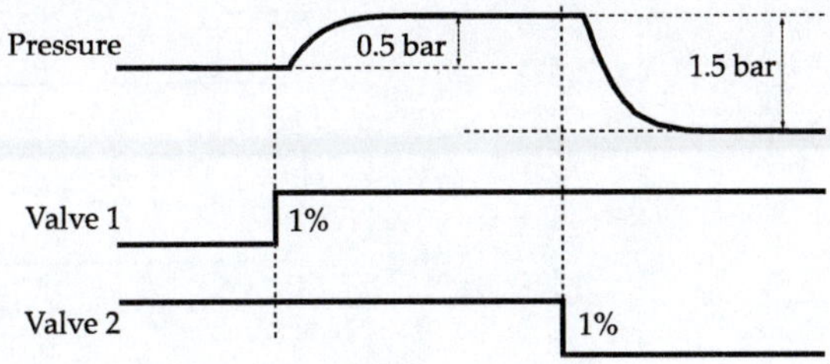

Fig. 4.18 Examples of split-point calculation with gains

because valve PC-101B is in the 70–100% leg of the split range of PC-101. When PCV-101A is later re-instated to CAS, the output to PCV-101A would see a jump from 0 to 100%. For mission-critical applications, this type of bumpy transition cannot be tolerated. It is critical to specify in the operating procedure that PCV-101A should be put back to CAS before PCV-101B, even though the bumpless transfer feature may already be built-in in the control module.

3. Slow action: Another issue with split-range control is that the pressure controller output must transverse the range of one valve before acting on the other valve. In emergency conditions, this slows down the speed of response. For example, in the split-range control in Fig. 4.15, if the controller output is at 35%, the two control valves, PCV-101A and PCV-101B, will have openings of 50 and 0%, respectively. Suppose a significant process upset is experienced, and the pressure abruptly increases; In that case, we expect the pressure control PCV-101B to open as quickly as possible to stop the pressure from going higher. With the split-range control scheme, the controller output would increase from 35 to 70 to cause the normal regulatory control valve PCV-101A to open 100% before opening the flare valve PCV-101B. The slow reaction may be unacceptable.

A simple solution for the above problems is to replace the split-range controller with two independent pressure controllers. This arrangement is the dual-controller control scheme. Both controllers read from the same pressure transmitter, but each independently manipulates one of the two valves. The two independent controllers can have independent tunings and are thus straightforward to achieve optimal tuning for each. This dual pressure control scheme is depicted in Fig. 4.19.

The vital requirement for a dual-controller control scheme to work is that the two controllers must have sufficiently different setpoints. Otherwise, the two controllers would "fight" each other to achieve a similar setpoint. The requirement of different setpoints is the primary drawback of the dual-controller scheme. For most control applications, this is not an issue. However, for applications that require a single control setpoint, split-range control is still the preferred scheme.

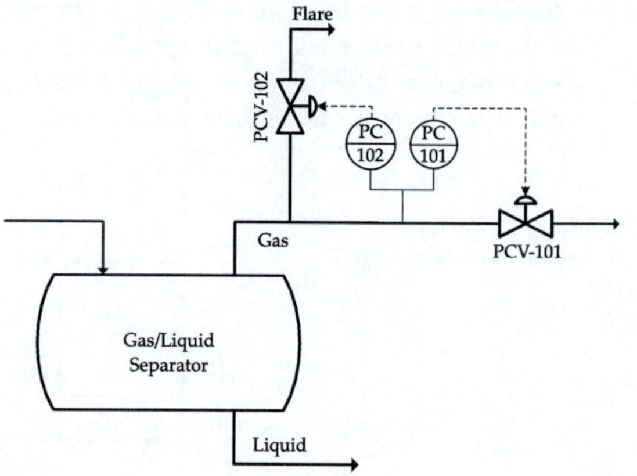

Fig. 4.19 Dual controller versus split-range

In general, a dual-controller scheme is preferred over split-range control due to its simplicity in tunings and configuration. The dual-controller scheme can even be extended to a multi-controller control scheme to achieve more layers in protective control. However, this is not recommended for processes with significant noises and fluctuations.

4.2 Structure of Complex Control Loops

A basic PID controller comprises one controlled variable and one manipulated variable (see Fig. 2.2) and is sufficient for most real-world applications. However, there are also many applications with multiple interacting variables and require more complex control schemes than a single PID controller.

There are two approaches to deal with multi-input multi-output (MIMO) problems. One approach is to use multiple PID controllers and make them work together to address the multivariable interactions. The result is the *complex PID control* schemes discussed in this chapter. The other approach resorts to a single advanced controller consisting of sophisticated calculations that implicitly address the multivariable interactions through models and calculations. This approach is represented by *model predictive control* (MPC) that will be discussed in Chap. 5. To the operator, a complex PID loop is a "white-box" solution that is fully transparent, while MPC is a "black-box" solution with the details hidden.

For clarity, we will use the terms *controller*, *control loop*, and *control scheme* to describe the increasing level of complexity for PID-based control solutions, such as a standalone PID controller, a standard complex PID loop, and a more complex control scheme. For a complete control design, simple or complex, we will call it a process control *solution*. These terminologies are solely for convenience and clarity and may be used interchangeably in a particular context. For example, a model predictive control is a complex control scheme but may be called an advanced control loop in control performance monitoring.

4.2.1 Common Function Blocks

A PID-based complex control scheme consists of one or more PID controllers and some supporting control function blocks. These function blocks are built-in modules in modern control systems such as distributed control systems (DCS) and supervisory control and data acquisition (SCADA) systems. They are the building blocks for process control solutions, both simple and complex. Table 4.4 provides a list of the commonly available function blocks.

For example, in the Foundation Fieldbus control system and Emerson DeltaV system, 10 "basic" blocks, 19 "advanced" blocks, and five other complex function blocks are specified, as listed in Table 4.5.

Table 4.4 PID-based complex control loops

Control blocks	Description	Back-calculation support
PID controller	PID control	✓
Ratio control	Ratio control	✓
Manual loader	Manual control	✓
Automatic switch	Two-way or multi-way switch	✓
Control selector	High/low selector	✓
Control splitter	Split-range, fan-out	✓
AO/DO	Analog output/digital output	✓
AI/DI	Analog input/digital input	
Gain/bias block	Gain and bias	
Lead/lag	Lead/lag calculator	
Dead-time	Time delay block	
Manual switch	Two-way or multi-way switch	
Signal selector	High/low selector	
Signal characterizer	Conditioning, selecting, filtering	
Calculation block	General-purpose calculation	

It is beyond the scope of this book to go into the details of all the function blocks. They can be looked up in the specific control systems manuals when needed, and the IEC standard (IEC-61804) defines the common requirements. However, a reasonable familiarity with their general functionality and typical configuration is essential for designing and maintaining complex control solutions.

For practical reasons, the function blocks offered by different control systems vendors are not standardized, even in their naming. We will use generic names in this description, and focus only on the general functionality.

One crucial aspect is the tracking and initialization (Sect. 3.4) of the PID controllers in a complex control scheme. For complex control schemes with multiple function blocks, the tracking and initialization become more critical and complicated. In consideration of loop integrity, we divide the function blocks into two categories: *control blocks* and *non-control blocks*. Control function blocks provide built-in support for tracking and initialization. Non-control function blocks, such as arithmetic and logic blocks, do not typically support tracking and initialization.

Control Function Blocks

The control function blocks provide `BKCALI` and `BKCALO` parameters to propagate the initialization values for the control element's output, which constitutes the back-calculation data flow (see Sect. 3.4.2):

Table 4.5 Function blocks in fieldbus control system

Control blocks	Description
AI	Analog input
AO	Analog output
B	Bias/gain
CS	Control selector
DI	Digital input
DO	Digital output
ML	Manual loader
PD	Proportional/derivative control
PID	Proportional/integral/derivative control
RA	Ratio station
	Pulse input
	Complex analog output
	Complex discrete output
	Step output PID
	Device control
	Setpoint ramp
	Splitter
	Input selector
	Signal characterizer
	Dead time
	Calculate
	Lead/lag
	Arithmetic
	Integrator (Totalizer)
	Timer
	Analog alarm
	Discrete alarm
	Analog human interface
	Discrete human interface
	Multiple analog input
	Multiple analog output
	Multiple digital input
	Multiple digital output
	Flexible function block

1. PID Controller: A basic PID controller block has two inputs (`PV` and `SP`) and one output (`OP`). The process value (`PV`) is typically pulled from another function block, such as an analog input `AI` block or a calculation block. The setpoint (`SP`) is typically from the operator (`AUTO` mode), another function block (`CAS` mode), or a user program (`RCAS` mode). The basic PID control block provides such standard features as input signal scaling and limiting, PID calculation, output signaling processing, alarming, and output limiting. Some implementations may have additional control functions built into the PID function block to perform additional functions such as feedforward and nonlinear gain scheduling.
   PID controller typically offers full support for back-calculation through `BKCALI` and `BKCALO` propagation (see Figs. 3.13 and 4.20a).
   However, suppose the secondary element does not propagate the windup status or initialization data back to the primary block, explicit external feedback would have to be provided in the control design to help the PID control. The feedback value is typically the process value or setpoint of the secondary block, which is used to initialize the output value of the primary block. That is, the controller output is then given by

$$\mathtt{CV} = \mathtt{EXTFBK} + \mathtt{GAIN} \cdot \mathtt{ERROR} \tag{4.20}$$

   where `CV` is the controller output, `EXTFBK` is the external feedback, and `ERROR` is the control error. See Fig. 3.7 for an illustration of using the valve position as external feedback.
2. Control Selector: A control selector receives two or more[2] control signals (typically controller output) and selects one of them based on the following configurations:

   - High selection: A high selector selects the largest value of the inputs.
   - Low selection: A low selector selects the smallest value of the inputs.
   - Mid-of-three: With three inputs, the median value of the three is selected.
   - High median: The median value is selected if the number of inputs is odd, or the larger median value if the number of inputs is even. A high median selection is the same as a mid-of-three selection in the case of three inputs.
   - Low median: The median value is selected if the number of inputs is odd, or the lower median value is selected if the number of inputs is even.
   - Average: The average selector calculates the average value of all the inputs.

   Figure 4.20b is an illustration of a selector with two inputs. A control selector (e.g., an override control block) typically supports the following back-calculation features:

   - One `BKCALI` connection for the input, for receiving back-calculation information from the downstream control element.

[2] DeltaV supports two inputs while Experion supports up to four.

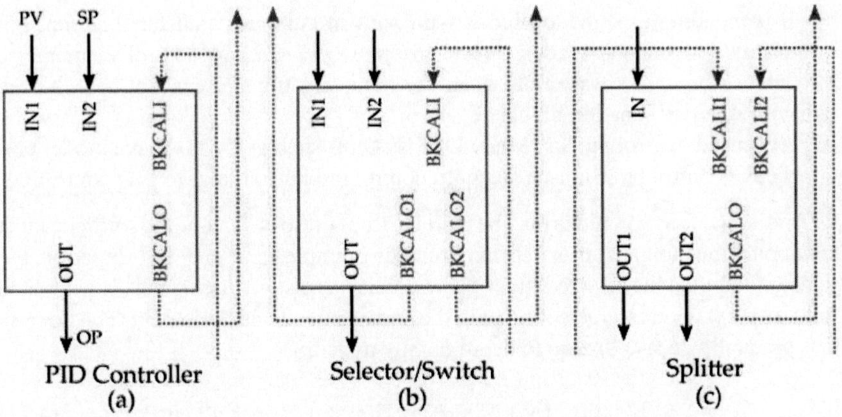

Fig. 4.20 Function blocks: PID controller, control splitter, and control selector

- Multiple BKCALO connections, one for each output, to send the back-calculation information to the upstream control elements (e.g., controllers).

Contrary to a control selector, a signal selector is an arithmetic function block performing the straightforward signal comparison and selection without support for tracking and initialization.

3. Control splitter: A control splitter splits the control signal into multiple ranges and sends them to multiple downstream control elements. The control splitter is commonly used for split-range control and fan-out control. Depending on the control system, the splitter may be offered as one-to-two, one-to-three, or one-to-many split.
   Figure 4.20c shows a one-to-two control splitter. A control splitter (e.g., a fan-out block) typically supports one or more of the following back-calculation features:
   - Multiple BKCALI connections, one for each input, to receive back-calculation value from the downstream control elements.
   - One BKCALO connection, to send back-calculation value to the upstream control element.
4. Switch control block: A switch block can be positioned to assign a different primary element to a secondary element. The switching can be initiated by the operator, a user program, or another function block. A switch block typically supports back-calculation, similar to a control selector block. The difference is that a selector automatically makes the selection based on the internal comparison, while a switch forces the selection externally.
5. Manual loader: A manual loader allows the operator to adjust the manipulated parameter, i.e., for manual control. A manual loader is also called a *hand-indication control* (HIC). Another common application of the manual loader block is providing "bumpless" output following initialization or mode changes.

It is the simplest function block with built-in bumpless transfer function. For example, a manual loader can be inserted between two control elements to achieve bumpless transition when the bumpless transfer can not be achieved with standard function blocks.

6. Advanced control blocks: Many DCS systems such as DeltaV have model predictive control built-in as a standard control module, similar to PID controller.

Additional functionalities may be built into the control selectors and splitters, most commonly the bumpless transfer function. For example, in Honeywell Experion, the selector output is the selected input plus two bias terms. A floating bias is generated for the output upon re-connection with the downstream control element. The floating bias gradually decays to zero to avoid bumpy transition:

$$\texttt{OP} = \texttt{SELECTED\_INPUT} + \texttt{BIAS.FIX} + \texttt{BIAS.FLOAT} \tag{4.21}$$

where `BIAS.FIX` is a fixed bias term and `BIAS.FLOAT` is a floating bias term. A gap between the primary element's output and the secondary element's setpoint may exist when a disconnected control loop is re-connected. The floating bias `BIAS.FLOAT` is initialized to this gap value and gradually ramps to zero:

$$\texttt{BIAS.FLOAT} = \texttt{BIAS.FLOAT} - \texttt{DECAY\_RATE}. \tag{4.22}$$

This floating bias is a straightforward yet effective approach to achieving bumpless transfer (see Sect. 4.3.4).

Some control systems choose to use a first-order filter to decay the initial bias, which serves the same purpose.

Non-control Function Blocks

Non-control function blocks, such as arithmetic or logic calculation blocks, do not typically support tracking and initialization. Therefore, it is critical to double-check the tracking and initialization capability if the control loop containing any non-control function blocks:

1. Gain block: A gain block is typically implemented as a gain plus bias function:

$$G(s) = K_c. \tag{4.23}$$

2. Lead/lag block: A basic lead/lag block has a transfer function of the following form:

$$G(s) = \frac{1}{\tau_1 s + 1}, \qquad G(s) = \frac{\tau_2 s + 1}{\tau_1 s + 1}. \tag{4.24}$$

Since the denominator has a phase lag and the numerator causes a phase lead, thus the name of lead/lag. It is a primitive building block for more complex functions.

3. Time delay block: As the name suggests, a pure time delay block delays a signal by the prescribed duration. The transfer function is given by

$$G(s) = e^{-\theta s}. \tag{4.25}$$

4. Calculation block: The general-purpose calculation block (also called Characterizer in some control systems) provides both logical functions and arithmetic computational capability within one integrated environment. A calculation block provides the ultimate flexibility in signal processing. However, it typically does not support tracking and initialization, and thus should be used with caution.

The gain, lead/lag, and delay blocks are the primitive function blocks. A first-order-plus-time-delay (FOPTD) transfer function can be readily built from the three primitive function blocks:

$$G(s) = K_c \cdot \frac{1}{\tau_1 s + 1} \cdot e^{-\theta s}. \tag{4.26}$$

The PI control and I-only controller shown in Sect. 3.1.9 can be built with these primitive blocks.

Table 4.6 provides a summary of the complex control loops and the function blocks they need.

The configuration support for tracking and initialization varies from one control system to another. For example, in Honeywell Experion and Yokogawa Centum, the back-calculation data flow is automatically established upon the connection of the control blocks when the control scheme is built. In Schneider Foxboro and Emerson DeltaV, an explicit configuration is required to connect `BKCALO` with `BKCALI`. Explicit connection is an extra step in configuration but offers added flexibility.

Table 4.6 Complex control loops and function blocks

Complex control loop	Function block	Initialization concerns
Ratio control	PID + Ratio	
Override control	PID + Control selector	Reset-windup
Split-range control	PID + Control splitter	Back-initialization
Fan-out control	PID + Control splitter	Back-initialization
Dual-controller control	PID + PID	
Feedforward control	PID + Lead/lag + Delay + Gain/bias	Bump transition
Decoupling control	PID + Calc	
Balancing control with offset	PID + Gain/bias	Saturation
Model predictive control		

4.2.2 Built-in Function Versus Custom Function

Modern control systems provide a rich collection of function blocks to fulfill the different control needs, from simple control loops to complex control schemes. Different designs for complex control solutions can achieve the same control objective, but the resulting control performance can be substantially different. Some general recommendations are given below to help achieve better control performance:

1. Use the built-in control blocks wherever possible: These built-in function blocks are carefully designed, feature-rich, and field-proven. There is no need to re-invent the wheel unless absolutely necessary.
   For the same reason, use the function blocks that offer the full features for the need. For example, a PID can be built from primitive control function blocks such as a gain/bias, lead/lag, and delay block, as shown in Fig. 3.5 or Fig. 3.6. However, there is no reason not to use a single standard PID function block if available. Similarly, a ratio control can be constructed with a calculation block and PID controller, but a standard ratio control block should be used instead, if available.
2. Prefer control function block to non-control function blocks: Control function blocks have built-in support for tracking and initialization to protect the loop integrity under different operating scenarios. On the other hand, a function block that does not support the back-calculation, e.g., a calculation block, interrupts the data flow and sometimes may need extra help to bridge the "gap" in the data flow. For example, a control selector supports tracking and initialization, and the loop integrity is protected against a broken connection or limited values. A signal selector may be used in place of the control selector. However, the back-calculation data flow is broken at the signal selector since tracking and initialization are not supported.
   Consider a practical example. In a water injection application, a pressure controller controls the pressure of the injection header pressure. The output of the pressure controller is fanned out to dozens of flow controllers; each controls the flow rate to an individual injection well. Control splitter has back-calculation support, and using the built-in control splitter is the preferred approach. However, the control splitter on some control systems is limited to a 1-to-2 splitting.[3] If there are 32 injection wells connected to this manifold, it will take five layers and a total of $1 + 2 + 4 + 8 + 16 = 31$ splitters to fan out the controller output to all the 32 wells. The multiple layers of selectors seem to be a very messy implementation. Thus, it may be tempting to utilize a single calculation block to distribute the process controller output to the 32 flow controllers. In reality, as we discussed before, a calculation block does not support tracking and initialization, therefore to ensure proper back-propagation of the value and status of the secondary controllers to the pressure controller, some other calculation and logic blocks are needed to

[3] In Honeywell Experion, a fan-out block can have up to eight secondary elements, and with back-calculation support.

mimic the back-calculation data flow and the handshake process. Designing and configuring the calculation and logic blocks can be very involved and are also prone to errors. In addition, the custom design and implementation also pose a significant challenge for maintenance if not adequately documented. Therefore, it is better to accept the superficial complexities and use the built-in control splitters to achieve this 1-to-32 splitting of the control signal.

3. Custom control logic: Sometimes back-end control logic needs to be implemented to meet special requirements not provided by the standard control function blocks. However, experience shows that the back-end logic can be another challenge for operation and troubleshooting since it can be easily overlooked due to its low visibility. Treat back-end logic as a necessary evil, and use it with discretion.

4.3 Configurations of Complex PID Loops

Although PID configuration is the responsibility of control system engineers or technicians, experience shows that many configuration options are incorrectly configured or simply left to the system defaults. Many configuration options are dictated by the control strategy and should be specified by process control engineers.

The PID configurations that may significantly impact the control performance are discussed in Sect. 3.5. However, for complex PID control loops comprising multiple PID controllers and supporting function blocks, the loop configuration can be dauntingly complex and demands many additional practical considerations.

4.3.1 Data Flow in Complex PID Loops

A complex control problem typically involves multiple control handles and multiple control targets. The data flow from the controllers to the final control elements is no longer a one-to-one fixed path as in simple PID control but can be many-to-many. Suppose selectors or splitters are utilized in the control scheme, the control action produced by a particular controller may travel a different data flow path and reaches a different final control element under different operating condition. In addition, the data flow can be disrupted or limited at many locations and for many reasons. Similarly, the changes in the final control element are fed back to the controllers through back-calculation, and the back-calculation channel can be disconnected or disrupted.

As discussed in Sect. 3.3, the four abnormal scenarios with the data flow become exacerbated for complex control schemes:

1. Bad data: The control data is bad or missing due to loss of communication, out of service control devices, or bad measurements.

2. Data flow is broken. The data flow is blocked, or the flow path disconnected, e.g., for the following reasons:
   - A control element such as a controller or analog output block on the data flow path has switched away from cascade mode.
   - The data is re-routed by a selector, a switch, or a splitter.
3. Data value is limited, saturated, or constrained. For example,
   - Controller output is high or low limited.
   - Setpoint sent down has reached the setpoint high or low limit of the secondary element.
4. The loop dynamics have changed. Many reasons can cause this to happen, such as the following:
   - Process dynamics change due to aging, fouling, or process modifications.
   - Valve problems such as stiction, hysteresis, and leaking.
   - Measurement errors such as lack of calibration.

   Reliable tracking and detecting gradual dynamics change is difficult and remains a challenge for both theory and practice.

The tracking and initialization of the PID control implementation is a critical consideration to ensure loop integrity. With tracking and initialization, the controller can take appropriate actions to initialize itself when an abnormal condition is detected. All control elements on the data flow path must have the same support to make the complex control loop work. Unfortunately, not all the function blocks (see Table 4.4) fully supports tracking and initialization.

4.3.2 *Common Configuration Considerations*

DCS implementation of a PID controller provides many configuration options to meet the various operating requirements. Although PID configuration is the scope of work of control system engineers, several configuration options can significantly impact control performance and need to be addressed by process control engineers.

Section 3.5 has provided a general introduction to PID loop configuration from the perspective of the PID controller itself. However, for complex control schemes, the interaction and inter-operation of the controllers and the supporting control function blocks deserve special attention. For example, an essential configuration item for overriding control is anti-reset windup, while bumpless transfer is a common consideration for a fan-out control. In addition, for all complex control schemes, careful consideration should be given to crippled mode operation and PV tracking. Table 4.7 lists the common configuration requirements that should be considered for each type of standard complex PID control loop.

Table 4.7 Common configurations for complex PID loops

	Crippled mode Operation	PV Tracking	Anti-reset Windup	Bumpless Transfer	Gain Scheduling
Cascade control	X	X			
Override control	X	X	X	X	
Switching control	X	X	X	X	X
Fan-out control	X	X		X	X
Split-range control	X	X		X	X
Custom calculation	X		X	X	

The selection of PID equation type depends on the specific control scheme. See Table 4.8 for a list of the recommended equation types. As mentioned in Sect. 3.1.4, most PID control in practice uses PI algorithm; thus, the difference between `PID` and `PI-D` is negligible.

The mode of operations of a PID control can significantly impact the loop dynamics (see Sect. 3.2). A controller in manual mode breaks the PID loop, and a cascade controller in local override mode simply bypasses the controller. For PID controllers in a complex control scheme, the impact of the controller mode on the loop dynamics is even more significant, especially when other control elements such as selectors and splitters are involved. It is thus critical to review the requirements on controller mode, setpoint value, and controller output value during the different operation modes as listed in Table 4.9. All the question marks in the table need to be considered during design.

There are many beautiful theories and techniques for control stability analysis. However, they are of limited value to practical process control because these theories and techniques assume the process dynamics are known and fixed. For process control, the stability analysis must be based on the loop dynamics rather than the process dynamics alone. In a complex PID control scheme, the more detrimental factors affecting control stability are often other components in the loop, e.g., the control valve with stiction or saturation, the splitter or selector in the control that alters the control structure online.

Table 4.8 Equation type for complex PID loops

	Equation type	SP tracking PV
Standalone PID	I-PD	NO
Protective PID	I-PD	NO
Overriding PID	I-PD	NO
Cascade (primary)	I-PD	NO
Cascade (secondary)	PI-D	YES
Fan-out (primary)	I-PD	NO
Fan-out (secondary)	PI-D	YES

Table 4.9 Mode of operation for complex PID loops

	Controller mode	Controller setpoint	Controller output
During shutdown	?	?	?
During startup	?	?	?
During normal operation	?	?	?
Entering crippled mode	?	?	?
During crippled mode	?	?	?
Leaving crippled mode	?	?	?
Emergency shutdown	?	?	?

In summary, it is highly desirable to reach the highest level of automation and minimize the need for operator intervention, especially during busy startups and emergency shutdowns. For this purpose, all the potential abnormal conditions should be considered, and the PID controllers and supporting functional blocks should be configured to be "fool-proof" and "fail-safe" as much as possible. On the other hand, the practical reality limits how far we can push in automating the operation. The best decision is always a trade-off between desire and practicality.

4.3.3 Anti-reset Windup in Complex Control Loops

As a closed-loop feedback control technique, PID control assumes that moving the controller output (OP) in the right direction would bring the process value (PV) closer to the desired setpoint (SP) and eventually eliminate the control error. If the control error does not go away, the controller's integral term will continue to change the output in the same direction, regardless of whether the feedback is lost or compromised. For example, override occurs when another controller takes over control of a particular loop (e.g., for safety reasons). If the controller is unaware that its output has been overridden, it will continue to act on the control error it can no longer influence. Eventually, the output will be driven to saturation.

This phenomenon is called *reset windup* or *integral windup* (Marlin 2015) and has been briefly explained in Sect. 3.5. The risk of reset windup is that when the overridden controller needs to take control, the output may take some time to come out of windup before becoming effective again. Reset windup degrades control performance and causes an acceptable delay in action.

Reset windup can be made clear by revisiting the PID algorithm in Eqs. 3.7 and 3.25. The PID controller produces the incremental change in control action $u(k)$ for eliminating the current control error $e(k)$:

$$u(t) = K_c \left(e(t) + \frac{1}{T_i} \int_0^t e(\tau) d\tau + T_d \frac{d\,e(t)}{dt} \right) \tag{4.27}$$

$$u(k) = K_c \left(e(k) + \frac{T_s}{T_i} \sum_i e(i) + \frac{T_d}{T_s} (e(k) - e(k-1)) \right). \tag{4.28}$$

The control $u(k)$ is added to the current process output and becomes the new control output

$$U_{\text{new}} = U_{\text{curr}} + u(k) = U_{\text{curr}} + f\,(\text{SP} - \text{PV}) \tag{4.29}$$

where U_{curr} is the current control output, and U_{new} is the new control output. U is the actual output that drives the final control element, and $u(k)$ is the increment change in output produced by PID to eliminate the current control error. In other words, $u(k)$ represents the delta change in the control action, and U is determined by the steady-state condition, which is the bias term in Eq. 3.24.

The integrating action in the PID algorithm is a powerful feature of PID control for eliminating control offset, but it also generates many practical issues. The integrating action in the PID controller does not know, nor does it care, if the control error is caused by its own control action or other reasons. In addition, the PID control calculation produces a change in the output, and does not know, nor does it care, what the actual value of the output should be. As long as there is a control error $e(k)$, the integral action will continue integrating on the error and producing a non-zero incremental change $u(k)$ added to the current control output.

In practice, there are built-in default limits for most of the variables in a controller. A valve opening is limited to 0~ 100%, and a controlled variable (controller PV) is limited to its measurement range. If the control error persists, the output would continue to change in the same direction until it reaches the limits.

The standard solution is to stop the integrating action on the error once the control action is limited. The PID controller then becomes a simple P-only controller. That is

$$u(k) = K_c \cdot e(k) = K_c \cdot \frac{\text{SP}(k) - \text{PV}(k)}{[\text{PV Range}]} \cdot [\text{OP Range}]. \tag{4.30}$$

In a practical implementation, the back-calculation provides the actual output U_{curr} via the `BKCALO` to `BKCALI` propagation. For example, in Fig. 3.14, the setpoint for the AO block is the controller output, and the back-calculation value `BKCALI` for the PID controller is the `BKCALO` value of AO block output. The `BKCALO` value for the AO block is usually set to the value of the AO output, which is the actual control action propagated down from the PID controller. This `BKCALI` value serves as the current controller output U_{curr} in Eq. 4.29. Therefore, when the control data is limited, the output of the PID control becomes

$$\begin{aligned} U_{\text{now}} &= U_{\text{curr}} + u(k) \\ &= \text{BKCALI} + u(k) \\ &= \text{BKCALI} + K_c \cdot \frac{\text{SP}(k) - \text{PV}(k)}{[\text{PV Range}] \cdot [\text{OP Range}]}. \end{aligned} \tag{4.31}$$

Similarly, the `BKCALO` value of the PID block is set to its process value (`PV`) and is sent back to its upstream control block as its `BKCALI` input.

If `BKCALI` is not available, the controller uses its own output from the last execution as U_{curr}, not knowing whether the output has reached the next control element on the data flow.

Example 4.2 Anti-Reset Windup. Consider the example in Fig. 4.21. The level controller `LC-101` is for normal regulatory control, while `PC-101` is for the high-pressure protective override.

Without anti-reset windup protection, the `SP`, `PV`, and `OP` trends of both controllers, `LC-101` and `PC-101`, are shown in Fig. 4.22.

In the beginning, when the level controller `LC-101` is in control, its output is selected by the low selector `FY-101` and sent down to the flow controller `FC-101`. The output of the pressure controller `PC-101` is ignored (overridden). The pressure controller `PC-101`, not knowing that its output is not selected, continues increasing its output, attempting to minimize the control error. This effort is fruitless since the loop is effectively open at the selector. The integrating action eventually drives the output to 100% at time 10 and remains at 100%.

At time 20, the pressure starts to rise for some process reasons. At time 33, the pressure rises above its setpoint (the high-pressure limit). High pressure is an unsafe condition, and the pressure controller is expected to act immediately to override the level controller and reduce the flow. However, the saturated pressure controller will have to un-wind the saturation first. The plot shows that the pressure controller output (`PC101.OP`) takes 22 seconds to come down and become lower than the level controller output (i.e., to have `PC101.OP` ≤ `LC101.OP`). Only after that point (time 55) does the pressure controller output (`PC101.OP`) get selected by the low selector

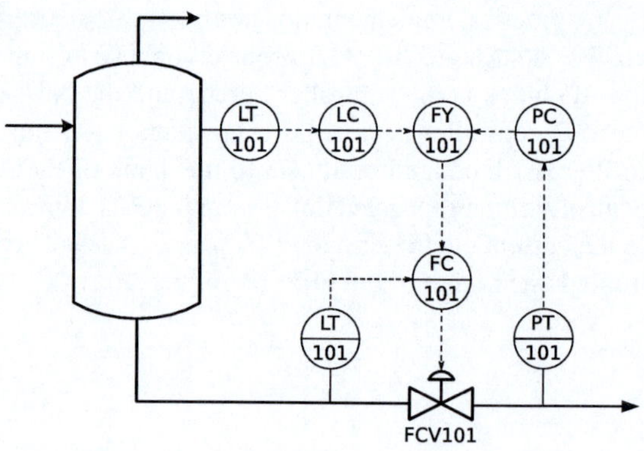

Fig. 4.21 Examples of override control

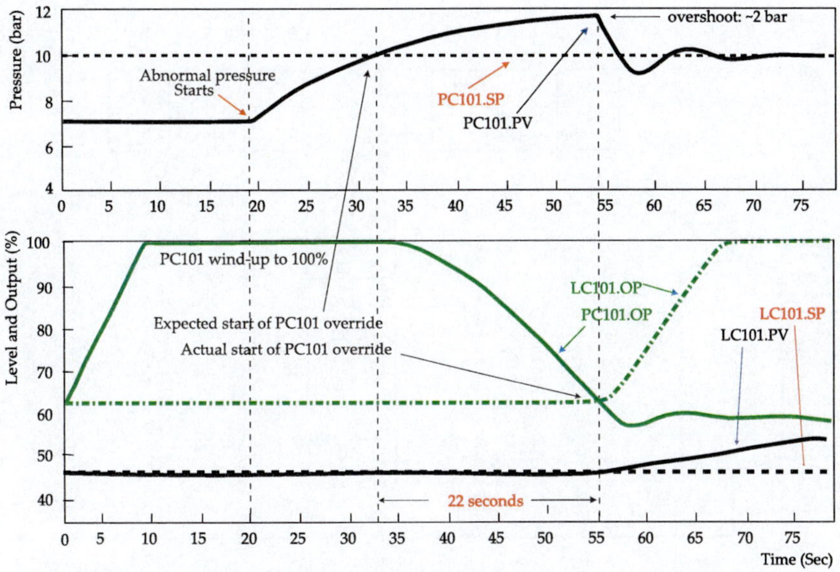

Fig. 4.22 Override control without anti-reset windup enabled

FY-101, and the flow starts to decrease. At the same time, the pressure continues to rise and reaches 2 bars above the pressure limit. This delay in action is unacceptable for a protective control!

Now let us look closely into the anti-reset windup configuration. With anti-reset windup configured, the information flow of this level-pressure override control loop is shown in Fig. 4.23.

Assume that the level controller is in control, then the pressure controller (an overriding protective controller) is open-loop, and its output is not "consumed." For the information flow, note the following:

1. First, note that the back-calculation path is from the valve to the two top-level PID controllers. If the actual valve opening is not available, then the back-calculation input for the AO block will default to its output.
2. Secondly, at the selector FY-101, the two data flow streams (going forward) from the two PID controllers merge into one. Naturally, the back-calculation splits into two paths at the selector. The back-calculated value at the selector is the actual output of the selector.

A controller in the override control must always remain ready to act even when not selected. Anti-reset windup aims to prevent the controller from going into saturation and is achieved through two actions:

1. When the pressure controller output is not selected, i.e., the loop is open, the pressure controller tracks the downstream status and value through back-calculation. The valve position is back-propagated to the AO block, the flow controller

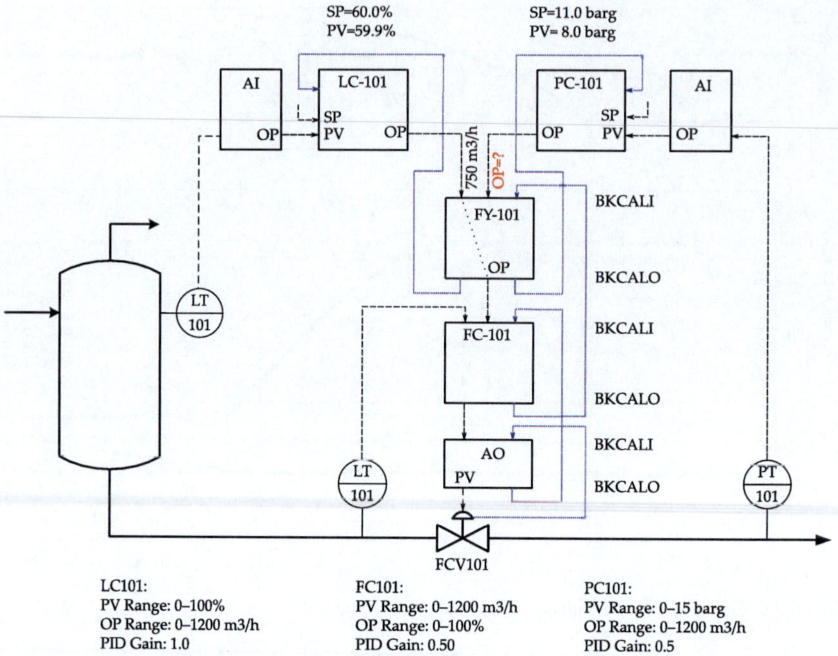

Fig. 4.23 Data flow in an override control loop

`FC-101`, and then to the selector block (`FY 101`). The selector "informs" both the level controller `LC-101` and the pressure controller `PC-101` which controller output is currently selected and the actual value of the selector output.

2. The controller algorithm is temporarily switched to a P-only algorithm to avoid integrating action, which is

$$\texttt{OP} = [\text{BKCALI}] + \text{GAIN} \cdot \text{ERROR}. \tag{4.32}$$

Considering the dimensionless calculation in the PID algorithm discussed in Fig. 3.4 of Sect. 3.1.5, the control output with proportional-only calculation is given by

$$\texttt{OP} = [\text{BKCALI}] + \text{GAIN} \cdot \frac{\text{PV} - \text{SP}}{[\text{PV\_SPAN}]} \cdot [\text{OP\_SPAN}] \tag{4.33}$$

$$= 750 + 0.5 \cdot \frac{11-8}{15-0} \cdot 1200 = 870\ m^3/h \tag{4.34}$$

where 750 is the output that is sent down to the AO block by the level controller `LC-101`.

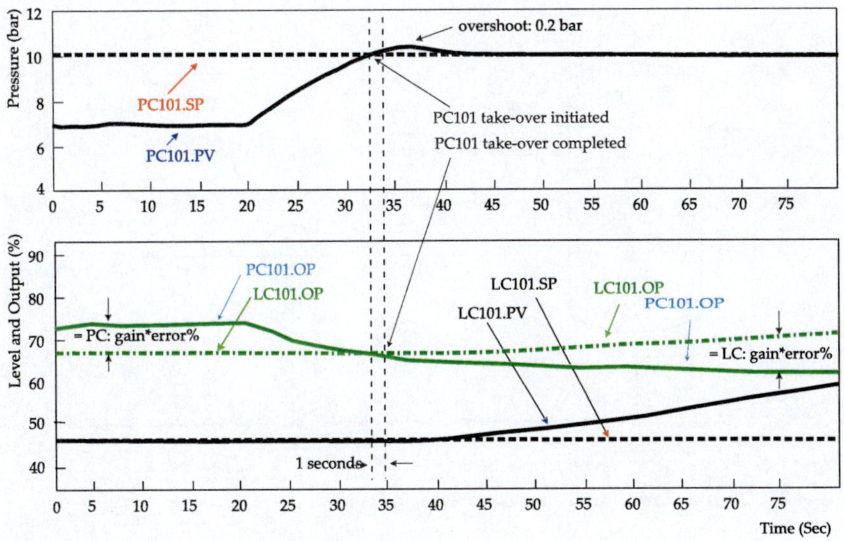

Fig. 4.24 Override control with anti-reset windup enabled

Figure 4.24 shows the process response when anti-reset windup is enabled. Instead of 22 s, the pressure control takes control almost immediately, one second (one scan time) after the process value rises above the pressure setpoint. The overshoot in pressure is only 0.2 bar!

Note that in Fig. 3.4, since the control error is the difference between the setpoint and process value, it decreases to zero when the pressure goes up to the same value as the setpoint. The PID calculated output becomes zero as well (Eq. 4.30). From the above equation, it is clear that the output decreases from 870 m^3/h to 750 m^3/h and equals the level control output! At this point, the control error is zero, the two outputs are equal, and the calculated output change is zero. So the cross-over and transition (the takeover) are completely bumpless! If the pressure value increases further and becomes higher than its setpoint, the pressure controller output would become lower than 750 and take over the control from the level controller.

Figure 4.24 also clearly shows that the level `LC-101` starts to rise above its setpoint because the pressure controller `PC-101` takes over the level controller, and the level is basically off control. Its output, however, will not wind up thanks to the anti-reset windup configuration on the level controller.

The override control scheme in Fig. 4.21 can also be implemented between the flow and pressure controllers, as shown in Fig. 4.25.

Compared to Fig. 4.21, the data flow will differ, but the same principle and analysis apply.

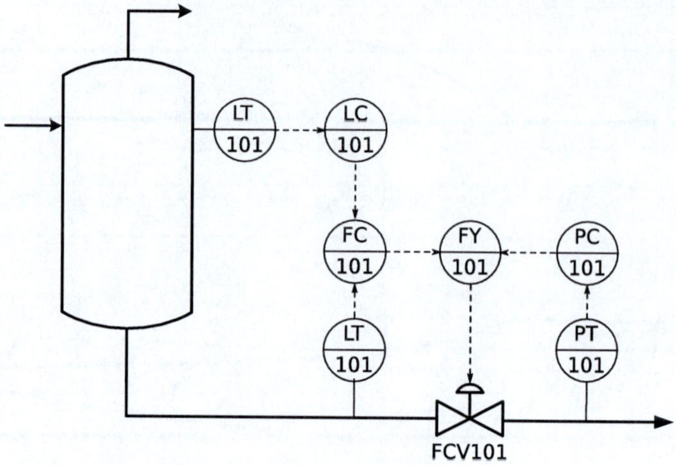

Fig. 4.25 Override control with an alternative configuration

As discussed in Sect. 4.2.1, modern control systems have the anti-reset windup (ARWU) function built into their basic control modules. Most control function blocks support tracking and initialization that can sense the changes in the data flow and communicate it back to the controllers. However, function blocks such as arithmetic calculation and logic processing do not support them. Besides, most final control elements (values, switches) may not support position read-back to report its actual position. In control design, it is crucial to pick the right function blocks to build the control scheme and know where extra configuration work is needed to assist the function blocks that do not have built-in back-calculation.

As explained in Sect. 4.3.3, reset windup occurs when the data flow is broken or limited by such events as controller mode change, selective control, switching, valve not responding, frozen signal with measurement. The output can be overridden by a signal selector, a switch block, or be limited by high/low limits on the output itself, the setpoint of the secondary element. As a general rule, anti-reset windup protection must be configured for any PID controller if its output may be limited or overridden by other applications.

In a complex control loop, the data flow can become overwhelmingly complicated. Therefore, the probability of the data flow being disrupted or compromised is exponentially higher. As a result, the anti-reset windup configuration becomes nontrivial. The anti-reset windup requirement is due to the reset (integrating) action in the PID controller. All PID controllers with I-action configured must consider anti-reset windup protection. For a P-only controller, reset windup is not a concern.

ARWU is preferably done with DCS built-in function wherever possible. If not, external feedback shall be provided based on calculated feedback values.

4.3.4 Bumpless Transfer in Complex Control Loops

Generally speaking, a bumpless transfer is required whenever this is a potential interruption in the data flow. When the data flow is re-connected, abrupt changes ("bump") in the control output to the final control elements should not occur.

PID control tracks the data flow (status and value) for disconnection and re-connection and tries to initialize the controller to avoid transfer bumps (Sect. 3.3). The tracking may be trivial for a simple control loop but can become overwhelmingly intricate when a complex PID control scheme is concerned.

Transfer Bump

As briefly discussed in Sect. 3.4, the data flow in a control loop may be interrupted for many reasons such as mode change, override, selection, and switching. In a cascaded connection where a PID control is the primary element and an AO block is secondary, the primary element's output is the secondary element's setpoint. When the connection is broken because the secondary element is switched away from the cascade mode, a gap will develop between the primary element's output and the secondary element's setpoint. When the data flow is re-connected, this gap can cause a transfer "kick" or "bump" if proper initialization is not performed.

Standard complex control loops that may experience transfer bumps include cascade control, fan-out control, ratio control, and split range controls (see Table 4.7 for a summary):

1. Cascade control: In a cascade control loop (see Fig. 2.18), if the secondary controller (inner loop) has switched away from cascade (CAS) mode, the primary controller output is "unconsumed" and dangling. When the secondary controller is switched back to cascade, its setpoint no longer matches the primary controller's output. The difference or gap may cause a "bump" in the secondary loop's setpoint if re-connected without adjustment of the gap.
2. Fan-out control: The primary controller's output is fanned out to more than one secondary control elements in a fan-out control scheme, as shown in Fig. 2.17. If one of the control elements is disconnected (e.g., switched to manual) and then re-connected (switch back to cascade) after some time, the setpoint of the secondary element may no longer match the output of the primary controller. Again, this gap can potentially cause a transfer bump if not correctly configured.
3. Split-range control: In a split-range control setup such as in Fig. 2.15, the controller output is sent to two secondary control elements (valves). Assume that one leg in the split-range control is broken by switching to manual control, and the other leg of the split-range control is connected, the output of the primary control is still consumed, and the loop is still closed. When the disconnected leg is re-connected, the difference between the restored valve position and the controller output may have already developed a gap, and the re-connection may cause a bump.

There may be other scenarios where the disconnection and re-connection of the data flow will cause a transfer bump. Some scenarios may be quite complex. For example, consider the split-range control in Fig. 2.15 that has one controller and two valves, with a split range of 0∼70% and 70%∼100%:

1. Assume that the controller output is 35%. Based on the 70% split point, the first valve (in the lower leg) will be 50%[4] open, and the second valve (in the upper leg) is at 0%. Now assume that the first valve is switched to manual mode and a persistent control error has driven the controller output to 73%.[5] The 73% output causes the second valve to open to 10%, and the first valve remains manual at 50%. If the first valve is switched back to cascade mode (re-connected), the valve opening will jump from 50 to 100% because the controller output is higher than 70%. This sudden change in the valve is a "bump" that may be unacceptable for some operations.
2. Similarly, assume that the controller output is at 85% and that both valves are in cascade mode. With a 70% split point, the first valve should be 100% and the second valve 50%. Suppose the second valve is switched to manual mode, and the control error drives the controller output to 35% due to process condition change. Now assume the second valve is now re-connected, the valve position will experience a step-change from 50 to 0% since the controller output is now in the 0∼70% range. This 50% bump in the valve opening may not be acceptable.
3. Now assume both valves are disconnected. In this case, which valve should the controller track and which valve position should the controller output re-initialize to? The controller will typically back-initialize itself to the first control element switched to cascade, and this re-connection is thus bumpless. When the remaining control elements are re-connected, the primary control can no longer back-initialize. Each newly re-connected control element will have to initialize its setpoint and take care of any gaps to ensure bumpless transfer.

Complex control schemes require a careful analysis of the bumpless transfer scenarios during control design and implementation.

Bumpless Transfer

Bumpless transition is an essential control requirement. In modern control systems such as DCS, most control function blocks have the bumpless transfer feature built-in. For example, mode switching between manual, automatic, and cascade mode is typically bumpless by design. Typically there are three options to respond to this transfer bump:

[4] The split-range of the first leg is 0%∼70% $\mapsto$ 0%∼100%, therefore, 35% of controller output drives the first valve to an opening of $\frac{35-0}{70-0} \times (100 - 0) = 50\%$.

[5] Similarly, for the second leg with a mapping of 70%∼100% $\mapsto$ 0∼100%, a controller output of 73% is $\frac{73-70}{100-70} \times (100 - 0) = 10\%$ in the opening of the second valve.

1. Do nothing: If the secondary element is a properly tuned PID controller, the bump implies a step-change in the setpoint of the secondary controller at the time of re-connection. The PID loop as a dynamic controller should be able to provide the required damping in the response, which is acceptable by many applications. In addition, the `I-PD` equation in the PID algorithm (Sect. 3.1.4) can be used to eliminate the transition bump caused by setpoint changes.
2. The primary control loop adjusts its output to match the setpoint of the secondary control elements, be it another controller or a valve: This type of initialization is the so-called *back-initialization*, and has been the standard feature in most control systems. It is automatically configured by default (see also `PV Tracking` in Sect. 3.5). For example, a standard cascaded control loop relies on back initialization to ensure bumpless transfer. With back-initialization, when the secondary element is switched back to cascade, there will be no bump experienced since the primary controller's output has already lined up with the setpoint of the secondary controller, resulting in a zero control error.
3. Secondary elements to decay the mismatch to avoid "bump": Sometimes, the primary control loop can not align its output to match the secondary elements' setpoint because the primary controller's output is already connected to other control elements. In this case, each secondary element to be connected must ensure bumpless transfer on its own if a bump is not acceptable. This type of initialization is typically achieved with a built-in bumpless transfer function. With the bumpless transfer function, the initial error at the time of re-connection is gradually reduced and eventually eliminated over multiple scan cycles through a ramping function or first-order filter.
   For example, in a control scheme where the controller output is fanned out to two valves, one valve is with 60% opening, and the other is offline and fully closed. If the offline valve is put back to control, the valve opening will jump to 60% in one shot, creating an unacceptable disturbance. An explicit function block such as a manual loader (Sect. 4.2.1) can be inserted in front of the control valve to provide a decaying function on the mismatch and achieve bumpless transfer. With a bumpless transfer configuration, the 60% initial error will be gradually reduced. See Fig. 4.26.
   Several function blocks (Sect. 4.2.1) have the bumpless transfer function built-in, including the manual loader block, control splitter, and the control selector. Consult the DCS manual for accurate information.

Note that the bumpless transfer functionality is not the same as PV filtering. The bumpless transfer function is activated only when the control path is re-connected (e.g., mode change). The damping function applies only to the initial error at the time of re-connecting the loop. Any errors developed after the re-connection will not be subject to the damping function.

The bumpless transfer function is also different from controller back-initialization. Back initialization is for the primary controller to align its output to the secondary control elements and is the preferred approach since the ultimate goal is to eliminate the transfer bump in the final control elements. A bumpless transfer function block

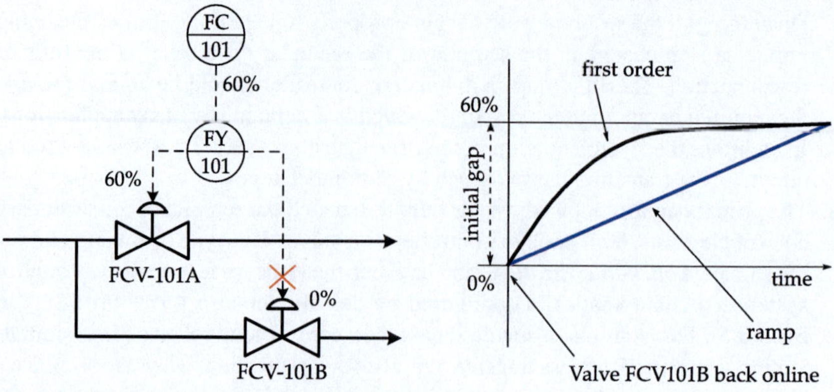

Fig. 4.26 A PID control loop requiring bumpless transfer

is used when back-initialization can not be achieved, and the downstream control element must take care of itself to prevent the transfer bump.

If none of the above standard options is applicable, an alternative control design may be considered. For example, a dual-controller control is preferred over a split-range control in practical applications for its simplicity and robustness in control performance.

4.3.5 Bumpless Transfer Versus Anti-reset Windup

A transfer bump is caused by a mismatch in the data value between the primary and secondary control elements due to the disconnection and re-connection of the loop. On the other hand, reset windup is due to the data flow being limited or saturated at some point in the control loop while the control loop is still closed.

Figure 4.27 is an example of an overriding control loop with a selector. The level controller `LC-101` is the primary controller for normal operation, while the pressure controller `PC-102` is an overriding protective controller to avoid high back-pressure downstream the valve. The low selector selects the smaller of the two controller outputs and sends it down to the flow controller `FC-103`. The flow controller output is sent down to two control valves through a split-range control arrangement.

For loop integrity, a PID controller (`LC-101`, `PC-102`, or `FC-103`) in the complex control loop not only has to send the controller output down to the valves; it also must know whether its output reaches the valve and which one. Due to the branching and merging of the control signal through selector and splitter, the output generated from a controller may not be the one that is received by the target valve.

Based on the previous discussion, this control scheme is vulnerable to reset windup due to the override control. It is also prone to transfer bumps because of the use of

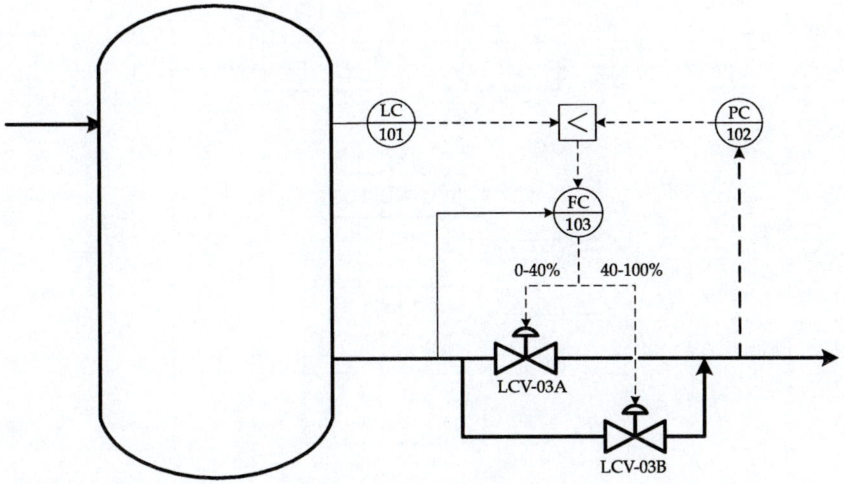

Fig. 4.27 A PID control loop for back-calculation

split-range control. Figure 4.28 shows the simplified information flow, including both the data flow and back-calculation. The back-calculation path propagates the valve position information from the valves, through the splitter and selector, back to the level controller `LC-101`, using `BKCALO` and `BKCALI` daisy-chain connections. As a result, the level controller knows whether its output is "consumed" by the downstream control elements. When the controller is "informed" by back-calculation that the pressure controller has overridden its output, it simply stops the integrating action to prevent output windup.

The function and operation of the back-calculation can become very perplexing for complex control loops. Some DCS implementations, such as Honeywell TDC, simply offer no back-calculation support for this specific control structure, i.e., a splitter after a selector (Honeywell 2003).

4.4 Summary

A complex PID control scheme typically consists of multiple PID controllers and other supporting function blocks. It can be surprisingly versatile in addressing complicated practical problems such as multivariable interactions, constraint handling, preferential flows, and limit pushing. However, effective use of complex PID control schemes require a good understanding of the operating requirements and also intimate familiarity with the functionality of the building blocks.

Many application scenarios can be addressed by either complex PID schemes or model predictive control, with different pros and cons. Complex PID schemes are

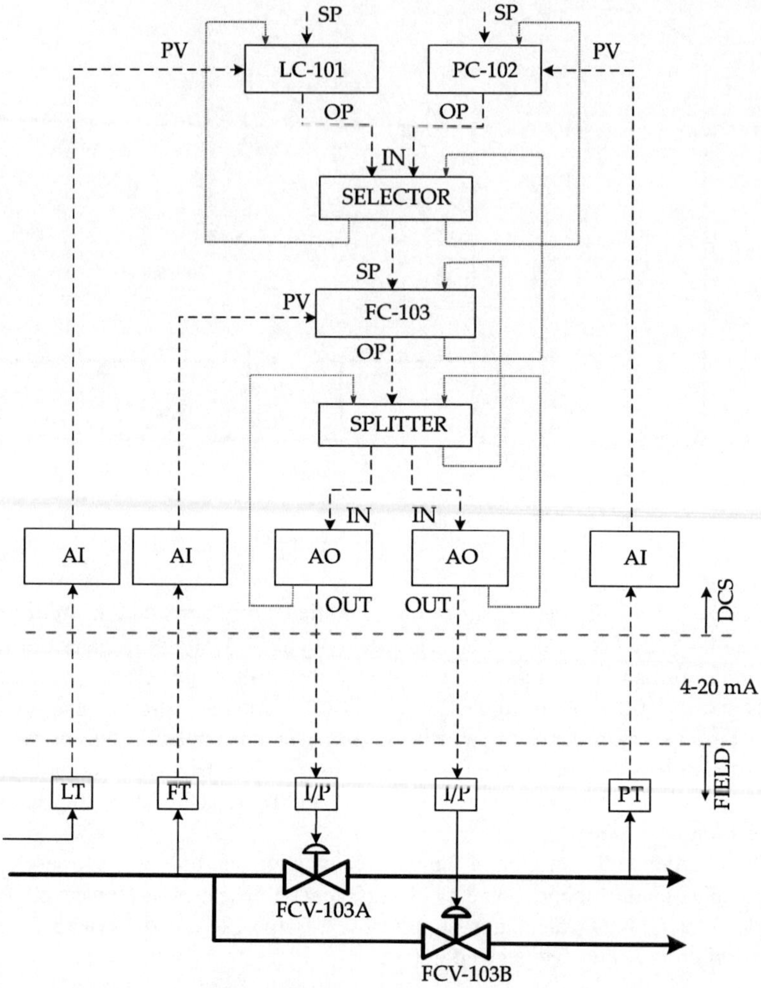

Fig. 4.28 Back-calculation in a PID control loop

native DCS applications and are entirely transparent to operators and control system engineers. However, the complexity in implementation can easily overwhelm the operator and the maintenance personnel.

Feedback control relies on a known and stable data flow to close the loop. With complex control loops, the data flow can become overly complicated. At each execution, the output of a controller in a complex control may travel a different path to reach the final control elements. Many scenarios can cause the control signal to be re-routed, disconnected, or limited along the data flow path. How to protect the integrity of the data flow is crucial to the function of the control scheme. Modern DCS implementation of the control scheme provides rich functionality in protecting the loop integrity, with tracking and initialization being the standard features. However, there are many scenarios for a complex control scheme that the built-in protection is incomplete or insufficient. It remains a challenge to provide "fool-proof" protection against loop integrity problems with DCS default functionality.

The loop integrity can be a determining factor in deciding the control strategy. Process control engineers play a central role in ensuring that the control design has considered all the possible scenarios. A process control engineer must know when the built-in functions are sufficient to protect the loop integrity and when external help must be provided to help the PID controllers.

References

Gordon L (2005) Advanced regulatory control: decoupling. Control Eng

Honeywell (2003) Advanced process manager. Control function and algorithm, Honeywell

Luyben WL, Lyuben ML (1997) Essentials of process control. McGraw-Hill

Marlin TE (2015) Process control—designing processes and control systems for dynamic performance, 2nd edn. McGraw-Hill

Shinskey FG (1978) Control systems can save energy. Chem Eng Progress 74:43–46

Shinskey FG (1994) Feedback controllers in the process industries. McGraw-Hill Publishing, New York

Smith CL (2010) Advanced process control—beyond single-loop control. Wiley

Stephanopoulos G (1984) Chemical process control—an introduction to theory and practice. Prentice-Hall, New York

Chapter 5
Advanced Process Control

Most practical applications are more complex than simple PID controllers can handle. Although complex PID control schemes are capable enough to address many common needs, their inherent limitations and implementation challenges call for more robust and elegant solutions (Sect. 5.1). Advanced process control technologies, such as inferential property estimation and model predictive control, result from these practical needs. Most advanced control technologies are model-based (Sect. 5.2), with inferential property estimation (IPE, Sect. 5.3) and model predictive control (MPC, Sect. 5.4) being the most successful. MPC has also changed the landscape of practical process control and has become the most important alternative or replacement of complex PID control solutions. MPC represents a mindset change from localized control to large-area control. Its success depends on a good understanding of the process and a holistic view of the control problem, which is reflected by an adequate control structure and high-quality process models. A well-developed work process (Sect. 5.6) is critical to enforce this mindset change and ensure the adequacy of the model quality and control structure.

5.1 Why Advanced Process Control?

As the dominating technology in process control for over 100 years, PID control has been very successful. At the same time, being a simple feedback control algorithm, PID also has severe limitations. There have been countless efforts to break these limitations; Some enhance the existing PID algorithms, while others bank on entirely different technologies. See Samad et al. (2020) for the result of a recent survey conducted by the IFAC industry committee in 2018, where the impact of the current and future control technologies are polled.

S. S. Niu and D. Xiao, *Process Control*, Advances in Industrial Control,
https://doi.org/10.1007/978-3-030-97067-3_5

5.1.1 Process Challenges and PID Control Limitations

The process industry has the following significant challenges for process control that are not common in most other industries:

Challenge #1: Strong multivariable interactions: Process control often involves multiple inputs (control handles) and outputs (control targets) that heavily interact and interfere with each other. Changes in one variable may cause more than one variable to change; Similarly, there are multiple handles to use to effect a change in one variable.

Challenge #2: Large time delay: Many processes exhibit long-time delays. Delay dominant processes are particularly challenging for feedback control.

Challenge #3: Operating constraints: As process operation relentlessly strives for efficiency and profitability, the operating points move closer and closer to their limits and constraints, continuously raising the bar on the process control requirements.

Challenge #4: Noises and disturbances: Many sources of noises and disturbances with unpredictable characteristics affect the process's dynamic behavior. Many complexities in the new control technologies are introduced just for dealing with noises and disturbances.

Challenge #5: Unreliable or infrequent measurements such as quality measurements based on online analyzers or lab samples: As process control is shifting from isolated process variable control to higher-level quality control, the reliability and response speed of most online quality measurements do not yet meet the requirements for online closed-loop control.

Challenge #6: Nonlinear and time-varying dynamics: Most real-world processes are inherently nonlinear. Besides, process behavior changes as the process operating point moves from one region to another or from time to time. Since most controllers are designed for linear and time-invariant (LTI) processes, the control performance may substantially degrade when the process behavior changes with operating condition or time.

A PID control loop, with a simple algorithm and pure feedback mechanism, struggles to face most of the above challenges. In the oil & gas industry, the downstream operations have taken the lead in adopting advanced control technologies, mainly due to process complexity (interactions), constraints, and optimization opportunities from frequent changes in product demand, pricing, and economics. On the other hand, the upstream exploration & production (E&P) have relatively simple process flows but faces different challenges such as geographically dispersed assets, higher variations in feed flow and quality, and measurement reliability.

5.1.2 Process Control Solutions Before MPC

The interaction and coordination among all the concerned process variables and controllers are critical for controlling a wider operation area. A PID control loop controls a specific variable in isolation and does not necessarily care about the interaction and coordination with other controllers. Complex PID control schemes discussed in Chap. 4 provide some capabilities to handle multivariable problems (Challenge #1 in Sect. 5.1.1). However, the inherent limitations of the PID feedback control leave the rest of the above challenges unanswered. Besides, as the number of PID controllers and supporting functional blocks increases, the control scheme's complexity increases exponentially. The implementation and maintenance become a significant challenge (see Sect. 4.2 for complex loop structure).

Decoupling control (Sect. 4.1.2) (Gordon 2005; Shinskey 1994) is another approach to address the interaction in multivariable control explicitly (challenge #1). However, its complexity has limited its widespread implementation.

The Smith predictor (Smith 1957) is an early effort tackling the long time delay problem (Challenge #2). The Smith predictor introduces an excellent control structure for compensating long dead time. However, due to the modeling requirement and non-trivial tuning, the Smith predictor did not find widespread application in the process industry (Åström and Hägglund 1995).

Nonlinear process dynamics is a more challenging problem, and there are not yet mature theories or standard technologies for practical process control. Nonlinear dynamics are usually addressed in one of the three ways:

1. Limiting the operating point to a local linear region.
2. Applying a nonlinear transformation on process variables.
3. Using a nonlinear control algorithm.

Most linear control solutions assume local linearity in the process dynamics. Signal characterization (nonlinear transformation) is a popular technique for transforming nonlinear variables to linear and is widely used. However, it is mainly on a case-by-case basis. Gain scheduling (Sect. 4.1.5) has offered some help for managing nonlinear process dynamics (Challenge #6), but it is for individual PID controllers and is thus more or less *ad hoc*.

5.1.3 Advanced Process Control Technologies

Most process control technologies originate from the classical control theory. Over the past decades, the unique challenges facing the process industry, as mentioned in Sect. 5.1.1, forced the process control researchers and practitioners to select from the rich pool of control technologies for those that can meet the process control needs, or adapt them to address process control requirements. However, this effort has not been easy due to the unique challenges of the process industry. The prevalent process control techniques, from simple PID controller, complex PID schemes, to model predictive control, all originated from the process control practice, developed by process control practitioners, and propagate back to academia and then formalized and improved.

The success of the advanced process control technologies hinges on the tight coupling between the control theory and practical needs. Model predictive control (MPC) has been a significant breakthrough in process control technology (Cutler and Ramaker 1980) that was developed by process control and for process control, with a clear goal of addressing the specific process challenges (Sect. 5.1.1). The process model at the center of MPC has provided the capability to handle multivariable problems, compensate for long time delays, and apply optimization techniques to honor constraints (Challenges #1, #2, and #3), all in a single controller. It is thus a much more capable and elegant solution than complex PID schemes discussed in Chap. 4.

The process control hierarchy (Fig. 1.10) shows that process control follows a layered approach, with MPC as one layer above the baselayer control. MPC typically runs at a lower frequency than baselayer control by a difference of one to two orders of magnitude. Therefore, MPC is typically implemented on top of a baselayer control rather than replacing baselayer control.

Figure 5.1 provides an outline of the process control architecture. At the bottom is the PID-based baselayer control, which directly interacts with the process. At the baselayer level, the control objective is primarily about maintaining individual process variables such as temperature, pressure, flow, or level at their targets. MPC is built on top of the baselayer and is concerned about variables or properties such as

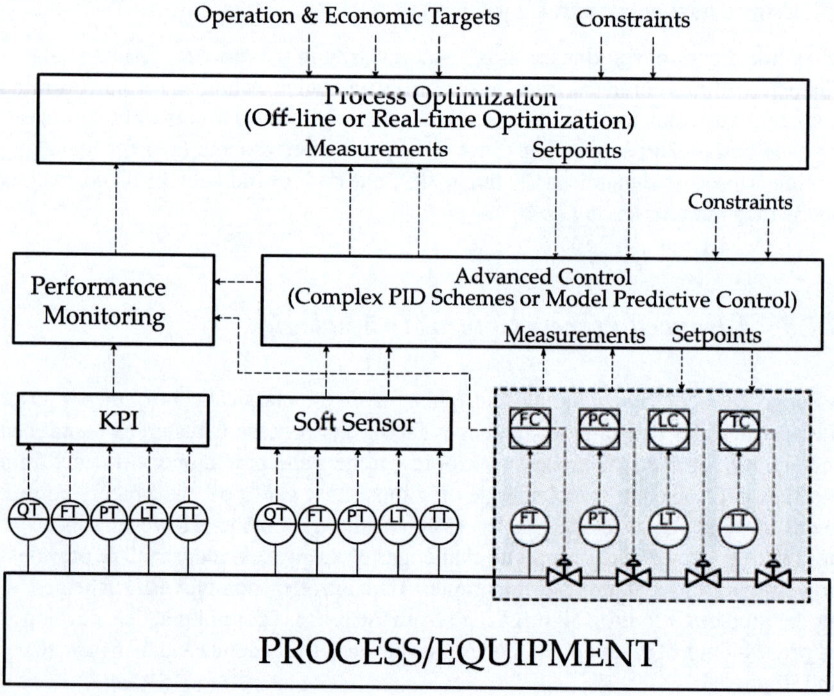

Fig. 5.1 Architecture of advanced process control

quality, capacity, and yield. These higher-level control targets can only be achieved by a coordinated effort of multiple baselayer control loops.

Many process variables cannot be directly, timely, or reliably measured using currently available process measurement technology such as online analyzers or lab tests. In this case, *inferential property estimator* (IPE) or *soft sensors* can provide the missing measurements or improve the measurement quality. There is another type of calculated value called *key performance indicators* (KPI) that provide the performance indications for various process properties, including controllers for monitoring purposes.

MPC has been widely used in the process industry and is currently the *de facto* standard for advanced process control. However, MPC is not a panacea. Nonlinear time-varying dynamics and unreliable measurements (Challenges #5 and #6) are still open challenges in both theory and practice.

5.2 Model-Based Applications

For any process, be it industrial, economic, social, or human behavioral, there must be a basic understanding of its behavior in order to control it. Usually, the more we understand it, the better chance we can control it.

A *process model* is a popular approach to represent our understanding of the process. The process model extracts the cause and effect relationships between the process input and output and abstracts the behavior with a mathematical representation (see Fig. 2.23). Advanced calculations can then be applied to the mathematical model rather than the actual process, thus opening up the opportunity for sophisticated data processing including real-time control and optimization. It is not exaggerating to say that all advanced process control applications will require a process model of a particular type.

5.2.1 *Process Analytics and Process Models*

There are many types of models available for various purposes. See Fig. 5.2 for an illustration of the different applications in advanced control and their model requirements. The model type can range from an implicit heuristic model for PID to rigorous first-principle models for an operator training simulator (OTS). Typically the more sophisticated the application is, the higher the model requirement. For example,

- PID controller tuning makes implicit assumptions on the process model. As discussed in Chap. 9, the trial-and-error tuning method requires a basic understanding of the process dynamics in terms of high/low gain, slow/fast dynamics, stable/unstable response.[1] On the other hand, model-based PID tuning relies on an

[1] A fuzzy logic model is a more structured way of formulating this implicit process model practice.

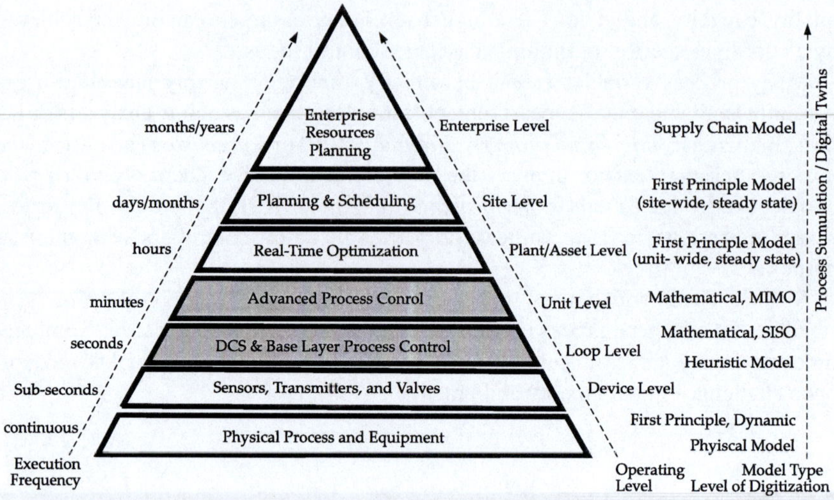

Fig. 5.2 Type of models for control applications

explicit process model to calculate the tuning parameters. The utilization of an explicit model results in more confidence in the tuning parameters and leads to more consistent performance than the empirical trial-and-error approach.

- Model predictive control (MPC) requires linear dynamic models, while real-time optimization (RTO) is based on steady-state first-principle models.
- Dynamic simulation and operator training simulators (OTS) require the highest rigorousness and fidelity in the models. Consequently, the dynamic simulation model is also the most costly and time-consuming to develop and maintain.

See Table 5.1 for a list of the most common model types. As a general guideline, the model type is application-dependent, and the best model type is the one that is fit-for-purpose. A process model merely provides a simplified approximation to the actual processes. The challenge is where and how to simplify it. "All models are wrong, but some are useful" (Box 1976).

For advanced control applications, there are two major categories of models: static and dynamic. A *static model* does not consider the time effect in the dynamics and

Table 5.1 Type of process dynamics

Linear	vs. Nonlinear
Dynamic	vs. Static
Deterministic	vs. Stochastic
Parametric	vs. Non-parametric
First-principle based	vs. Empirical
Time-domain	vs. Frequency domain
Continuous	vs. Discrete

focuses only on the steady-state behavior. The model output is related to the current model input and does not consider any effect from the past values. For example, the flow calculation from differential pressure is a simple static model:

$$F_v = C\sqrt{\Delta P/\rho} \tag{5.1}$$

where ΔP is the measured pressure difference across an orifice, and ρ is the density of the fluid. C is the parameter to be determined and specific to each orifice plate.[2]

A *dynamic model* has a memory of time. The model output is related to both the current and the past input values. For example, a simple first-order linear dynamic model is represented as

$$G(s) = \frac{K_p}{\tau s + 1} e^{-\theta s} \tag{5.2}$$

$$y(t) = K_p \left(1 - e^{-\frac{t-\theta}{\tau}}\right) u(t-\theta) \tag{5.3}$$

where u is the process input, $y(t)$ is the process output. The current output $y(t)$ is determined by $u(t-\theta)$, which is θ minutes in the past.

A process model has two critical aspects: the *model structure* and the *model parameters*. The model structure describes the type of (or lack of) cause and effect relationships among the process variables; and the model parameters representing the exact dynamic behavior for each relationship. For example, Eq. 5.1 defines a square-root-type causal relationship between the input pressure ΔP and the output flow F_v, while Eq. 5.2 shows a linear dynamic relationship between the input u and output y. The model parameters, C in Eq. 5.1 and K_p, τ, θ in Eq. 5.2 define the exact relationship.

When the past inputs and outputs are known, a dynamic model can predict the future outputs.[3] Conceptually, the future output is given by

$$\begin{cases} y(t+1) & = k_p \left(1 - e^{-\frac{t-\theta+1}{\tau}}\right) u(t-\theta+1) \\ & \vdots \\ y(t+\theta) & = k_p \left(1 - e^{-\frac{t}{\tau}}\right) u(t). \end{cases} \tag{5.4}$$

If we know the *desired* future output values, we can somehow "back-calculate" the required inputs through matrix "inversion." This back-calculation of the process inputs from the desired future outputs is the basic principle of MPC (see Fig. 2.23). Of course, the actual implementation is much more complicated.

---

2 With the recent development in data science, the static model can also be called process analytics or descriptive analytics.

3 Dynamic models can also be called predictive analytics because of the capability of prediction.

5.2.2 Data Analysis and Process Identification

As there are many types of process models, there are also many ways of obtaining the process models. Static models are typically developed based on physical knowledge and engineering insight, supplemented by trial-and-error experiments to decide the model structure (i.e., the equation type). The equation's coefficients are typically determined by experimental data via data analysis techniques such as linear regression, nonlinear equation solvers.

Dynamic models can be developed via theoretical modeling or process identification. Theoretical modeling constructs the model based on physical knowledge of the process behavior, while *process identification* builds the model from process data. The former is also called *white-box modeling*, and the latter *black-box modeling*.

The dynamic model used by MPC is typically a linear dynamic model, and the technique for obtaining the model is *process identification*. Process identification or system identification is a very active research area with a vast scope (Ljung 1999; Niu et al. 1992, 1994). In a nutshell, process identification is the methodology and techniques to select, given the process input/output data and subject to a specific criterion, the mathematical model of a given type that can best represent the actual process (or rather, the process data). The three ingredients for process identification are thus as follows:

1. Process data: The process data can be from the physical understanding of the process, a well-designed plant test, or simply from the normal plant operation. Since process identification is a data analysis technique, the data quality determines the final model quality. The goal of (black-box) process identification is "data-in, model-out." However, if the process data does not contain sufficiently rich information, process identification can often become "garbage in, garbage out."
2. Process model: The process can be represented with various models, such as linear versus nonlinear, time-variant versus time-invariant, deterministic versus stochastic, discrete-time versus continuous-time, time-domain versus frequency domain, lumped-parameters versus distributed-parameters, difference equation versus differential equation, input–output versus state-space (Corriou 2018). However, the most commonly used model type in process identification is the linear, time-invariant (LTI) difference equations.
3. Criteria: Process identification aims to find the mathematical model to "best" approximate the process input/output relationship. The criterion defines in what sense the model is the "best." The most commonly used criterion for process identification is the least-squares criterion, a quadratic function of the estimation errors.

In other words, process identification is a data-driven technique where the "black-box" process is systemically disturbed, and the response data is gathered. The dynamic cause and effect behavior is then abstracted to a mathematical model based on the input/output response data under the assumption of a specific model structure and particular criteria for the goodness-of-fit (Ljung 1999).

Linear time-invariant dynamic empirical models provide a good trade-off between complexity and fidelity and are preferred for advanced process control. Within this category, there are still many types, such as the following:

1. Finite step response model (FSR) and finite-impulse-response model (FIR).
2. Transfer function model, including first-order-plus-time-delay model (FOPTD) and higher-order models.
3. State-space models.

Except for the simplest model structure with clean data (e.g., Eq. 5.2), specialized process identification software tools are needed to analyze the data and produce the model, which consists of two major steps: *structure determination* and *parameter estimation*. The quality of the data determines the quality of the model, which in turn dictates the final control performance. Process identification is thus the vitally important step in the MPC applications.

5.2.3 First-Order-Plus-Delay-Time (FOPDT) Model

For process control purposes, the first-order-plus-delay-time (FOPDT) model plays a vital role because of its simplicity. Most real-world processes can be approximated with this simple FOPTD model for most control applications. The FOPTD model uses three parameters to describe *when*, *how much*, and *how fast* the output variable $y(t)$ responds to a change in the input variable $u(t)$:

1. Process gain K_p: By moving the input variable by one unit, *how much* change (direction and magnitude) will be eventually caused on the output variable.
2. Time delay or dead time θ: After a change is made to the input (the manipulated variable, or MV), when will the output variable (the controlled variable, or CV) start to move?
3. Time constant or lag time τ: The amount of time that takes from the moment that the output variable starts to respond until the output value reaches 63.2% of the new steady-state value. The time constant describes how fast the process dynamics are.

The transfer functions for the first-order-plus-dead-time (FOPDT) model, in both continuous-time and discrete-time, are given in as follows:

$$G(s) = \frac{K_p}{\tau s + 1} e^{-\theta s} \quad \mapsto \quad y(t) = K_p \left(1 - e^{-\frac{t-\theta}{\tau}}\right) u(t - \theta) \tag{5.5}$$

$$G(z^{-1}) = \frac{b z^{-d-1}}{1 - a z^{-1}} \quad \mapsto \quad y(k) = a\, y(k-1) + b\, u(k - d - 1) \tag{5.6}$$

$$\text{with } \; a = e^{-T_s/\tau}, \; b = K_p/\tau, \; d = \theta/T_s \tag{5.7}$$

where k denotes the current sampling time, y is process output, u is the process input, and θ is the time delay. a, b, and d are the model parameters to be determined. Note that θ is the time delay in continuous time, and $d = \lfloor \theta / T_s \rfloor$ is the same time delay in discrete time, with T_s being the sampling time.

The FOPTD model can be conveniently and intuitively related to the step response. When the process noise level is low, the three model parameters K_p, τ, θ can often be calculated from the step response with sufficient accuracy, as shown in Fig. 5.3. For example, the process gain can be readily calculated as the change in output (Δy) over the change in input (Δu) $K_P = \Delta y / \Delta u$, the dead time is given by θ, and the time constant by τ. The three parameters in continuous-time are related to discrete-time by Eq. 5.7. Modeling from a clean step response is the most commonly used method for obtaining a FOPDT model in practice. Note that the gain is calculated based on changes in input and output and does not concern the steady-state values. In PID control, the control output is calculated as the change in control action rather than the actual value (see Sect. 3.1.6).

A second method is via solving a set of linear simultaneous equations. From Eq. 5.6, there are three parameters to determine a, b and d. The time delay d is nonlinear to the output $y(t)$, so it must be determined first either visually or via trial-and-error. Once the time delay d is known or assumed, the parameters a and b can be calculated by solving the following equations:

$$\begin{cases} a\,y(k-1) + b\,u(k-d-1) & = y(k) \\ a\,y(k-2) + b\,u(k-d-2) & = y(k-1) \\ a\,y(k-3) + b\,u(k-d-3) & = y(k-2) \\ & \vdots \\ a\,y(k-n) + b\,u(k-d-n) & = y(k-n). \end{cases} \tag{5.8}$$

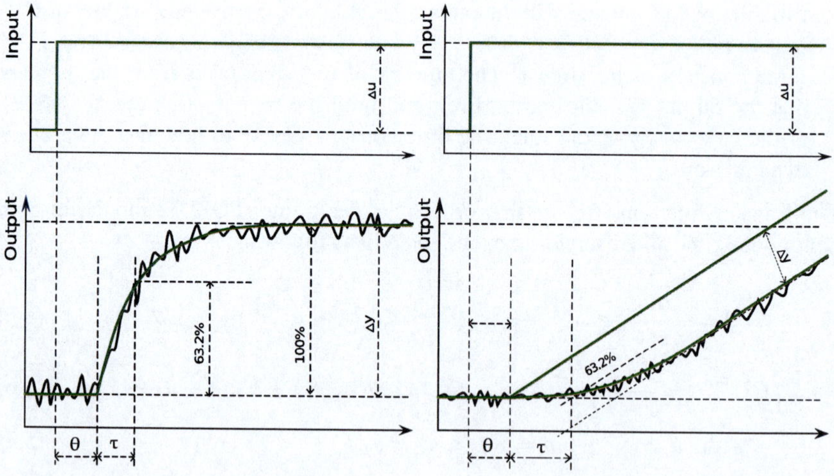

Fig. 5.3 Step response and step response model. Left: stable process, right: ramp process

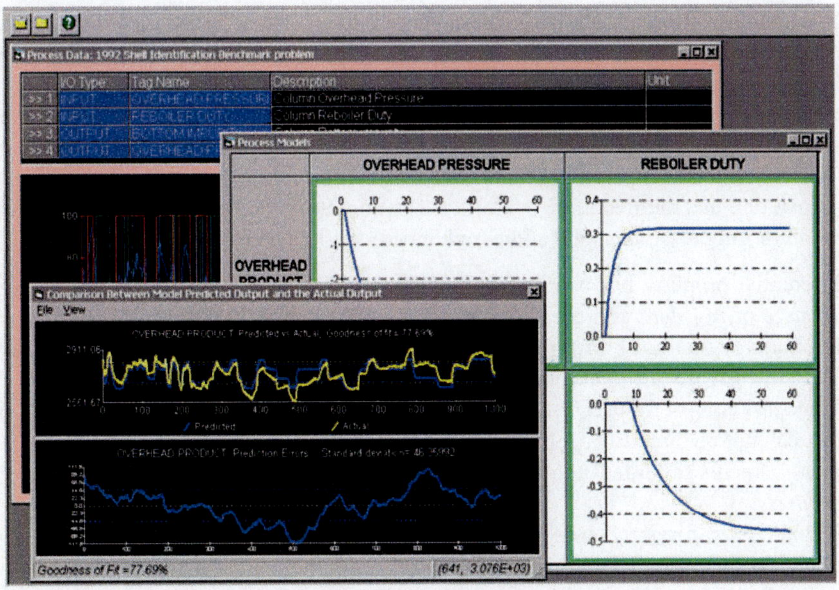

Fig. 5.4 An example of process identification (modeling) software tool

Equation 5.8 is a set of overdetermined simultaneous equations with two unknowns, a and b. There are many tools, even spreadsheet software, for solving equations.

The generic approach for modeling is to use a dedicated process identification tool. A process identification software tool can work on a wide variety of models, including the first-order-plus-time-delay models and the finite-impulse-response model (FIR) in Eq. 5.18. Figure 5.4 is an example of a generic process identification software.

For completeness, a first-order integrating model is typically used as follows:

$$G(s) = \frac{K_p}{s\,(\tau s + 1)} e^{-d s} \quad \mapsto \quad y(t) = K_p \left[t - \left(1 - e^{-\frac{t-d}{\tau}} \right) \right] u(t-d). \qquad (5.9)$$

5.3 Inferential Property Estimation

Inferential property is one of the most successful model-based applications and has been used for multiple purposes (Blevins et al. 2003; Marlin 2015), including the following:

1. Improving process measurements: Poor process measurement is Challenge #5 in Sect. 5.1.1. Many desired measurements are either not available or unreliable. Using inferential property based on some more reliable measurements is one practical and effective approach to improve.

2. Nonlinear transformations: Most advanced control techniques, including MPC, are based on linear models. To directly control nonlinear process dynamics would need an entirely different control technology. However, using nonlinear inferential property can isolate the nonlinear part of the process dynamics and results in a more straightforward and cleaner solution. For example, controlling a pressure-compensated temperature (PCT) is better than controlling both pressure and temperature in a multivariable controller. The nonlinear correlation between temperature and pressure can be addressed outside the linear controller.

Inferential property has been more popularly known as a *soft sensor*. The recent upsurge of *big data* and the resurgence of *artificial intelligence* bring some new terminology such as *descriptive analytics*, *predictive analytics*, and *prescriptive analytics*, which are different forms of inferential properties that have been in existence in process control for many years. The fundamental idea is that for process variables that can not be directly or reliably measured, we can use other commonly available measurements to "infer" their values. The inferred value is thus called *inferential property*.

Inferential property estimate (IPE) can be used for open-loop monitoring or closed-loop control. Monitoring has a lower requirement on reliability than control, and thus has found more widespread use for IPE. Many recent technologies such as performance monitoring, abnormal situation management, equipment health monitoring, fault detection, data validation are wholly or partly dependent on inferential property estimates (process analytics).

Inferential property can be broadly categorized into *static* and *dynamic* (see Fig. 5.5). The static inferential property is usually based on *spatial* information where several variables are used to infer another, while the dynamic inferential property uses both *spatial* and *causal* information. For the causal information, both *current* and *past* values are used to predict future values. The final form and technique of the IPE are mainly dependent on the peculiarities of the specific application.

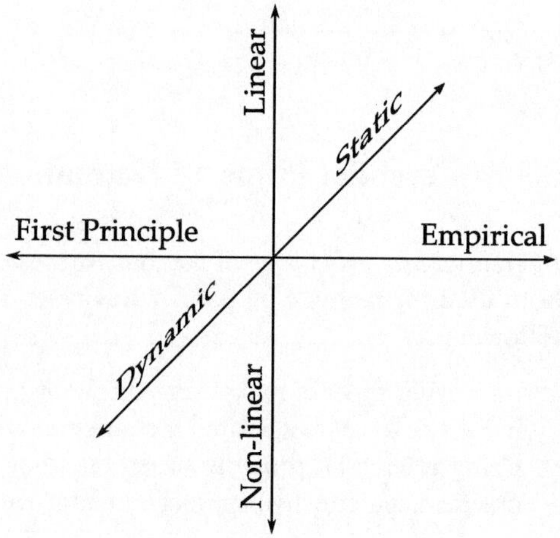

Fig. 5.5 Common model types

Models based on first principles relate vital physical features such as pressure and temperature to the process dynamic behaviors. They are thus valuable in establishing relationships between the process variables. Unfortunately, this approach has severe limitations due to the complexities of physical processes. For example, a typical distillation column may need several hundreds of differential equations and an overwhelming number of parameters to characterize the thermodynamic relationships. It is thus a significant effort to formulate the equations and solve them (Marlin 2015). For many applications, it is not worth the effort and complexity to go with first-principle models. In practical applications, many decisions are a result of a trade-off between *desire* and *practicality*.

5.3.1 Static Inferential Property Estimation

Static IPE is typically based on a mathematical formula or *static model*. Most of the time, the model is a nonlinear equation derived from physical understanding. For this reason, the IPE can be highly flexible and powerful. This approach is also called the *analytical* approach.

Static IPE has been in use for a long time under different names. Many examples of static IPE exist even in this book. Below is a list of some IPE calculations in various chapters:

1. Flow calculation (Sect. 10.3): Flow is never directly measurable. Instead, the flow rate is inferred from other variables such as differential pressure, fluid velocity, or density. See Table 2.3 for the measurement principles and Table 10.3 for the inferential formulas. In fact, even the flow velocity is inferred from other directly measurable variables such as frequency, elapsed time, or voltage. Most process variables are not the direct measurement from the sensors. The process variable is inferred from sensor output through a correlation governed by physical principles (Marlin 2015).
2. Pressure-compensated temperature (PCT) (Sect. 10.4): Pressure and temperature are highly correlated for gas and steam (Process challenge #1). The correlation can be embedded into one new variable called *pressure-compensated temperature* (PCT) (Eq. 10.20) and be used as one controlled variable instead of two. This PCT is a typical static IPE.
3. Surge margin calculation (Sect. 11.3): The compressor surge margin can not be measured and is typically inferred from pressure and flow (see Eqs. 11.36 and 11.37).
4. Signal characterization (Sect. 10.1): Signal characterization is a common approach to transform a signal, typically from nonlinear behavior to linear. Signal characterization is supported by most control systems, typically as a look-up table. With the table look-up approach, the signal characterization can perform a generic free-form transformation.

5. Key Performance Indicator (KPI): Most KPIs are inferred from other measurements or values.

Although in different shapes and forms, the static inferential properties have the following characteristics in common:

- Based on engineering insight: The static inferential property estimators are typically backed by deep physical insight and engineering principles.
- Flexible in format: The estimator is highly flexible in format and can use any mathematical formulas supported by the target system.
- Low cost: An inferential property estimator is merely a calculation in the control system.
- Reliable: The inferential property estimator for basic measurements such as pressure, temperature, flow, and level are regarded as direct measurements because they are sufficiently accurate and reliable.

We here provide an example showing how a static IPE is developed from first-principle engineering knowledge.

Example 5.1 Flow meter with an orifice plate. An orifice plate is a thin plate with a hole in the middle (Fig. 5.6). The orifice is placed in a pipe in which fluid flows.

When the fluid reaches the orifice plate, the fluid is forced to converge to go through the smaller hole; the point of maximum convergence occurs shortly downstream of the physical orifice at the so-called *vena contracta* point. As it does so, the velocity and the pressure change. Beyond the vena contracta, the fluid expands, and the velocity and pressure change once again. By measuring the difference in fluid pressure between the normal pipe section and at the vena contracta, the volumetric and mass flow rates can be inferred.

We can use Bernoulli's equation to calculate the flow rate. Between point ① and ②, we have

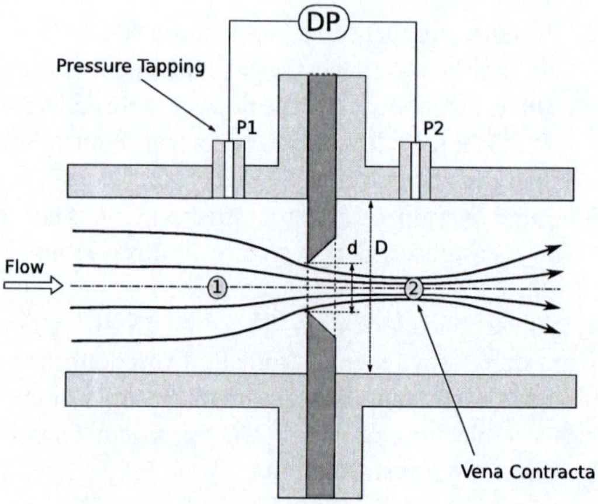

Fig. 5.6 An orifice plate for DP measurement

$$P_1 + \frac{1}{2}\rho_1 v_1^2 = P_2 + \frac{1}{2}\rho_2 v_2^2. \tag{5.10}$$

Assuming the change of fluid density across the orifice is negligible ($\rho_1 = \rho_2$), and by continuity of mass, $A_1 v_1 = A v_2 = F_v$, where A_1 and A_2 are the cross area of the orifice and the vena contracta, and F_v is the volumetric flow, we then have

$$P_1 - P_2 = \frac{1}{2}\rho\left(v_2^2 - v_1^2\right) = \frac{1}{2}\rho\left(\frac{F_v^2}{A_2^2} - \frac{F_v^2}{A_1^2}\right) = \frac{1}{2}\rho F_v^2\left(\frac{1}{A_2^2} - \frac{1}{A_1^2}\right). \tag{5.11}$$

Define $\Delta P = P_1 - P_2$,

$$F_v = \frac{A_2}{\sqrt{1-(A_2/A_1)^2}}\sqrt{2\frac{\Delta P}{\rho}} = \frac{\pi}{4}\frac{d^2}{\sqrt{1-\beta^4}}\sqrt{\frac{2\Delta P}{\rho}} = C_v\sqrt{\frac{\Delta P}{\rho}} \tag{5.12}$$

where d is the diameter of the orifice and D is the diameter of pipe, and with

$$A_1 = \frac{1}{4}\pi D^2, \quad A_2 = \frac{1}{4}\pi d^2, \quad \frac{A_2}{A_1} = \left(\frac{d}{D}\right)^2 = \beta^2, \quad \beta = \frac{d}{D}.$$

C_v is called the *flow coefficient*. Once an orifice is built, the flow coefficient remains unchanged. The mass flow rate F_m is given by

$$F_m = \rho F_v = C\sqrt{\Delta P \cdot \rho}. \tag{5.13}$$

The generic and precise equation for mass flow rate measurement is given by ISO5167 (2003) standard[4] as

$$F_m = \frac{C_d}{\sqrt{1-\beta^4}}\varepsilon\frac{\pi}{4}d^2\sqrt{2\Delta P\rho} \tag{5.14}$$

where C_d is the discharge coefficient, ε is the expansion factor, d is the bore diameter, and D is the pipe diameter. They are flow constants that are experimentally determined after the orifice is built.

5.3.2 *Dynamics Inferential Property Estimation*

Inferential property has become popular with MPC. With MPC, the attention has shifted from individual process variable control (flow, pressure, temperature, and

[4] Note that ISO 5167 (all parts) is applicable only to flow that remains subsonic throughout the measuring section and where the fluid can be considered as single-phase. It does not apply to the measurement of pulsating flow.

level) to product quality control. However, the product quality is not as readily and reliably available as the conventional process variables. Many properties are not measurable, and those that can be measured via online analyzers and lab sampling suffer from the problems of low availability and poor reliability, which are severe challenges for online control:

1. Availability: The process variable is not directly measurable. Many process variables that need to be controlled are simply not measurable for technical or economic reasons. In this case, it is often possible to use existing measurements to infer the value of the desired process variable.
2. Reliability: The process measurement is not sufficiently reliable for real-time control. Conventional process measurements such as flow, pressure, temperature, and level have been widely used for closed-loop control, although they occasionally suffer from measurement failures. Quality measurements through online analyzers are notorious for poor reliability and low up-time. Frequent maintenance such as calibration is required but often not received.
3. Speed of response: The online process measurements are too slow for real-time control. Many online analyzers, such as gas chromatography (GC), have a slow response time producing the result. Lab sampling is even less responsive. The long time delay is a significant challenge for feedback control.

With the advance in technology, a modern operating plant usually has abundant and redundant process measurements. It is thus possible to utilize the more reliable and more responsive measurements to infer other process variables.

Static inferential property estimation typically uses static, nonlinear, first-principle models based on engineering insight of the process variables. Dynamic inferential property estimation is typically data-driven and limited to linear dynamic empirical models. For example, the overhead composition in a distillation column can be related to the temperature and pressure via a linear dynamic empirical model. On the other hand, for a high purity distillation tower, the impurity level is typically a nonlinear function of the inputs (e.g., temperature). A linear model is insufficient to support the process operation within a wide range of variations in purity. A more sophisticated nonlinear model such as a neural network or a linear model with a nonlinear transformation may become necessary.

Building a dynamic inferential property estimator is relatively straightforward among all model-based applications. Although data-driven, a good understanding of the physical property and cause and effect relationship is essential. The data used for parameter estimation can be historical data if sufficient excitation is present to characterize the dynamic behavior. Otherwise, plant tests should be conducted to obtain high-quality data.

Example 5.2 Inferential property estimator for a stabilizer column. For the stabilizer column in Fig. 2.24, the total vapor pressure (TVP) is a control target, with an online analyzer installed downstream of the column. However, the analyzer readings often were not available or not credible due to calibration or maintenance issues.

A PID controller was configured to maintain the TVP value on target. However, the challenge was that the control target was the TVP value inside the column, but the online analyzer was located a long distance downstream. The transport delay and the analyzer response time added up to 40 minutes, which was impossible for a feedback PID controller to handle. The analyzer's long time delay and low up-time had prevented the quality controller (`AC-101`) from running in closed-loop ever since the plant startup.

An inferred TVP value from the pressure and temperature was developed later and used as the `PV` for the `AC-101` controller. The inferential property was estimated from two process variables, MV1 (tray 9 temperature) and MV2 (column pressure), with the transfer function as follows:

$$\widehat{\mathrm{PV}}(s) = \frac{K_{p,1}\, e^{-\theta_1 s}}{\tau_1 s + 1} \mathrm{MV}_1(s) + \frac{K_{p,2}\, e^{-\theta_2 s}}{\tau_2 s + 2} \mathrm{MV}_2(s) \tag{5.15}$$

where $\widehat{\mathrm{PV}}$ was the estimated TVP value. MV1 and MV2 are the two process variables used for the estimation. θ_1 and θ_2 were the time delays between the two inputs and the measured property value. Keeping in mind the two issues we try to address with online analyzers or lab samples are as follows:

1. Long time delay: The long delay time was a significant challenge for control. In this example, it took about 40 minutes for the TVP analyzer to produce the response, partly because of the long response time of the analyzer itself but mainly because the analyzer was located a long distance downstream of the stabilizer column for design reasons.
2. Low availability: The online analyzer had chronic problems with calibration and maintenance. More than half of the time, the results were either not available or could not be trusted.

One of the goals for the inferential property estimator was to "remove" the time delay so that the dynamics between the TVP and the temperature/pressure become more responsive. The simple approach was to remove the common delay $\theta_0 = \min(\theta_1, \theta_2) \approx 36$ min from the transfer function in Eq. 5.15:

$$\widehat{\mathrm{TVP}}(s) = \frac{K_{p,1}\, e^{-(\theta_1-\theta_0) s}}{\tau_1 s + 1} \mathrm{MV}_1(s) + \frac{K_{p,2}\, e^{-(\theta_2-\theta_0) s}}{\tau_2 s + 2} \mathrm{MV}_2(s) + \widehat{\mathrm{offset}}. \tag{5.16}$$

The IPE value with the common delay removed was then used as the controlled variable for real-time control, either PID or MPC.

The term $\widehat{\mathrm{offset}}$ in Eq. 5.16 is the static error that may develop between the predicted value and the analyzer reading over time. The actual value from the analyzer or lab sampling is fed back to the estimator to update the $\widehat{\mathrm{offset}}$. The inclusion of feedback is a practical yet effective approach to compensate for the discrepancy between the process model and the actual plant. The main reasons for this model-plant mismatch include the following:

1. The process model is always over-simplified, especially in the presence of non-linearity.
2. There are always unmeasured disturbances that can not be captured.
3. The process dynamics may drift away from the region defined by the data that the model is derived from.

The vital step in taking the feedback from online analyzers is the timestamp alignment. That is, the offset is a deviation between the predicted value and the actual value. Since the predicted value has removed the common delay θ_0 from the model, to compare the predicted value with the actual value, the predicted value needs to add the delay back to align with the analyzer value in time. See Fig. 5.7 for an illustration of IPE for feedback control, and compare it with standard feedback control in Fig. 2.2.

A sample implementation of the inferential property estimator for TVP is given in Fig. 5.8, where the gain, lag, and delay are primitive function blocks in standard DCS. Depending on the target system for implementation, there may be more advanced function blocks to lump the calculations together. For example, one function block may be able to perform all three functions.

Figure 5.9 shows the inferred and the actual TVP values based on historical data, with feedback correction based on the online TVP analyzer value once every hour.

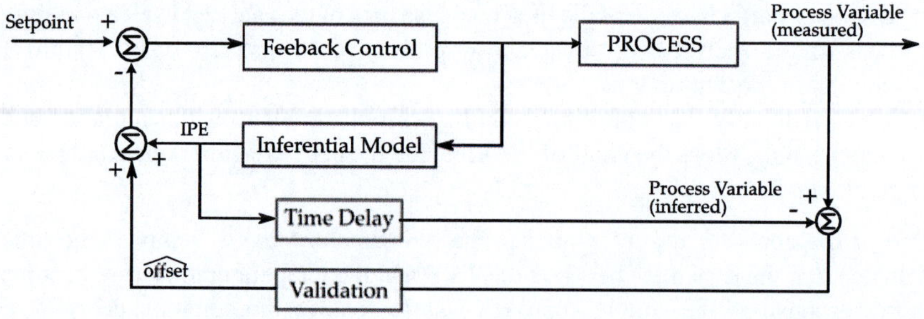

Fig. 5.7 IPE for feedback control

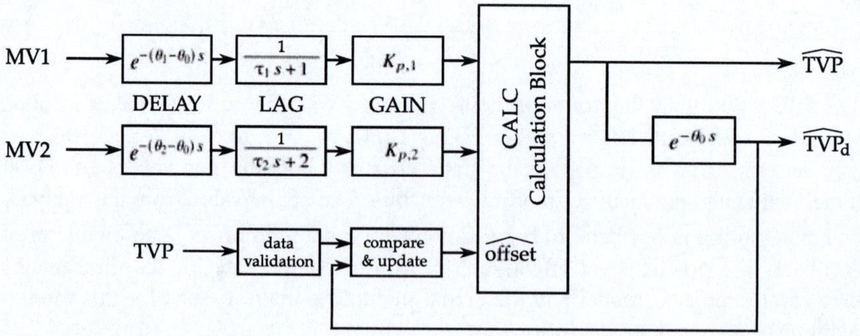

Fig. 5.8 A sample implementation of IPE

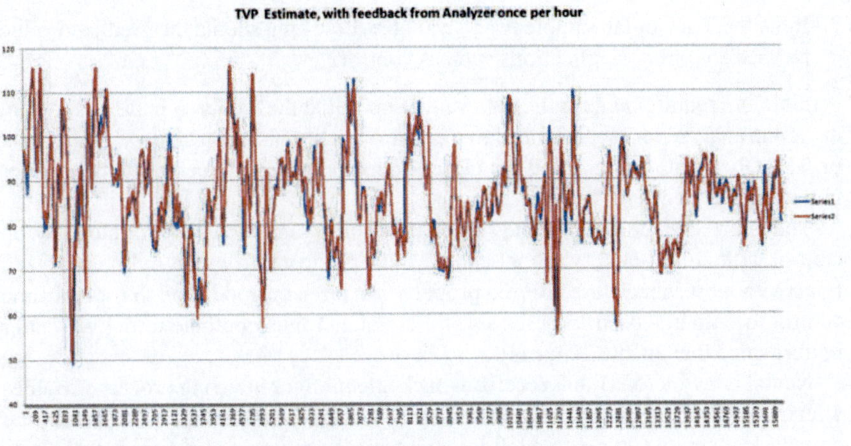

Fig. 5.9 Inferred TVP values, with hourly feedback from analyzer

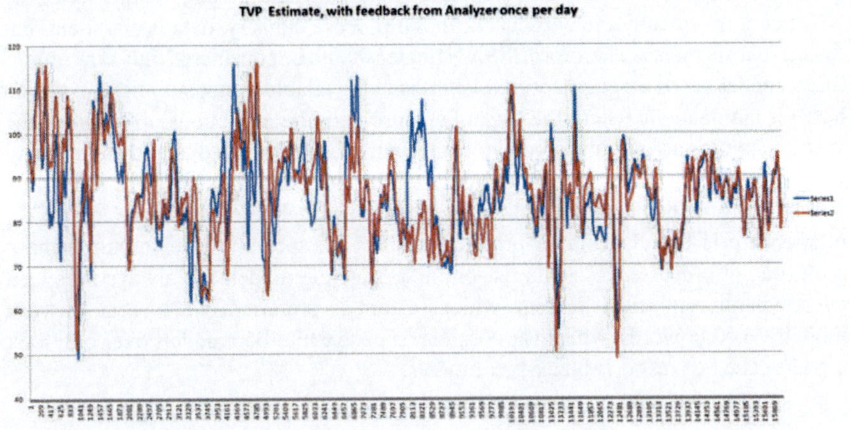

Fig. 5.10 Inferred TVP values, with daily feedback from analyzer

Figure 5.10 shows the inferred TVP values with the same IPE but with feedback from the TVP analyzer once every day. The IPE provides a decent estimate with analyzer feedback once an hour. Even if the feedback input is reduced to once a day, the IPE still offers valuable information for monitoring.

An open challenge for the inclusion of feedback in the IPE is the online quality assessment on the feedback:

1. Is the analyzer feedback credible and reliable for correcting the predicted value?
2. If the feedback is from lab samples, how often is the lap sampling, and how to reliably match the timestamp of the lab result with the actual time the sample is taken?

3. If the feedback or lab sample is not available, how long should the predicted value be used for online applications such as control?

A minimum requirement for the data validation is that the feedback data, either from an online analyzer or lab test, need to provide a quality indicator as *good*, *uncertain*, or *bad*. Good data can be used for closed-loop control and uncertain data only for monitoring.

The inferred value can provide sufficient accuracy and reliability for closed-loop control with a high-quality model and periodic feedback from the online analyzer. Even without feedback, the inferred property can often be good enough for real-time control to remain closed-loop for several hours and thus continue to deliver better performance than manual control.

Neural network (NN) has received much attention for inferring process variables with a significant nonlinear relationship with the independent inputs. However, the drawback of neural network models is that it typically requires a lot more "clean" data to train the network than parsimony models do. For IPE applications where the "good" measurement data from the online analyzer or lab are insufficient, the quality of the NN model is typically the bottleneck. In the process industry, data is abundant, but "good" data is scarce. Therefore, the NN model should be considered only if a simpler linear model based on engineering insight is not available or not sufficient. It works best for applications where the dynamic cause and effect relationship between the specific inputs and outputs is known from fundamental knowledge and engineering insight, with only the exact form of nonlinearity and the parameters to be estimated.

Modeling without understanding the physical principle of the cause and effect relationship is usually counter-productive. This is especially true for multi-input multi-output processes. For this reason, first-principle models are always preferred over dynamic empirical models, which in turn is preferred over neural network models. In other words, whenever possible, choose white-box models over gray-box models. The last resort is black-box models.

5.4 Model Predictive Control—The Concept

Model predictive control (MPC) mainly refers to dynamic matrix control (DMC) (Cutler and Ramaker 1980) and generalized predictive control (GPC) (Clarke et al. 1987). MPC is a vast topic deserving separate books for systematic coverage (Corriou 2018; Garcia et al. 1989; Lee 2011; Morari and Lee 1999; Qin and Badgwell 2003; Rawlings et al. 2017). Here we will skip the theoretical derivations and focus only on the fundamental concept and the best practice work process in implementation and applications.

5.4.1 MPC as Multivariable Control

Model predictive control is inherently a multivariable control algorithm. It is capable of addressing process challenge #1 in Sect. 5.1.1 elegantly as compared to a complex PID control scheme in Sect. 4.1.2. Unlike complex PID schemes where the multivariable interaction is addressed explicitly by separate function blocks (thus a white-box approach), MPC relies on multi-input multi-output (MIMO) models to implicitly handle the multivariable interactions (a black-box approach). The efficient and reliable handling of multivariable interactions empowers APC to control a large process area, such as an entire column or unit, with a single controller.

Figure 5.11 shows a generic distillation column, along with the key process variables. A distillation column is the most widely used unit operation for separating mixtures. It typically consists of a tray section, a condenser, a reflux drum, and a reboiler. The section above the feed tray is called the rectifying section (or enriching section), and the part below is called the stripping section. In a distillation column, the overhead temperatures and pressures are highly correlated, and all have a substantial impact on the composition of the top product. The overhead conditions also heavily interact with the bottom, affecting the bottom product quality. The feed temperature and flow rate are significant disturbances to the entire column operation.

The interaction between the process inputs and outputs can be illustrated in Fig. 5.12, where G_{12} designates the transfer function between input #1 and output #2. As the number of inputs and outputs increases, the number of transfer functions can grow rapidly.

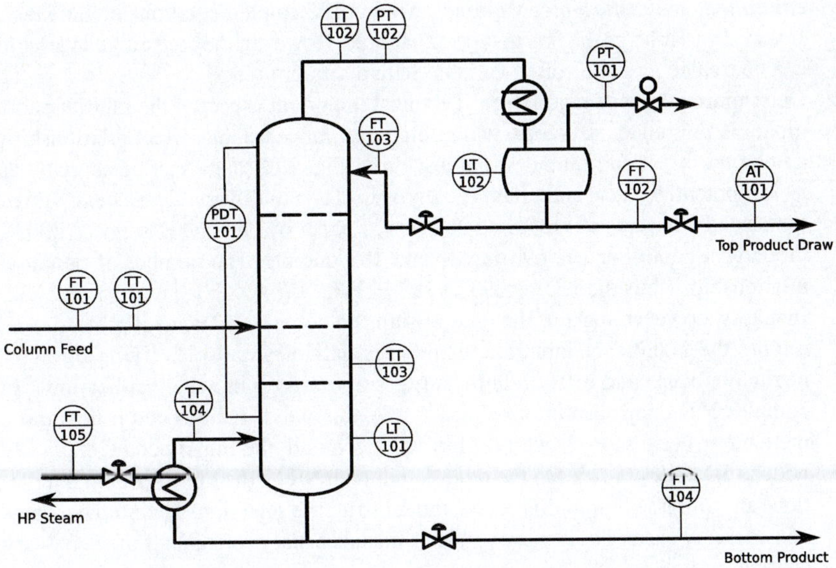

Fig. 5.11 A generic fractionator column with typical measurements

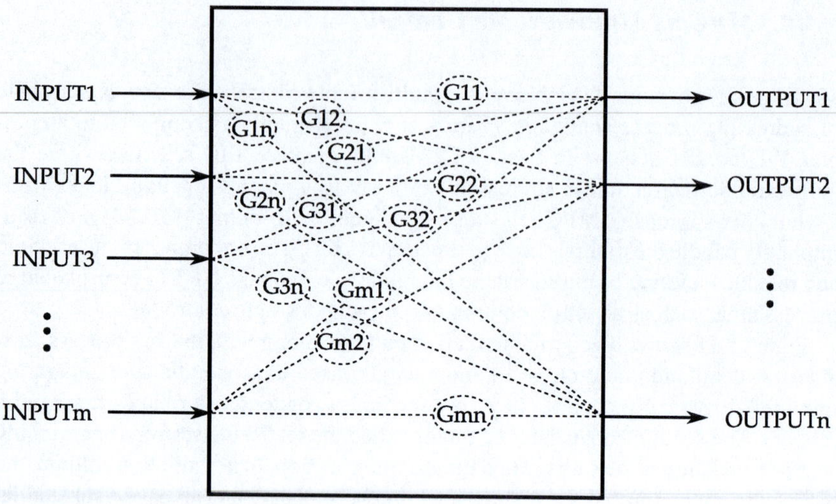

Fig. 5.12 Interactions in a MIMO process

There are several critical aspects in the multivariable representations as follows:

1. Selecting the process variables: Selection of the process inputs (including manipulated variables and disturbance variables) and the process outputs (the controlled variables) is the first step and often goes back and forth before finalization. MPC typically includes three types of variables in the controller: *manipulated*, *controlled*, and *disturbance* variables. Some MPC implementations include *constrained* variables as the fourth type of variable. However, the constrained variable can be treated as a controlled variable with a setpoint range.
2. Determining the model structure: The most important aspect of the multivariable model is the model structure, which includes the cause and effect relationships among the input and output variables. From Fig. 5.12, there can be as many as $m \times n$ potential cause and effect relationships for a multivariable process with m inputs and n outputs. As reported in a survey paper by Qin and Badgwell (2003), the largest controller has 693 inputs and 283 outputs. The number of potential relationships (thus models) is $693 \times 283 = 196{,}119$!
In reality, however, most of the relationships do not exist or are negligible. Determining the significant input/output process variables and identifying the non-negligible cause and effect relationships is a vital step in MPC applications. In complex PID control solutions (Sect. 4.1.2), this step is tedious and is thus more meticulous because each relationship will eventually be implemented explicitly with a PID control loop. On the other hand, with MPC, the multivariable interactions are addressed implicitly with a model structure selection. This step becomes considerably easier but is more prone to mistakes and oversights if the necessary cautions are not taken.

3. Estimating the model parameters: The determination of the model structure is still an art to a certain extent. After the model structure is decided, the parameter estimation is relatively more straightforward. There are many mature techniques for parameter estimation, so it is more science than art. Most commercial software tools are capable of producing the model parameters once the model structure is given.

In practical applications, the multivariable interactions are usually represented in an intuitive and convenient table format called the cause and effect table (CET), as illustrated in Table 5.2.

The most challenging step in MPC is to formulate operating requirements into a *control strategy*. The knowledge and skills required for controller variable selection and model structure determination are the same for complex PID control discussed in Sect. 4.1 and MPC. An in-depth understanding of the operating objectives and the process dynamics is essential, and a systematic process analysis is vital to select the appropriate model structure to and ensure that the critical process characteristics are captured in the control design. Chapter 6 is dedicated to a structured approach in process analysis applicable to both complex PID control and MPC.

Inferential property is often an integral part of MPC. Once the inferential property is built, it can be used the same way as with other directly measurable process variables. Many controlled variables are inferential properties instead of direct measurements. It is vital to have the inferential properties developed and commissioned before MPC step testing starts.

As an example, Fig. 5.13 shows all the models produced by the process identification tool as a full matrix, with both the step response model and the parametric model superimposed on each other. Unfortunately, many models are not of high qual-

Table 5.2 Multivariable interactions

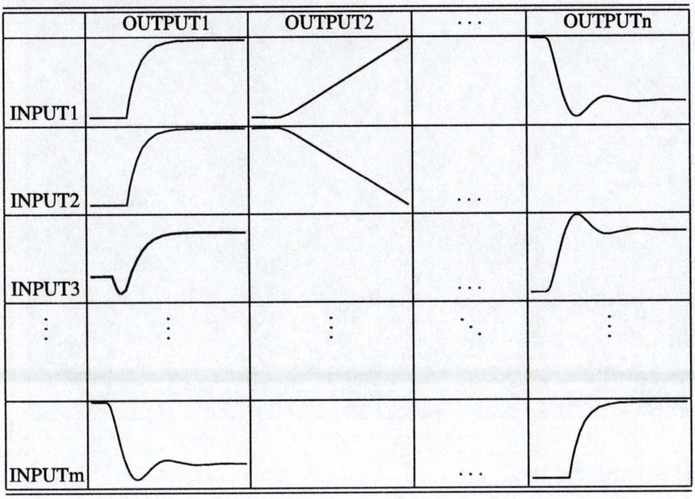

	OUTPUT1	OUTPUT2	...	OUTPUTn
INPUT1				
INPUT2			...	
INPUT3			...	
⋮	⋮	⋮	⋱	⋮
INPUTm			...	

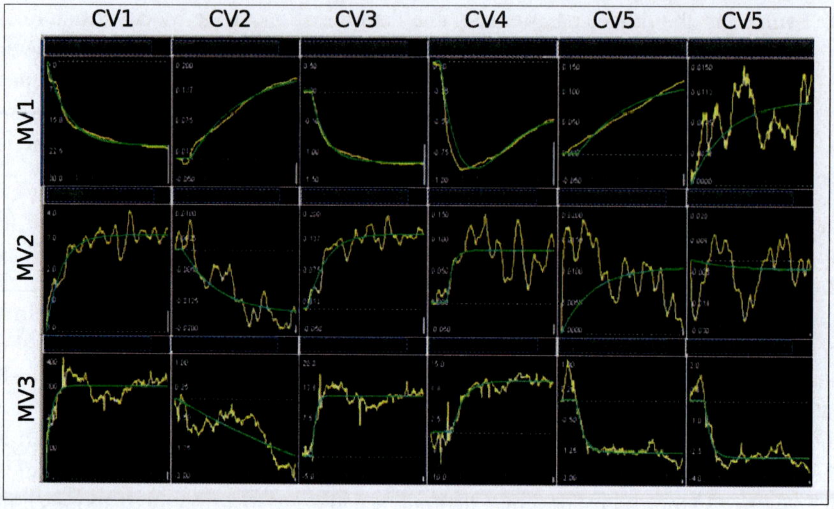

Fig. 5.13 Raw model matrix for MPC

ity (low confidence). The decision for the final model matrix is somewhat subjective, depending on many factors such as the understanding of the process dynamics and the trust level on the plant test data. The final model matrix may be like Fig. 5.14 or different depending on the preference and competency of the engineer designing the MPC controller.

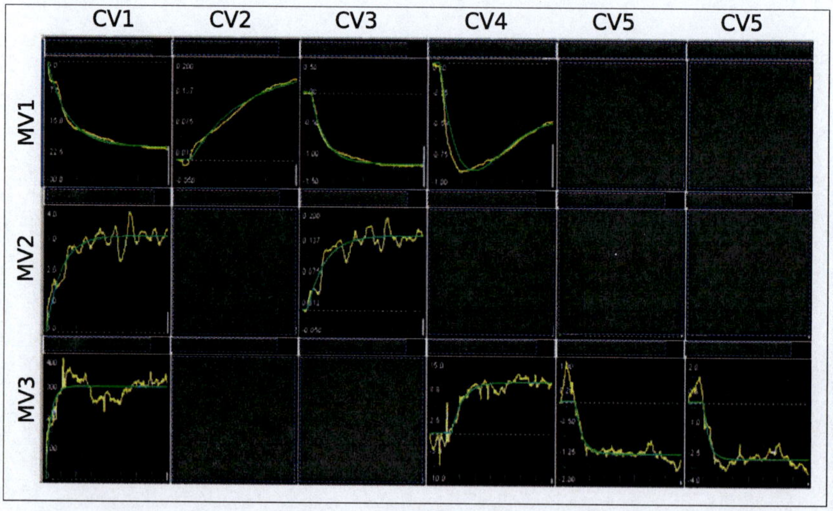

Fig. 5.14 Final model matrix for MPC

One of the weaknesses of most MPC software is that the calculation does not differentiate between good and bad models. All models in the controller are given equal trust. Discriminative treatment of the models based on the model quality is an open challenge for improving MPC. One approach is to separate the process models into prediction models and control models. The control model is a subset of prediction models that represent the key cause and effect relationships with higher quality. The implicit "matrix inversion" inside the MPC only involves the control models.

All commercial software has the built-in feature of model validation or model quality assessment (the goodness-of-fit), which is an essential checkpoint on the model quality. No corners should be cut related to model quality.

5.4.2 MPC as Predictive Control

In addition to decoupling multivariable interactions, the dynamic process models can predict future process behavior (Rossitier 2003).

Unlike purely reactive feedback control, MPC can "predict" the future process output with the process model. The dynamic nature of the model means that the current output is a function of the past inputs and outputs. The future outputs can be predicted based on the present and past process inputs with the same model. With this predictive capability, APC can predict where the controlled variables are moving to, then back-calculates *which*, *when*, and *how much* to move the manipulated variables to correct the undesired deviations from control targets. As a result, the MPC control actions are calculated based not only on where the process is currently running but also on where it is predicted to be moving to in the next few minutes or even hours. This predictive capability, a vast improvement over PID controllers, makes the control performance much smoother than purely feedback-based control. A reactive control compared to a predictive control is like a blind person walking with a stick (one-step-ahead prediction) compared to a normal person walking in daylight (multi-step-ahead long-range prediction).

With Eq. 5.2, the future prediction of the output y are given by

$$\begin{cases} b\,u(k-d-1) - a\,y(k-1) & = y(k) \\ b\,u(k-d) \quad - a\,y(k) & = y(k+1) \\ b\,u(k-d+1) - a\,y(k+1) & = y(k+2) \\ b\,u(k-d+2) - a\,y(k+2) & = y(k+3) \\ & \vdots \\ b\,u(k-d+m-1) - a\,y(k+m-1) & = y(k+m). \end{cases} \tag{5.17}$$

The corresponding finite-impulse-response (FIR) model with n parameters is given by

$$\begin{cases} b_n\, u(k-d-n) \quad +\cdots+ b_2\, u(k-d-2) + b_1\, u(k-d-1) & = y(k) \\ b_n\, u(k-d-n+1) + \cdots + b_2\, u(k-d-1) + b_1\, u(k-d) & = y(k+1) \\ b_n\, u(k-d-n+2) + \cdots + b_2\, u(k-d-2) + b_1\, u(k-d+1) & = y(k+2) \\ & \vdots \\ b_n\, u(k-d-n+m) + \cdots + b_2\, u(k-d+m-2) + b_1\, u(k-d+m-1) & = y(k+m) \end{cases} \tag{5.18}$$

where the three parameters a, b, d in the first-order-plus-delay-time (FOPDT) are expanded to $n+1$ parameters $b_n, \ldots, b_2, b_1$ and d in the finite-impulse-response model.

A crucial design parameter (sometimes exposed as a tuning parameter) is the so-called *prediction horizon*. The prediction defines how far MPC predicts into the future on the process response. The farther it predicts, the smoother (or more sluggish) the control performance is. An analogy is driving at night versus driving in daylight. The former has a short prediction horizon, while the latter has a longer one. Night driving must be slower since the car can easily overrun the range of the headlights at a higher speed.

5.4.3 MPC as Constrained Control

One of the distinctive MPC advantages is the ability to handle multivariable interactions and constraints in a unified manner, thanks to the explicit process model. The internal mathematical model opens the opportunity to utilize optimization technologies such as quadratic programming (QP) to produce control actions that meet operating limits and constraints. There are several types of constraints as follows:

1. Constraints on manipulated variables: These are the hard limits on the manipulated variables, such as valve saturation limits.
2. Constraints on the move size of the manipulated variables: These are the limits on the rate of change in the manipulated variables for stability.
3. Constraints on controlled variables: These are the limits imposed on controlled variables to avoid overshoots or minimize deviations from setpoints.

An MPC algorithm without considering constraints is a straightforward linear algebra problem of solving matrix equations. However, MPC needs an optimization engine to solve the equations with constraints. The *quadratic programming* (QP) algorithm is a popular choice.

Many a time, it is not feasible to meet all the constraints specified. For this reason, the constraints are further classified as *hard* constraints and *soft* constraints. Hard constraints, such as valve limits, cannot be violated, while soft constraints are allowed to be broken with a penalty. The overall objectives are thus two folds:

1. Fast response for setpoint tracking and disturbance rejection.
2. Minimum moves (typically valves) to conserve energy and reduce wear and tear.

The two objectives are often contradictory, and thus the optimal behavior is merely a compromise between the two. If represented as an objective function, it would be like the following:

[Objective Function] = [Total Control Error] + [Total Control Energy]

$$J = \sum_k \left[r^2 \left(y(k) - y_{sp}(k) \right)^2 + s^2 \left(u(k) - u(k-1) \right)^2 \right].$$

The tuning aims to strike a good balance between fast response and minimum effort. Mathematically, this balance is determined by the relative weighting factors r and s.

As an example, consider the four reactors in Fig. 5.15. The reactors are exothermic. The operating requirement is to maintain the temperatures in all four reactors at 152 °C. However, at the start of the catalyst life cycle, the reaction heat generated by reactor R-103 is more than enough to raise the T-104 temperature above its setpoint, even after the heating medium flow FC104 is reduced to zero.

Without changing process design and the type of catalyst, one option for correcting the temperature profile is distributing the control error among the four reactors, especially between T-103 and T-104. That is, within the allowable range of the temperature variation, lowering the temperature setpoints of T-101 to T-103 to below the desired 152 °C so that less heat is generated and T-104 temperature can come closer to 152 °C. This error distribution is a coordinated effort among all four reactors. The temperature profiles with and without MPC are illustrated in Fig. 5.16.

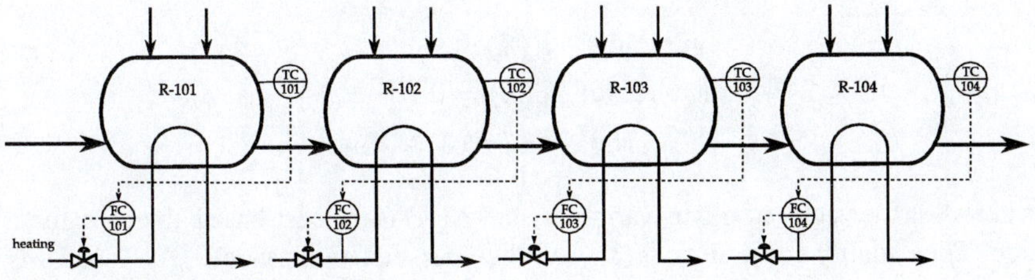

Fig. 5.15 Example of constrained control

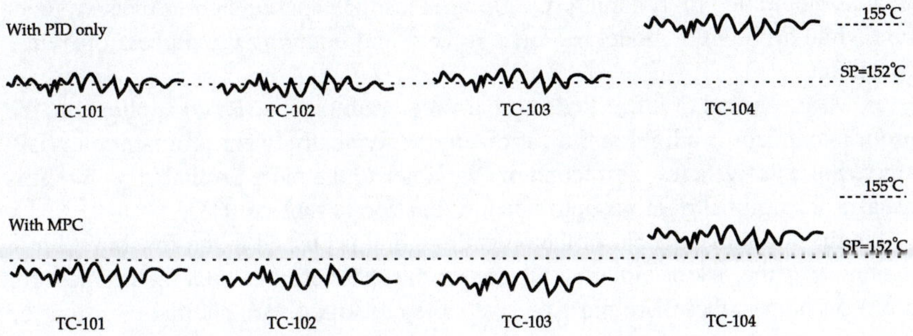

Fig. 5.16 Constrained control with MPC

An alternative PID-based approach is an override control on R-103 temperature control by R-104 temperature. Another option is to use valve position control (see Sect. 4.1.4) to prevent the R-104 control valve from saturation. However, MPC is a better approach to achieve a balanced temperature profile reliably. The optimization engine in MPC can optimally distribute the control error among the four reactors based on the temperature setpoint range of each reactor.

5.4.4 MPC as Optimizing Control

MPC provides a unified solution for addressing several process challenges (see Sect. 5.1.1) elegantly in a single controller and can replace the complex PID control schemes in most applications. However, the more deep-rooted reasons for MPC's massive success in the petrochemical industry are attributed to its capability to push for more profitable operations (See Fig. 1.9).

MPC stabilizes the process operation and thus creates the opportunity to push the operating point closer to the most profitable operation region to maximize economic benefits. Usually, another optimization layer is above MPC to calculate the more profitable operation points and passes them down to MPC as setpoints. The optimization is based on some economic targets. The role of this profit optimization layer is the same as the real-time optimization layer in Fig. 2.25. The optimization problem can be formulated as follows:

$$\begin{aligned} \text{minimize:} \quad & P(\mathbf{X}) \\ \text{subject to:} \quad & f(\mathbf{X}) = 0 \\ \text{and:} \quad & \mathbf{X}_{\text{low}} < \mathbf{X} < \mathbf{X}_{\text{high}} \end{aligned} \tag{5.19}$$

where $\mathbf{X}$ is the vector of model variables, and $P(\mathbf{X})$ is the cost-based objective function. The equality constraints, $f(\mathbf{X})$, are the open equation models. The inequality constraints on the model variables, $\mathbf{X}$, are the high and low constraints placed on the system to represent plant operating constraints. Mathematical optimization software such as quadratic programming (QP) minimizes the economic cost objective function while driving the model residuals to zero and honoring the process operating constraints.

In MPC, instead of a full-fledged real-time optimization (RTO) application, the profit maximizer is a light optimization engine, typically linear programming (LP) algorithm, to drive a few key economic variables to the more profitable region. It is usually not a global rigorous optimization function to replace RTO.

On the downside, the closer the operating point is pushed to the constraint limit, the higher is the risk of violating the constraints. Therefore, a vital requirement for profit pushing is that there must be a reliable constraint variable that can stop the pushing. For every *pusher*, there must be a reliable *stopper*!

5.5 Model Predictive Control—Practical Considerations

MPC application is more about the work process than technology. Compared with traditional PID control, MPC is a change in the "control mindset." This mindset change is made through the exercise of rigorously formalizing and analyzing the underlying problems, selecting input and output variables, representing complex physical phenomena by descriptive yet simplified mathematical models, and using mathematical tools for design and analysis.

Generally speaking, MPC has been an enormous success over the last few decades. However, not every control problem is suitable for MPC (see Chap. 7). There has been a tendency of overuse and even abuse of MPC technology. "To a man with a hammer, everything looks like a nail." There are many examples of disastrous failures. The quality of implementation is highly dependent on the competence of the implementors, much more so for MPC than for PID controls.

One of the crucial reasons for the failures is that most commercial MPC products have packaged the MPC theory and algorithm into the software tools and usually are not sufficiently transparent to the user. Many practitioners overly depend on the tools and have treated the MPC application like a software application. The importance of good process knowledge and the best practice work process is grossly undervalued or conveniently bypassed. Under the pressure of project delivery, they tend to cut corners on some critical steps in the MPC work process. Instead of spending sufficient effort to obtain a high-quality process model and deciding on a sound controls structure, they often prematurely rush to controller design with questionable model structure and parameters. Late in the project, they end up spending much time tinkering and twisting the controller tuning parameters betting on good luck to arrive at the "best" performance and thus fall into the same trap as in the "trial-and-error" approach in PID loop tuning (Sect. 9.2.2).

The process aspects of process control are the key to success; the systems aspects (the bits and bytes) are largely irrelevant. If we can do an application with one commercial system, then we can do it with any of them (Smith 2009).

5.5.1 *Process Noises and Disturbances*

Process noises and disturbances come from many sources and exhibit different frequency, magnitude, repeatability, and stochastic properties. Most process noises and disturbances do not meet the textbook descriptions, such as Gaussian distribution or stationary trends. Thus a large part of the control effort is to handle the varying and uncertain process dynamics caused by noises and disturbances. Table 5.3 lists some of the familiar sources of disturbances in a typical operating plant.

The long-range model prediction in MPC has successfully addressed the dynamic uncertainty caused by noises and disturbances. The control action is calculated based on the overall long-range prediction trajectory and is not un-proportionally biased by individual noise or disturbance along this trajectory.

Table 5.3 Examples of process disturbances

Source	Examples of noises and disturbances
Process	Steam pressure changes, recycle loops
Measurement	Measurement noises, failures
Valves	Stiction, slip, deadband, saturation
Controllers	Tuning, interactions, windup
Operator	Manual interventions, operator "Sweet Spot"
Environment	Daily and seasonal changes in temperature/pressure
Market	Supply and demand balancing, feed/product prices

5.5.2 *Baselayer Control Versus Advanced Control*

For multivariable process control, there is a significant overlap between baselayer control and MPC in terms of applicability. The decision between baselayer control and MPC solution is often a difficult one, and sometimes even a "political" one. One crucial lesson in process control is, the best solution is not the most sophisticated one, but the one that is simple and fit-for-purpose. As a result, it is not always good to go with MPC when a PID control scheme can serve the purpose. There is a threshold to pass to start using MPC, typically with three prerequisites as follows:

1. A commercial software license to apply for MPC: Most real-world MPC applications are based on commercial software packages. It is not as trivial as it sounds to choose the right vendor.
2. Hardware infrastructure to run the application: Some MPC software is built in DCS modules, while others require dedicated machines.
3. Engineering expertise to build or maintain the MPC application and the necessary training to run (and the mentality to accept) MPC applications by operations.

In an operating site, the first MPC application should be a carefully selected one with a sufficiently high return on investment (ROI) to justify the application and showcase the benefits. A failure in the first application may set back the MPC acceptance by many years (Smith 2009).

There are two schools of thought regarding the division between baselayer control and advanced control. One advocates keeping as much as possible baselayer PID controllers since PID control is more reliable in performance, and with multiple PID controllers, the risk is distributed. The role of MPC is primarily for handling constraints and pushing for profits. The other school of thought is to put everything under MPC to the extent that MPC directly manipulates all the control valves and other final control elements. All the PID controllers will be bypassed. The argument is that this approach provides total control and eliminates the inefficiency caused by the PID controllers in between. For example, a PID controller introduces extra dynamics between setpoints and controlled variables and thus slows the response or introduces limitations.

There are pros and cons to both approaches. In practice, it often becomes a personal preference on where to draw the line. The final control strategy is often a healthy mix of baselayer PID and advanced MPC for most applications: the baselayer PID controller keeps the individual process variables on target, while MPC supplies the setpoints for these PID controllers.

Considering the number of valves and control variables in the fractionator example (Fig. 5.11), it would be very challenging, albeit not impossible, to bypass baselayer PID loops, including the level controls, and put everything under MPC. A realistic approach is to use PID controllers to close those loops that are not very interactive with other loops. For example, the levels `LI-101` and `LI-102` are better off being put under PID control, so is the accumulator pressure `PI-101`. The top product draw, `FI-102`, can become a flow control loop. The baselayer control scheme will then be as shown in Fig. 5.17.

If the operating conditions require that MPC be taken offline, such as during a significant process upset, it would safely and seamlessly "shed" to the next layer down, i.e., baselayer control.

After the MPC technology becomes widely available, many existing PID control schemes are upgraded to MPC for better performance. Although the two control technologies are significantly different, they are based on the same process analysis and operating requirements. The difference is that the control objectives may be fulfilled more fully or more elegantly with the upgrade.

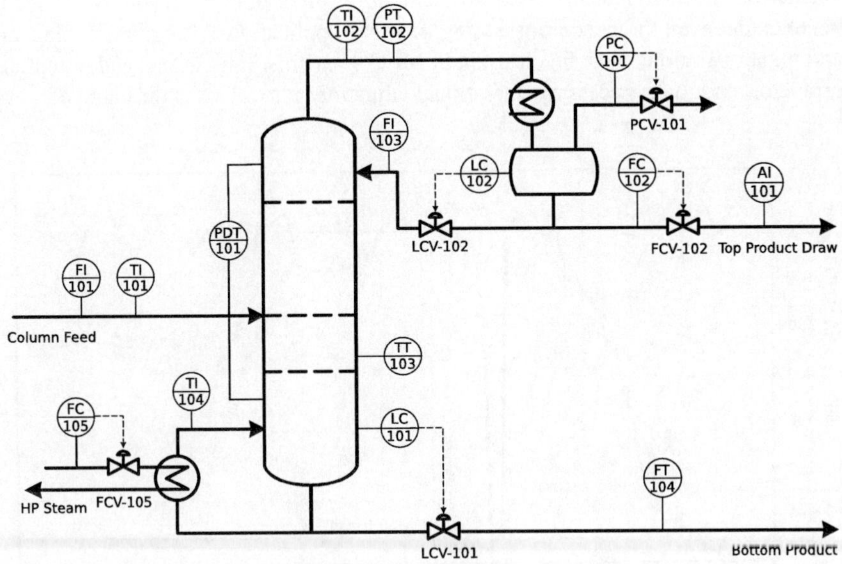

Fig. 5.17 A fractionator column, baselayer control scheme

5.5.3 MPC Maintenance and Sustainability

As with any sophisticated application, sustained performance and benefit are always desired but not easy to achieve. Experience shows that the controller performance will deteriorate after a year or two, and the deterioration would accelerate if proper maintenance measure is not taken to turn the trend around. The reasons that regular MPC maintenance is required include the following:

1. Change in process dynamics requires review and potential update of process models or even control structure.
2. Change in operating limits or constraints may require an update of the control structure.
3. Wear and tear in control valves or other final control elements may demand model updates.

An example of the MPC benefit prediction based on well-maintained and poorly-maintained MPC applications is given in Fig. 5.18. The data is based on a 3-year APC implementation project in an ethylene plant comprising over a dozen APC applications. Without proper maintenance, the APC benefits are expected to diminish after a year or so. The loss of benefits will accelerate until all MPC controllers are taken out of service for good.

The ongoing monitoring and maintenance of the MPC controllers is as crucial as the initial implementation but are often ignored or undervalued. The lack of maintenance has been an industry-wide problem and is often the key reason for reduced benefits and even the decommissioning of MPC applications. The plant managers and plant operations must be clear about MPC's benefits, limitations, and potential problems, and make necessary operational adjustments to adapt to the change.

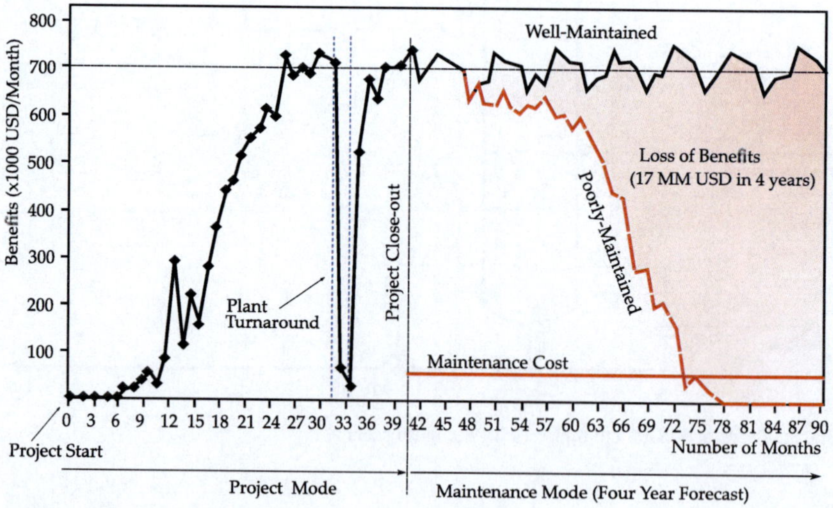

Fig. 5.18 MPC benefits forecast based on maintenance level

The first step toward good maintenance is proper monitoring of the controller performance. It is recommended that a *controller performance monitoring* system (CPM) be implemented to provide 24/7 monitoring on all the baselayer (PID) controllers and advanced process control (MPC) applications. See Sect. 8.5.2 for more discussions on CPM.

The CPM system should be proactive, systematic, and automated to provide the basis for both predictive and reactive maintenance. The CPM system typically produces daily, weekly, monthly reports on the controller performance indicator (KPI) targets such as the on-stream factor (e.g., 90%) and CV compliance (e.g., 95%). The KPIs should be regularly reviewed as part of the maintenance routine:

1. Daily monitoring: The goal of daily MPC performance monitoring is to timely troubleshoot any problems that may have arisen during the previous day and ensure that MPC applications are not unnecessarily turned off.
2. Weekly performance reporting: A weekly performance review should be performed and include the following actions:
   a. Review the weekly trend of the on-stream factors for all the applications.
   b. If any of the on-stream factors is lower than expected or displays an abnormal pattern, check the MPC maintenance logbook and talk to the operators to determine why.
   c. If deemed necessary, retrieve history data to take a closer look at the trends.
   d. If necessary, call a meeting with operations and engineers to decide on actions to take.
3. Annual performance auditing: MPC performance deteriorates over time. As part of the MPC maintenance, it is recommended that proactive performance evaluation be conducted on every controller every six months to a year by performing the following tasks:
   a. Meet with engineers and operations and discuss if the controller performance is still acceptable and if there is a need to improve the controller.
   b. If any controller performance is in question, collect two weeks of normal operating data, import them into the modeling tools, and check for significant model-plant-mismatch.
   c. The maintenance could be as simple as an update of model gain or as complex as completely revamping the control strategy. Depending on the above outcome, perform an update of the MPC controller following the standard work process.
   d. Determine the maintenance budget for the coming year.

5.5.4 Knowledge Management and Training

Although the MPC penetration rate in refining, petrochemical, and chemical plants has been pretty high, there is still resistance in accepting MPC implementation in many plants, especially in other industries. The key concerns include:

1. Operations feel that MPC is too complex and does not trust the moves that it makes.
2. Operations love MPC and rely on MPC so much that they lose familiarity with the process operation over time.

MPC focuses on control objectives such as throughput, production specifications, and economics, and less on the smooth and stable operation. For this reason, MPC can be unpopular among operation personnel since MPC moves the process point away from their "comfort zone." The operators often switch MPC off (or part of it) due to concerns about stability over profitability.

The ultimate solution is to deliver high-quality MPC applications that are intuitive, reliable, and transparent, to facilitate understanding and operation. However, MPC applications are inherently more complicated than baselayer control applications both in the understanding and operation. Therefore, adequate training and education is crucial.

1. Knowledge Management. The MPC knowledge and skills must be appropriately retained and managed within the support team, including proper management and maintenance of

   a. software manuals;
   b. training manuals;
   c. controller documentation (MPC narratives and Operating Manuals);
   d. work processes;
   e. best practices and lessons learned.

2. Training. Training is the primary means to maintain the MPC expertise. A good interest level and learning potential will help to motivate and retain MPC engineers and operators.
   Another common concern on MPC is that the operators may gradually lose their intimate familiarity with the plant operation with MPC controlling the plants. This gradual loss of skills is a general problem across the industry, and there have been no standard or effective solutions developed yet. One mitigating option is to make full use of an operator training simulator (OTS) system to train (or regularly refresh) the operator on handling abnormal conditions.

5.5.5 Future Developments

Adaptive control is the hope of the future for addressing the time-varying process dynamics and some nonlinearity (Challenge #6). The foundation of the adaptive control is the continuous online update of the process model, including the model parameters and potentially the model structure, to adapt to the loop dynamics changes. Although academic research has been active, a genuinely reliable, practical, and general-purpose algorithm similar to MPC is still unavailable. Nevertheless, some emerging exciting progress in MPC development includes the following:

1. Periodic model updates: A locally linearized model may become inadequate over time due to operating condition changes, a common maintenance challenge for both PID and MPC applications.
   The self-tuning PID has been available for decades, where the tuning parameters are automatically updated via a one-button click. However, self-tuning has not been widely accepted because there is still much to expect in the performance. PID tuning is still an art requiring engineering insight and practical experience (see Chap. 9).
   For MPC applications, the recommendation is to review and update the MPC models every six months to sustain the MPC performance. However, most MPC applications do not receive the model update as recommended due to budget limitations. As seen in the MPC work process in Sect. 5.6, MPC modeling is usually the most challenging and time-consuming part of MPC implementation and is thus the most costly. It is usually easier to get approval for starting a new MPC implementation in many operating plants, often with much fanfare, but challenging to receive the continuing budget support for regular maintenance. The best practice is to include the annual maintenance in the initial budget planning. One recent development is the automated step testing, which allows online model updates behind the scene while MPC is in control. Automated step testing provides some limited adaptive control capability, although it is still manually initiated and requiring a high level of expertise to carry out.
2. Nonlinear Models: The MPC model is inherently linear and time-invariant, mainly because of its simplicity and low computation load. Some MPC implementation includes nonlinear gain (e.g., via neural network models) while keeping the rest of the model in linear form. There are incremental advantages over a purely linear model.
3. First-principle models: Most industrial processes are inherently nonlinear and time-varying. A very sophisticated and complex model would be required to capture this complicated dynamic behavior fully. The current state-of-the-art model for process description is dynamic first-principle models based mainly on partial differential equations (PDE). It has been used for dynamic simulation studies and operator training simulators (OTS).
   The complexity of first-principle models makes it impossible for real-time control, at least for now. However, as the online computational power increases, the next generation of model-based process control may be based on high-fidelity first-principle models.

Another direction of advancement is simplifying the process dynamics and improving the measurement reliability (Challenge #5) via inferential properties, a.k.a. process analytics. Inferential property estimation is capable and cost-effective for many practical problems and deserves a lot more attention than it is now receiving. Static inferential properties calculation has been in extensive use. Dynamic inferential property estimation has been a crucial supplement to MPC for simplifying or enhancing the MPC models. In addition to the soft sensor, many nonlinear problems are transformed into linear ones through transformations or characterization, backed by sound engineering insight.

5.6 Model Predictive Control—The Work Process

Model predictive control has been the *de facto* standard for process control solutions. On the one hand, MPC is taken as the panacea for almost all control problems. On the other hand, many practitioners are sent to the field without adequate background and training on process control fundamentals. As a result, the results of MPC are a mixed bag. There are many failure examples due to inappropriate implementation. As discussed in Sect. 5.4, MPC is just another power tool in the process control toolbox. It can solve more control problems than the conventional PID-based feedback control; it can also cause more damage than a PID-based solution can if not correctly implemented.

The applications are typically based on commercial MPC packages (Qin and Badgwell 2003) that contain all the necessary tools for model identification, controller design, performance tuning, and real-time monitoring. The theory and technology are built into the software tools and are typically not transparent to the user. Compared with baselayer control, MPC is inherently multivariable and concerns a much larger area of the process; it is thus even more vital to understand the underlying process. In addition, the control design requires more careful considerations due to constraint handling and economic target pushing. It is thus essential to follow a well-defined methodology and procedure to ensure the proper execution of MPC projects. The methods of process analysis in Chap. 6 and control design in Chap. 7 are all applicable. As a result, the level of success of MPC applications highly depends on the competence of the implementors and their compliance with the standard methodology and best practice procedures. A proven work process is one way of guiding and ensuring that the necessary steps are followed, and no dangerous shortcuts are taken.

Some key points that need to pay close attention to are highlighted as follows:

1. MPC is built on top of the baselayer controllers. Improving the baselayer control loops is the prerequisite for MPC applications. More than half of the MPC benefits come from improved baselayer control.
2. The quality of the process model is the most critical factor for MPC's success. A good model depends on the engineering insight and the best practice procedure (Marlin 2015). The cause and effect analysis and dynamic response analysis in Chap. 6 is essential in arriving at the correct model structure and parameters. High-quality step-test data is instrumental in confirming the model structure and estimating the model parameters.

The rest of the section provides an MPC best practice work process for MPC applications of various complexities. For some tasks, such as model update during regular maintenance, only part of the work process needs to be followed.

5.6.1 Project Preparation

Before starting an MPC application implementation, it is vital to understand the project definition, the process to be controlled, and the objectives to achieve. The following is a checklist of the things to know before calling for a kickoff meeting:

1. Understand the project scope, schedule, budget, deliverables, and benefit expectations. There are typically some documents available such as feasibility studies or master-plan studies that define the MPC project from a high-level perspective. Collect and study these documents to get a basic understanding of the commercial side of the project.
2. Check if there is any software and hardware to be procured before the project starts. Since the procurement of hardware and software often requires considerable lead time, plan well in advance to allow sufficient time for delivery.
3. Obtain a copy of the following documents where available, and spend as much time as necessary to better understand the process flow and plant operating philosophy.
   a. Process control narratives (PCN) document.
   b. Process description and operator training manuals.
   c. Latest process PFD and P&ID documents.
   d. Vessel sizing information and level transmitter data sheets to be used for tuning level controllers.
   e. For product quality variables to be controlled, collect at least six months of lab data if online analyzers are not available.
   f. For revamping-project, obtain all documentation from previous MPC implementation.
4. Define, or validate if already defined, both the control and economic objectives.
5. Identify the stakeholders of the project that need to be included in meetings and communication. These typically include process engineers, economists, instrument technicians (analyzer team), and operations.

5.6.2 Project Kickoff

MPC application is a multi-team effort and always requires close collaboration between several parties, including process engineers, economists, control system engineers, instrumentation technicians, and panel operators.

A project kickoff meeting is essential for team mobilization and alignment. A recommended agenda for an MPC kickoff meeting is listed as follows:

1. Team introduction, including roles and responsibilities.
2. Review the MPC scope, schedule, and deliverables.
3. Review the control and economic objectives.

4. Explain project execution methodology (schedule, staffing, key project steps, milestones, reporting, interim review points).
5. Conduct a review and discussion on the following:
   a. Major operation and control challenges.
   b. Operating limits, constraints, and bottlenecks.
   c. Known problems in instrumentation and equipment.
   d. Significant disturbances to the unit (e.g., feed change, ambient temperature).
6. Check and note the unit blackout dates, such as scheduled shutdowns and turndown operating mode. Also, check and record the vacation plans of key players.
7. Agree on MPC benefit expectations. Discuss and agree on a baseline period for benefit-review purposes, if not yet done.
8. Issue meeting minutes with action items (with item owners and due dates).

5.6.3 System Installation

The MPC infrastructure should be set up as soon as feasible since certain activities may need a long lead time to complete (e.g., procurement). A checklist of the activities related to infrastructure is listed as follows:

1. Set up the MPC server machines if MPC runs on a separate and dedicated machine.
2. Install and configure MPC software.
3. Configure and test the communication between DCS and the MPC machines.
4. Install and configure the MPC communication watchdog on DCS, if applicable.
5. Configure MPC offline design tools to gain access to the site historian database. If a direct connection to the database is not possible, make sure that alternative means of data acquisition are set up and functional. For example, the historical data can be exported to Excel or CSV files and imported into the MPC modeling tool.
6. Set up the project file structure on a shared location and make it accessible to all concerned parties.
7. Determine how the process data will be collected. If a historian database is used, check the data compression to see if the compression limit needs to be reduced or removed.

5.6.4 Baselayer Review and Improvement

MPC relies on the baselayer controllers to function. A properly designed and well-tuned baselayer is a critical requirement for the success of MPC. As part of the MPC activities, a thorough review and improvement of the baselayer control are necessary. A recommended list of items to check is as follows:

1. Check the required analyzers' availability, reliability, and accuracy and determine whether inferential property estimator (IPE) applications are needed.
2. Create any new MPC-required calculation tags on DCS as needed and historize them for both DCS and historian database.
3. Obtain a complete list of the relevant DCS tags for data collection. Identify all the candidates for manipulated variables (MV) and disturbance variables (DV).
4. Obtain a complete list of relevant PID controller tags, including PV, SP, OP, and Mode. Also, include the essential properties such as tuning parameters, equation type, filter parameters, engineering units, and measurement ranges.
5. Check history data of the process variables and determine if excessive measurement noises are present. If so, consider adding measurement filtering to smooth out the noises.
6. Find out why any controller that has run in the manual (MAN) mode for an extended period.
7. Identify the controllers whose performance is not satisfactory and re-tune them by following the standard troubleshooting procedures.
8. Define and document the baseline condition of operation conditions based on the agreed baseline period.
9. Implement and commission the inferential property estimator (IPE) applications if required. Historize the inferred property values.

5.6.5 *Pre-testing Activities*

Pre-testing is essential for understanding the underlying process dynamics and identifying potential problems in baselayer control, instrumentation, equipment. Besides, coupled with historical data, the pre-test can help optimize the step testing plan, substantially reduce the time spent on step testing, or help avoid some unnecessary step tests. A checklist of things to watch out for during pre-testing is provided as follows:

1. Make a list of all parameters for which historical information will benefit MPC implementation, and check to ensure that these parameters are configured and historized in the historian database.
2. Check and make sure that the data compression setting in the historian database is appropriate. The historian database's default compression band (typically 0.1%) is typically too high for the desired data resolution; it is often necessary to change to a smaller deadband (e.g., 0.01%) for process modeling. Compile a list of variables whose compression settings need to be changed, and ask the database administrator to make the changes.
3. Obtain and document the product specifications and operating targets relevant to the MPC application.
4. Determine the approximate time to steady-state (TTSS) and the move size of the step changes for each MV and DV variable based on historical data. The general principle is to use move sizes as large as possible to achieve the highest

possible signal-to-noise ratio. However, many factors, primarily operating safety and local linearity, limit the size of step change that can be made. Always consult the operator before making any moves.

5. Operating condition: Before step test, check the operating condition against the condition that MPC intends to run. If needed, ask the operator to make necessary changes to bring the unit to the desired condition (e.g., unloading the column).
6. Controller modes: Make sure the baselayer controllers are in the correct modes intended for MPC. For example, if one of the PID controllers in cascade mode will be run with the cascade loop open when MPC is implemented, the cascade loop should be kept open while conducting step tests.
7. PID algorithm type: For specific DCS systems, check and ensure that control algorithms of the PID controllers to be manipulated by MPC are set to the right algorithm type. For example, PID with algorithm "`I-PD`" shall be changed to "`PID`" or "`PI-D`" to ensure proper setpoint tracking (see Sect. 3.2.1).
8. Ambient temperature: Ensure that a reliable ambient temperature tag is included in data collection, as the ambient temperature is commonly a vital disturbance variable (DV).
9. Step changes: Make at least one step change in each direction on every MV/DV variable that is expected to move during formal step testing. Validate/confirm the move sizes and TTSS with operators before making any moves.
10. Interaction with operators: Engage operators as much as possible in the test. Ask, record, and cross-validate their opinions as follows:

    a. Are there any valve, transmitter, or tuning problems?
    b. What handles do they usually use to control each CV, and what the control ranges are?
    c. What do they typically move each MV for?
    d. How much grace time does the operation allow for off-spec products?

11. Inferential properties: Build inferential property estimator (IPE) models and applications if required.
12. Modeling tool: Setup the offline modeling tool and import pre-test data and baseline data as necessary.
13. Preliminary cause and effect table: With process knowledge and test data, identify the cause and effect relationship between the MVs, DVs, and CVs. Show them in a cause and effect matrix, and mark the confidence on the relationship with certain symbols. For example, if a strong relationship with a positive gain is expected, mark the cell with +++. If a reasonably strong relationship is expected, mark it as ++. If a positive relationship may exist, mark it as +. Do the same for relationships with negative gains. If uncertain, mark the cell as `?`.

After the pre-test is completed and the data is analyzed, a pre-test report or summary should be issued to the operations and engineers to prepare the unit for formal step testing. Depending on the nature of the problems discovered, this may take some time to complete all the fixes. The pre-test report should include the following content at a minimum:

1. Equipment and instrumentation problems, especially valve problems.
2. Any changes in the baselayer control strategy to be implemented. For example, the column's sensitive tray for temperature may be shifted after column unloading, and the temperature controller may need to be relocated and re-configured.
3. A list of the controllers that have been re-tuned and the controllers that are planned to be re-tuned.
4. Any new instrumentation that is required to be added.
5. Evaluation of the analyzers and the decision on whether inferential property estimators (IPE) need to be implemented.
6. Key control objectives and control handles that are identified for MPC.
7. A preliminary model structure showing the expected cause and effect relationship and dynamic response characteristic (approximate gain, delay, and lag) between each cause and effect variable pair.
8. Detailed step-test plan including the variables to be tested, such as the step size, test duration, variables to watch during step testing.

5.6.6 *Plant Test*

A formal step-test practice aims to systematically and meticulously disturb the process to observe the dynamic responses of the interested variables. Step test is a critical step in MPC implementation and is also the most time-consuming part of an MPC project. A carefully designed and conducted step test yields high-quality process data and consequently leads to high-fidelity models. As emphasized already, the model quality (both the structure and the parameters) is the most important determining factor for MPC's success (or failure).

Some recommendations on step testing are provided as follows:

1. Check and make sure that the inferential property estimators (IPE), if any, are functioning with acceptable reliability and accuracy.
2. Check if analyzers need to be calibrated. Request for calibration if necessary.
3. Ensure the data collection mechanism is appropriately set up and reliably working, and data compression is removed or reduced.
4. Use a standard template to create a test log file with a list of the MVs and DVs and their move plans.
5. Ensure that the unit is in the same operating condition as that MPC intends to run at, including process values and PID controller modes. If not, ask the operator to make changes to bring the unit to the desired operating condition.
6. Check and request extra lab samples during plant tests as needed.
7. Brief the panel operators on what will be done and request their cooperation to
   a. inform the MPC team of any significant upcoming disturbances, e.g., feed change, fieldwork on equipment that could impact the test results and
   b. inform the MPC team of any key control moves they make that could impact the test results.

8. In the test log file, record all key disturbances and operator-made moves that may affect test results.
9. Frequently review the test data in offline modeling tools (if feasible) and make adjustments on the move plan such as move size, move frequency, and test duration.

5.6.7 Process Modeling

The power of MPC comes from the process models, which capture the cause and effect relationships and dynamic response behavior between the manipulated variables and controlled variables. Although the quality of the models is primarily determined by the data available, experience and skills are essential in extracting the maximum information from the data at hand. Some general recommendations regarding process modeling are provided as follows:

1. Collect test data. Note that data can be from various sources, including historical data, pre-test data, and step-test data. Step test data is obtained under controlled conditions and is typically most trusted. However, pre-test and historical data can often be used as supplementary information or for cross-validation of the models.
2. Build models. Use process understanding and engineer knowledge in all steps, especially in data pre-processing, e.g., slicing and dicing.
3. Assign a quality indicator to each model identified, e.g., A for good models, B for fair models, and C for low-quality models.
4. Use the simplest models (e.g., first-order plus time delay) whenever possible, even if a more complex model (e.g., second-order model) can offer marginally better performance.
5. If a critical model (based on process understanding) is expected, but the model quality is questionable, re-do the step test to improve the model. Otherwise, leave it out of the controller. A flawed model may be worse than no model.
6. Decide whether a nonlinear transform such as a logarithm function is needed on certain variables.
7. Perform a relative gain array (RGA) analysis to determine that the resulting control matrix is not singular or ill-conditioned (Skogestad and Phostlethwaite 2005).

5.6.8 Controller Design

The controller design part is relatively straightforward with a sound control strategy (following the analytical steps in Chap. 6) and high-fidelity models (see Chap. 7). The following are some of the things to watch out for during controller design:

1. Obtain economic values (market values) on the following streams for economic function (EF) calculation:

   a. Feed Stocks.
   b. Products.
   c. Utilities such as steam, fuel oil, and electricity.
   d. Waste (disposal cost).

2. Calculate the economic values of the variables if available. Otherwise, assign economic values based on their relative economic significance.
3. Decide and define sub-controllers. Note that change in sub-controller definition at a later time may require more re-work, such as the update of the watchdog.
4. Simulate, test, and fine-tune the controller offline with unconstrained move size limits.
5. Simulate the controller offline in different cripple modes, e.g., removing some MVs and CVs from control.
6. Update the control matrix to show the following information at a minimum:

   a. All the MVs, DVs, and CVs.
   b. The cause and effect relationships between the MV/DV and CVs, including the gains.
   c. Model Quality rating.
   d. Setpoint values or setpoint ranges.

5.6.9 *Model and Controller Review*

After the models are identified and the controller built, it is necessary to conduct a thorough review of the models and detailed control strategy with operations and process engineers before commissioning the controller. Some general guidelines for the review meeting are as follows:

1. Invite process engineers, operation representatives, and senior operators to the meeting.
2. Review the detailed control and economic objectives.
3. Review the CVs, MVs, DVs and their cause and effect relationship. Use the control matrix for convenient presentation.
4. When a CV is controlled by multiple MVs, decide which MV(s) is the principal handle for the CV. Prioritize the MVs if possible and tune the controller accordingly. Decide whether some of the MVs should be duplicated as DVs.
5. Review/confirm the product specifications, operating targets, and critical constraints.
6. Review the commissioning plan. Discuss and request operation support for commissioning.
7. Ask Operations to notify operators of the upcoming commissioning and training.
8. Issue meeting minutes with the following information at a minimum:

   a. The agreed changes to the controller strategy and models.
   b. The product specifications, operating targets, and critical constraint limits.
   c. Commissioning Plan.

5.6.10 Controller Commissioning

After the control strategy and controller models are reviewed and agreed upon, the controller can be tested online.

The controller commissioning typically consists of two steps: pre-commissioning and commissioning. Pre-commissioning aims to set up the proper communication between MPC software and DCS and identify any errors in controller configuration. Following the pre-commissioning is the formal commissioning. The main objective of the commissioning phase is to fine-tune the controller to achieve the desired control performance.

Below is a list of the general steps for controller commissioning:

1. Create a complete list of MPC MVs with the following modes: normal mode, MPC mode, and Shed Mode. Validate the list with operators. Note that the shed mode of a controller may not necessarily be the same as its normal mode.
2. Build and configure the MPC/DCS communication watchdog.
3. Run the MPC control in online open-loop mode, i.e., MPC calculates the next moves as if it is online, but the moves are not actually implemented in DCS. After every execution, do the following:

   a. Check the calculated MV moves. These are the control actions to be sent to DCS for all the MPC MVs. Based on the current CV errors, validate the sign and magnitude of the moves and make sure they agree to the process understanding and the values are in the expected range.
   b. Check the predicted steady-state values of all MVs if available. These are the total planned MV changes at the new steady state based on current CV errors. Make sure the direction and size of the MV changes are reasonable in comparison with current CV errors.
   c. Check the predicted steady-state values of all CVs if available. These are the predicted steady-state values for all CVs, if the planned moves above are all implemented. Check whether the predicted values are within expectation.
   d. Check the innovations on the PVs and make sure they are not too large. Each PV has an innovation value, which is the difference between the controller's prediction of the PV and the actual measurement of the plant. This innovation is an indicator of the "goodness" of the prediction model for the PV.
   If any of the above is not as expected, stop the commissioning and double-check on the controller tunings, DCS mappings, or even the controller models to find the cause of the discrepancies.

4. Before turning on the controller in the closed-loop mode for the first time, review/confirm again with the operations and engineers on the MPC MV limits and CV set ranges, MV/CV predictions, and next MV move sizes. Adjust the limits as necessary.
5. For critical MV, consider temporarily reducing their move limits to avoid unexpected large moves due to potentially too aggressive tunings. Remember to put

these limits back to normal ranges once the controller has been "proven" to be working correctly online.

6. Compare the CPU and network loads before and after the controller is turned on. Verify that the load is within the expected range.
7. During pre-commissioning (day-time-only), make an MPC screen capture showing all the current MV and CV values before turning on the MPC. Leave a copy to the panel operator in case the operator needs to bring the unit back to where it was.
8. If available, the online monitoring tool shall be configured and used parallel to monitoring the MPC status and performance.

5.6.11 Operator Training and Documentation

Before the MPC application is commissioned, it is necessary to introduce the application to all shifts of operators. Operator training typically involves the following activities:

1. Prepare the operating manual using the standard document template.
2. Prepare operator training presentation slides based on the operating manual.
3. Check operator shift schedule and work out a training plan (group training or individual training). Inform the operators in advance about the training schedule. Recommended time for operator training is when the controller is commissioned and relatively stable (for hand-on training). All operator training needs to be completed before running the controller round-the-clock.

The operator training should cover the following contents at a minimum, with an appropriate quiz to assess the training efficacy:

1. The purpose of the MPC application, along with historical data showing pre-MPC performance.
2. Overview of all MPC variables (MVs, CVs, DVs).
3. Overview of modifications to existing graphics, as well as new MPC graphics.
4. Criticality of variables in the controller, for example, which variables are essential for control.
5. Any special instructions for the MPC.
6. Operation of the controller, especially how to shed the MPC controller to the baselayer in abnormal conditions.

5.6.12 Controller Handover to Operators

After completing the operator training and a few weeks of stable operation, the controller can be officially handed over to operators. For doing that, make sure the following items are covered:

1. Make sure the operators have received a copy of the MPC operating manual.
2. Ensure the operators understand that they are assuming the responsibility of MPC control once it is on 24/7, even though engineers may be remotely monitoring.
3. Make sure operators have the emergency contact numbers of the MPC engineers.

Even after controller handover to operators, close monitoring and maintenance support is still necessary to correct any issues discovered, improve the controller performance, and in the long run, ensure the sustained performance of the controller. Typical maintenance duties include the following:

1. Conduct regular (e.g., daily) performance reviews of the controller.
2. Discuss with operations and engineers for feedback and make fine tunings as necessary.

5.6.13 Post-implementation Review (PIR)

After two to three months of stable operation, a post-implementation review (PIR) can be conducted, with the following recommended agenda:

1. Review of MPC benefits.
2. Unit performance comparison before and after MPC.
3. Feedback on control performance and added values.
4. Future value opportunities identified during the project.
5. Issue meeting minutes.

The benefits analysis for the plant will involve calculating the post-MPC economic performance against the baseline performance over a specific period. Since the baseline data were gathered over two weeks of operation, it makes sense that this same period is used for the PIR. An appropriate two-week period of process operation with MPC will need to be identified for each plant as the "MPC period."

5.6.14 Project Close-Out

To close out the project, ensure that all the required documents have been delivered and relevant data files have been appropriately archived for each MPC application. The checklist is to include the following items:

1. Organize and clean up the project file repository and make sure all the required project deliverables and application-related files are properly archived and in the correct format, including data files.
2. Finalize and distribute the MPC operating manual.
3. Finalize and distribute the MPC engineering manual (MPC narrative).

5.7 Summary

Advanced process control (APC) refers to any control scheme that is more complex than a simple PID control loop. Now that model predictive control (MPC) has become prevalent, APC is often used interchangeably with MPC. However, this book uses APC separately from MPC and reserves the word APC for all advanced technologies, including, but limited to, model predictive control, nonlinear control, adaptive control, and robust control.

Most advanced control applications are based on models of a specific type, from the heuristic model, input/output black-box model, state-space model, to the first-principle model (Sect. 5.2). The more advanced the technology is, the higher requirement it has on the underlying model. Inferential property estimator (Sect. 5.3) as an open-loop calculation has flexibility in terms of model requirement. It is a reliable and cost-effective approach to improve operation and control and is thus an area deserving more attention in both theory and practice.

Model predictive control (Sect. 5.4) is a step-change from traditional PID feedback control. Introducing explicitly and separately identifiable process models into the control algorithm has been a breakthrough that successfully breaks the limitation inherent to PID-dominated feedback control. MPC applications yield better and more reliable control performance and have gained widespread use in the process industries such as chemical plants and oil refineries since the 1980s.

One crucial change that MPC has brought about is that the MPC work process forces the practitioners to pay more attention to the process analysis and overall control strategy as part of the implementation requirement. MPC is inherently multivariable and covers a larger area of the process than the conventional PID control approach. The control design must see the "bigger picture," i.e., the overall control strategy. Chapters 6 and 7 will elaborate on the "bigger picture" while discussing the details of process analysis and control design. Due to the complexity of the MPC application, a field-proven work process (Sect. 5.6) is critical to ensure that the key application steps are not bypassed or compromised.

References

Åström KJ, Hägglund T (1995) PID controllers: theory, design, and tunings, 2nd edn. Instrument Society of America

Blevins TL, McMillan GK, Wojsznis WK, Brown MW (2003) Advanced control unleashed—plant performance management for optimum benefit. ISA

Box GEP (1976) Science and statistics. J Am Stat Assoc 71(356):791–799

Clarke DW, Mohtadi C, Tuffs PS (1987) Generalized predictive control—part I. The basic algorithm. Automatica 23(2).137–148

Corriou JP (2018) Process control—theory and applications, 2nd edn. Springer

Cutler CR, Ramaker BL (1980) Dynamic matrix control—a computer control algorithm. In: Proceedings of American control conference

Garcia CE, MPrett D, Morari M, (1989) Model predictive control: theory and practice—a survey. Automatica 25(3):335–348

Gordon L (2005) Advanced regulatory control: decoupling. Control Eng

ISO-5167 (2003) Measurement of fluid flow by means of pressure differential devices inserted in circular cross-section conduits running full, 2nd edn. ISO

Lee JH (2011) Model predictive control: review of the three decades of development. Int J Control Autom Syst 9(3):415–424

Ljung L (1999) System identification: theory for the user, 2nd edn. Prentice-Hall, Englewood Cliffs, NJ

Marlin TE (2015) Process control—designing processes and control systems for dynamic performance, 2nd edn. McGraw-Hill

Morari M, Lee JH (1999) Model predictive control: past, present and future. Comput Chem Eng 23(4–5):667–682

Niu S, Fisher DG, Xiao D (1992) An augmented UD identification algorithm. Int J Control 56(1):193–211

Niu S, Fisher DG, Ljung L, Shah SL (1994) A tutorial on multiple model least-squares and augmented UD identification. Technical report LiTH-ISY-R-1710, Department of Electrical Engineering, Linkoping University, Linkoping, S58183, Sweden

Qin SJ, Badgwell TA (2003) A survey of industrial model predictive control technology. Control Eng Pract 11(7):733–764

Rawlings JB, Mayne DQ, Diehl MM (2017) Model predictive control: theory, computation, and design, 2nd edn. Nob Hill Publishing, LLC

Rossitier JA (2003) Model predictive control—a practical approach. CRC Press

Samad T, Bauer M, Bortoff S, Cairano SD, Fagiano L, Odgaard PF, Rhinehart RR, Sanchez-Pena R, Serbezov A, Ankersen F, Goupil P, Grosman B, Heertjes M, Mareels I, Sosseh R (2020) Industry engagement with control research: perspective and messages. Ann Rev Control 49:1–14. open access

Shinskey FG (1994) Feedback controllers in the process industries. McGraw-Hill Publishing, New York

Skogestad S, Postlethwaite I (2005) Multivariable feedback control—analysis and design, 2nd edn. Wiley

Smith CL (2009) Practical process control—tuning and troubleshooting. Wiley

Smith OJM (1957) Closed control of loops with dead time. Chem Eng Prog 53(5)

Part III
Analytic Skills and Problem-Solving Methodologies

Process control is about the control of processes. The practical challenge to process control is that there are countless processes in numerous industries, and there is also a rich collection of control technologies. As a result, process control design is by no means a trivial job. For instance, a process control engineer in a service company implementing model predictive control (MPC) applications may have to deal with a completely different process in each new project. It is unrealistic to expect him or her to have the required process knowledge to start with, nor is it advisable for him or her to blindly copy and paste from existing projects. Even with the same process design, every process typically has very different operating requirements and constraints and must be treated differently. It is more practical to develop the essential and generic process analysis skills applicable to all new processes encountered. After all, the process knowledge required by process control engineers is different and less comprehensive than that of process engineers.

A good understanding of the process characteristics and in-depth knowledge of control technologies is essential for designing and executing the control solution. As a cross-disciplinary specialty, process control requires knowledge in many related but distinct areas such as process engineering, rotating equipment, instrumentation, control systems, operations, maintenance, and project execution, in addition to the core technologies of automatic control. For this reason, a process control practitioner must be very selective in the acquisition of knowledge and skills, and more importantly, pursue a more efficient approach to gain the necessary expertise.

For this purpose, we have recommended a learning process in two phases: build the core knowledge framework as the foundation (Part II of this book) and then develop the necessary skills to perform the job efficiently (Part III). This learning process includes the following steps:

1. Have the basic knowledge of the necessary tools and enabling platforms, such as control system infrastructure (DCS/PLC), sensors, transmitters, control valves, network and communication, and information technology.
2. Master the core process control technologies, including the baselayer control and advanced process control.

3. Develop the analytic skills for problem-solving such as process analysis, control design, troubleshooting, and performance monitoring and improvement.
4. Follow the best practice methodologies to tackle process control design or troubleshooting tasks.
5. Eventually, reach the mastery level of process control instinct or mindset.

Structured process analysis (Chap. 6) helps in gaining the necessary process insight. The process analysis skills such as *process flow analysis*, *supply and demand analysis*, *cause and effect analysis*, *degree-of-freedom analysis*, and *dynamic response analysis* provide a systematic approach to extract the necessary information about the process as required by process control. Tools such as process overview diagram (POD) and cause and effect table (CET) can help improve the consistency and efficiency of the process analysis.

The process control design is the next step following process analysis. A structured approach to analyzing the control needs and guiding the control technology selection and implementation is critical. The techniques such as overall control strategy for plant-wide control and protective/regulatory control for layered protection can help design fit-for-purpose control solutions (Chap. 7). Control overview diagram (COD) is an effective visual tool to assist the design.

The principles and skills of process analysis and control design can also apply to control problem troubleshooting (Chap. 8) and improvements (Chap. 9), which demand the most knowledge, skills, and experience. It requires a sound knowledge of how the control scheme works, and a good understanding of how the control scheme may fail. The skills are built through many years of hands-on experience that has no substitute for. The guidelines and procedure in Chaps. 6 and 7 can make the loop troubleshooting and improvement more focused and less painful.

Chapter 6
Methodology of Process Analysis

Process control is about process *and* control. The previous chapters have discussed the different control technologies, from simple PID control to advanced process control. However, they are just various tools in a process control engineer's toolbox. Effectively using these tools to perform meaningful work depends on a good understanding of the process design and operating requirements (Sect. 6.1). As there are countless processes in various industries, it is unrealistic for a process control practitioner to become intimately knowledgeable on all the processes encountered. The more practical approach is developing generally applicable skills and methodologies to help with process analysis. This chapter presents a structured work process for efficiently and consistently extracting the required process information into the form suitable for process control. These skills include process flow analysis (Sect. 6.2), supply and demand analysis (Sect. 6.3), cause and effect analysis (Sect. 6.4), degree of freedom analysis (Sect. 6.5), and dynamic response analysis (Sect. 6.6).

6.1 Operating Requirements

Process control serves the operation and production in two main areas: process safety and production efficiency. Familiarity with the operating objectives and operating requirements is essential for understanding the process operation and designing the control solution (Downs and Skogestad 2011; Marlin 2015; Seames 2018).

6.1.1 Operating Objectives

Generally speaking, the operating objectives are about producing the right products with the right quality and quantity at the right time (Sect. 1.2). For example, in the oil & gas industry,

S. S. Niu and D. Xiao, *Process Control*, Advances in Industrial Control,
https://doi.org/10.1007/978-3-030-97067-3_6

1. Upstream exploration & production (E&P) is primarily concerned about the movement of the various fluids (gas, oil, and water) for simple separation, light processing, and transport. So the operating requirements are to ensure the various streams flow to the right location with the right quantity. The general control requirement is thus *capacity* (quantity) control.
2. Downstream operations, such as petrochemical and chemical production, aim to produce specific products that meet the given specifications. So *quality* control is the top priority. The market determines the product rate and can be unpredictable. Large inventory buffers such as feed tanks and product storage tanks usually are utilized to facilitate capacity control.
3. Mid-stream operations such as refineries have a good mix of quality control and capacity control.

The following reference documents can provide a good start in understanding the operating objectives at either the plant or unit level:

1. Existing process control narratives (PCN) document, if available. A good PCN document is the best source of information to understand the process from a process control perspective.
2. Process flow diagram (PFD): The PFDs provide the details of the process streams and equipment. They typically also include each stream's process conditions and chemical compositions, such as the steady-state material balance data (Turton et al. 2018).
3. Piping & instrumentation diagram (P&ID): P&ID provides detailed information for the plant's construction, including instrumentation and control.
4. Operating philosophy document and operating manual.
5. DCS Schematics.

This chapter will use the following process example to explain the methodology and procedures of process analysis.

Example 6.1 Gas compression. Figure 6.1 shows a gas compression process in an upstream production facility. The produced gas from the production separators is sent to a produced gas compressor (PGC, `K-101`) for compression to the desired pressure. The compressed gas is used as fuel gas by four heat recovery steam generators (HRSG) as the top priority. The remaining gas is used as lift gas by multiple gas lift wells. The diagram in Fig. 6.1 is a process overview diagram (POD) simplified from the actual PFDs and P&IDs, with the non-control-relevant details removed for clarity.

The operating objective is to provide a continuous gas supply at the desired pressure (quality) and flow rate (quantity) for the steam generators and gas lift wells.

With Example 6.1, the process analysis methodology is explained step-by-step in the following sections.

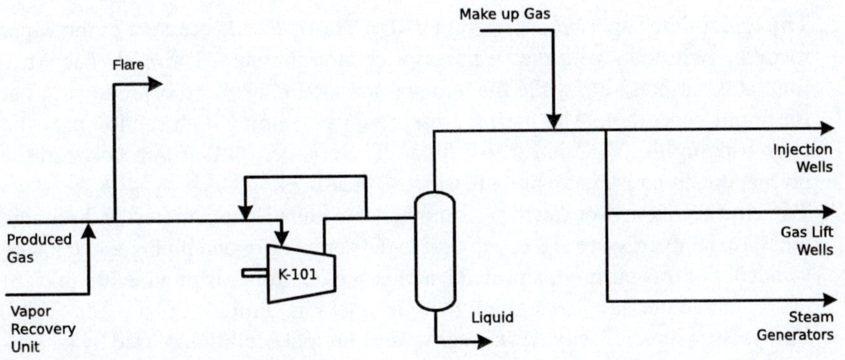

Fig. 6.1 A gas compression process

6.1.2 Product Specifications

All process operations aim at producing something useful, either as final products or as semi-products. Most of these products have specifications on their quality or quantity. Even the produced wastes would have to meet specific requirements before they can be disposed of. For example, the produced water in Example 6.1 must be sent to the injection wells or the deep disposal wells, as required by environmental regulations.

The product specifications are vital to process engineering and process control design; so it is crucial to get familiarized with the relevant product specifications and their impact on control design.

6.1.3 Limits and Constraints

Another critical aspect of the operation is the design limits and operating constraints that the process control scheme must honor. These limits and constraints may be related to product quality or operating safety. They can significantly influence the control design decisions. Each plant has its own operating limits and constraints, even if it is based on the same design. The difference in the limits and constraints is critical, and a control scheme designed for one plant cannot be copy-and-pasted to another plant without going through detailed process analysis. The differences must be considered in the overall control strategy rather than treated in a piecemeal approach. See Sect. 1.3.5 for an explanation of how the operating setpoints and limits stack up.

The control design for the compressor in Example 6.1 must observe the following constraints to protect the equipment or piping:

1. Trip limit of the vapor recovery unit (VRU): The operating pressure of the vapor recovery unit floats with the compressor suction pressure. The VRU has a trip limit at 2 bar pressure, while the compressor suction pressure is around 1.4 bar for normal operation. The flaring pressure for preventing high suction pressure from tripping the VRU is set to 1.8 bar. There is very little room between the normal operating pressure and the tripping pressures.
2. Trip limit at compressor discharge: Another constraint is the compressor discharge pressure. High pressure is a safety risk to the compressor and piping and must be avoided. For this purpose, a protective pressure controller is provided to prevent the discharge pressure from going over the operating limits.
3. Compressor surge: *Compressor surge* is an abnormal condition caused by insufficient flow through the compressor. It must be prevented because it can potentially lead to severe mechanical damages to the compressor or the associated piping.

6.1.4 Mode of Operations

It is essential to know all the modes of operation under which the plant/unit is expected to operate. Typical modes of operation include startup mode, normal operating mode, turndown mode, crippled mode, and shutdown mode. There are often other operating modes specific to the operational requirements. For example, different feed requires different operating conditions. Reduced capacity due to low demand or partial shutdown may force the operation into a turndown mode. Failure in specific equipment may cause the overall process into crippled mode operation.

The challenge is the safe and smooth transition between the different modes of operations. One example is polymer plant operation, where the production is a batch operation. Each product is based on a different "recipe." The transition from one product to another, called "recipe" change, happens frequently. The transition is expected to be quick and smooth to minimize off-spec products.

For the process control solution in Example 6.1, at least the following modes of operation need to be considered:

1. Normal operation: During normal operation, the wells and compressor have sufficient capacity to supply high-pressure gas needed by both the steam generators and gas lift wells.
2. Compressor down: If the wells are shut-in or the compressor goes down, the entire operation will enter crippled mode. The steam generators will draw gas from the main gas line.
3. Steam generators down: If one or more of the HRSGs are offline, the operation will transition to turndown mode with reduced capacity. The injection wells will take all the produced gas, and the production will be subject to the injection well capacity.

6.2 Process Flow Analysis

A modern operating plant design can often produce dozens or even hundreds of pages of process flow diagrams (PFD) and piping & instrumentation diagrams (P&ID). There are also a comparable number of DCS schematics. An inexperienced or unprepared engineer can easily be overwhelmed by the daunting number of schematics and fails to the "bigger picture."

Process operation is mainly about material movement and energy transfer, governed by physical laws such as conservation of mass, energy, and momentum. The mission of process control is to enforce these physical laws, not to defy them. Thus, it is essential to start with a structured process flow analysis to identify the system boundary, locate all the incoming and outgoing material and energy streams of interest, identify any recycle and loop-back flows, list all possible process flow line-ups, and enumerate all potential modes of operations (Arnold and Steward 1999; Campbell 2004; Marlin 2015).

6.2.1 *Process Flow and System Boundary*

The process flow analysis is about the process topology. The study starts with a process overview diagram (POD) derived from the PFDs and P&IDs. The POD focuses on the "bigger picture" and provides a "full view," in one single page, of all major process flows, essential devices, equipment, final control elements, and key process variables. The control-relevant content is extracted from the PFDs and P&IDs and included in the POD. All the details not relevant to process control are left out. The following information is generally expected in the POD:

1. The system boundary and the boundary conditions.
2. All the streams that may continuously flow into the system boundary (the *supplies*) during normal operation, along with the stream specifications.
3. All the streams that may continuously flow out of the system boundary (the *demands*) during normal operations, along with the stream specifications.
4. All the control elements such as switches and on/off valves potentially affecting the process flow line-up during normal operation.

In Example 6.1, the system boundary is around the compressor, with three streams coming in and four streams going out. It can serve as the initial POD and progressively update with more information as the analysis continues:

1. Gas supply from the separator is the main process flow feeding the compressor. The inlet pressure floats with the compressor suction.
2. Gas from the vapor recovery unit is a side stream with a much lower flow rate. The pressure also floats with the compressor suction pressure.
3. The primary consumers of the compressed gas are the steam generators, with a specification on the supply pressure.

4. The next consumers in line are the gas lift wells; each has a specification on the gas flow rate.
5. Any excess gas is sent to the injection wells.
6. Sweet gas from the main gas pipeline is provided as make-up gas in case the produced gas is insufficient.
7. Flaring is provided as an emergency measure to relieve excess gas during abnormal conditions, e.g., compressor shutdown.

6.2.2 *Key Disturbances*

Noises and disturbances are a matter of life in the process industry (see Sect. 5.5.1). Disturbances are the inputs or external conditions from the surrounding environment that the operator or the control system cannot control. Some typical process disturbances include ambient air temperature, feed temperature, feed flow rate, feed composition, steam pressure fluctuations, and cooling water temperature swings. For example, the feed (flow, temperature, composition) is usually a critical disturbance in distillation operation. The ambient temperature often has a significant impact on the operation as well.

Disturbances can be measurable or unmeasurable. Both can drastically affect the unit operation. For proper control design, all the primary sources of disturbances must be identified and considered in the control strategy.

In Example 6.1, the process flow may be disrupted or altered by the following events or devices:

1. Well shutdown: The compressor will run in turndown mode due to insufficient gas when some production wells are shut down. Make-up gas from the main gas line will be brought in to fill the gap in supply.
2. Compressor shutdown: When the compressor goes down, the gas will be drawn from the main gas line to meet the demands of the HRSG and gas lift wells.
3. HRSG shutdown: When the steam generators are offline, the excess gas will be diverted to the injection wells.

6.3 Supply and Demand Analysis

Process operation is about the continuous (or batch) movement of material, with energy added or removed along the way. Any stream coming to the system boundary is a supply, while any stream going out is a demand. Generally speaking, any stream coming to a point or *node* is a supply, while any stream going away from the point is a demand. At any point in the process flow, the material and energy must be balanced statically and dynamically; otherwise, the imbalance will be internally accumulated as inventory change or externally released abnormally, e.g., to flares.

A stream can be supply-driven or demand-driven, depending on how the flow rate is determined. Consequently, the control scheme can be drastically different for supply-driven or demand-driven operations. Some early discussions on supply and demand control can be found in Bequette (2002); Luyben et al. (1998).

6.3.1 Material and Energy Balance

Material and energy balance follows the fundamental physical laws. It is one of the primary considerations for process design, demanding the proper sizing and routing of the piping, vessels, and equipment:

$$\begin{aligned}\texttt{[Inventory Increase]} &= \texttt{[Material In]} - \texttt{[Material Out]}\\ &= \texttt{[Supply]} - \texttt{[Demand]}.\end{aligned} \tag{6.1}$$

Since gas is compressible, the imbalance between supply and demand in a gas processing unit impacts the pressure first, caused by the so-called *line-packing*. For incompressible liquid flow, the imbalance between supply and demand is reflected in liquid inventory, indicated by the fluid levels in tanks and vessels.

Sustained steady-state imbalance in material or energy flow is a process design mistake. A common consequence is continuous flaring in the gas operation, overflow/emptying of surge tank in liquid operation, or choked operation with reduced capacity. During process control design, it is essential to review the stream data, such as the design values of the flow, temperature, and pressure, and validate the steady-state material/energy balance against the different modes of operations, such as startup, normal operation, turndown operation, and depleted operation. The stream data can typically be found in the PFDs.

A crucial concept is the *swing stream*. Along any process flow path, the supplies and demands must not be all fixed. At least one supply or demand stream must have sufficient capacity to "absorb" the fluctuations or imbalances between supply and demand. That is, the flow rate must have room to swing up and down. For Example 6.1, the gases from the wells and the VRU are the supplies, with their flow rates determined by upstream operation. The gases to the wells and HRSG are demands, with their flow rates set by the respective wells and HRSGs. Based on the reservoir simulation results, the gas supply may be less than the demand for some operating scenarios; thus, the sweet gas stream is available as make-up gas when needed, which serves as a swing stream.

6.3.2 Supply-Driven and Demand-Driven Operation

The supply and demand model describes how the material and energy are moved with the process flow. Each stream can be a supply or a demand, and each node on the flow path can have a mix of supply and demand coming in and going out.

1. Supply-driven stream: A stream is supply-driven if the upstream operation dictates the flow rate, and the current node has no control over it. A supply-driven stream is also called a *supply-pushing* stream since it is pushed into the node. For example, the produced gas in Example 6.1 is a supply-driven stream since the production of the wells determines how much gas is produced and comes into the system boundary. The operating objective is to maximize the (oil) production, and the surface facility is expected to adapt to the production variations. Similarly, the gas from the vapor recovery unit (VRU) is supply-driven since the compression unit has no control over how much gas will come in from the VRU.
2. Demand-driven stream: If the consumer determines how much it takes from the system, it is a demand-driven or *demand-pulling* stream. The primary consumer of the compressed gas is the heat recovery steam generators (HRSG). The HRSGs operate at a pre-defined load. So the gas flow stream to HRSG is determined by the HRSG load and is thus demand-driven. Similarly, the gas lift determines the gas lift wells' gas flow rate; therefore, this stream is also demand-driven.
3. Reversible supply and demand model: A supply-driven ("pushing") stream can become a demand-driven ("pulling") stream, and vice versa, depending on the material balance. In Example 6.1, the HRSG pulls gas from the compression unit based on how much it needs. However, if the valve to the HRSG is already fully open, but HRSG still cannot get all it needs, then the HRSG will have to settle with whatever it can receive. It then becomes supply-driven. This reversal may be only temporary if the steady-state material balance is adequate. It may also become permanent if the supply and demand model has changed permanently.

In summary, for Example 6.1, the process inside the system boundary is a mixed supply-driven and demand-driven operation with two supply-driven streams (produced gas and VRU gas), two demand-driven streams (fuel gas for HRSG and lift gas for wells), and two swing streams.

6.3.3 Supply-Driven and Demand-Driven Control

Process control enforces the supply and demand model. The control strategy can be supply-driven control or demand-driven control corresponding to the supply-driven or demand-driven operation. It is essential to know how to recognize or design a supply-driven control or a demand-driven control. Typical control configurations are shown in Fig. 6.2 and may include the following common scenarios:

1. No control or controller out of service ⇒ supply-driven.
2. Controller in manual mode or control output saturated ⇒ supply-driven.
3. Standalone flow controller upstream of system boundary ⇒ supply-driven (Case 1a in Fig. 6.2).
4. Standalone flow controller downstream of system boundary ⇒ demand-driven (Case 2a).
5. Pressure controller upstream of a control valve ⇒ supply-driven (Case 1b).

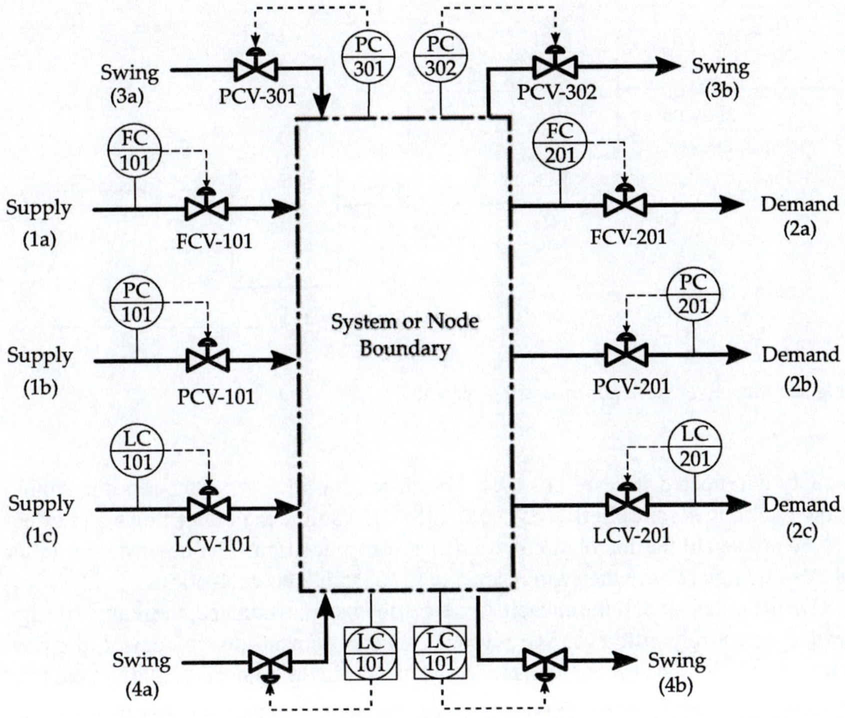

Fig. 6.2 Control options for supply and demand operation

6. Pressure controller downstream of a control valve ⇒ demand-driven (Case 2b).
7. Level controller upstream of a control valve ⇒ supply-driven (Case 1c).
8. Level controller downstream of a control valve ⇒ demand-driven (Case 2c).
9. Standalone flow controller inside system boundary ⇒ attention! There cannot be two active flow controllers (over-specified!) on any continuous flow path unless there is also a swing stream.

The above rules can be used as a guide for placing process measurements and final control elements. These rules can also be followed when analyzing an existing process design.

The imbalance between supply and demand is sensed by pressure or level. Capacity control for gas processing aims to maintain a desired *pressure–flow profile* along the process flow path, and capacity control for liquid processing maintains a healthy *level–flow profile*.

The supply and demand analysis starts from the incoming streams at the system boundary and propagates downstream for supply-driven operation. Conversely, the analysis starts from the demand point at the system boundary and propagates backward to upstream for demand-driven operation. For example, for the supply-driven operation Fig. 6.3, the flow and level control are so placed that the fluctuation in

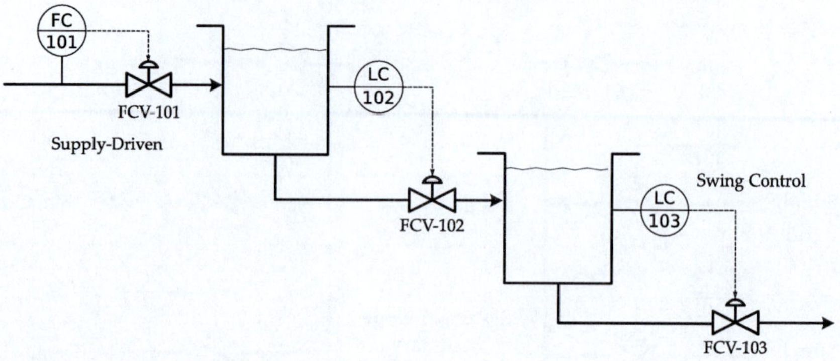

Fig. 6.3 Supply-driven liquid processing operation

supply is reflected in level changes. The flow controller `FC-101` sets the supply flow rate. Level controllers (`LC-102`, `LC-103`) adjust the outlet flows and effectively propagate the fluctuations to downstream operation. The downstream (after `FCV-103`) must have the swing capacity to "absorb" the fluctuations.

On the other hand, if the capacity is set by downstream demand, the control design would be entirely different (See Fig. 6.4). In this demand-driven control, the flow controller `FC-104` sets the demand. Any changes in the demand will affect the tank levels, and the level controllers (`LC-102`, `LC-103`) would have to adjust the inlet flows to keep the level constant. This level–flow interaction propagates backward all the way to the inlet flow of the first tank, which must have sufficient swing capability to absorb the demand fluctuations.

For gas processing, the focus is the *pressure–flow profile* instead of the *level–flow profile*. Figure 6.5 is an example of a supply-driven gas processing operation. The fluctuation in gas supply to the compressor is reflected in the pressures. The pressure controllers force more flows through the compressor and piping. This flow–

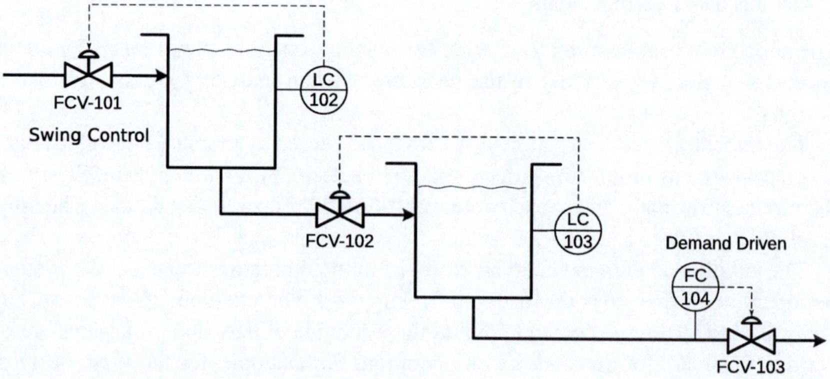

Fig. 6.4 Demand-driven liquid processing operation

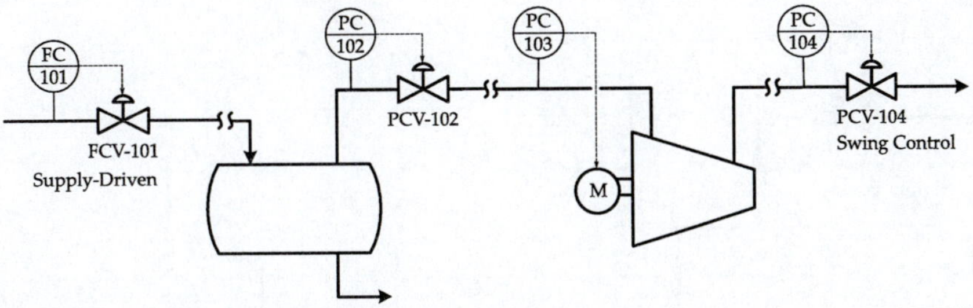

Fig. 6.5 Supply-driven gas processing operation

pressure–flow–pressure–flow propagation goes on until reaching the final consumer. The consumer of the gas needs to have the capacity to absorb the swing in supply.

For the same process, if driven by demand, the control strategy would have to be changed to that shown in Fig. 6.6. The critical requirement is that the fluctuation in demand must be propagated to, and accommodated by, the supply.

A noteworthy scenario is that a supply-driven operation can become demand-driven temporarily or permanently, and vice versa. In this case, the control strategy will need to be changed to *supply-driven control with demand override* or *demand-driven control with supply override*. This concept is illustrated in Fig. 6.7. Overriding control is prevalent in processing facilities in E&P operation (See Sect. 2.2.7).

Adding override protection against supply and demand reversal, Fig. 6.3 would become Fig. 6.8. In case the demand side cannot take all the flows (e.g., the valve is already fully open), the operation automatically reverts to demand-driven, and the

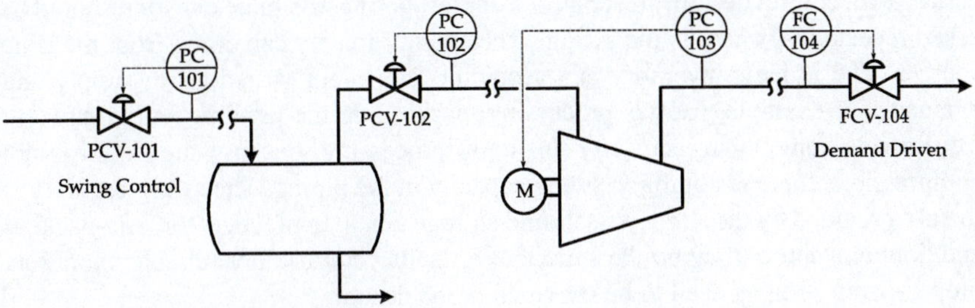

Fig. 6.6 Demand-driven gas processing operation

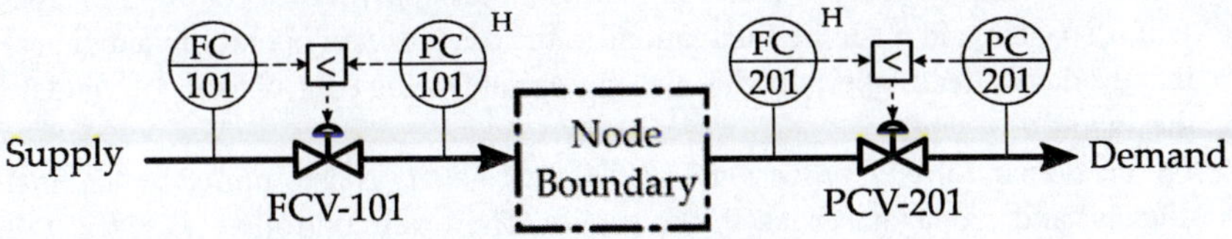

Fig. 6.7 Supply and demand operation with override protection

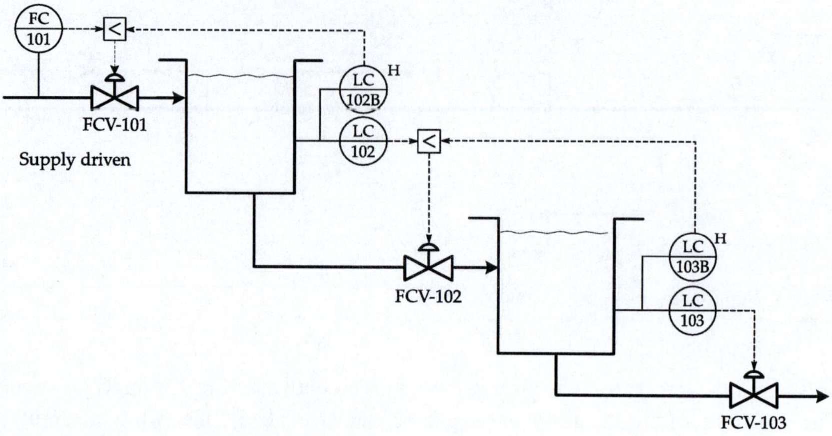

Fig. 6.8 Supply-driven control with demand override

overriding controllers (`LC-103B` and `LC-102B`) take over, propagating the demand back to the supply side. Again, the upstream operation must be able to reduce supply to adapt to the demand change.

6.3.4 Swing Capacity and Swing Control

The swing capacity is critical in balancing supply and demand dynamically. In process design, control design, and control troubleshooting, the same question should be asked repeatedly: where is the swing? The swing capacity can come from multiple sources. The process inventory can accommodate minor variations in supply and demand. For example, for gas processing, packing up the pipeline and equipment can provide some swing capacity. In liquid processing, because the liquid is not compressible, there is not much swing capacity in the piping. The swing capacity is mainly provided by the surge vessels and storage tanks. In practice, the line-packing and liquid inventory may not be sufficient for swing control. For reliable operation, proper swing streams need to be provided in the design:

1. For a supply-driven process, such as in Fig. 6.3, the final demand must have sufficient swing capacity to take the fluctuation in supply. Level control `LC-103` will propagate and push the fluctuations to the downstream operation and expect that the downstream operation has the appropriate swing control to accommodate the swing.
2. For the demand-driven process in Fig. 6.4, the capacity is determined by the final demand and propagates back to the supply. The level controller `LC-102` will back-propagate the demand variations upstream of the `FCV-101` and expect that upstream has swing control to adjust the supply.

3. For a process where the supply and demand model may reverse, both the supply and demand will have to be able to swing. The case with Fig. 6.8 is an example. Although seemingly complex, this reversal in supply and demand is common in practical applications.
4. For a mixed model with both supply-driven and demand-driven streams, as in Fig. 6.1, dedicated swing streams such as the flows to injection wells, the make-up gas, and the gas to flare are provided.

6.4 Cause and Effect Analysis

Process control, either open-loop or closed-loop, is based on the assumption that adjusting some variables (control handles) can cause changes to some other variables. The cause and effect analysis aims to identify all key process variables (the *effect*) and ensures the necessary control handles (the *cause*) are available to affect the desired changes.

6.4.1 Controlled Variables

Although the number of process variables in scope can be overwhelming, a small subset of the variables can often provide a sufficiently reliable indication of the overall operating health. Keeping these process variables at their desired values ensures process stability (Juliani and Garcia 2017). We will call these variables the *key process variables* (KPV). Most of them will become the controlled variables for the process control.

The key process variables typically include the following:

1. Quality control variables: All measurements with commercial or operational specifications on their quality or quantity, such as composition, impurity, dew point, vapor pressure, qualify as quality control variables. Pressure, temperature, and flow rate can also be product specifications. In Example 6.1, the gas supply pressure (`PT-105`) for the HRSGs and gas lift wells is a quality control variable.
2. Capacity control variable: Capacity control is for maintaining the dynamic supply and demand balance. Some critical variables are the relay points for the control action propagation, typically pressures (for gas) or levels (for liquid). In Example 6.1, the suction pressure `PT-101` is a capacity control point since the pressure is an indicator of the supply and demand imbalance and can be used to adjust the compressor speed to achieve the required capacity. The discharge header pressure `PT-105` is another capacity control point that needs monitoring and control.
3. Constraint variables: These are the limits and constraints that define the operating and safety envelope. They can lead to the shutdown of the operation if not controlled or limited. In Example 6.1, the suction pressure `PT-101` is a constraint variable since an excessively high pressure may trip the vapor recovery unit.

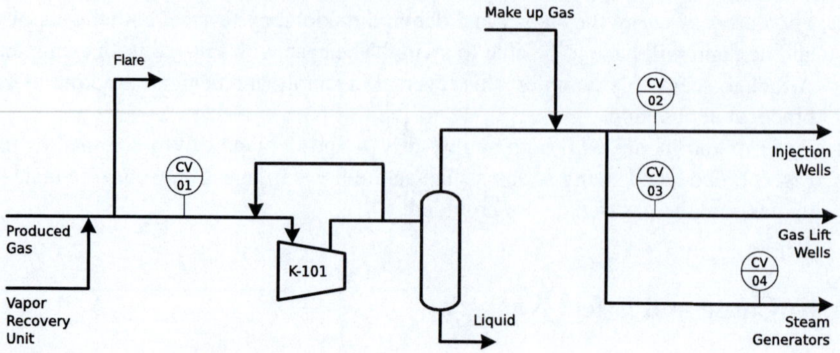

Fig. 6.9 A gas compression process: key process variables

The compression process is shown again in Fig. 6.9, with the key process variables added.

6.4.2 *Manipulated Variables and Disturbance Variables*

The causes for process changes could be active or passive. A final control element such as a valve or electric switch can actively cause a change to the process, while ambient air temperature may passively affect the process variables. The causes that we can manipulate are called the *manipulated variables* (MV), and those that we cannot manipulate are called the *disturbance variables* (DV).

Let us examine all the control handles in Example 6.1. Temporary tag names are used for now. They will be assigned permanent tag names later once the control strategy is finalized. The manipulated variables include the following:

1. `MV-01`, Flaring valve: A control valve to send excess gas to flares.
2. `MV-02`, Compressor speed: Usually, an electric motor drives the compressor via a variable speed gearbox (VSGB) to adjust the compressor speed.
3. `MV-03`, Compressor recycle valve: This valve is primarily for anti-surge control (see Sect. 11.3).
4. `MV-04`, Make-up gas valve: A control valve to bring in sweet gas as make-up gas.
5. `MV-05`, Supply valve for injection wells.
6. `MV-06`, Supply valve for gas lift wells.
7. `MV-07`, Supply valve to HRSG.

Figure 6.10 shows the process flow with the control elements (valves and motor drives). The placement of the control elements is based on the supply and demand analysis and is in line with supply-driven or demand-driven operating requirements.

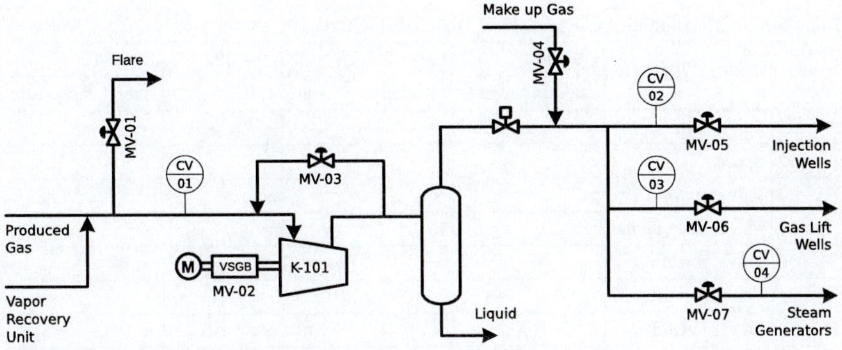

Fig. 6.10 A gas compression process: with control valves

6.4.3 Cause and Effect Relationships

We use the manipulated variables (MV) to keep the controlled variables (CV) at the targeted values to maintain the material and energy balance. The vital requirement is that the cause and effect relationship between the MVs and CVs must exist and must be known. When designing or analyzing a process control scheme, it is crucial to identify all the key *manipulated* and *controlled* variables and examine the cause and effect relationships.

The relationship may be single-input single-output (SISO) or multi-input and multi-output (MIMO). Most of the cause and effect relationship is simple and obvious. For example, it is evident that the flow control valve is the cause for a flow controller, and the flow rate is the effect. However, in many cases, the flow rate may be the cause for other effects, e.g., in a level-to-flow cascade loop, the flow is a cause and an effect simultaneously. In complex control schemes with multiple MVs and multiple CVs, the cause and effect relationship can become quite complicated and need to follow a structured approach for the analysis:

1. Focus on the "bigger picture." The challenge for a multivariable process is in the interactions between process variables. A change in one MV can cause changes in multiple CVs. Similarly, the desired CV change can be effected by adjusting multiple MVs. In the compressor example (Fig. 6.10), the suction pressure `PC-102` can be controlled by changing the compressor speed or adjusting the recycle valves. Using two MVs to control one CV is a design decision, as discussed in Sects. 2.2 and 4.1.
2. Simplify. The purpose is to examine the interaction between the MVs and CVs, and certain localized variables can be simplified for better clarity. For example, in a temperature–flow–valve cascade control, the cause and effect relationship between the control valve and the flow rate is well-known and localized to the flow. The effect of the valve on other variables goes through the flow; therefore, the flow is considered a manipulated variable instead of the control valve.

Table 6.1 PGC example: cause and effect table, preliminary

		CV-01	CV-02	CV-03	CV-04
		Suction pressure	Discharge pressure	Lift-gas flow	Fuel gas pressure
MV-01	Valve, flare	×			
MV-02	Speed	×	×	×	
MV-03	Valve, recycle	×	×	×	
MV-04	Valve, makeup gas		×	×	
MV-05	Valve, injection		×		
MV-06	Valve, gas lift			×	
MV-07	Valve, HRSG		×	×	×

Following the supply and demand analysis in Sect. 6.3, let us examine the cause and effect relationship among the manipulated variables and the key process variables in Fig. 6.9. A cause and effect table (CET) is produced in Table 6.1, where "×" indicates that a cause and effect relationship exists between the corresponding variables.

From the table, it is clear that a final control element (e.g., compressor speed MV-02) can cause multiple process variables (CV-01, CV-02, and CV-03) to change, while a process variable (e.g., CV-02) may be affected by several control elements.

"Correlation does not imply causation" (Aldrich 1995). Many times the cause and effect relationship is just a correlation in process input/output data. The cause and the effect must be further examined and validated from the physical principles of the process dynamics.

6.5 Degree of Freedom Analysis

Degree of freedom (DOF) is the number of variables that must be specified to define a process completely; for process control design, every manipulated variable (MV) or control handles, such as a valve, motor speed, or a slave controller, is counted as one DOF.

6.5.1 Degree of Freedom Requirements

Depending on the number of controlled variables and manipulated variables, the control problem can be as follows:

1. Over-specified: The number of controlled variables is more than the number of manipulated variables. All the controlled variables cannot be maintained, and some must be neglected or compromised.

2. Under-specified: The number of manipulated is more than the number of controlled variables. There may not be a unique solution in terms of control action, and some of the manipulated variables will eventually be driven to saturation to exhaust the DOFs.
3. Fully specified: The number of manipulated variables is the same as the number of controlled variables.

A DOF must be a variable that can be independently manipulated and is usually provided by a final control element such as a valve or a motor. Typically the DOF has a cost in terms of CAPEX and OPEX, and an unconsumed DOF is a waste. It is expected that process control design will fully utilize all DOFs. A control valve that remains saturated (fully open or fully closed) renders this DOF unavailable. Leaving a valve in the manual mode also removes the valve as a DOF. On the other hand, sometimes multiple DOFs are required to control one process variable, such as in a split-range control or fan-out control scheme where multiple valves are counted as one DOF.

For proper control design, all DOFs must be consumed by pairing up all the variables and controlled variables. A controlled variable, once paired with a manipulated variable, consumes one DOF. Two controlled variables must be sufficiently independent of each other to consume two DOFs. If they are strongly correlated, they may be considered one controlled variable and consume only one DOF:

1. For simple control (Chap. 3), DOF analysis is usually straightforward. For example, a standalone flow controller has the flow rate as the controlled variable and the control valve as the manipulated variable and is thus fully and properly paired up.
2. In complex PID control schemes (Chap. 4), such as those containing split-range control, or selective control, the DOF condition is less obvious and often overlooked. The consequence is poor control performance.
3. In model predictive control (MPC, Chap. 5), the DOF condition is handled implicitly in the model structure and needs to be very careful. Relative gain analysis (RGA) is typical to verify the DOF condition and variable pairing are adequate.

6.5.2 *Pairing of Process Variables*

The manipulated variables (MV) and controlled variables (CV) must be fully paired up to leave no DOF unconsumed. The MV and CV pairing needs to consider many factors, with the supply and demand model and the cause and effect relationship as the critical considerations. In the multivariable control scheme, the following questions are required to be answered:

1. Which MV to control which CV? Generally speaking, a CV should be controlled by the MV with the strongest and most direct effect.
2. What if there are not enough MVs to control all the CVs? When there are more CVs than MVs, the typical approach is to reduce the number of CVs via cascade control, override control, selective control.

3. What if there are more MVs than CVs? When there are more MVs than CVs, the typical approach is complex PID schemes with split-range or fan-out control.
4. What if multiple MVs affect multiple CVs? Strongly interacting MVs and CVs constitute a multivariable control problem, and explicit or implicit decoupling is typically needed.

Let us continue the compressor example. Table 6.1 has provided all the critical cause and effect relationships between the final control elements and the key process variable. Some of the variables will pair up as manipulated variables and controlled variables through a DOF analysis. Others may need further actions to ensure the DOF conditions are met. We will go through all the variables one by one as follows:

1. `CV-01`, the suction pressure: The suction pressure is a key process variable. It is the primary indicator for supply and demand balance and thus should be upgraded to a controlled variable. During normal operation, the compressor is supply-driven by design. Therefore, this pressure controller serves as the capacity controller. The compressor speed `MV-02` is the obvious choice to pair with this CV based on engineering insight. In terms of supply and demand control, this is Case 2a in Fig. 6.2.
2. `CV-02`, the discharge pressure: The discharge header supplies the compressed gas to multiple consumers and should be another controlled variable. By design, the compressed gas is supply-driven, and the injection gas is the swing stream. As a result, the discharge header pressure `CV-02` is paired with the injection gas flow valve `MV-05`. The supply and demand model is Case 3b in Fig. 6.2 for supply and demand balancing.
3. `CV-03`, the lift gas flow: Per operating requirements, the lift gas flow is demand-driven. Therefore, the gas flow `CV-03` pairs up with the flow control valve `MV-06`. The supply and demand model is Case 2a in Fig. 6.2.
4. `CV-04`, fuel gas to HRSG: The make-up gas is another demand-driven stream, and the flow rate is dictated by gas pressure. So the CV-MV pair-up is `CV-04` with `MV-07`. The supply and demand model is Case 2b in Fig. 6.2.

With the above CVs and MVs paired up, the cause and effect diagram is now shown in Table 6.2, where the symbol ⊗ indicates the paired variables.

Now let us look into the remaining variables that are not yet paired:

1. `MV-01`, flare valve: This valve sends excess gas to flare in case of imbalance caused by more supply than demand. The controlled variable to pair with should be the suction header pressure, which has already been paired with compressor speed `MV-02`. The flare valve is an extra DOF to be consumed.
2. `MV-03`, recycle valve: The recycle valve is dedicated by design for anti-surge control (see Sect. 11.3). It also has a strong effect on the suction header pressure `CV-01`. From engineering insight, the compressor speed also strongly affects the surge condition of the compressor. This control problem thus has two MVs and two CVs to pair up, and a multivariable control scheme is needed to handle the complex cause and effect relationship.

Table 6.2 PGC example: cause and effect table, updated

		CV-01	CV-02	CV-03	CV-04
		Suction pressure	Discharge pressure	Lift-gas flow	Fuel gas pressure
MV-01	Valve, flare	×			
MV-02	Speed	⊗	×	×	
MV-03	Valve, recycle	×	×	×	
MV-04	Valve, makeup gas		×	×	
MV-05	Valve, injection		⊗		
MV-06	Valve, gas lift			⊗	
MV-07	Valve, HRSG		×	×	⊗

3. MV-04, make-up gas valve: This make-up gas flow valve is open only when the overall supply is less than demand, and in normal operation, it is fully closed by design. The effect of this valve is on the discharge pressure header CV-02, which has already been paired up with injection valve MV-05. This make-up gas valve is an extra DOF to be consumed.

The different control options will need to be evaluated to pair up all the process variables completely. We will defer the remaining DOF analysis to Chap. 7.

6.6 Dynamics Response Analysis

The cause and effect analysis in Sect. 6.4 reveals the cause and effect relationships between the process variables. For process control, the bare existence of a cause and effect relationship is not enough. A quantitative description of the relationship is essential to process control design.

Most process control technologies assume that the process dynamics are relatively linear and do not (significantly) change with time. However, most industrial processes are inherently nonlinear and constantly change with time. In addition, this nonlinearity and time-variance are often unpredictable. Therefore, it is critical to know the process dynamic characteristics and have a good idea of the applicability range of the control solution.

6.6.1 Process Disturbances

Disturbances and fluctuations are inevitable in process operation! They are the main reasons for the need to control. There are many types of process disturbances from different sources and at different levels.

Disturbances can be *measurable* or *unmeasurable*. Process control can reduce the impact of measurable disturbance by resorting to feedforward control or model predictive control. For unmeasureable disturbance, the only means to combat it is probably feedback control, after it has impacted our controlled variables.

It is important to remember that process control cannot reject disturbances. It simply transfers the noise and disturbances from one place to another, from a critical variable to a less critical one. For example,

1. A flow is stabilized by shifting the fluctuations to valve movements. As per the design, if the valve is driven by instrument air, the flow fluctuations are shifted to the instrument air consumption.
2. A temperature in a heater is maintained constant by shifting the temperature variations to the fuel oil flow.
3. With a surge tank, inlet disturbance is absorbed by level (less critical) in exchange for a smoother outlet flow (critical variable); conversely, the tank level can be maintained stable by shifting the level fluctuations to the outlet flow and then to downstream (e.g., a storage tank).
4. Poor control itself can generate disturbances as a result of, for example, valve malfunction and poor controller tunings.

When a disturbance is included in the control design, it is crucial to consider where it is expected to be transferred to.

6.6.2 Dynamics Response

Process dynamics refer to the time trajectory of a variable in response to a change in another variable. All control actions are based on known and predictable process dynamics. The type of dynamic response determines the applicability of the different process control technologies and affects the control strategy.

We all have the inherent appreciation that the effect of a cause takes time to manifest itself. It takes seconds to see the flow change after a valve position change, but it may take many minutes or even hours before its impact on product quality can be observed. In general, the dynamic response of a stable (self-regulating) process has three key characteristics, that is, once there is a change in input variable $u(t)$, *when*, *how much*, and *how fast* the output variable $y(t)$ will respond:

1. Process gain (K_p): By moving the input variable by one unit, how many units of change (direction and magnitude) will be eventually caused on the output variable. A stable process will eventually reach a new steady-state, while an integral (ramp) process does not have a new steady-state.
2. Time delay or dead time (θ): After a change is made to the input (the manipulated variable, or MV), when will the output variable (the controlled variable, or CV) start to move? Time delay is typically caused by transport delays. Large time delay is one of the most significant challenges for control.

3. Transient behavior: After the time delay elapses and the output starts to move, how fast will the output reach a new steady-state, and what trajectory will it follow?

Although a first-order-plus-delay-time (FOPDT) model can approximate most dynamic response behaviors, many processes exhibit higher-order or nonlinear characteristics. The different process dynamics differ mainly in the transient behavior, including the response speed and trajectory. Figure 6.11 shows the commonly encountered simple process dynamics:

1. First-order dynamics (Fig. 6.11a): Most industrial processes can be sufficiently approximated by a first-order-plus-dead-time (FOPTD) characteristic, where the transient behavior is defined by process gain K_p, delay θ, and time constant τ, and d is the time delay in the number of samples:

$$G(s) = \frac{K_P}{\tau\, s + 1} e^{-\theta\, s} \quad \mapsto \quad y(t) = K_p\, e^{-\frac{t-\theta}{T_s}} \tag{6.2}$$

$$G(z^{-1}) = \frac{b_1\, z^{-\theta-1}}{1 + a_1\, z^{-1}} \quad \mapsto \quad y(k) = -a_1 y(k-1) + b_1\, u(k-d-1). \tag{6.3}$$

   The time constant is the amount of time that takes from the moment that the output variable starts to respond until the output value reaches 63.2% of the new steady-state value.

2. Higher-order dynamics (Fig. 6.11b): Although most process dynamics can be approximated with a first-order-plus-dead-time, some processes exhibit a truly high-order characteristic (characterized by the "S" shaped rising curve). Second-order process dynamics may be sufficient to approximate most high-order dynamics:

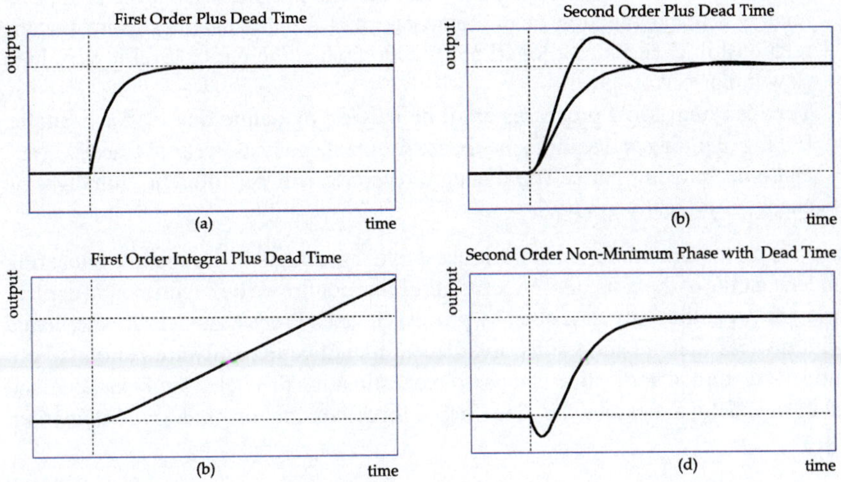

Fig. 6.11 Step response of common process dynamics

$$G(s) = \frac{K_p}{(\tau_1 s + 1)(\tau_2 s + 1)} e^{-\theta s} \quad \mapsto \quad G(z^{-1}) = \frac{b_1 z^{-1} + b_2 z^{-2}}{1 + a_1 z^{-1} + a_2 z^{-2}} z^{-d} \tag{6.4}$$

$$y(k) = -a_1 y(k-1) - a_2 y(k-2) + b_1 u(k-d-1) + b_2 u(k-d-2). \tag{6.5}$$

3. Integrating process (Fig. 6.11c): Some processes, such as the liquid level in a tank or the pressure in a fixed volume, are not self-stabilizing. They behave like a ramp, and there is no steady-state in step response:

$$G(s) = \frac{K_p}{s} e^{-\theta s} \quad \text{or} \quad G(s) = \frac{K_p}{s\,(\tau s + 1)} e^{-\theta s}. \tag{6.6}$$

4. Inverse response (non-minimum phase, Fig. 6.11d): Some processes may show strange behaviors and pose a significant challenge to control. For example, when cold feed-water is added to a steam boiler, the water level may go in the "wrong" direction before returning to the "right" direction. The control solution thus needs to address this behavior.
5. Unstable: Some processes, such as an exothermic reaction process, are inherently unstable. If not adequately controlled, it can diverge quickly and lead to loss of containment and possible explosion.
6. Nonlinear process: A nonlinear process exhibits different dynamics at different operating conditions. A notorious example is pH control, where a flow of acid or base (called the reactant) is added to the product stream to control the pH value to a given value. The process is very nonlinear, and an example of the process gain is illustrated in Fig. 6.12. A simple PID control is almost impossible for proper pH control. A feedforward control, ratio control, or preferably model predictive control is often needed to battle the nonlinearity.
   Another example is the high purity distillation. The impurity level is approximately a linear function of the temperature at a particular tray. However, the relationship changes to a logarithm function when the purity level is very high (low impurity).
7. Time-varying: Most processes are time-varying by nature due to, for example, fouling of piping and equipment, de-activation of catalysts, wear of tear of valves, and other moving parts. The dynamic response will be different, although the change is typically gradual.

In practice, the determination of the process dynamics starts with a sound engineering understanding of the cause and effect relationship, confirmed by experimental or plant tests, and quantitatively described by process models. Process models are facilitated with process identification software. Decent identification software supports both model structure identification and parameter estimation that help supply and demand analysis and dynamics analysis. See Chap. 5 for discussions on *process identification*.

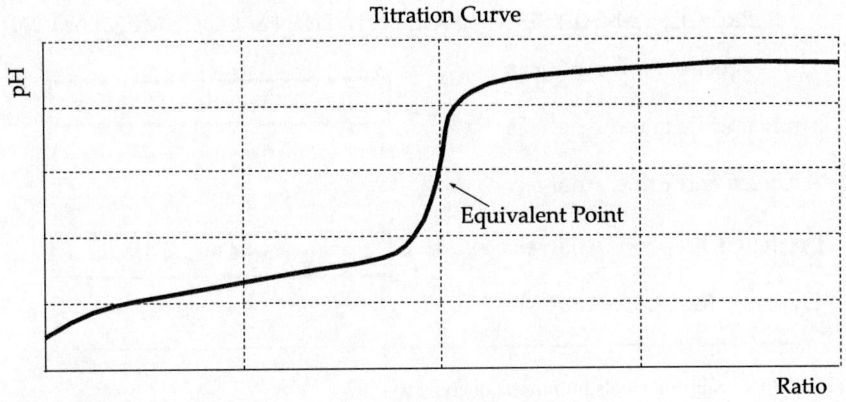

Fig. 6.12 Example of nonlinear process: pH process gain

Depending on the type of dynamics, different models may be required. Subsequently, other control technologies may have to be considered. More discussions will follow in Chap. 7.

6.7 Process Analysis Documentation

We conclude this chapter by summarizing the best practice for process analysis and formalizing the two simple visual design tools for the visualization and documentation of the analysis results.

When an unfamiliar process is encountered, it is helpful to follow a structured process analysis methodology to familiarize oneself with the process quickly. The information required by process control is quite different from that required by process engineering; therefore, it is vital to know what to look for:

1. Understand the operating objectives and requirements.
2. Identify all the flow streams that may affect material and energy balance.
3. Decide on the key process variables to be controlled.
4. Identify all the significant disturbances and constraints.
5. Determine the supply and demand model for each stream.
6. Decide on the placement of final control elements based on the supply and demand model.
7. Perform a cause and effect analysis to pair up the controlled variables and manipulated variables. Add or relocate the variables as necessary.
8. Gain a basic idea of the dynamic response of each cause and effect relationship, and decide on the appropriate type of control needed.
9. Repeat the above steps if the process design is modified or new information is obtained.

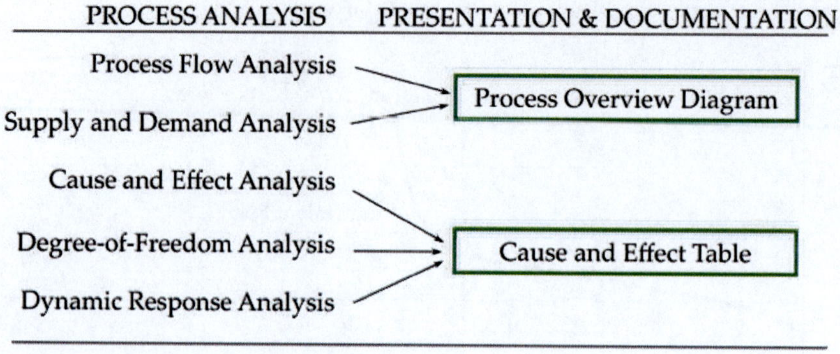

Fig. 6.13 Visualization tools for process analyses

The process engineers are the best resources for in-depth discussions during process analysis.

Two visual design tools are helpful to document the analysis results and for more effective communication with other teams. One is the process overview diagram (POD) for providing a holistic view of the process flow, and the other is the cause and effect table (CET) for visualizing the cause and effect relationship and optionally the dynamic response characteristics. Both have been extensively used in this chapter (Fig. 6.13).

Figures 6.1, 6.9 and 6.10 are examples of POD for unit operation. Figures 1.3 and 12.27 are POD examples for large-scale plant operations.

The CET can be simple or complex. Tables 6.1 and 6.2 are simple CET examples showing the model structure and the cause and effect relationships without the process dynamics. Table 5.2 and Fig. 5.14 are examples of the CET with dynamic responses. The crucial information in the CET include the following:

1. The controlled variables (CV), the manipulated variables (MV), and the disturbance variables (DV).
2. The existence of a cause and effect relationship between an MV/DV and CV.
3. The approximate characteristics of each identified cause and effect relationship.
4. The quality (credibility) of each identified cause and effect relationship.

Both the POD and CET are progressively developed as the process analysis proceeds.

6.8 Summary

There is an infinite number of processes in various process industries. Even the same process design may have different operating requirements and constraints in a different plant (Sect. 6.1). Although we strive for simplification and standardization, we must validate each control solution to ensure it is fit-for-purpose. On the other hand, it is unrealistic, nor necessary, to expect a process control person to achieve the same level of process understanding as a process engineer does. As a result, one of the main challenges facing process control engineers or researchers is, given a new and unfamiliar process unit, where to start with the process analysis to quickly reach the level of understanding required to perform process control duties.

Fortunately, the process knowledge required by a process control is very different from that of a process engineer. We can quickly and systematically gather what we need for the underlying process following well-defined procedures and guidelines with the basic knowledge in process engineering such as thermodynamics, heat transfer, fluid mechanics, mass transfer, and chemical reaction.

Chapter 5 already provides a glimpse into the process flow analysis methodology for model predictive control. This chapter formalizes this work process and presents the essential skills and practical methodology for understanding the process dynamics. These include process flow analysis (Sect. 6.2), cause and effect analysis (Sect. 6.4), supply and demand analysis (Sect. 6.3), DOF analysis (Sect. 6.5), and dynamic responses analysis (Sect. 6.6).

The scope and requirements of process analysis are summarized in Fig. 6.14. Based on the complexity and specificity, all or part of the procedure can be fol-

- Process Analysis
  - Operating Requirements
    - Operating Objectives
    - Product Specifications
    - Modes of Opertions
    - Limits and Constraints
  - Process Flow Analysis
    - System Boundary
    - Key Material and Energy Flows
    - Major Disturances
    - Interaction with Other Units
  - Supply and Demand Analysis
    - Material and Energy Balance
    - Supply-Driven and Demand-Driven Operation
    - Swing Stream and Swing Capacity
  - Cause and Effect Analysis
    - Controlled Variables
    - Manipulated and Disturbance Variables
    - Cause and Effect Relationships
    - Cause and Effect Table
  - Degree-of-Freedom Analysis
    - Degree-of-Freedom Requirements
    - Cause and Effect Variable Pairing
    - Degree-of-Freedom Balancing
  - Dynamic Response Analysis
    - Types of Dynamic Response
    - Quality of Dynamic Response
    - Simplification of Dynamic Response

Fig. 6.14 Scope and checklist for process analysis

lowed when an unfamiliar process is encountered. Although the process engineers are always the best source for in-depth knowledge when needed, we need to know enough to know what we do not know.

References

Aldrich J (1995) Correlations genuine and spurious in Pearson and Yule. Stat Sci 10(4):364–376

Arnold K, Steward M (1999) Surface production operations—design of oil-handling systems and facilities, vol 1, 2nd edn. Imperial College Press

Bequette BW (2002) Process control-modeling, design, and simulation. Prentice Hall

Campbell J (2004) Gas conditioning and processing—the equipment modules, vol 2, 8th edn. John M, Campbell and Company

Downs J, Skogestad S (2011) An industrial and academic perspective on plantwide control. Ann Rev Control 99–110

Juliani RCG, Garcia C (2017) Plantwide control: a review of design techniques, benchmarks, and challenges. Ind Eng Chem Res 56(28):7877–7887

Luyben WL, Tyréus BD, Luyben ML (1998) Plant-wide process control. McGraw Hill, New York

Marlin TE (2015) Process control—designing processes and control systems for dynamic performance, 2nd edn. McGraw-Hill

Seames W (2018) Designing controls for the process industries. CRC Press

Turton R, Shaeiwitz JA, Bhattacharya D (2018) Analysis, synthesis, and design of chemical processes. Prentice Hall

Chapter 7
Methodology of Process Control Design

Experience shows that a large percentage of operational difficulties originate from poor process control design. Once in operation, the issues caused by flawed design are typically difficult and expensive to correct. A modern operating plant is usually very complex, so is the control solution (Sect. 7.1). The flawed control design is often caused by the lack of a coherent overall control strategy (Sect. 7.2). It is not unusual that all the individual control loops perform well in silos but fail to work in concert when put together. So a structured design methodology should be followed to achieve optimality at all plant levels (Sect. 7.3). The control solution is an integral part of the plant operation; the documentation and visualization of the control scheme (Sect. 7.4) are critically important but have been long neglected or undervalued. This chapter provides some general discussions and guidelines to address these issues.

7.1 Process Hierarchy and Process Control Scope

A control solution must serve a clearly defined control objective. A modern production process can be overwhelmingly complex, so is the control solution. It is essential to follow a "divide and conquer" approach and break the process and control into more manageable chunks while not losing sight of the "bigger picture." (Åström and Kumar 2014; Marlin 2015; Seames 2018).

This design methodology and the link between process analysis and control design are illustrated in Fig. 7.1. The critical requirement for process control design is the strong connection and interaction between process design and control design. Process design must keep the process control strategy in mind so that the appropriate control handles and measurements are available and with adequate capacity or rangeability (Marlin 2015). The control design may have to go back to process analysis frequently to validate and improve the design. In control troubleshooting and improvement, the same methodology applies.

This tight integration between process and control is exemplified by model predictive control (Chap. 5). It is worth re-emphasizing that the process analysis is the basis for control design. Whether selecting a base layer control strategy or an advanced control strategy depends on the result of process analysis. As discussed

S. S. Niu and D. Xiao, *Process Control*, Advances in Industrial Control,
https://doi.org/10.1007/978-3-030-97067-3_7

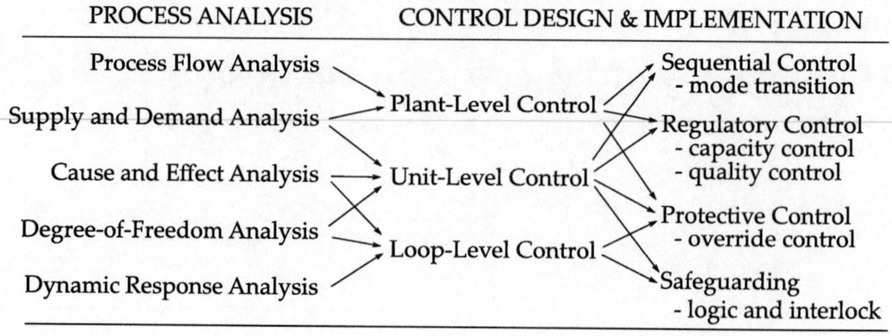

Fig. 7.1 From process analysis to control design

before, control design is always a trade-off of many factors, and the process analysis result is the basis for the decision.

7.1.1 *Process Hierarchy*

A production plant can be viewed as a *process hierarchy*. The ISA-88 standards define the physical model of operation as an enterprise, site, area, process cell, unit, equipment module, and control module. However, we will only consider three levels for control design, i.e., *plant-level*, *unit-level*, and *loop-level* operations.

An operating plant typically consists of one or more process units. Each unit comprises one or more control loops, and each loop may be a simple PID controller, a complex PID loop, or a model predictive control scheme. At each level, the operating requirements are different, and the control objectives are thus also different:

1. Plant level: The control objectives at the plant level are typically related to maintaining adequate material and energy balance. The material balance is reflected by the *pressure–flow profile* (for gas) or the *level–flow profile* (for liquid). The material balance analysis may run across multiple operating units and sometimes even cross the battery limit. Maintaining the correct supply and demand profile, both static and dynamic, is the primary consideration in the overall control strategy.
2. Unit level: At the unit operation level, product quality control is prevalent. Pressure, temperature, and quality control are commonplace. Energy integration and optimization have resulted in more compact and complex process design and thus made process control more challenging. Most of the process challenges discussed in Sect. 5.1.1 are the unit-level considerations, which have the most significant impact on the control structure.
3. Loop level: A control loop can be a simple standalone PID loop, a complex PID scheme, or a model predictive controller (MPC). The control loop performs a specific control task that is part of the overall unit-level control strategy, which in turn is part of the plant-level control strategy. Whether a PID or MPC, the

controller is an integral part of the control loop and must always be treated that way.

A PID controller is like a well-trained soldier. One soldier may be good enough to guard a post. To hold a fortress, it needs a team of well-trained soldiers, while to protect a city, it would need an army of well-organized and highly-coordinated teams. Controlling a process unit is like guarding a fortress while controlling a plant is like protecting a city. Individual well-trained soldiers are not enough. It needs plans, strategy, tactics, and most importantly, coordination to achieve a higher goal. The best process control solution is always a fit-for-purpose solution based on process design and meets the operating requirements.

7.1.2 Plant-Wide Control

An operating plant typically consists of multiple interconnected units. The material moves from upstream to downstream, often with recycles. Along with material movements, there are energy changes and exchanges.

For example, in the E&P production plant illustrated in Fig. 1.3, the operating objective is to send the separated gas, oil, and water to the different destinations, subject to the capacity (quantity) constraints. Purity requirements (quality control) are also becoming increasingly common. The plant operation comprises multiple processing units, including gas processing, condensate processing, and water treatment. Similarly, there are many separation and fractionation columns in a refinery or olefin plant, and each is a separate unit operation.

Traditionally, control textbooks and academic teachings provide very little guidance on achieving a sound design for the overall control at the plant level. The study of plant-wide control started three or four decades ago (Foss 1973; Stephanopoulos 1983; Luyben, Tyréus, and Luyben 1998) and has received more attention lately (Downs and Skogestad 2011; Juliani and Garcia 2017; Rangalah and Kariwala 2012). However, the literature on plant-wide control is still scarce compared to the importance of this subject, and most of them focus on plant-wide steady-state optimization (corresponding to the real-time optimization and planning layers in the automation pyramid in Fig. 1.10) rather than dynamic control.

HAZOP (HAZard + OPerability) review is compulsory at multiple phases of new or revamping process design projects. It is an ideal opportunity to catch flaws in plant-level control design, including safety hazards and plant operability. However, in reality, the HAZOP sessions typically devote most effort to hazard analysis and spare very little time and attention to operability reviews. The primary reason is the lack of process control expertise or awareness. Many practical challenges in operation can be traced back to poor operability in design. Therefore, operability analysis at the plant level deserves more attention than it currently receives.

The plant-level control design is based on process flow analysis, supply and demand analysis, and degree of freedom (DOF) analysis in Chap. 6.

Process Topology and Process Flows

The supply and demand model is one of the plant-level concerns that can cross the plant or unit boundary. A change in process flow in one process unit may alter the supply and demand relationship and require changes in other units. Similarly, a change in the piping layout in another plant may force a change in the process design or control design in the current plant.

Plant-wide control requires a holistic view of the plant operation, comprising all the interconnected unit operations. The job of plant-wide control design is often customarily assigned to process engineers alone. However, as mentioned in Sect. 1.1.4, a process engineer's control knowledge and mindset are significantly different, flaws are common in the overall control strategy designed by process engineers. Many of them are even against basic process control principles. It is usually too late to correct the flaws after the control solution is finalized or the plant is built; a more elaborated control solution must be designed to accommodate or compensate for the controllability issues in the process design. Many times plant operations have to live with the flaws until the next major turnaround. Therefore, process control should be involved as early as possible in the process design to ensure the controllability requirements are adequately addressed (see Sect. 1.1.3).

Plant-Wide Supply and Demand Control

The primary control consideration at the plant level is the dynamic balance between supply and demand. Take the typical E&P production process in Fig. 1.3 as an example. The process illustrates an entire surface production plant, from the production wellhead all the way to the final sales point. The operating objective is to maximize the oil production; therefore, the well production rate determines the overall throughput (capacity). The separator separates the multi-phase fluid into gas, oil, and water, and sends them to the respective processing units under pressure or level controls.

Following the guidelines on supply and demand analysis in Sect. 6.3, a supply-driven control design is illustrated in Fig. 7.2. The gas processing unit compresses the gas ① to the desired pressure, and sends the gas to downstream users ②, following a supply-driven propagation. Similarly, the oil ③ and water ④ processing is supply-driven as well. The supply and demand model is maintained through the *pressure–flow* or *level–flow* profiles.

Now assume that the consumer of the produced gas has a flow or pressure specification on the gas. A flow controller is installed to adjust the intake of the supplied gas. This change effectively turns the supply-driven operation into demand-driven and would cause significant changes in the process design and control design.

By requirement, at least one swing stream must exist along the gas flow path from the separator to the sale point, to absorb the supply and demand fluctuations. The gas stream to sales used to be the swing stream in the supply-driven operation model. It is now dictated by downstream consumption and can no longer freely "swing." The supply-driven model has effectively changed to demand-driven. With the demand-driven operation, the demand change must propagate back to the separator and then to the wells; the production rate must swing to adapt to the demand changes. The re-designed overall control strategy is shown in Fig. 7.3, with stream ② being demand-driven and stream ① a swing.

Swings in the well flow affect the overall oil production. If the oil production is a target to be maximized, choking back on the oil production at the wells would not be an option. Another gas consumer must be added to serve as the swing stream to ensure the supply-driven operation. Otherwise, the imbalance in gas will have to be *continuously* flared, which is against environmental regulations.

Similarly, a change in environmental regulation has banned the wastewater disposal to shallow water wells. All the processed water has to be disposed of in deep-water wells. Before completing the deepwater disposal (DWD) facility, the plant operation is constrained by water disposal and becomes demand-driven. Temporary changes to the overall control strategy are required to adapt to the reversal of the supply and demand relationship. Once the DWD facility is complete, the supply and

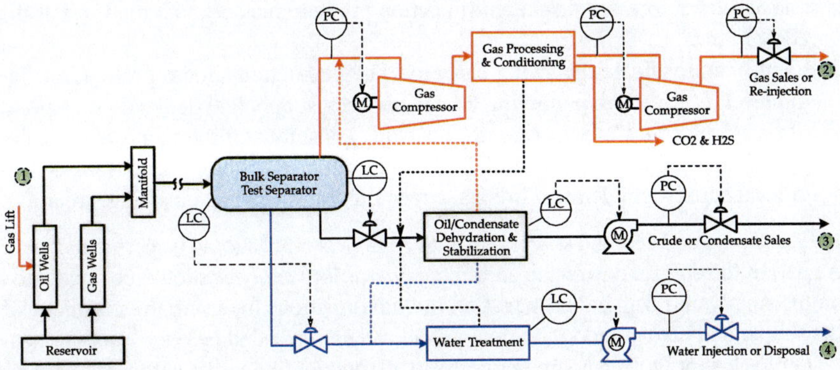

Fig. 7.2 Overall control strategy for E&P process, supply-driven

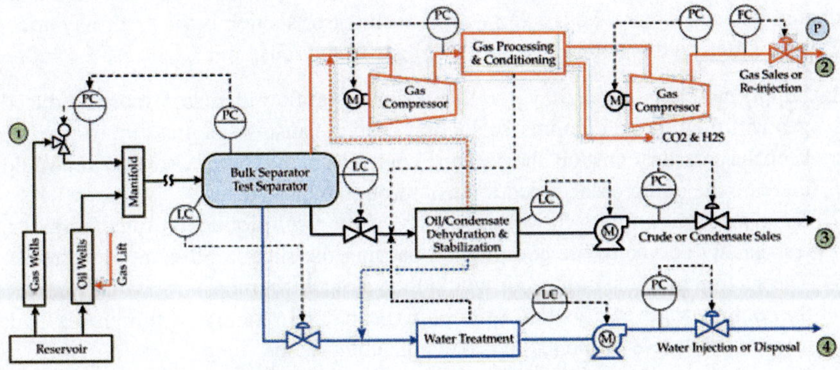

Fig. 7.3 Overall control strategy for E&P process, demand-driven

demand analysis must be carried out again to assess if the control design needs to be modified again.

The above discussion shows that the controllability at the plant level can significantly impact both the control design and the process design. The actual process flow might be more complex than what is shown above.

Some general procedures are provided below for determining or selecting supply-driven or demand-driven control strategies:

1. For each incoming and outgoing stream, determine if they are supply-driven or demand-driven, or both. Whether a stream should be supply-driven or demand-driven is primarily dictated by operating requirements.
2. Along the process flow path, identify/define some propagation nodes. These nodes are typically the *key process variables* and *key control points*. The supply and demand imbalance is "sensed" by these process variables and propagates forward via control loops. Thus we can also call these variables the *relay points*, around which the supply and demand analysis will be performed. For the compression process in Example 6.1, we can have three nodes, one at the compressor suction, one at the compressor discharge (injection pressure header), and the third at the HRSG inlet.
3. Identify and validate the swing capacity. The fluctuations in gas pressure or liquid level reflect the dynamic imbalance between supply and demand at the point of pressure or flow measurement. The swing capacity at these points is typically provided by gas line-packing or liquid surge volume for minor fluctuations. A significant supply and demand imbalance would require dedicated swing stream(s).

The supply and demand analysis discussed in Sect 6.3.2 should be performed first. As part of the control design, an additional *operability analysis* should be conducted on the supply and demand propagation in both directions to ensure the adequacy of the supply and demand model. Some guidelines are provided below.

For each supply-driven stream, perform a structured hypothetical *what-if analysis*. This analysis is similar to what is done in a HAZOP session, where the standard scenarios such as "more flow," "less flow," "reverse flow," and "no flow" are conducted for each stream. For Example 6.1, compressor suction is the first relay node. The gas entering this node is the produced gas from wells:

1. Supply increase ("more flow"): If the incoming gas flow increases, more produced gas will flow to the compressor suction header, causing the pressure to rise. To keep the pressure constant, the pressure controller at the compressor suction would increase the compressor speed to push through more gas.
2. Supply decrease ("less flow"): Similarly, if the well production rate decreases, less gas will come to the compressor, causing the suction pressure to decrease. The suction pressure controller would then slow down the compressor to reduce the compressor capacity. If the minimum speed or minimum flow limit is reached, then recycle valve is opened to protect the compressor.
3. Limited demand ("reversed supply and demand model"): If the compressor throughput is limited and cannot take all the gas coming in, then the supply

and demand model at this node will reverse to demand-riven. The capacity of the compressor now sets the demand. The pressure at the suction will increase. The excessive gas will have to be sent to flare. Otherwise, well production has to be reduced.

At the compressor discharge, which is the second node of our analysis, the inlet is supply-driven, while the outlet is a mix of supply-driven (injection gas) and demand-driven (lift gas and HRSG gas). More flow from the compressor will cause the pressure to go higher, and the discharge header pressure controller would then force more gas to the injection wells. Less flow from the compressor will cause the pressure header to decrease; the pressure controller will reduce the gas flow to the injection wells. If the gas to injection is reduced to zero, and the pressure still goes down further due to insufficient gas from the compressor side, the make-up gas stream will open and fill the gap.

For each demand-driven stream, perform similar hypothetical *what-if analyses* on all streams and check the capacity changes from the demand point propagating backward all the way to the supply point. For example, the fuel gas to HRSG (the third node in Example 6.1) is demand-driven. More demand will force the pressure controller to open the make-up gas valve and cause the header pressure to decrease. The pressure controller detects the pressure change and compensates it by reducing the gas flow to injection wells.

Transition Between Different Mode of Operations

Changes in operation mode are common in all operations. For example, during plant startup, the operation transitions from shutdown mode to normal operating mode. This transition has impacts on all plant levels. For a typical operating plant, the startup and shutdown is a complicated process with significant risks. Coordination at the plant level is essential. Sequential control with sequence and logic, coupling with regulatory control, is a practical yet challenging need.

7.1.3 Unit-Level Control

Unit-level control is typically focused more on quality control than capacity control. Since all commercial operations are for-profit, quality control is the highest priority, which can in turn affect or dictate the capacity control strategy at the plant level. For example, in the fractionator example in Fig. 5.17, the top product draw has a quality requirement on the composition. The quality controller will determine the product flow and forces the control to be demand-driven. The feed is supply-driven; therefore, the bottom product flow must be supply-driven to meet the dynamic supply and demand balance. The bottom product has no specification, and thus a simple level control is sufficient to meet the need for this supply-driven stream.

Stringent HSE regulations and energy optimization have led to leaner and more integrated process designs, reflected by higher purity specifications, higher yields/selectivity, more recycle streams, tight energy integration, and reduced swinging capacities. Control of material and energy flow at one location will more likely cause changes and disturbances to other locations due to interconnection and recycles. They all make process control more challenging.

Unit-level control design relies on the results of the cause and effect analysis (Sect. 6.4), DOF analysis (Sect. 6.5), and dynamic response analysis (Sect. 6.6). The challenge is often the large number of possible control options, which can grow combinationally as the number of controlled targets and DOFs increase. Therefore, the critical control concern at the unit level is the control strategy or the control structure. It is vital to focus the attention on the unit-level issues and not be distracted by local details. "Even though control engineering is well developed in terms of providing optimal control algorithms, it is clear that most of the existing theories provide little help when it comes to making decisions about control structure" (Skogestad 2000).

Example 7.1 Control of an Acetylene Hydrogenation Unit. Figure 7.4 shows an acetylene hydrogenation process discussed in Al-Dawery and Dakhil (2012), a process unit in a polyethylene plant. The feed to the polymerization unit is a mixture of hydrocarbons from the olefin plant containing the undesired impurity of acetylene (C_2H_2), which should be controlled to below 2–3 ppm. Acetylene is a harmful contaminant in high-purity polymer-grade ethylene and must be removed. The typical process to remove the acetylene is via selective hydrogenation over the catalyst in multi-bed adiabatic reactors.

From a quick process flow analysis, it is clear that the unit operation is strictly supply-driven. The supply is from the C2/C3 separator going straight through the heat exchangers and reactors. The material balance should have been addressed in

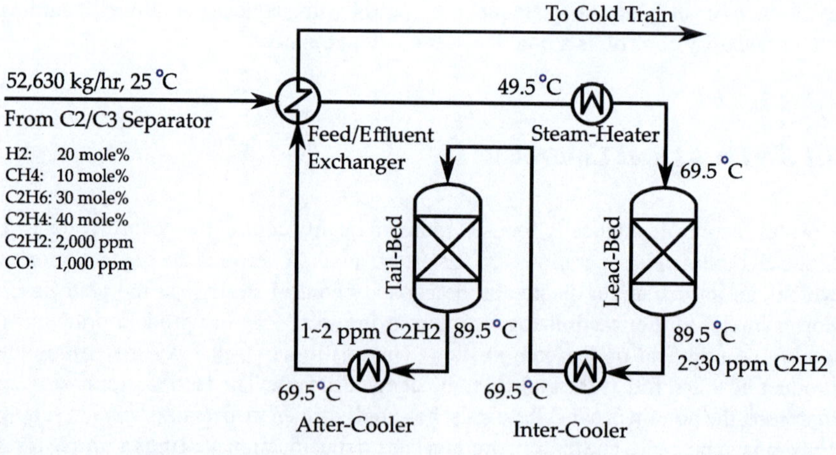

Fig. 7.4 An acetylene hydrogenation process

the plant-level control design. There is no capacity control requirement in this unit operation, and thus, there is no need for any flow control elements along the process flow.

On the other hand, it is also clear that the primary goal for the unit operation is quality control, i.e., on the impurity level of the acetylene. The control strategy manipulates the heat exchange at the heaters and coolers to achieve the desired temperature and composition. Therefore, this unit control is strictly an energy balance problem, and energy optimization can be a consideration in the control design.

7.1.4 Loop-Level Control

At the loop level, a critical decision is in choosing the appropriate control algorithm. A control algorithm can be a simple standalone PID control (Chap. 3), a complex control loop consisting of multiple PID controllers and supporting functional blocks (Chap. 4), and can also be an advanced process control scheme like MPC (Chap. 5). A simple standalone PID loop is sufficient for most applications, and the complex PID control loops provide an economical solution for multivariable problems. Model predictive control (MPC) provides a new technology that is more versatile than PID-based control.

For the example in Fig. 7.4, the options for the control strategy can include the following:

- Simple PID temperature controls after each heater/cooler.
- Cascaded temperature controllers from reactor outlet to heater/cooler outlet temperatures.
- Model predictive control where one single controller is used to control the reactor outlet composition, and perform energy optimization.

The process analysis and control design will be elaborated in Sect. 12.5.

The best control is always the fit-for-purpose control. It is crucial to keep the control solution as simple as possible. Simpler solutions tend to be more reliable. A simple solution is also easier to understand and thus more acceptable by the operators and plant managers, at least conceptually. The control solution that the operators do not accept will likely be turned off and eventually abandoned (also see the discussions in Sect. 5.5.2).

7.2 Control Hierarchy and Control Strategy

The control design is divided into multiple layers, including continuously regulatory control for normal operation, protective control for handling abnormal conditions, sequential control for transition and mode changes, and instrumented safeguarding as necessary (see Sect. 1.3.4). The purpose and function of each layer typically include the following:

1. Normal operation: Maintain the operating point at the desired location or within an acceptable region.
2. Abnormal condition: Protect the operating point from going outside the operating or safety envelope during abnormal situations.
3. Mode transition: Ensure a safe and smooth transition during the change of operating modes.

In line with the above operating requirements, the process control function can be classified into three main categories: normal regulatory control, abnormal protective control, and sequential control. Each type of control may have close interaction and integration with instrumented safeguarding (SIS):

$$\text{Process Control}\begin{cases}\text{Normal Regulatory Control} & \begin{cases}\text{Quality Control}\\ \text{Capacity Control}\end{cases}\\ \text{Abnormal Protective Control} & \begin{cases}\text{Protective Control}\\ \text{Overriding Control}\end{cases}\\ \text{Sequential Control} & \begin{cases}\text{Control Sequence}\\ \text{Logic and interlock.}\end{cases}\end{cases} \tag{7.1}$$

This layered control solution can be achieved by either baselayer PID control or model predictive control (MPC). The targets for regulatory control loops are usually the controlled variables in MPC. The targets for the protective control loops are often included in MPC as constrained variables. Although the implementation differs between PID based control and model predictive control, the control objectives and control strategy are the same.

7.2.1 *Overall Control Strategy*

The control objectives can typically be finalized after the process analysis. However, the control strategy can have many options for the same control objectives, depending on many practical factors. When designing new control solutions or troubleshooting existing ones, it is essential to always have the "full picture" in mind. The high-level control objectives and overall control strategy determine how the next level of details should be laid out. For this reason, a top-down approach for process control design is recommended following the different phases of project execution (Rangalah and Kariwala 2012).

The control strategy is dictated by the control objectives and is based on a thorough process analysis discussed in Chap. 6. There is no "one size fits all" solution for process control. PID control is very effective, economical, and robust but has limitations for more complex process dynamics. Model predictive control (MPC) has been very successful for complex process control problems but has a higher requirement on expertise and resources. There are always compromises to be made in process control design.

7.2.2 *Normal Regulatory Control*

Normal Regulatory Control is responsible for the continuous adjustment of the manipulated variables to keep the controlled variables at the desired setpoint values. The normal regulatory control is also the control we commonly refer to.

Normal regulatory control consists of *quality control* and *capacity control* (quantity control). Quality control is prevalent in downstream operations, while capacity control is more commonplace in upstream E&P operations. As the HSE requirements become more stringent, quality control in E&P is also becoming a common requirement.

Quality Control

All production processes mean to produce something useful, either as final finished products for sale or as semi-product to send to other units or plants for further processing. They all have certain specifications or quality standards to meet, typically represented by product quality such as composition, impurity level, dew points, and pH value. Any stream flowing out of the system boundary is considered a product, and the requirement on pressure, temperature, or flow rate are also considered product specifications. The acetylene concentration in Fig. 7.4 is a typical example of quality control.

Although already a common practice in downstream operations, moving from *process variable control* (temperature, pressure, level, and flow) to *product quality control* is still a value proposition in many upstream operations. Significant challenges for quality control include the following:

1. Availability: Many product qualities are not easily measurable online, while others depend solely on lab samples. Most of them are not available at the frequency required by online control.
2. Credibility: The accuracy of most online measurements is not on par with the conventional process measurements such as flow and pressure. Common problems such as analyzer freezing ("flat-lining"), spikes, drifting render it not viable for online control.
3. Reliability: Online quality measurement devices are prone to malfunctions and require regular calibrations and maintenance. They typically have a much lower *uptime* than conventional devices.

The slow and unreliable online quality measurements can be enhanced with inferential property estimation ("soft sensors"). The product quality can be "inferred" from the fast and reliable conventional measurements based on the cause and effect relationship and dynamic response characteristics. The inferred values are regularly corrected by feedback from online analyzer or lab results to improve accuracy. See Sect. 5.3 for inferential property estimation (IPE).

Capacity Control

Capacity control is a vital aspect of the process control design in the general plant-wide control, especially in E&P operations. By definition, a process is the continuous or semi-continuous (batch) processing of certain materials. Maintaining the inlet/outlet material and energy balancing is an essential requirement. Capacity control is the continuous regulatory control of the plant/unit throughput, ensuring the dynamic material and energy balance. Capacity control design follows the supply and demand analysis in Sect. 6.3, and maintains the desired supply-driven operation, demand-driven operation, or mixed-operation.

For gas processing, the imbalance in supply and demand is reflected in the gas inventory in the process and can be sensed by the gas pressure. The capacity control is thus focused on maintaining the desired *pressure–flow profile*. For liquid handling, capacity control is concerned with maintaining the proper *level–flow profile* along the process path, as the liquid levels in the vessels reflect the liquid inventory. For example, the suction pressure controllers in Fig. 7.2 are responsible for capacity control for the supply-driven operation. If the operation is demand-driven, as in Fig. 7.3, the discharge pressure controllers will take care of capacity control.

For energy balance, maintaining the proper *temperature–flow profile* is the objective of capacity control. However, since the energy balance is typically managed by heat exchange, and the heating or cooling media are typically part of the utility flow that is not directly coupled with the process flow, the plant-wide energy balance is not a significant concern. The typical challenge with energy balance is the saturation in heating or cooling supply, especially if associated with exothermic processes.

For the compression process in Example 6.1, the cause and effect table (CET) is updated and shown in Table 7.1, where $\oplus$ indicates a positive process gain and $\ominus$ a negative gain.

7.2.3 Protective Control

The optimal operating points are often located at or near the boundaries of the operating envelope, which results in more profitable production but also increases the risk of going out of control and trip the unit to shut down. A protective control layer is often included in the control design to avoid unnecessarily tripping the operation. This protective layer stays dormant during normal operation but becomes active if the operation approaches too close to the boundary of the operating envelope. The abnormal protective control layer is typically tuned much more aggressive than the normal regulatory controllers for fast action.

Table 7.1 The PGC example: cause and effect table, with process gain

		CV-01	CV-02	CV-03	CV-04
		Suction press	Discharge press	Liftgas flow	Fuel gas press
MV-01	Valve, flare	×			
MV-02	Speed	⊖	×	×	
MV-03	Valve, recycle	×	×	×	
MV-04	Valve, make-up		×	×	
MV-05	Valve, injection		⊖		
MV-06	Valve, gas lift			⊕	
MV-07	Valve, HRSG		×	×	⊕

As discussed in Sect. 1.3.5 on the layered approach for control and safeguarding, a protective control layer should be considered, if feasible, to avoid unnecessary trips whenever there is a safeguarding trip limit for a particular control point. The protective control can be a standalone PID controller or an overriding PID controller (see Sect. 4.1.3 for detailed discussions on protective control).

Now let us continue from where we left in Sect. 6.5.2 and finish the MV-CV pairing for the compressor control example (Example 6.1). From the cause and effect relationships in Table 6.2, the MV-CV pairs marked with ⊗ are all normal regulatory control loops for capacity control. The MV-01 (flare valve) is evidently for protective control of the suction pressure CV-01 during normal operation, which is already paired with MV-02. We now have a control problem with one controlled variable and more than two manipulated variables. From Sect. 4.1.6, the extra DOF can be consumed by split-range control or dual-pressure control design. A dual-pressure control scheme is ideal because flaring pressure and normal operating pressure are different by design. As a result, a second pressure controller from the same pressure measurement CV-01 is created as a protective controller for the flaring. Let us name it CV-01B for now.

The same analysis applies to the make-up gas valve (MV-04). The make-up gas valve is for protective control. The make-up intake is only expected when the discharge pressure is below the normal operating pressure and reaches the pre-specified setpoint. Again, a dual-pressure control loop is selected over split-range control. A protective pressure controller CV-03B is created to pair up with the make-up gas valve MV-04.

The recycle valve is designed explicitly for anti-surge control (see Sect. 11.3 for details on anti-surge control), so another controlled variable, CT-01C, is created to pair up with MV-03, the compressor anti-surge control valve. The anti-surge control is a standalone protective controller.

Now that all the DOF are exhausted, let us assign the permanent tag names to the controlled variables and manipulated values following the naming conventions, as shown in the updated CET (Table 7.2).

Table 7.2 PGC example: cause and effect table, final

		PC-101 Flare gas	PV-102 Suction P.	UC-103 Surge	PV-104 Discharge	PV-105 Injection gas	FC-106 Lift gas	PC-107 HRSG gas
PCV-101	Flare valve	⊗						
SCV-102	Speed		⊗	×	×			
UCV-103	Recycle		×	⊗	×			
PCV-104	Make-up valve			×	⊗			
PCV-105	Injection valve					⊗		
FCV-106	Lift gas valve						⊗	
PCV-107	HRSG					×	×	⊗

7.2.4 *Sequential Control*

Sequential control such as automated startup, shutdown, or transition is an integral part of process control and is a vast topic by itself. We will provide some brief discussions on this topic, emphasizing its interaction and integration with regulatory control. Sections 12.3 and 12.4 provide two unit-control examples where regulatory control and sequential control are integrated to meet the unique operating requirements.

Another type of sequential control is the transition between the different operating modes. For example, a feed change in an ethylene plant may require many adjustments to the operating points and operating envelope. The safe and smooth transition between the two operating conditions is crucial.

Crippled mode operation is another common scenario. A piece of equipment or device may malfunction, forcing the operation to shut down partially and enter the so-called *crippled mode operation*. Critical considerations include the smooth transition to the crippled mode, the safe operating in the crippled mode, and the quick transition back to the normal operating mode. For example, with a complex integrated compressor control scheme (Sect. 11.3) consisting of multiple compressor trains, a compressor automatically opens the recycle flow valve to stay in operation when another compressor in parallel trips.

Conventionally, dedicated control sequence or logic handles the sequential control scenarios. However, it is preferable if a control solution can gracefully handle the transition.

7.2.5 *Instrumented Safeguarding*

Instrumented safeguarding interacts with process control in two prominent areas:

1. The proper configuration of setpoints and limits: There is typically a trip limit for every key process variable (KPV) to prevent the process from going out of the

operating envelope or safety envelope in case of abnormal operating conditions. The normal regulatory control loops maintain the operating point during normal conditions. Protective control loops are implemented to keep the operating points inside the operating envelope in case of disturbance. The normal control setpoint, the protective control setpoint, the alarm limit, and the trip limit must be appropriately validated and rationalized for stable operation. For example, a setpoint for protective control close to the trip limit does not offer much protection.
The rationalization of the setpoints and limits is a joint effort between multiple teams, including process engineering, operations, rotating equipment, and process control, based on the specific operating conditions and requirements.

2. Control logic embedded in the control loops that interact with safeguarding: Although back-end logic is strongly discouraged for the sake of clarity in design and ease of maintenance, behind-the-scene control logic does exist as the "necessary evil" and can not be avoided entirely. They are a part of the control scheme and need to be carefully considered during process control design and adequately documented.

The level of sophistication in automation, control, and safeguarding is dictated by the nature of operating units and the associated economic benefits. The safest plant would be a plant in stopped condition; however, such a plant does not make money. Implementing lots of safety features may cause the plant to trip too frequently without a particular reason (nuisance trips), which leads to unnecessary production loss or deferment. So reliable safeguarding (more trips) and high availability (fewer nuisance trips) are often conflicting goals. A careful compromise is needed to reach the "optimal" balance.

7.3 Process Control Functional Design

The *process hierarchy* and the *control hierarchy* present the "big picture" of process control design. This representation is illustrated in Table 7.3, where "x" indicates the prevalence of the type of control. For example, capacity control is typically considered at the plat level, while sequential control is often a requirement at the unit level.

Table 7.3 Process control hierarchy

	Process control requirements					
	Quality	Capacity	Override	Protect	Sequence	Logic and interlock
Plant-level control		x	x	x		
Unit-level control	x			x	x	x
Loop-level control						x

7.3.1 Variable Pairing and Degree of Freedom Validation

A vital step in control design is determining the control structure, which involves variable pairing and DOF analysis. The DOF analysis is typically an explicit step with PID-based complex control. For model predictive control, the DOF analysis is often built into the software tools. Although the control structure may seem entirely different between complex PID control and model predictive control, the analytic skills and results are the same.

As discussed in Sect. 4.1.2, if there are more control targets than control handles, the typical approach is to reduce the number of control targets through selective control such as overriding control, ratio control, or cascade control. On the other hand, if there are fewer targets than handles, extra DOFs are not consumed. The solution is typically to reduce the number of control handles by making them non-independent via split-range, fan-out, and switching techniques. The extra DOFs can also be consumed by increasing the number of targets by making the same variable as the control target of a new controller (but with a different setpoint).

Now let us come back to the compressor example in Example 6.1 and validate the adequacy of the manipulated variables against the supply and demand model and check on the DOF requirements:

1. Flaring: Since gas flaring is supply-driven, a pressure controller upstream of the control valve is appropriate. Pressure measurement `PT-101` will become pressure controller `PC-101`.
2. Compressor capacity: Overall production is supply-driven; therefore, the compressor is also supply-driven. The suction pressure is an indicator of supply change, and thus a suction pressure controller `PC-102` controls the compressor capacity by manipulating the compressor speed.
   High load on the compressor may cause the discharge pressure `PT-103` to go excessively high and over the safety limit. A pressure controller `PC-103` based on discharge pressure measurement `PT-103` is added to override the capacity controller `PC-102` in case of high discharge pressure.
3. Recycle: When compressor flow (capacity) goes below a low limit, the compressor may enter into an unsafe condition of *compressor surge*. A minimum flow through the compressor is maintained by recirculating part of the gas from the discharge side back to the suction to prevent the surge from happening. Therefore, the recycle valve is reserved for anti-surge control (more on anti-surge control in Sect. 11.3).
4. Gas supply to HRSG: The gas supply to HRSG is demand-driven; thus, a pressure controller `PC-106` downstream the valve meets the requirement.
5. Gas supply to gas lift wells: The gas lift determines the gas demand; thus, it is demand-driven. A flow controller upstream of the valve fulfills the demand.
6. For the gas header pressure `PT-104`, the gas flow to injection provides the swing capacity, clearly a supply-driven operation. A pressure controller `PC-104` upstream of the valve is appropriate.
7. Make-up gas flow is demand-driven. Based on the rules, a pressure controller downstream of the valve is an obvious choice. Since the header pressure `PT-104`

has been used to manipulate the valve to injection wells, another pressure controller can be spawned from the same measurement, but with a higher setpoint, to adjust the makeup gas in-take.

The final challenge with the control design is the interaction between multiple MVs and CVs. The cause and effect relationship is solid for all the controllers except the suction pressure and surge, and the dynamics are very fast, so the interactions are not a serious concern. However, the compressor speed and recycle valve both affect the suction pressure. From the compressor dynamics analysis, the speed alone also limits the controllability range of the capacity control, so a split-range control is used for the suction pressure control.

The final control structure is shown in Table 7.4, considering the supply and demand model, the cause and effect relationship, and the dynamic response characteristic of each causal relationship.

The overall control strategy is presented in a control overview diagram shown in Fig. 7.5.

Table 7.4 PGC example: cause and effect table, updated

		PC-101	PV-102	UC-103	PV-104	PV-105	FC-106	PC-107
		Flare gas	Suction	Surge	Discharge	Injection gas	Lift gas	HRSG
PCV-101	Flare valve	⊗						
SCV-102	Speed		⊗	×	×			
UCV-103	Recycle		⊗	⊗	×			
PCV-104	Make-up			×	⊗			
PCV-105	Injection					⊗		
FCV-106	Lift gas						⊗	
PCV-107	HRSG					×	×	⊗

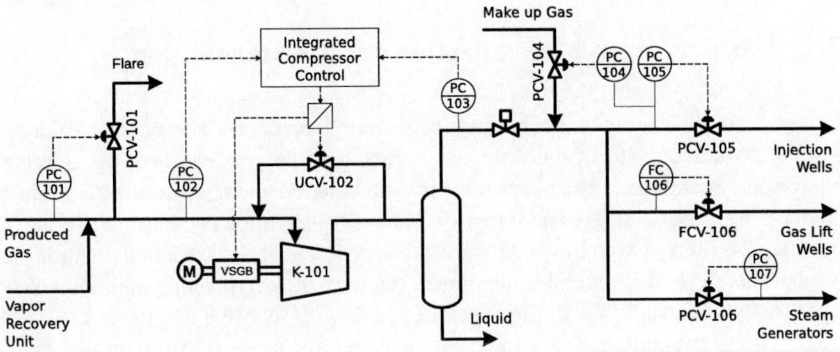

Fig. 7.5 A gas compression process: control strategy

7.3.2 Control Algorithms

With the widespread adoption of model predictive control, a critical decision in control strategy selection is the control algorithm, typically between PID-based baselayer control and MPC-based advanced process control. PID control is based on existing infrastructure and expertise and is thus cost-effective. The downside is that PID-based solutions have limited capability in handling very complex control problems (Sect. 5.1.1). Model predictive control is significantly more versatile and powerful but has a higher requirement on expertise and resources, especially for the first application.

As already mentioned, even for the simple example in Fig. 7.4, there can have several different control strategies, including simple PID control loops, complex PID control schemes, and full-blown MPC solutions. In addition, there can be many ways to pair up the variables. For example, Chin (1979) shows 21 ways to control the pressure of a fractionator; each has advantages and disadvantages.

Riding with the vast success of MPC, there is a tendency of over-use or abuse in MPC applications, mainly in two aspects:

1. Treating MPC as a panacea for all control needs. Many times a baselayer control scheme can do equally well or even better than an MPC solution.
2. Blindly applying the MPC technology without sufficient process insight. For many MPC applications driven by tight project schedules, the proper process analysis discussed in Chap. 6 is often bypassed or short-cut.

Careful considerations of all the factors need to be taken when deciding between baselayer control and advanced control. Once the decision is made, the discussions and guidelines in Chap. 4 can be followed for complex PID control loops and Chap. 5 for advanced process control.

7.4 Process Control Visualization and Documentation

Process control is a cross-disciplinary engineering specialty. The control solution is a collaborative result of multiple teams, including process engineering, rotating equipment, operations, control systems, instrumentation and measurements, project engineering, process safety and integrity. The control solution is expected to control the plant 24/7 throughout the plant's life cycle, operated and maintained by multiple teams. Therefore, it is crucial to document the control design *simply and concisely* to be easily understood by all the concerned parties of diverse backgrounds.

Based on the author's experience with countless projects, from minor troubleshooting tasks to multi-billion dollar mega-projects, one of the most challenging tasks, which is also a task that most people have no interest in doing, is the process control documentation. When performing process control maintenance and troubleshooting, the most needed document is the process control document. However, most of the time, it is challenging to find a decent control document. The control document most likely lies in the brain of the original designer, who has left the company a long time ago.[1] Even if there is a control document, it typically does not have the expected quality to be very useful.

The *process control narrative* (PCN) document is a critical delivery from process control. The PCN document is authoritative on the process control design, implementation, operation, and maintenance. It is also an essential communication tool among all the teams regarding the control solution. Therefore, the quality of the PCN document has a profound impact on the plant operation.

However, the development of a high-quality PCN is a highly challenging task. It requires a holistic view and intimate knowledge of the process, operation, and control to cover all the critical information. It requires an even higher skill to present the information in a *simple* and *concise* fashion.

This section provides some guidelines on standardizing and simplifying this challenging task of process control documentation. A *control overview diagram* (COD) is recommended to facilitate the understanding of the control solution. A standard document structure is provided to improve the consistency and completeness of the information presented.

7.4.1 Process Control Narrative Document

The PCN is the authoritative document on the design and implementation of the control solution. It typically consists of the process description, overall control strategy, narratives of complex control loops, implementation considerations, and operation recommendations. The following contents are recommended to be included in a PCN document.

1. Process description: It is vital to include a concise process description to provide the background for the process control design. The guidelines on developing the process overview diagram (POD) in Sect. 6.7 can be followed as the starting point for COD:

    a. The design intent and operating requirements: A brief description of the operating objectives and requirements is essential in providing the background of the process operation.
    b. The process streams and specifications: The major process flow streams coming into and going out of the system boundary, with the critical specifications for each stream.
    c. The design limits and operating constraints: The limits and constraints can have a significant impact on the control design.

[1] In the USA, in the industries that are subject to FDA (food and drug administration) review, it is commonly said that "if you didn't document it, you didn't do it".

2. Plant-level control design (if applicable):
   a. The plant control objectives.
   b. The narratives on the overall control strategy at the plant level.
   c. The control overview diagram at plant level.
3. Unit-level control design (if applicable):
   a. Unit control objectives.
   b. Narratives of the overall control strategy at the unit level.
   c. Control overview diagram (for complex units).
   d. Control narratives are required for any complex loop whose intent or design cannot be unambiguously understood from the control diagrams such as COD or P&ID.

   A common mistake in writing control narratives is elaborating on the simple control loops but neglecting the more complex control schemes. For example, it is very often to see detailed descriptions of how each simple flow feedback controller works but lacks any description of the override control scheme comprised of these very flow controllers.
4. Implementation and Configuration:
   a. Implementation aspects: Many implementation considerations require a good understanding of the process design, operation requirements, control strategy, and the capability of the control systems. Process control is arguably the only specialty in the operating plant that has a holistic view of all these aspects.
   b. Controller configuration: The controller implementation is the task of control systems engineers/technicians. However, many configuration details such as anti-reset windup and bumpless transfer are best specified by the process control engineers (Sect. 3.5). MPC configuration is the sole responsibility of process control engineers. Including all the configuration details in PCN is unnecessary, but some critical or unique configurations must be adequately documented.
5. Special Considerations: Many practical scenarios must be taken into account.
   a. What if something goes wrong? The control design challenge is not how to make the control solution work under normal conditions but how the control system should respond to the different operating scenarios including abnormal situations. Special considerations must be taken to prevent or mitigate the negative impact of these abnormal scenarios.
   b. Special features.

The PCN document is progressively developed through the different phases of the project. The expected content from different phases of a project is shown in Table 7.5.

The PCN can be elaborated or succinct, depending on the actual requirement and the preparer's preference. However, the following three components are strongly recommended to be part of the PCN document:

Table 7.5 The phased process control deliverables

Project phase	Process control narratives	Check point
Conceptual design	Process flow analysis	Design review
	+ Overall control strategy	
	+ Control overview diagrams	
Front-end engineering (FEED)	+ Key process control loops	Design review
	+ Process control schematics	HAZOP review
Detailed engineering	+ Detailed function specifications	Design review
	+ Loop configuration details	HAZOP review
Implementation	+ DCS control modules/sequences	Factory acceptance
	+ Logic flow diagrams	
Commissioning	+ "As-Built" updates	Site acceptance
Maintenance/improvement	+ Management of change (MOC)	

1. Control overview diagram (COD). A one-page overview to provide a holistic view of the overall control strategy.
2. Narratives of overall control strategy and complex control loops.
3. Key configuration parameters of the controllers.

7.4.2 *Control Overview Diagram*

A modern operating plant/facility design can often produce dozens or even hundreds of pages of process flow diagrams (PFD) and piping and instrumentation diagrams (P&ID), along with a comparable number of DCS schematics. An inexperienced engineer can easily be overwhelmed by the daunting number of schematics and fails to see the forest for the trees, thus loses the "bigger picture" (the overall control strategy). A flawed process control design is often caused by the lack of a sound overall control strategy. It is not unusual that all the individual control loops perform well in silos but fail to work when put together.

A control overview diagram (COD) is one way to force the design engineers to think about the bigger picture and thus make sound decisions. The COD is a one-page "visual summary" of the *process flow* and *control strategy*, evolved from the process overview diagram (POD) discussed in Chap. 6. The COD captures the process analysis results in Chap. 6 and control design in this chapter and aims to provide a simplified yet concise high-level summary of the overall process design intent and essential operating requirements, focusing on the dynamic material/energy balance. The COD focuses on the *overall control strategy*, emphasizing the integrity of the control scheme and interaction among different parts of the operation.

The COD should include the following essential contents:

1. The critical process flow streams with stream specifications, including all the process streams, equipment, and devices that may affect the dynamic material and energy balance.
2. Key process variables (KPVs): The *key process variables* are a subset of process variables whose values and statuses provide a sufficiently complete and reliable indication of the overall control status and operating health when viewed together. The KPVs can include, for example, the quality control variables, capacity control variables, and critical protective control variables.
3. Key process control loops: The key control loops are the control loops that represent the overall control strategy at the plant level and unit level. They reflect the supply and demand model and cause and effect relationships among the key process variables.

The COD should include only the minimum and essential details for an uncluttered view of the process flow and control strategy ("the big picture"). Any details that do not contribute to improving the presentation clarity of the control scheme should be excluded. The details of process engineering and control systems should refer to the P&IDs or individual control schematics in the PCN document. Figures 7.2 and 7.5 in this chapter are examples of COD. All the application examples in Chap. 12 include a COD.

Due to the cross-disciplinary nature, process control engineers need to collaborate closely with several other teams during control design and maintenance. The COD serves as a very effective *visual* communication tool when working with process engineers, control systems engineers, and instrumentation engineers, operations personnel, and operation manager, especially for complex control strategies. From the authors' experience with hundreds of applications, from small brownfield projects to multi-billion dollar greenfield projects, the COD has been an indispensable tool for visualization, communication, and documentation of the process control strategy.

7.4.3 Control Loop Narratives

Simply put, the PCN provides the answers for the essential questions on all the control schemes or control loops whose intent or function is not immediately clear from the control schematics such as PFD or P&ID. The questions and purposes are listed in Table 7.6.

The same questions apply to a controller, a control loop, or a control scheme, and should be asked at different plant levels. For example,

1. At the plant level, the questions are typically about the overall control strategy.
2. At the unit level, the questions are often on the control design, either a complex baselayer scheme or a model predictive control.

Table 7.6 Essentials elements of control narratives

Intent	Sample narratives
WHERE?	Where is the control scheme or loop located/placed? The placement of the measurements and valves are based on process design, and is a result of supply and demand analysis, cause and effect analysis, and dynamic response analysis
WHY?	Why is the control scheme/loop needed? This is the process reason for justifying the needs for this particular scheme/loop, such as maintaining product quality, enforcing material balance, addressing process non-linearity
WHAT?	What does the control scheme/loop provide? This is the control strategy and functional specification for the control scheme, based on the WHY and WHERE analysis. The control strategy addresses the design intent and operating requirements, while the functional specification is the implementation details that can be implementable by control system engineers
HOW?	How is the control scheme/loop implemented?
WHEN?	When is the control expected to be in operation, and when should it be taken offline?
WHAT IF?	What if things go wrong? In case of an abnormal operating condition such as measurement failures, emergency unit shutdowns

3. At the control loop level, the questions are more about implementation and configuration, whether a PID loop, an on/off control loop, model predictive control, or a logic sequence.

All control loops, sequences, or logic whose purpose or function are not immediately evident from the COD or P&ID require a narrative in the PCN document. For example, for the compressor anti-surge control loop in Fig. 4.2, a minimum control narrative can be as brief as shown in Table 7.7.

The control narratives are expected to be short and to the point. Engineers in the field do not have the time and patience to flip through a 500 page PCN document to find the information they urgently need. It is much more effective to document the control design in drawing such as a COD or bullet-point list of the critical information listed above. For example, the complex capacity control scheme for the entire gas compression process unit in Fig. 4.2 can still be as brief as shown in Table 7.8, as long as the main points are captured.

7.4.4 Loop Configuration Details

A PID controller has many configuration parameters, all having system default values. However, the default values are rarely adequate for the intended application and must be reviewed and assigned with appropriate values per control design.

Usually, the PID configuration is the scope of work of control system engineers or technicians. However, a subset of the configuration parameters are closely related to the operating requirements and overall control strategy, typically beyond the knowl-

Table 7.7 Sample control narratives for compressor anti-surge control

Intent	Sample narratives
WHERE?	An anti-surge controller (`UC-111`) shall be implemented at each stage of the compressor …
WHY?	…to prevent the compressor from going into surge
WHAT?	It calculates a surge parameter with `UY-111` based on suction pressure `PT-111`, discharge pressure `PT-112`, and discharge flow `FT-111`, and use the anti-surge control valves `ASCV-111` to regulate the recycle flow and keep the compressor flow above the minimum flow requirement
HOW?	The control loop is a PID controller with I-PD setting and in DIRECT action, and runs at 100 ms …
WHO?	… and shall be operated in `AUTO` mode during all operation operation modes except for crippled mode. The controller mode and setpoint are locked from the operators
WHAT IF?	In the case of `BAD PV`, the output shall be ramped up to the fail-safe position of the valve; during ESD/PSD, the control valve shall open to the fully open position in one shot

Table 7.8 Sample control narratives for compressor capacity control

Intent	Sample narratives
WHERE?	A pressure controller `PC-002` is implemented at the compressor suction header …
WHY?	…as the capacity control loop to enforce the supply-driven operation
WHAT?	It manipulates the compressor speed `SC-101` and recycle valve `ASCV` in split-range to maintain the pressure setpoint `PC-002`
HOW?	The pressure controller manipulates the compressor speed via a cascade control through the load control `XC-101` and is the first leg of the split-range control. If the compressor speed is insufficient to achieve the desired capacity turndown, the recycle valve `ASCV-111` and `ASCV-121` are used as the second leg in the split-range control setting
WHO?	The capacity control should normally operate in `AUTO`. Mode change is part of the compressor startup logic. Operators are allowed to change control mode after normal startup
WHAT IF?	The capacity control `PC-002` can be overridden by the discharge header pressure protective controller `PC-003` via a low selector. The capacity control output `PC-002.OP` to the speed control can be overridden by motor power protective controller `EC-101` via a low selector. The pressure control output to the recycle valve can be overridden by the anti-surge controller `UC-111` and the discharge pressure controller `PC-112` via a high selector; both are protective controllers

edge of control system engineers, and must be decided by process control engineers. For this reason, it is recommended to include a table of the loop configuration details in the PCN document to guide the control systems engineer on the configuration of those critical configuration parameters. This table contains the typical and critical configuration details for the PID controllers, either specified by the process control engineers or extracted from other master sources.

A sample table is provided in Table 7.9 and the guidelines can be found in Sect. 3.5. Some additional remarks on the PID loop configurations are provided here:

Table 7.9 Control loop configuration (sample)

Properties	Example	
Loop name on P&ID	15FICA-111	…
Loop name in DCS	15FC111	
Loop description	K1510 stage 1 suction flow control	
SP/PV engineering unit	M3/HR	
OP engineering unit	%	
Loop type (regulatory or protective)	Regulatory	
PV measurement range high	500	
PV trip limit high (if applicable)	–	
PV alarm limit high (if applicable)	450	
SP range limit high (if applicable)	300	
Typical PV value (e.g., nominal value)	250	
SP range limit low (if applicable)	100	
PV alarm limit low (if applicable)	–	
PV trip limit low (if applicable)	–	
PV measurement range low	0	
OP range limit high	100	
OP range limit low	0	
Algorithm (P, PI, PID, PID, I-PD)	PI-D	
Direction of action (direct/reverse)	REVERSE	
SP tracking PV when in MAN	YES	
Anti-reset windup (Y/N)	YES	
Explicit bumpless transfer (Y/N)	NO	
Initial PID tuning: gain	3.0	
Initial PID tuning: integral (minute)	0.5	
Initial PID tuning: derivative (minute)	0.0	
Has custom calculations/logic? (Y/N)	NO	
Operator access (MODE/SP/OP)	YES/YES/YES	
Mode during normal operation	AUTO	
Mode during startup	MAN	
Mode during shutdown/OP = 0%	MAN	
Mode when entering crippled operation	MAN/OP = Last good	
Mode when resuming from crippled mode	AUTO, by Operator	

1. Tag in DCS often has a shorter name than that on P&ID (see Sect. 1.3.3).
2. For the process values (PV), the typical value, alarm value, trip value, and measurement range, along with the setpoint range limits and output range limits, are collected and shown in the same table to facilitate the rationalization of the setpoints and limits (see Fig. 1.14).
3. Anti-reset windup: Anti-reset windup is required if the controller output may be limited or disconnected (see Sect. 3.5). The standard built-in anti-reset windup configuration should be used whenever possible. If the built-in anti-reset function is not sufficient, an external anti-reset windup is required, detailed narratives must be provided in the main context of the document (see Sect. 3.5).
4. Explicit bumpless transfer support: In most modern control systems, the bumpless transfer function has been a standard feature for most control blocks. However, if the standard built-in functions do not support the bumpless transfer requirement, an explicit bumpless transfer block must be added and configured (see Sect. 3.5).
5. Initial PID settings: The system default for the PID settings is typically not adequate for most of the PID loops. Initial PID tuning should be provided in the table to avoid defaulting to the system values.
6. If custom calculations, sequences, or logic are present, indicate YES in the table and provide narratives in the relevant section of the PCN document. A common source of frustration is the lack of documentation on back-end logic in the controller.
7. Operator access: Many PID controllers may need to be locked from the operator to avoid mal-operation. The lock-up can include `MODE`, `SP`, or `OP`. See Sect. 3.6 for more information.
8. Mode of operation: The controller `MODE`, `SP` (in `AUTO`), or `OP` (in `MAN`) shall be specified for the following operation modes if the default behavior is not acceptable or the standard practice is not adequate:

   - Normal operation.
   - Startup.
   - Shutdown.
   - Entering crippled mode.
   - Recovering from crippled mode.
   - Any other major operation modes.

7.5 Building Control Overview Diagram—The Work Process

Visualization and documentation of the process control solution are essential requirements but are often neglected. One reason is that good visualization and documentation require that the developer have both a thorough understanding of the overall control strategy and also intimate knowledge of the details.

The control overview diagram (COD) is one recommendation for simplifying the job. A COD should typically be a one-page diagram with a top-down layout. The

COD must provide the top-level holistic view of the entire facility within the process control scope. If there is room for more details, the responsible engineer can decide what additional details to include at his/her discretion, provided that the added details will not inadvertently affect the clarity of the overview.

A sample procedure for COD development is recommended as below. Typically, this procedure needs to be repeated multiple times before reaching a satisfactory outcome.

7.5.1 Collect Relevant Information

The following information or documents are essential for the process analysis and control design. Lack of reliable documentation is a major headache to process control engineers, so it is beneficial to collect as much as possible for completeness and cross-validation of the information:

1. Existing PCN document, if available.
2. Process flow diagrams (PFD) (with the material balance sheets).
3. Piping & instrumentation diagrams (P&ID).
4. Operating philosophy document and operating manual.
5. DCS schematics.

7.5.2 Build Process Overview Diagram

Follow the process analysis process and build the process overview diagram (POD) as the backdrop of the COD:

1. Define the project or process boundary.
2. Identify and add all the major streams flowing into the system boundary.
3. Identify and add all the major streams flowing out of the system boundary.
4. Identify and add the piping and equipment essential for showing the process flow and field instrumentation.
5. Identify and add all the final control elements such as valves and motors that can affect the product quality or production capacity.
6. Identify and add all the control elements such as switches and on/off valves that can affect the process flow line-up during normal operation.

7.5.3 Add Regulatory Control Loops

1. Identify and add the essential quality control and capacity control loops to the diagram.

2. Perform a *cause and effect analysis* between all final control elements and key process variables, and verify that the control PV and OP selection are appropriate.
3. Perform a *dynamic response analysis* of the control loop. Assuming a step-change in the *controller output*, through the final control elements and the process itself, assess the dynamic response of the *controller PV* of the same controller in terms of gain, lag, and delay. Note that transient response characteristics from the controller `OP` to the controller `PV` give the *loop dynamics*.
4. Perform a DOF analysis to verify that there is no surplus or deficit in the DOF for the current selection of manipulated variables and controlled variables, e.g., multiple controllers (implicitly) controlling the same variable or one valve attempting to control multiple variables.

7.5.4 Add Overriding and Protective Control Loops

1. Identify and add overriding protection control to complement the capacity control loops in case the supply and demand relationship may change, such as the following:

    a. Supply-driven operation may temporarily become demand-driven due to process upset or operator intervention.
    b. Demand-driven operation may permanently become supply-driven over time due to depletion of supply or process modifications.
    c. Operation may be unexpectedly limited by critical equipment (e.g., compressor becomes the bottleneck), and the capacity becomes equipment-driven.

2. Identify and add any critical protective control loops as needed. Generally, for any key process variables with safeguarding trip configuration, it is essential to evaluate the necessity and feasibility of adding a protective control layer between normal regulator control and safeguarding trip.
3. For each control loop, determine if it requires special configurations such as anti-reset windup and bumpless transfer. Start from the controller output and trace the control signal to the final control elements, reset windup protection is needed if the signal could be disconnected or limited anywhere along the path. If the change due to disconnection or re-connection of the signal can cause an undesirable bump in the control components, then an explicit bumpless transfer block should be considered.
4. Perform a rationalization check on the operating ranges for each key control variable, including normal regulatory control setpoint, protective control setpoint, alarm limit, trip limit, and measurement range.

7.5.5 Add Special Configuration Requirements

1. Add a "#" mark to the controller if the controller has any of the following special requirements:
   a. Supporting calculations (other than standard built-in functions such as filtering, gap control).
   b. Back-end logic that affects the operation of this controller or other controllers.
   c. External logic that may affect the operation of this controller.
   d. Anti-reset windup (ARWU) via external feedback.
   e. Explicit bumpless transfer blocks.

   Any controllers with the "#" mark should have more detailed control narratives in the PCN document.
2. Add notes to indicate any other additional information required to facilitate the design or understanding. For example, for level control of a large vessel, show the vessel dimension (height and diameter) and the flow range to help vessel-size-based controller tuning (see Sect. 9.5).

The progression from Figs. 6.1, 6.9, 6.10, and finally Fig. 7.5 illustrates the work process. The plant or unit-control examples in Chap. 12 all have a COD as the primary means of presentation. After all, "a picture is worth a thousand words."

7.6 Summary

The process analysis in Chap. 6 helps build a solid understanding of the process flow, design intent, operating requirements, and operating challenges. It is the prerequisite for a sound process control design, either baselayer control or advanced control.

This chapter continues from the process analysis and shows how to perform process control design based on the process analysis results. Because of the diversity of process flows, providing ready-to-copy control solutions for various process control problems it is not practical nor efficient. Instead, this chapter provides some essential analytical skills and generic processes that can be followed by any process control design. The same skill and methodology also apply to troubleshooting and improvement of existing control solutions. The key considerations in process control design are summarized in Fig. 7.6.

Section 7.1 discusses the process hierarchy and the corresponding control scope at the different operating levels. While in Sect. 7.2, the overall control strategy and the layered control structure are explained. Section 7.3 provides some guidelines on control functional design, with emphasis on the control algorithm selection. Finally, Sect. 7.4 discusses the typical process control deliverables in a new project and emphasizes the critical importance of the visualization and documentation of the process control design.

- Control Design
  - Control Objectives
    - Plant-wide Control
    - Unit-Level Control
    - Loop-Level Control
  - Control Strategy
    - Normal Regulatory Control
    - Protective Control
    - Sequential Control
    - Instrumented Safeguarding
  - Functional Design
    - Simple Standalone Control
    - Complex PID Control Schemes
    - Model Predictive Control
    - Other Advanced Process Control
  - Control Implementation
    - Control Configurations
    - Setpoints/Limits Rationalization
    - Back-End Logic and Interlocks
  - Control Deliverables
    - Process Control Narrative
    - Control Overview Diagram (COD)
    - Loop Configuration Details

Fig. 7.6 Scope and checklist for process control design

References

Al-Dawery SK, Dakhil HM (2012) Plantwide control of acetylene hydrogenation process. J Eng Sci Technol 7(1):11–24

Åström KJ, Kumar PR (2014) Control: a perspective. Automatica 50(1):3–43

Chin TG (1979) Guide to distillation pressure control methods. Hydrocarbon Processing

Downs J, Skogestad S (2011) An industrial and academic perspective on plantwide control. Ann Rev Control 99–110

Foss AS (1973) Critique of chemical process control theory. AIChE J 19:209–214

Juliani RCG, Garcia C (2017) Plantwide control: a review of design techniques, benchmarks, and challenges. Ind Eng Chem Res 56(28):7877–7887

Luyben WL, Tyréus BD, Luyben ML (1998) Plant-wide process control. McGraw Hill, New York

Marlin TE (2015) Process control - designing processes and control systems for dynamic performance, 2nd edn. McGraw-Hill

Rangalah GP, Kariwala V (eds) (2012) Plantwide control: recent developments and applications. Wiley

Seames W (2018) Designing controls for the process industries. CRC Press

Skogestad S (2000) Plantwide control: the search for the self-optimizing control structure. J Process Control 10:487–507

Stephanopoulos G (1983) Synthesis of control systems for chemical plants - a challenge for creativity. Comput Chem Eng 7(4):331–365

Chapter 8
Methodology of Control Problem Troubleshooting

Control problems can range from sub-optimal performance, crippled operation, to complete breakdown (Sect. 8.1). The causes may be simple instrumentation failures, dynamics changes, and the control design flaws (Sect. 8.2). For this reason, troubleshooting is typically the most challenging work for a process control engineer demanding the highest level of expertise and experience. The troubleshooter not only must understand how things work but also needs to know how things may fail (Sect. 8.3). Consequently, the solution can be simple instrument repair, controller tuning improvement, controller configuration, to a control scheme re-design (Sect. 8.4). Overall, control problem troubleshooting is a mix of art and technology with no fixed recipes. Nevertheless, some general guidelines and best-practice procedures make the troubleshooting effort more technical and less artistic.

8.1 Common Control Problems

Process control is a "hidden" technology (Åström 1999). When the control scheme works well, nobody notices. However, when it fails, the results can be catastrophic! Some major accidents include Three Mile Island nuclear reactor (PA, USA, March 1979), Chernobyl nuclear reactor (April 1986), and the Texas City Refinery fire (March 2005).

A typical operating plant can have hundreds to even thousands of control loops monitored by only a small and shrinking number of operators. The staffing level might be sufficient for normal operation but can be easily overwhelmed during major plant upsets. It is critical to minimize the abnormal operating conditions and detect the process control problems early before it deteriorates from a performance issue into a safety problem.

Some commonly reported problems (or symptoms) on process operation include the following:

- Frequent process upsets, trips, or breakdowns.

S. S. Niu and D. Xiao, *Process Control*, Advances in Industrial Control,
https://doi.org/10.1007/978-3-030-97067-3_8

- Excessive number of alarms leading to frequent operator interventions.
- Controller not in the desired normal mode most of the time.
- Persistent cycling in one or more process variables.
- Difficulties in achieving product specifications.
- Excessive flaring and recycling.
- An excessive number of starts/stops of critical equipment such as pumps or compressors.
- Difficult startups.
- Process values diverge and become unstable, and thus controller has to be taken out of control.
- A controller takes too long to bring PV to its setpoint or exhibits excessive PV deviation from the setpoint at steady-state.
- Controller outputs in saturation for a prolonged period.
- Excessive valve movements leading to excessive wear & tear.

For operation and troubleshooting, it is crucial to know what the normal performance is to be expected. This normal performance serves as the baseline when identifying abnormal performance issues.

It is also critical to make the initial judgment if the operating problem is indeed a control problem. Many of the problems are clearly control problems, while many others may not be that obvious. In an operating plant with a well-established process control team, many operating problems are customarily reported to the process control team for troubleshooting because of trust. Thus, the process control engineer needs to be aware that many alleged control problems may turn out to be non-control related at all.

Most control textbooks provide detailed theoretical analysis on the control stability using advanced tools such as Nyquist plots, root locus analysis, Lyapunov criteria. However, these analyses are of minimal practical use because they all assume a known linear time-invariant process dynamics. In reality, however, the process dynamics are rarely linear or time-invariant. Besides, the control design is based on loop dynamics rather than process dynamics alone. Other changes in the control loop, such as the valve characteristics, measurement noise, disturbances, and especially the control structure, play a more detrimental role in the loop stability. Table 8.1 lists the common issues with a complex control scheme, from a broken loop, bad data, constrained data, to drifting dynamics. These issues may be caused by process dynamics, measurements, valves, or control design.

Table 8.1 Common control issues and possible causes

Issues	Process	Measurement	Valve	Objective	Controller
Broken data flow	X	X	X		X
Bad data value		X			
Limited data value		X	X	X	X
Drifting dynamics	X		X	X	X

A control loop has five essential components (Sect. 2.1.2). Issues with the process dynamics are only part of the potential causes of control problems.

8.2 General Troubleshooting Procedure

Control scheme troubleshooting is a mixed bag of arts and technologies, with no cookbook recipes to follow. The best weapons are engineering common sense, process insight, control knowledge, practical experience, and enough patience. Despite the uncertainties, some general guidelines and procedures can help make the troubleshooting effort less painful.

The control problems can be divided into several levels, corresponding to the analysis in Chaps. 6 and 7. The general troubleshooting procedure follows this order:

$$\texttt{design} \rightarrow \texttt{implementation} \rightarrow \texttt{operation}. \tag{8.1}$$

1. Control design: Flaws in the control strategy or control structure have a significant impact on the overall operations. It requires a higher skill to locate and is typically expensive to correct.
2. Control implementation: For complex PID control loops or model predictive control, implementation is another area that can often go wrong. A thorough understanding of the control scheme and some help from control systems engineers are most useful.
3. Control operation: During operation, the loop dynamics may change for many reasons. System troubleshooting procedures can help to detect and diagnose the issues.

A general troubleshooting procedure is illustrated as a flow diagram in Fig. 8.1 for easy visualization. The following sections discuss the critical steps of loop troubleshooting and provide guidelines on what to check for each component of the PID loop.

8.2.1 Obtain a Problem Statement

Troubleshooting a process control problem is very much like a doctor seeing a patient. The first step is to obtain a problem statement, i.e., the symptoms or complaints. Control problems are typically reported by operating personnel but may also come from other sources, such as a proactive control performance monitoring (CPM) application (see Sect. 2.3.6).

To have a good understanding of the complaints, try to find out from operations (e.g., the panel operators) and instrumentation engineer for answers to the following questions:

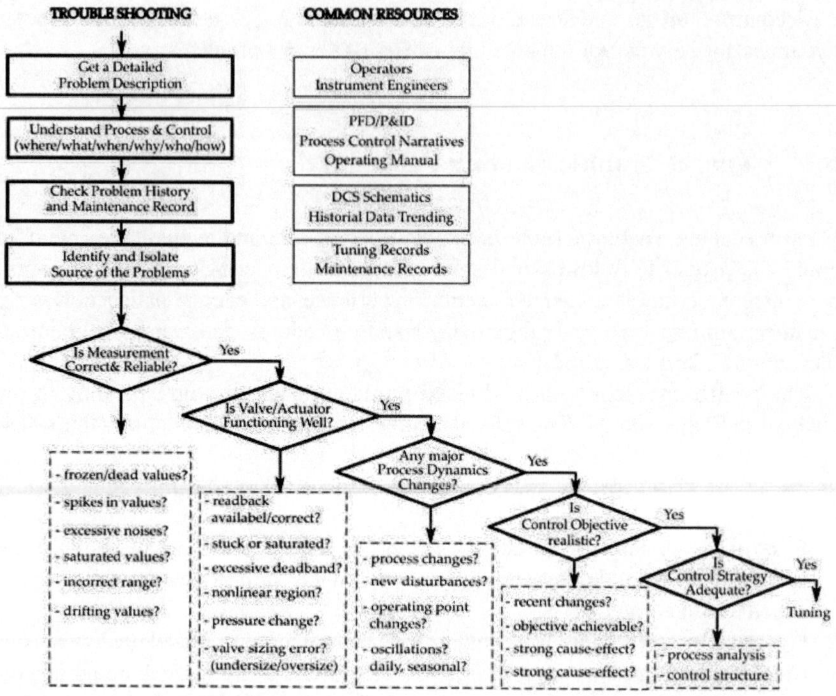

Fig. 8.1 PID loop troubleshooting procedure

1. What is wrong with the loop? That is, what are the complaints and symptoms of the reported control problems?
2. When did the problem start? Is this the first time happening, or is it a recurring issue?
   - If this is a new problem, what has been changed recently to process, equipment, operating condition, control valve, or instrumentation?
   - If this is a pre-existing problem, is it just a simple repetition of the last occurrence, or is it getting worse with new symptoms?
3. Have there been any significant changes to the plant or operation requirements? For example, are there any recent process modifications, feed/load changes, turndown operations?
4. Have there been any major repairs or maintenance work on the instrumentation or control valves?
5. What is the severity of the problem? That is, how critical is this control loop to safety, production quality, or process operability?
6. How urgent is a solution for the problem? That is, what is the risk if no action is to be taken immediately?

The problem statement provides a basic understanding of the severity and urgency of the reported problem and serves as a starting point for planning the troubleshooting effort. However, the description of the control problems from the operational personnel is often ambiguous, sometimes exaggerated, or understated. On the other hand, alerts from control performance monitoring (CPM) applications are typically more objective and accurate. Depending on the location and nature of the problem, the statements may need to be double-checked and cross-validated. The troubleshooter needs to research further into the reported problem to obtain a complete picture and verify it before starting troubleshooting:

1. Collect the process control narrative (PCN) document, process flow diagrams (PFD), and piping & instrumentation diagrams (P&ID) to understand the process design, operating philosophy, the control objective, and the interactions with other control loops.
2. Check history data and the maintenance record to verify the problem history.
3. Evaluate the criticality of the loop relating to process safety, product quality, and process operability.
4. Verify if there have been any significant process changes that can affect the loop(s) in question.
5. Verify if any operating condition changes (e.g., a significant shift in control setpoints) have moved the operating points outside the normal operating envelope.
6. Verify if multiple loops are suffering from the same or similar problems (e.g., oscillations).

8.2.2 *Identify Human Errors*

Human error is a common cause of process control problems. The consequence of human errors can be from poor control performance, crippled operation, to sometimes catastrophic incidents. For example, the Chernobyl nuclear reactor (April 1986) incident was a combination of mechanical and human error. The Texas City Refinery incident is caused by instrumentation error and human error (CSHI 2005).

Human error is more related to violations of standard operating procedure (SOP) and is thus beyond the scope of this book. However, it is essential to rule out human error preliminarily before immersing oneself in technical troubleshooting too quickly. Besides, if human errors have caused the issues, many can be prevented or mitigated by improved process control design:

1. Operator mal-operation: Normal plant operation is commonly repetitive and boring, and over time the operators can become less vigilant, and human errors start to occur. As an example from personal experience, while doing step tests for a model predictive control, a request was made to the operator to step change a temperature from 110°C to 100°C. When entering the new setpoint, the operator carelessly entered 10 instead of 100. Although he realized the error immediately and re-entered the correct setpoint right away, the vast step changes from 110 to

10 and back to 100 already caused a significant upset to the column operation, which took a few hours to re-stabilize. The small incident definitely ruined the step-test plan for that day. Fortunately, it did not result in a costly shutdown.
If this part of historical data is investigated as part of a troubleshooting effort for something else, it might be challenging to find out that the unit-wide fluctuation in the data was caused by a human error rather than any control problems.
Naturally, this problem could have been avoided if the operator stayed more vigilant when operating the plant. However, on the process control side, a high and low setpoint limit could have been configured for the PID controller (see Sect. 3.5) to prevent such an abnormally large setpoint change from the operator.

2. Controller not in the desired mode: Another common cause of poor performance is that operating personnel switches controllers away from their intended mode of operation (the normal mode). For example, many PID controllers are put in `MAN` mode instead of the desired `AUTO` or `CAS` mode for no good reasons. Each experienced operator usually has his or her own "comfort zone," which may not always be the best operating choice. It is common to see that after a shift change, the first thing the operators do is to make many setpoints changes to bring the process operation to their own "comfort zone." Model predictive control (MPC) is often switched off due to unfamiliarity with the technology and lack of understanding of the moves produced by MPC. For complex control loops involving multiple controllers, a poorly performing controller could impact the performance of the entire control scheme or even an entire unit and results in many PID controller running in manual mode.
The best solution is to improve the control performance so that the operators will become enthusiastic about keeping the controllers running in automatic mode since automated control typically makes their work easier after all. At the same time, the operating procedure should include a tracking mechanism to record why the operator chooses to take the controller offline. For example, control performance monitoring (CPM, Sect. 2.3.6) can produce automated reports on the statistical information on the number of changes made to PID controller mode, setpoint, or output.

Problems resulted from human error can be very difficult to troubleshoot because, by human nature, it is common to delay, forget, or even cover up the true cause of an incident or abnormal condition. Troubleshooting needs to be based on a good understanding of the process design and a thorough investigation of the historical data to make informed decisions.

8.2.3 Isolate the Problem

The next step is to narrow down the problem to a specific area or a particular loop. For example, it is crucial to decide if the problem is a plant-wide problem, a unit-level problem, or is local to a control loop.

A plant-level problem is typically caused by a dynamic imbalance in supply and demand. The inventory fluctuation is reflected by gas pressure or liquid level. Excessive flaring or overflow/underflow of vessels are the consequences. Understanding the process flow (Sect. 6.2) and the cause and effect relationships (Sect. 6.4) is vital in isolating the problems.

A common symptom of unit-level or loop-level control problems is the oscillation in the controlled variables. The oscillation often occurs to multiple loops and even to a whole unit. Consequently, the problem may often show up as a "chicken or egg" dilemma, challenging to determine which comes first and which comes after. The cause and effect relationships (Sect. 6.4) and dynamic response characteristics (Sect. 6.6) between the key control variables are crucial for troubleshooting.

A common approach to isolate the problem to a particular control loop is to put the most suspicious control loop in manual (`MAN`) mode for a few minutes to see if the oscillations in key process variables start to go away. Continue monitoring the loop to find out the following:

1. Is the variability in the controlled variables (`PV`) less than when the loop is in `AUTO`? If yes, it indicates that the controller is doing more harm than good.
2. Does the variability diminish or decrease in other loops when the loop is in manual? If yes, that means the interaction exists, and the loop under investigation may be the root cause.

8.2.4 *Identify the Root Cause*

The root cause of a control problem can be pretty challenging to determine, depending on the complexity of the control scheme. For example, oscillating loops can be caused by many reasons, including the following:

1. Interaction between the loops, where the output of one controller affects another controller and vice versa.
2. Valve stick–slip from excess friction in the packing or sealing surfaces.
3. Incorrect PID tuning.

Although the causes of the control problems can be many, process control problems can be classified into three categories or levels: design, implementation, and operation. It is often a combination of two or more problems lined up to cause a major incident. See the Swiss cheese model (Euro-Control 2006):

1. Design problems: Many chronicle control problems are due to flaws in control design, especially in the overall control strategy (see Sect. 1.1.3). Once in operation, these types of problems are difficult or expensive to correct. For example, a supply-driven control scheme for a demand-driven operation may result in continuous swings in process variables or permanent offset in steady-state values. A flawed control strategy is unlikely to achieve satisfactory performance even with all the individual controllers perfectly functioning. The troubleshooting thus

should start with a good understanding and careful validation of the overall control strategy.

2. Implementation problems: The configuration and tuning of a complex PID control loop or a model predictive control scheme are not trivial and often require many careful considerations. Incorrect configuration is a common source of performance issues. Reset windup or bumpy transition are examples of loop integrity problems in complex PID control schemes.
3. Operation problems: Operation problems are typically related to the changes in the loop dynamics. As discussed in Sect. 4.3.1, loop dynamics are determined by the data flow, subject to four types of common problems:

   a. Broken connection of the data flow.
   b. Bad data in the data flow.
   c. Limited or constrained data values.
   d. Drifting in loop dynamics.

   The control loop comprises five essential components, and each can contribute to the problem, with the final control elements, the measurements, and the process dynamics being the most common culprits. See Table 8.1. A structured review of the five essential components should be carried out if the problem is pinpointed to a specific PID control loop. Process measurement can be erroneous but can typically be fixed easily. Final control elements, especially control valves, are the most common causes of control problems and are often hard to resolve without plant shutdown.

   Controller tuning is less likely the cause for existing control loops. A control loop is designed and configured for specific loop dynamics. Unless there is a significant change in the loop dynamics, the controller is expected to deliver the same performance. As explained in Chap. 4, many factors affect the loop dynamics; it is crucial to determine which part of the loop dynamics has changed before jumping to control loop tuning.

8.2.5 *Propose a Solution*

For complex control schemes, proposing a solution to fix the existing problem can follow the same process control design methodology discussed in Chaps. 6 and 7. The solution can be a permanent fix or a temporary workaround. In response to an urgent request from operations, a temporary workaround is often proposed and implemented first to allow the operation to continue. In reality, however, it is often found that once the temporary workaround is in place, operations are content with it and stop pursuing a proper permanent solution. As a result, a temporary workaround becomes a permanent solution. As a lesson learned, when the temporary workaround is implemented, make sure that the permanent solution will follow through.

For standalone control loops, depending on the specific cause, process control may propose solutions ranging from modification of the process, cleaning of equipment, calibration of instrumentation, maintenance of control valves, or fine-tuning PID controls.

Except for PID tuning, all other remedies need to be carried out by other departments, a proof that process control is a cross-disciplinary engineering specialty (Sect. 1.1.1), and close collaboration with other teams is a must. On the other hand, only process control offers this cross-disciplinary view of the operating problems (Sect. 1.1.4).

8.3 Troubleshooting of Complex Control Scheme

The crucial step for a complex control scheme consisting of multiple variables and multiple PID controllers is to ensure that the overall control strategy is correct and fit-for-purpose.

8.3.1 Overall Control Strategy

The overall control strategy must meet the control objectives, which in turn serve the operating objectives. It is vital to have a good understanding of the operating objectives and operating requirements. The different components in a properly designed control scheme must perform their duties; they must work together seamlessly with other components to achieve the high-level goals. Even the best configured or tuned loops will not deliver the desired performance with a flawed control strategy.

Chapters 6 and 7 have provided a systematic discussion on reaching a sound overall control strategy. The same process analysis and control design methodology discussed in Chaps. 6 and 7 also apply to control scheme troubleshooting and improvement. For any control loop, whether a simple standalone PID loop or complex control scheme, and for either design or troubleshooting purposes, the following questions can be asked to gain the necessary knowledge about the loop.

WHERE	Is the control scheme/loop located?
WHY	Is the control scheme/loop needed (process reason)?
WHAT	Functionality does (should) the control scheme/loop provide?
WHEN	Is the control scheme/loop expected to respond?
HOW	Was (will be) the control scheme/loop implemented in DCS?
WHAT IF	The operating condition is not met?

The troubleshooting should include the following reviews and analysis to understand and validate the control strategy:

1. Process flow analysis.
2. Supply and demand analysis.

3. Cause and effect analysis.
4. Degree of freedom analysis.
5. Dynamic response analysis.
6. Layered control design.

Plant-level troubleshooting depends on a good analysis of the process flow and the supply and demand model; unit-level control relies on a good understanding of the cause and effect relationship and dynamic characteristics; the loop-level troubleshooting depends on the type of control scheme, whether PID-based control or advanced process control.

8.3.2 Dynamic Supply and Demand Balance

In upstream E&P operation, the most common operating requirement is the physical separation and movement of the produced fluid. Maintaining a healthy balance between the supplies and demands, both static and dynamic, is mandatory at any point on the process flow path.

For gas processing, a prolonged imbalance between gas supply and demand will result in continuous flaring. Environmental regulation has been calling for zero-flaring during normal operation, and thus flare reduction is a pressing target. Constant flaring indicates poor process design, while temporary flaring indicates process upset that has exceeded the controllability range of the normal regulatory control loops.[1] For liquid processing, imbalance in the supply and demand is typically buffered in surge tanks but up to a limit. Over the limit, the unit operation has to turn down the throughput; otherwise, the unit will be tripped to a shutdown.

Example 8.1 Continuous flaring. Let us re-visit the produced gas compression example from Sect. 6.3. The original process design is as shown in Fig. 8.2.

There is one gas supply from the gas/liquid separator, and there are two consumers for the compressed gas, namely the gas lift wells and four heat recovery steam generators (HRSG). The demand is forecast to be greater than the supply, so a make-up gas stream is provided from a sweet gas source if needed.

Following the analysis in Sect. 6.3, the supply side is production-driven, thus supply-driven. The two consumers are either on flow control or pressure control, so they are both demand-driven. The make-up gas provides the swing control to regulate the dynamic supply and demand balance.

After the plant was built and brought online, many operating problems were encountered, including frequent compressor tripping, severe and continuous flaring, and failure to maintain compressor suction pressure at the setpoint. The heavy flaring went on for two years. At the peak time, the flaring rate reached one million standard

[1] It is common to see large flares in a refining or ethylene plant during startup or shutdown, when the supply and demand balance cannot be met.

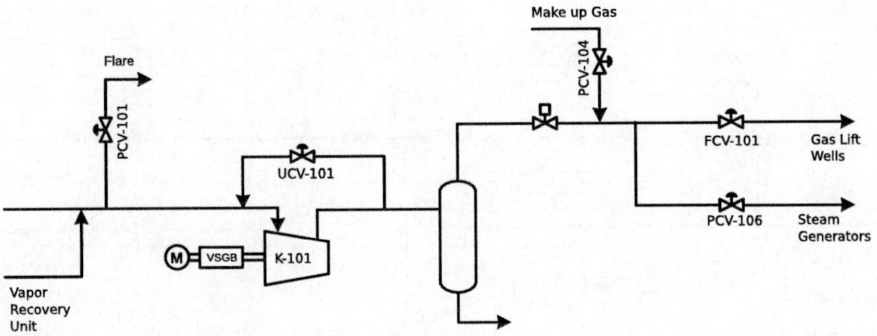

Fig. 8.2 Bad process design: gas compression

cubic meters per day! The operation had to rely on repeated renewal of government waiver on flaring to continue operation.

The process control group was approached to troubleshoot the flaring problem. After a careful review of the process flow, material balance, and operating condition, it was found that the root cause is the imbalance in supply and demand, due to the following causes:

1. The original design was based on the assumption that the produced gas is insufficient to meet the needs of the gas lift wells and the steam generators. A continuous make-up gas stream was expected to fill in for the supply shortage and serve as the swing stream. However, after startup, difficulties with the steam generators delayed the commissioning of two of the four steam generators. The gas demand was lower than the forecast and resulted in gas oversupply.
2. Subsurface study has underestimated the GLR (gas–liquid ratio) of the reservoir. For every barrel of oil produced, there was higher associated gas production than expected. This higher GLR made the gas oversupply problem worse.

The control overview diagram (COD) for the original control design is illustrated in Fig. 8.3.

The control scheme clearly shows that the excess gas has nowhere to go but to flares. The over-supply in the produced gas causes the compressor discharge pressure to increase. The elevated pressure causes the pressure controller `PC-103` to take charge and overrides the capacity controller `PC-102`, which leads to higher pressure at suction. The suction pressure reaches the setpoint of the protective controller `PC-101`, which kicks in and opens the valve to flare.

In other words, due to the supply and demand imbalance, the supply-driven operation was reversed to a demand-driven operation at the suction side of the compressor. The protective controller `PC-101` overrode the normal regulatory controller `PC-102` and became the normal regulatory controller. What makes things more challenging is that the operating pressure of the vapor recovery unit (VRU) floats with the compressor suction. The suction pressure operated much closer to the trip

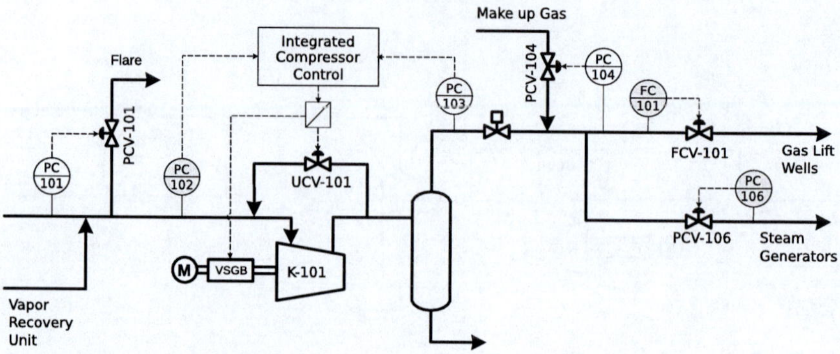

Fig. 8.3 Bad process design: supply/demand imbalance

limit of the VRU. Process upsets in the compressor train had caused several trip incidents in the VRU.

Plant operation was under pressure to maximize oil production, and thus, scaled down the overall capacity was not an option. This steady-state supply and demand imbalance could only be corrected by process modification. Based on the recommendation from process control, a plant shutdown was scheduled, and an extra stream was added to send excess gas to gas injection wells. The new process design and control scheme has been discussed in Sect. 6.3.

Flaws in plant-level control design are often related to inadequate supply and demand handling, and the fix is typically costly once the plant is built.

8.3.3 Dynamic Cause and Effect Relationship

As shown in Fig. 7.1, unit-level problems are often caused by incorrect or unreliable cause and effect relationships. Feedback control is based on a known and credible causal relationship between the controlled and manipulated variables. If this causal relationship is weak or changing, the control performance would be unpredictable.

Let us use a unit operation example to illustrate the criticality of the cause and effect relationship.

Example 8.2 Figure 8.4 is a simplified process flow for the true vapor pressure (TVP) control loop in a crude oil stabilizer. TVP is an important quality criterion for crude oil, and too high a TVP value is a safety hazard, while too low a TVP value means loss of oil to gas.

The corresponding control scheme is illustrated in Fig. 8.4. The control scheme shows that a PID control loop is implemented to control the TVP by manipulating the stabilizer bottom temperature.

However, the complaints about the stabilizer operation are multiple:

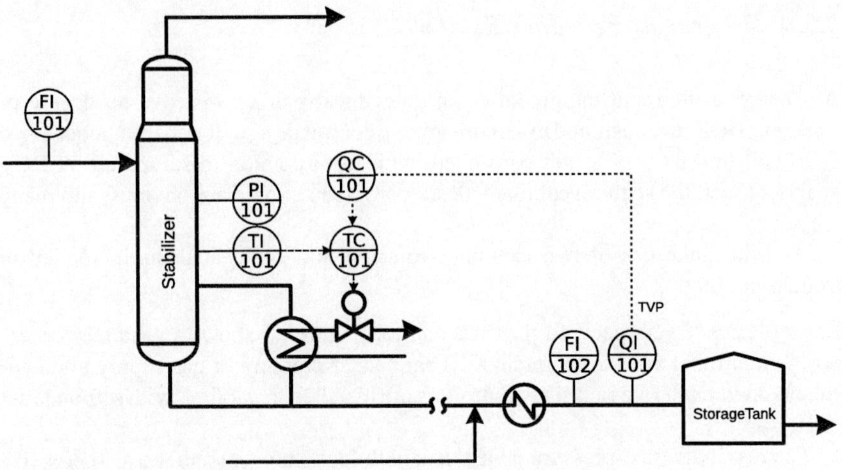

Fig. 8.4 Bad process design: unreliable cause and effect

1. The TVP controller can never be put into automatic AUTO operation for an extended time due to inferior control performance.
2. The TVP is manually controlled by manipulating the bottom temperature, and the TVP of the crude oil in the storage tank fluctuates so much and at one time even popped up the floating roof and caused a gas leak and oil spill.

Again, the problem was brought to the attention of process control. A systematic troubleshooting effort arrived at the following findings, among others:

1. The TVP online analyzer is highly unreliable due to technology limitations and poor maintenance. It is far from acceptable for closed-loop control.
2. The TVP analyzer measurement lags the actual changes in the stabilizer by about 40 min due to poor choice of location. As mentioned in Sect. 2.3.1, a long time delay is the number one challenge for PID control, and a 40-minute delay is beyond impossible for PID control.
3. The process variables that affect the TVP are not just temperature. The pressure in the stabilizer also significantly impacts the TVP value and is thus an essential component in the cause and effect relationship.

Based on the discussions in Sect. 8.3.3, to properly control the TVP value in the stabilizer, the control scheme must include all the critical cause and effect relationships related to the controlled variable, and the dynamic response corresponding to each cause and effect relationship must be fast, predictable, and reliable. The above control scheme has violated all the above requirements.

The improvement is implementing an inferential property estimator (soft sensor), where faster and more reliable temperature and pressure measurements are used to predict the TVP value. A PID controller controls TVP based on the predicted TVP value instead of the value from the online analyzer.

8.3.4 Degree of Freedom Review

All the controlled and manipulated variables must pair up to leave no degree of freedom (DOF) unconsumed in a multivariable control design. If not, the unconsumed DOF will find its way to get consumed, typically by going to saturation. A DOF analysis (Sect. 6.5) can reveal many design problems. Some are obvious, and many are not.

We will make use of two examples to show how DOF analysis is applied in troubleshooting.

Example 8.3 Over-specified pressure control. Figure 8.5 shows a gas pressure let-down circuit that aims to maintain a 10 bar header pressure of gas supply to downstream operation. Two pressure controllers are installed in parallel, with setpoints set to 10 bar.

The two individual pressure controllers both look innocent and are also properly configured and tuned. The two valves offer two DOFs, and there are two pressures as controlled variables.

However, this control scheme will not perform well. The problem is that the two pressure variables are the same header pressure. They are not sufficiently independent and can only consume one DOF. The overall control strategy of using two controllers to control the same header pressure is flawed. There are two valves for one pressure, which is one extra DOF unconsumed. As expected, at the steady-state, there is always one valve at saturation.

The solution to this control scheme is to re-do the control design. The improved design should have the header pressure as the only controlled variable and use either a split-range control or a fan-out control to manipulate the two control valves.

Sometimes this DOF over-specification is not apparent. For example, if the two controllers are far apart along the path, they may not appear on the same P&ID page or DCS screen (a key reason that a COD is necessary).

Example 8.4 Over-specified flow control. Figure 8.6 shows a water treatment process where the entrained sand is removed by educers upstream and a downstream hydro-cyclone.

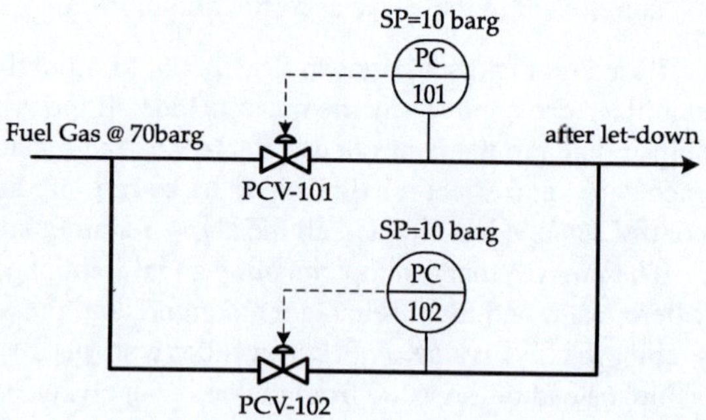

Fig. 8.5 Bad control design: over-specified pressure control

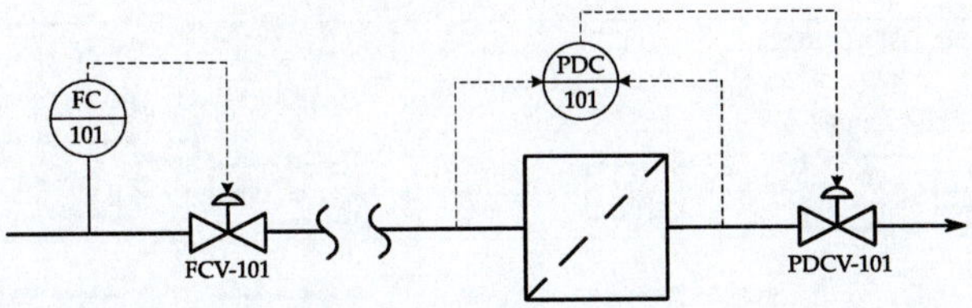

Fig. 8.6 Bad control design: over-specified flow control

In an early design, flow controllers are implemented at the educers in upstream. While downstream at the hydro-cyclone, the differential pressure (DP) across the hydro-cyclone is controlled to ensure the proper operation of the hydro-cyclone.

The educer and hydro-cyclone are supplied by two different vendors. Both vendors insist that the control scheme for their equipment is adequate with a proven track record.

Since the two systems are far apart, they seem to be okay when viewed in isolation.

However, flow rate is a function of differential pressure across the hydro-cyclone. The DP control is the same as flow control. This design has two flow controllers on the same stream. One extra DOF is not consumed, and one of the valves will end up in saturation (fully open).

8.4 Troubleshooting of PID Control Loop

If the control scheme is a simple standalone PID control, or a complex control scheme where the control problem has been pinpointed to a specific PID loop, then some systematic PID troubleshooting procedure can be followed.

As explained in Chap. 3, a PID control loop comprises five essential components, process dynamics, process measurements, final control elements, controller setpoint (objective), and the controller. The loop dynamics, determined by all the components on the data flow, dictate the control performance (King 2016; Luyben and Lyuben 1997; Marlin 2015). When troubleshooting, always keep the complete loop in mind and not take any component out of its loop context.

The most vulnerable component in the data path is the final control elements, specifically, the control valves. Process measurement also creates many troubles for control.

8.4.1 Process Dynamics

Control performance is determined by the loop dynamics instead of the process dynamics alone. However, as mentioned in Sect. 3.3.2, process dynamics are the

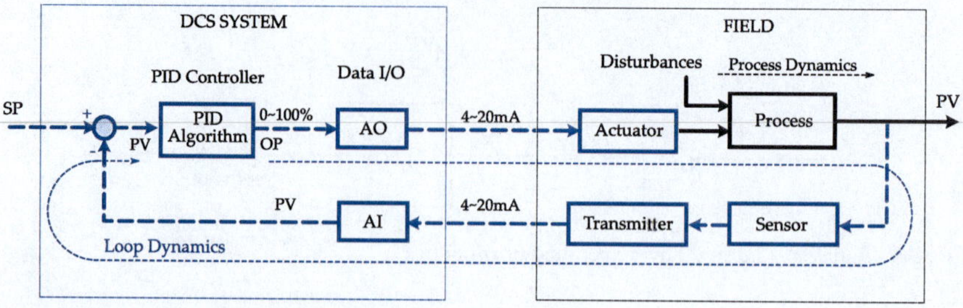

Fig. 8.7 Process dynamics verses loop dynamics

most critical part of the loop dynamics. Change in process dynamics can render the process control configuration inadequate, especially the PID tuning parameters. Figure 8.7 shows how the loop dynamics and process dynamics relate to each other.

Depending on the type of change to process dynamics, from slow aging/fouling of equipment to significant modification of process flow, its impact on control performance can vary. Some of the changes in dynamics are obvious, while others may be subtle.

Example 8.5 Control tuning for fan-out control. Suppose a water tank level is controlled by two pumps running in parallel at normal conditions, and the controller is adequately tuned. If the processing capacity reduces and only one pump needs to be in operation, the original PID tuning would be no longer adequate because the process gain has changed. Depending on how much performance degradation is acceptable, adjustment of the control gain may be required.

This example shows that although the individual pump dynamics remain the same, the process flow has caused the process dynamics (two pumps vs. one pump) to change and thus requires a change in PID tuning.

A more subtle example is with a distillation column where the process dynamics for control can vary with the operating condition and significantly affect the control design.

Example 8.6 Fractionator bottom temperature control. For the bottom impurity control, a tray temperature is typically selected as the manipulated variable. That is, the process dynamics are from the tray temperature to the bottom impurity level.

It is well known that the relationship between the tray temperature and the impurity is very nonlinear at a high purity level. It is also known that the most sensitive temperature (at the so-called *sensitive tray*) shifts to a lower tray if a higher purity (lower impurity target) is desired. When implementing advanced process control, which is supposed to improve the separation and lower the impurity level, a different tray temperature often needs to be selected to control the bottom impurity.

Heat exchanger fouling, filters/strainers blockage, distillation column flooding, and catalyst deactivation can all cause the process dynamics to change over time.

Some of the changes are within the range of robustness of the PID controller, while others may degrade the control performance so much that adjustment of the PID tuning or change of control strategy is needed.

8.4.2 *Process Measurement*

Process measurement typically includes a primary device (e.g., a sensor) and a secondary device (e.g., a transmitter). The sensor senses the changes in the process variables while the transmitter converts and conditions the sensor output to an electric signal appropriate for transmission to other devices such as DCS or PLC. Signals for transmission are typically in 4–20 mA current signals.

The analog input (AI) block in DCS receives the transmitter signal (e.g., 4–20 mA). It converts the signal to a value with the appropriate engineer unit and scale, based on the design datasheet and via calibration and re-calibration. This value becomes the process value, or PV, for the control loop.

All measurements have errors. For example, the measurement resolution is limited by the analog/digital (A/D) converter, which is at least 0.05% of the span for a 12 -bit A/D conversion ($1/2^{12}$). Instrument error can occur due to various factors such as calibration, environment, electrical supply, the addition of components to the output loop, and process changes.

The drift in the zeros, incorrect span, and linearization errors are typical calibration errors and can be corrected by re-calibration. Most instruments are provided with a means of adjusting the zero and span of the instrument. Many instruments offer a linearization adjustment to correct linearization errors.

Some common troubleshooting steps on measurement problems include the following (Brannan 2018):

1. Check whether the transmitter "freezes." A frozen transmitter could drive the controller to saturation and thus out of control.
2. Check for any significant non-uniform fluctuations in the measurements. For example, a spike or step change in flow or temperature that is physically impossible indicates a measurement problem.
3. Calibration: Check whether the instruments, especially analyzers, are correctly calibrated and serviced. Transmitter zero shift is a common issue.
4. Check the ranges and scan times to make sure they are adequate.
5. Check the location of the transmitter. For example, for a flow controller, the flow measurement should be located in front of the valve to avoid the turbulent flow caused by the valve. For liquid temperature measurement, the sensor should be immersed in the liquid, not in the vapor or the splashing liquid.
6. If PV is very noisy, consider adding a small PV filter to smooth out the high-frequency measurement noises.

Some noises and disturbances with slow frequency can be corrected by PID control, while others, such as those with higher frequencies, must be corrected at their origin.

Although high-frequency noises can be smoothed out with a small filter, the filter time will become part of the process dynamics and must be considered when designing process control loops.

8.4.3 Final Control Elements

The feedback controller calculates the control action to eliminate or minimize the control error by bringing the process value closer to its setpoint at each execution. The control output is either in the percentage of scale (0–100%) or in the actual engineering unit. The output signal is then converted to 4–20 mA by the analog output (AO) block and sent to the final control element. The final control element, which typically is a valve consisting of a positioner, actuator, and the valve, receives the control signal in the range of 4–20 mA, converts it to 3–15 psi pressure signal to drive the air-operated valve.

The control valve is the component in a control loop with the most troubles in a closed-loop control system, primarily because of wear and tear. The valve problems degrade the integrity of the control loop and render the valve dynamics (which is the part of the loop dynamics) uncertain or time-variant. Common valve problems include the following:

1. Valve stiction: Stiction refers to stickiness and friction. With a sticky flow valve, there is no change in flow until the accumulated changes exceed a certain amount ("the amount of stiction"). Valve stiction adds delay to a loop for any changes in the controller output.
2. Valve overshoot (slip): The change in flow is more discrete than continuous and is much larger than demand. Valve slip causes the PV to pass by the setpoint.
3. Dead band: No PV response is observed until the manipulated variable has changed more than a certain amount. Dead band can be caused by many factors, such as backlash, excessive friction, or shaft deflection.
4. Step resolution: The minimum step change in the input signal to which the control valve will respond while moving in the same direction. This phenomenon is typically caused by the tendency for the control valve to stick after coming to a rest and takes a more remarkable change in controller output to initiate the correction at the valve. Degraded step resolution means more lag and error that lead to the stick–slip limiting cycle problem, characterized by continuous cycling at a steady-state under closed-loop control, even with correct tuning.
5. Hysteresis: Hysteresis indicates that the valve position can take any valve position within a range of values for a given input signal. There is no change in flow when the controller reverses direction until the change exceeds a certain amount.
6. Nonlinearity: All valves have nonlinearity. The valve characteristics are different from valve to valve. The nonlinearity becomes more severe towards both ends of the characteristic's curve. It is advised to limit the normal operating range to the middle section of the valve curve, typically 10% to 80%.

7. Valve Rangeability: Valve is too small or too large for the loop to be controlled. For example, a column temperature control with an undersized reboiler is very likely to find that the valve stays saturated most of the time and is effectively out of control. On the other hand, an oversized or undersized valve may force the operating point to operate outside the valve's normal operating range, becoming more nonlinear.
   The normal process operating range should typically fall between 10% and 80% of the valve operating range, and the valve should be sized accordingly. In practice, however, it is not uncommon to see the contrary. For example, in a reciprocating compressor, the recycle valve is typically used for capacity control. Since recycling is expensive, many control specifications place the normal operating point too close to the upper range and thus lose the controllability in that direction. It is crucial to allow the valve sufficient room to move around the normal operating point in both directions.

Valve hysteresis plus dead band is the total friction in the valve. All valves have hysteresis and resolution limits. For control, valve slip is typically worse than valve stiction, and valve stiction is worse than dead band. To check for valve hysteresis or stiction, make several controller output changes: at least two steps in one direction and two in the other, followed by several tiny steps to check stiction. Using the data, run a hysteresis check on the loop. If the hysteresis is more than 1% for valves with positioner and 3% for valves without positions, repair or replacement of the valve should be considered. Hysteresis of 1–4% degrades loop performance, while with tight tuning, hysteresis greater than 3% may start to cause oscillations.

Some valves (e.g., smart valves in Fieldbus control system) can provide the actual valve position as *read-back*. The controller can use the valve read-back value to effectively eliminate many of the control problems related to the valve. For example, the anti-reset windup configuration shall use the valve read-back where available.

The following checklist can help troubleshoot some common valve problems:

1. Check valve history data (`SP`, `PV`, `OP`) to look for unusual behaviors or patterns.
2. Check if the controller output is high or low limited when the loop problems occur.
3. Check valve read-back, if available, to see if the positioner follows the controller output well.
4. Put the valve in manual mode, and see if the variability in the controlled variable is less than when the loop is in automatic mode `AUTO`. If yes, it indicates that the control loop is doing more harm than good.
5. Is the valve oversized or undersized for the loop? An oversized valve reduces the sensitivity of the controller, while an undersized valve limits the controllable range.
6. Is the valve sufficiently linear within the normal operating range. Put the loop in manual mode (`MAN`) and make a few steps in controlled output in both directions. The linearity check includes a symmetry check to investigate if the valve responds uniformly in both directions. If the valve nonlinearity cannot be cor-

rected, then more conservative tuning or valve characterization can be considered to compensate for the nonlinearity.
7. Is the loop problem limited to only a specific range of valve output, or can it occur at the full range of output?
8. Check the PV response to control output changes and determine the severity of the valve slip, stiction, and dead band.
9. Check the bypass valve, if any, to see if it is in the appropriate position. The bypass valve opening can shift the controllable range of the control valve.

A scatter plot of the controller `PV` vs. `OP` can reveal many valve performance issues. See Sect. 8.5 for examples of using scatter plots for valve troubleshooting.

Note that although the ultimate solution is to fix the root cause, many times, modifications to the controller tuning (or the control strategy if necessary) can temporarily reduce the severity of the problem and allows the operation to carry on until the next scheduled shutdown or maintenance. However, ensure that the temporary solution will not be taken as a permanent solution!

8.4.4 Control Objectives

The operating objective dictates the control objective. For complex control loops that involve multiple controllers, it is imperative to view the control scheme and control loops in their entirety and understand the overall control strategy. With the whole picture in mind, it is easier to appreciate how the loops work together to achieve the overall control objectives and understand how they interact and interfere. It often requires validating the following to troubleshoot the control objectives at the PID control level:

1. Know what to expect of the control scheme or control loop. For instance, a level control on a surge drum has a very different objective than that on a knockout drum.
2. Check the control strategy to see if it serves the control objective. It is not uncommon in practice to find that the control loop design does not meet the operating need at all times.
3. The control target is realistic and achievable. For example, controlling the steam pressure at its saturation temperature is hard to achieve. Similarly, nobody would expect the room temperature to reach freezing in a hot summer by setting the A/C thermostat to 0°. As another example, tight control of a delay-dominant process with PID may not be realistic.
4. Maintain an adequate balance between loop performance, stability, and tuning effort. The tightest control setting is often the one that can lead the process to unstable conditions during a process upset. Besides, the extra effort spent on reaching the "optimal" tuning may not be justified because the "optimal" tuning often changes with time and operating conditions.

8.4.5 *The Controller*

Some of the control design and configuration problems with the control algorithm or control design are listed as follows:

1. Cause and effect relationship: Is the controlled variable sensitive enough to the manipulated variable for decent control? If not, consider controlling a different process variable. For example, for a distillation column, if the current tray temperature is not at control sensitive tray in a column, the control performance would not be good, and a different tray may need to be used for control.
2. Process dynamics: Is the dead time of the process short enough for control? Loops with long time delays are notoriously hard for PID loops to handle.
3. Control structure: Is the controller designed to handle all the operating modes, such as overriding, split range, fan-out? see Sect. 4.3. Check whether a complex PID control scheme such as feedforward control and cascade control can help improve the performance.
4. Measurement: Consider other controlled variables such as pressure compensated temperature, internal reflux, separation index, heat duty, pressure/temperature compensated flow, reflux/feed ratio, and reflux/draw ratio.
5. Nonlinearity: Adding nonlinear transformation can often improve the control performance if the relationship between the manipulated and the controlled variable is very nonlinear. More on signal characterization in Sect. 10.1.

If a controller worked well in the past and stopped working all of a sudden, the most likely cause is typically not the controller itself (including the tuning) but something else in the control loop.

Many preventive and protective features are included in controller design and implementation to deal with abnormal operating conditions. The most common and essential features include anti-reset windup and bumpless transfer. While these features provide the flexibility to handle abnormal conditions, they are also familiar sources of loop problems.

For each PID controller, troubleshooting should include the validation of the controller configuration listed in Table 3.9. Some of the typical configuration errors that can result in persistent poor performance of the control loops include the following:

1. Incorrect algorithm type: An `I-PD` algorithm may not deliver the desired performance for setpoint tracking, while a full-form `PID` algorithm for a noisy flow loop can produce many fluctuations.
2. Wrong tuning parameters: During project commissioning, many PID controllers are left with inadequate system default tunings.
3. Reset windup in complex control loops such as override control involving selectors or switches.
4. Bumpy transfer in complex control loops that involve splitters and split-range controls.

In the context of a complex control scheme that consists of multiple complex control loops, it is crucial to account for all the operating scenarios that the controllers are expected to run.

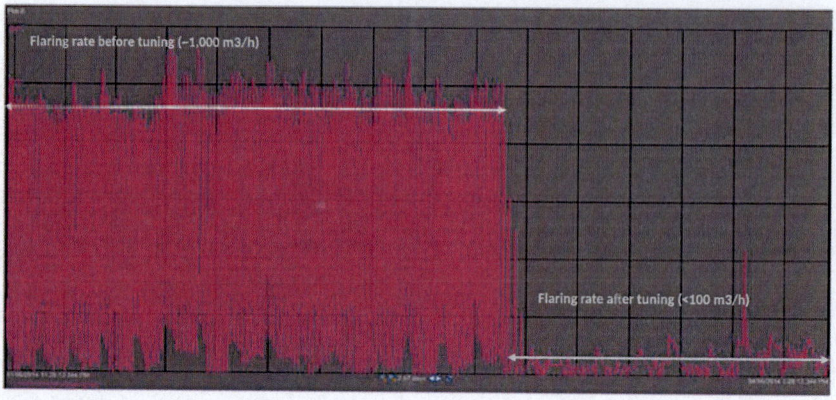

Fig. 8.8 Amount of flaring before and after PID tuning

Example 8.7 Improved control performance via tuning. Figure 8.8 shows an example of the difference in the amount of flaring in a production station before and after PID tuning improvement. The flaring rate is controlled by a protective pressure controller at a production separator. The separator level control caused severe oscillations due to highly aggressive tunings. Those tunings were probably left unchanged from the DCS system default (see Sect. 3.5.6). The improved tuning reduced the flaring from about 1,000 m^3/h to under 100 m^3/h, with additional room to improve.

Oscillations caused by poor loop tuning typically show sinusoidal behavior. Sawtooth-shaped oscillations are often caused by valve hysteresis, stiction, and nonlinearity. Level loop on a vessel that feeds another unit (downstream) is often a significant cause of variability. Similarly, a poorly performing pressure control loop is often a common source of oscillations in a large area.

For oscillations caused by valve stick–slip, the ultimate solution is to fix the valve. However, if the valve maintenance requires an unscheduled unit shutdown, it may have to be delayed unless the valve is in critical service. Meanwhile, de-tuning the loop (smaller controller gain or less integral action) can temporarily improve the control performance to an acceptable level to wait for the next scheduled shutdown.

8.5 Troubleshooting Tools

Software tools are becoming widely available to help monitor and diagnose loop performance problems. Most tools can continuously monitor all the control loops in a plant or unit and proactively report those under-performing loops based on predefined statistical or analytical criteria. Some software tools can even provide an excellent initial diagnosis.

Most of the tools, however, only detect and report the problems. To pinpoint and fix, the problem is the most challenging part of the effort. Knowledge, skills, experience, plus sufficient patience are still the key.

8.5.1 Data Trending and Scatter Plots

Process control engineer's most effective and frequently used tools are still the basic tools available on everyone's desktops:

1. Trending: Data visualization and trending is an essential feature in any decent data analysis software. Trending the setpoint (`SP`), process value (`PV`), and controller output (`OP`), sometimes also with controller mode (`MODE`), can reveal many problems in measurements and control valves.
2. X-Y Plot (scatter Plot): An X-Y plot of the process variable (`PV`) and the controller output (`OP`) with a spreadsheet software or most monitoring tools can often expose many control valve problems and controller nonlinearity.
3. Auto-correlation and cross-correlation analysis: Correlation analysis is a valuable tool for performance evaluation but is not as readily available as trending and x-y plots.

Below are a few examples of using scatter plots to identify valve problems.

Example 8.8 Normal control valve characteristics. Figure 8.9 shows a `PV` vs. `OP` scatter plot for a properly tuned PID control loop of flow. A good correlation exists

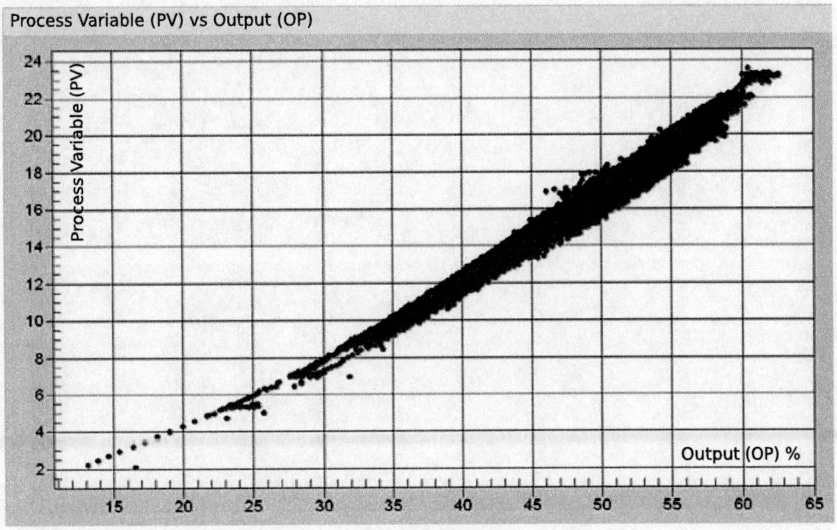

Fig. 8.9 Scatter plot of PV verses OP for a normal PID loop

between the PV and OP, and the line is almost straight over the normal operating range of the valve openings.

It is also observed that the operating range of the valves is between 30% and 60%. If the data represents the normal operating condition, the valve may have been a little oversized.

Example 8.9 Sticky control valve. Figure 8.10 depicts the real-time trending of the setpoint, process value, and output of a flow controller with a sticky valve. The setpoint (flat line) and process value are at the upper part of the chart and the output at the bottom.

The trending plot shows that the controller output is ramping up and down, but the process value does not follow the valve until the output change reaches a threshold. When the valve suddenly starts to move, the output change has already been too excessive, and causes the process value to jump to the other side of the setpoint, which immediately drives the output to move in the opposite direction.

Figure 8.11 is another example of a PID loop with a sticky valve. The trending plot shows that the valve stickiness is 8–9%, too much for a PID controller.

Plotting controller PV against its output in a scatter plot (xy-plot) immediately reveals the poor performance of the control loop. See Fig. 8.12 and compare it to the normal loop performance shown in Fig. 8.9.

Example 8.10 Valve Slip. A slippery valve can significantly degrade the control performance. For example, in a crude propylene oxide (PO) column in a styrene

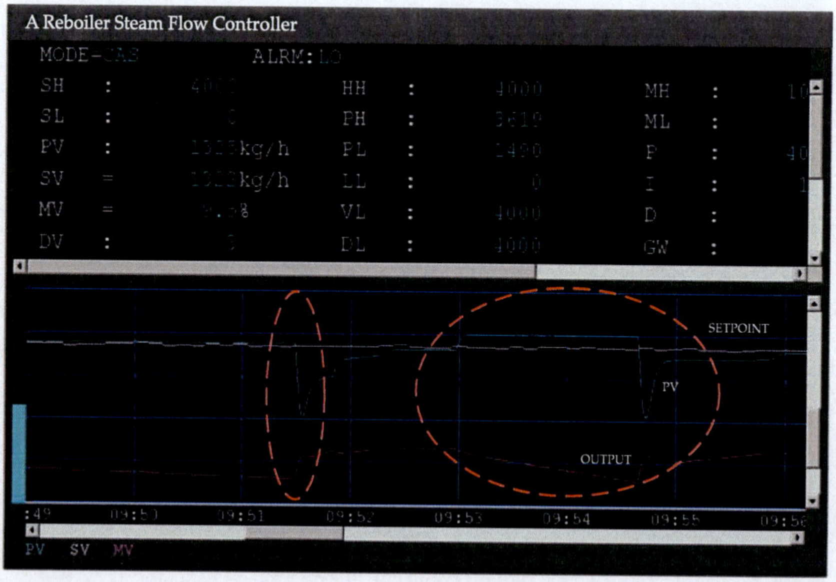

Fig. 8.10 PID data trending: sticky valve

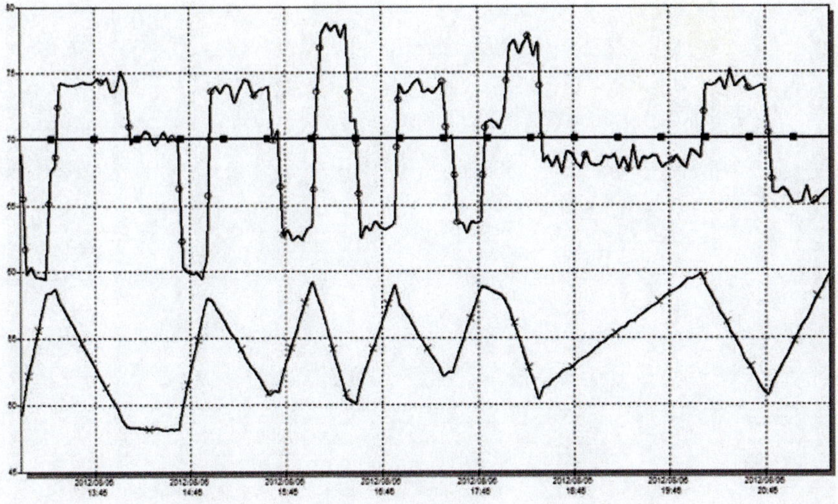

Fig. 8.11 PID data trending: severe stiction

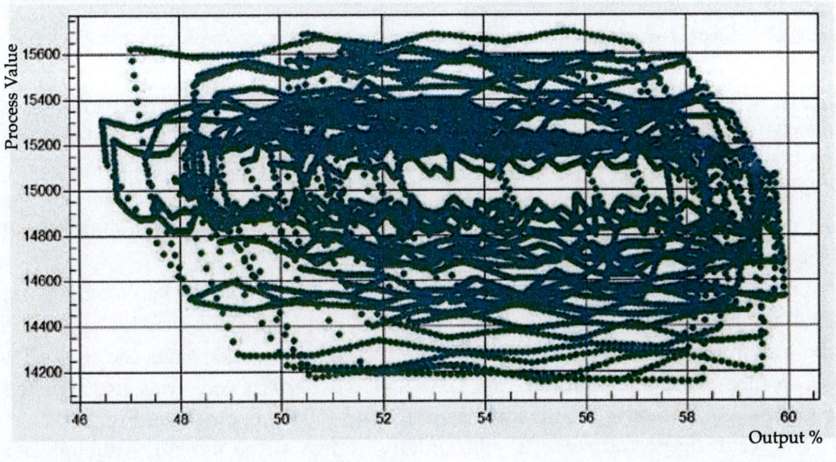

Fig. 8.12 PID data trending: poor performance

monomer and propylene oxide (SMPO) plant, the reboiler steam flow loop has inferior performance, leading to instability of the unit operation. The unstable operation of the crude PO column was believed to be the cause of the problems in polymerization in a downstream column, which once led to a costly plant shutdown. Figure 8.13 shows the trend of the history data.

The plot shows that the flow has been spiky by as much as 7%, causing unstable column operation (period ①). The flow loop was temporarily switched to manual (MAN) mode for troubleshooting purposes, and the flow immediately becomes more

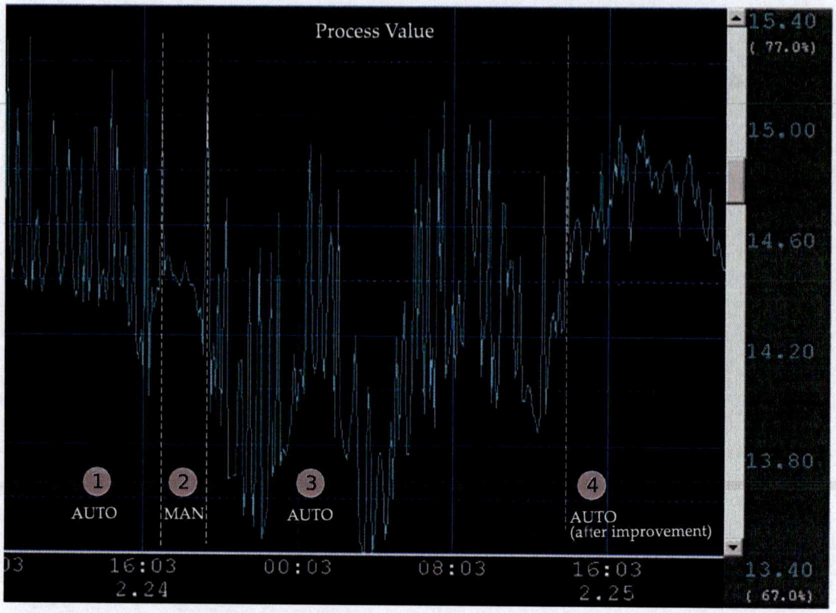

Fig. 8.13 PID data trending: slippery valve

steady. A few step changes on the valve output were performed, and the PV/OP data revealed that the valve has difficulties positioning itself. Valve is also found to be significantly oversized (period ②). When the loop was put back to the automatic model (AUTO), the spiking in flow re-appeared immediately. Consultation with the instrumentation engineer confirmed that the valve had difficulties positioning around 45% (period ③) due to wear and tear.

The ultimate solution for the valve slip problem was to take the valve offline and repair it. However, the valve could not be serviced without a unit shutdown, and the next turnaround was still one year away. As a temporary workaround, the controller was de-tuned with integral time from 16 to 45 s. The bypass valve was also adjusted so that the control valve would stay away from 45–50% opening (see Fig. 2.9).

The resulting performance is substantially improved (period ④). Although not optimal, the control performance is sufficiently well, awaiting the next plant turnaround.

8.5.2 *Control Performance Monitoring Tools*

A properly designed, implemented, and maintained process control solution is critical for safe and efficient plant operation. Once a plant is built and running, the challenge is tuning the control performance to the optimal level and sustaining it. For this reason, a *control performance monitoring* (CPM) system is becoming increasingly popular.

The CPM system provides a convenient, continuous, and reliable way of monitoring the performance of the control loops in operation. The objective of loop performance monitoring is to improve the control performance through the following steps:

$$\texttt{Monitor} \Rightarrow \texttt{Diagnose} \Rightarrow \texttt{Improve.} \qquad (8.2)$$

Proactive monitoring and troubleshooting are arguably more effective and efficient than reactive maintenance to achieve or retain higher control performance. A good CPM software application is valuable to help this transition from reactive to proactive maintenance.

Monitoring

Loop performance monitoring provides proactive detection and automated reporting of the loop performance issues. However, control loop performance has been an active area with many theories and practices but still lacks standard definitions or implementations. The practical requirements include at least the following:

1. Provide a statistical summary of the control loop status at any operational level and time period. For example, what percentage of control loops were running in their intended mode for a particular unit during the last month or the last eight hours? What is the daily up-time trending for all the controllers in the entire plant for this month? These are crucial information for management.
2. Detect the poorly performing control loops, or "bad actors", generate a "bad actor report." The bad actor report is a list of the worst-performing loops automatically generated by the system. It is typically sent to the mailbox of the responsible engineers for troubleshooting and improvement. As the worst-performing loops are fixed, the less severe ones bubble up and make into the bad actor report. Over time, the overall control performance is improved. Consequently, chasing the bad actors is a systematic and sustainable approach to improve loop performance.

The loop performance is usually represented as intuitive and reliable KPIs. For example, the statistical information may include the following KPIs:

1. Availability: A controller may not be in operation due to operational reasons but can be turned on if needed. For example, all the controllers on standby equipment are not actively controlling but are in ready-to-run condition, and thus, they are considered *available*.
2. Up-time: A controller running in its intended mode is called *in-control*. A controller in manual mode is not in control, while a controller in automatic mode or cascade mode is. The percentage of time a controller is in control is called the controller up-time (see Fig. 3.10 for PID controller modes).

3. Compliance: A controller that keeps the control error within the desired range is defined as complying with the control performance requirement. There are many ways of defining the compliance standard; a simple error band is the most common.

The definitions of loop performance differ between simple standalone PID controllers, complex PID control schemes, and model predictive controls. The control performance index (CPI) has been a popular KPI for simple PID controllers, comparing the closed-loop response with the ideal minimum-variance control algorithm. A good CPM tool is expected to include most of the following functions:

1. Proactively and continuously calculate the key performance indicators (KPI) for all control loops.
2. Detect any performance issues based on pre-defined performance criteria.
3. Support quick "roll-up" (aggregation) of the KPIs to any operational levels, from individual loop to a particular unit to the entire plant, and for any selected period.
4. Provide a convenient web-based interface to allow on-demand access to the various KPIs by any authorized user from anywhere and at any time.
5. Generate customized performance reports and automatically deliver them to the email or mobile device of the stakeholders or responsible persons. The report can be a high-level statistical view of the control loops for operation management or an exception-based report containing bad actor lists for process control engineers.
6. Provide relevant details ("drill-down") on the poorly performing control loops to facilitate root cause analysis.

Diagnosis

The diagnosis part of a control performance monitoring (CPM) tool provides accurate information and analytical tools to facilitate the root causes analysis for the problems detected and reported. For example, some software can pinpoint the problem to valve stiction or measurement error, while most software offers cross-correlation analysis to help isolate process variable oscillations. CPM software typically provides the following diagnostic information at the individual control loop level:

1. PID loop dashboard to show the running status and performance information, highlighting the detected problems.
2. Single-point access to the critical configuration information, performance history, and other dependent information to facilitate the root cause analysis.
3. Access to all the required loop configuration information.
4. Access to the change history of key PID configuration parameters, including the controller mode, setpoint, tuning parameters, and filter value.
5. Online access to the relevant P&IDs, PCN, and operating manuals.
6. Convenient analytical tools such as trending charts, cross-correlation analysis, and essential dynamic modeling capability.

Improvement

The ultimate purpose of the monitoring and diagnosis is to improve the control loop in question. Improvement is a highly complex process and can follow the same methodology and procedure for process analysis (Chap. 6) and control design (Chap. 7).

8.6 Summary

Depending on the complexity of the control scheme, the area of attention can also be quite different. For simple standalone PID controllers, the focus is on the loop dynamics, i.e., the five essential components of a PID loop. However, for complex control schemes, the troubleshooting should start with a good understanding and careful validation of the overall control design, the interaction between the different controllers, and the integrity of the control scheme configurations.

Troubleshooting is typically the most challenging work demanding the highest expertise. It requires that the troubleshooter not only understands how the control scheme work but also knows how it may fail.

Some general guidelines and best-practice procedures are discussed in this chapter. The basis is the knowledge presented in Chap. 3 for simple PID loops and Chaps. 4 and 5 for complex and advanced control schemes. The process analysis skills and control design methodologies in Chaps. 6 and 7 provide the methodology and guidelines.

When troubleshooting a control loop, the best support is typically from the operators. As a process control engineer, it is critically important to maintain a good work relationship with the panel operators. The operators can make the troubleshooting effort very enjoyable, and they can also make it miserable.

References

Åström KJ (1999) Automatic control - the hidden technology. In: Frank PM (ed) Advances in control - highlights of the ECC'99. Springer

Brannan C (2018) Rules of thumb for chemical engineers - a manual of quick, accurate solutions to everyday process engineering problems, 6th edn. Elsevier

CSHI (2005) Report no. 2005-04-I-TX: Investigation report – refinery explosion and fire. Technical report, U.S. Chemical Safety and Hazard Investigation Board. https://www.hsdl.org/?view&did=234995

Euro-Control (2006) Revisiting the Swiss Cheese model of accidents. Technical report, EuroControl Experimental Center

King M (2016) Process control: a practical approach, 2nd edn. Wiley

Luyben WL, Lyuben ML (1997) Essentials of process control. McGraw-Hill

Marlin TE (2015) Process control - designing processes and control systems for dynamic performance, 2nd edn. McGraw-Hill

Chapter 9
Control Loop Tuning and Improvement

There are many ways to tune a PID controller. PID loop tuning starts with a realistic expectation of the closed-loop control performance to achieve (Sect. 9.1). Trial-and-error and empirical methods (Sect. 9.2) were popular during the early days, but they rely heavily on the practitioner's experience and patience and the results are often hit-and-miss. Model-based PID tuning methods (Sect. 9.3), although a little more involved, significantly reduce the uncertainties in the PID tuning process and are gaining increased popularity to become the preferred methods. The work process of the model-based tuning method can be simplified for some control loops such as flow loops (Sect. 9.4) and vessel level control loops (Sect. 9.5), where the process gain can be calculated via some "shortcut" methods to save time and effort.

9.1 PID Control Performance

Different applications have different expectations of the ideal PID performance. The optimal PID response in robotic control may be unacceptable in industrial process control and vice versa. The desired control performance is also severely limited by the inherent weakness of the PID algorithm. Therefore, control performance is always a trade-off between desire and practicality.

9.1.1 Performance Criteria

The control performance refers to the closed-loop response (Fig. 9.1), typically gauged by setpoint tracking and disturbance rejection.

The quantitative measurements of the dynamic response characteristics of the closed-loop response typically include the following:

1. Dead time: After a control action is taken, the time it takes for the process output to respond.
2. Overshoot: How much the peak level is higher than the steady-state, normalized against the steady-state value.

S. S. Niu and D. Xiao, *Process Control*, Advances in Industrial Control,
https://doi.org/10.1007/978-3-030-97067-3_9

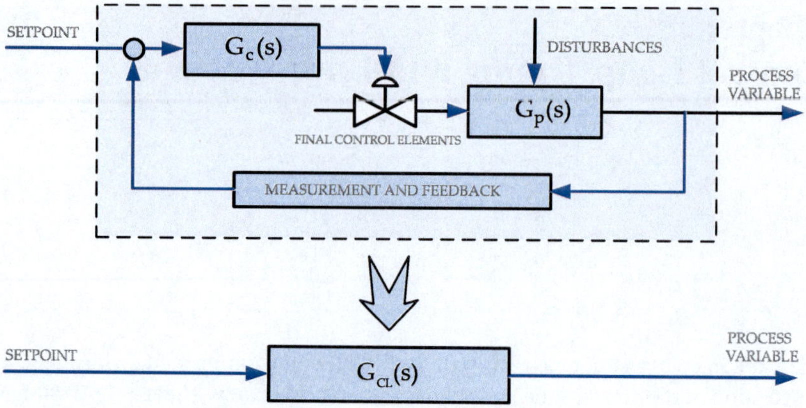

Fig. 9.1 Closed-loop transfer function

3. Steady-state error: The difference between the steady-state value and the desired value.
4. Settling time: The time it takes for the system to converge to its steady-state (within 2 or 5% of the final steady-state value).
5. Rise time: The time it takes for the process output (controlled variable) to reach the new steady-state value (typically 95%).
6. Peak time: The peak time is the time required for the response to reach the first peak of the overshoot.

In setting the control objectives, a good balance should be maintained between performance and simplicity. Simple often means reliable, and reliability is far more important than performance in the process industry. Ideally, we would like the closed-loop response to meet the following expectations:

1. Closed-loop stability: The closed-loop response must be stable.
2. Disturbance rejection: The effects of disturbances need to be minimized.
3. Setpoint tracking: Fast and smooth response to setpoint changes.
4. Steady-state error: There should be no steady-state offset.
5. Control actions: There are no excessive control actions.
6. Robustness: The control design is insensitive to a wide range of changes in process dynamics and operating conditions.

PID control performance tuning is achieved by the proper setting of the three parameters: proportional gain (K_c), integral time (T_i), and derivative time (T_d). However, due to the simplicity of the three-parameter formula, it is often difficult to achieve the desired *best* performance. The tunings that provide the fastest setpoint tracking are the settings most likely to cause instability if the *loop dynamics* change.

PID tuning thus always seeks a compromise between performance and robustness instead of performance only. Some processes must not allow an overshoot of the process variable beyond the setpoint if, for example, this would be unsafe. Other processes may require that the manipulated variable be limited to save energy.

Mathematically there are many criteria for deriving the tuning parameters and measuring the controller performance. We will not go into any details on these criteria. For industrial process control, most of the time, the desired PID control response has the following characteristics, similarly to line ② in Fig. 9.2:

1. Stable and smooth response in the controlled variable, with no overshoot or slight overshoot.
2. Relatively fast response to reject disturbances.
3. Relatively smooth response in the manipulated variable.

Quarter wave decay (Line ③ in Fig. 9.2) has long been accepted as the most popular performance criteria for PID response performance because of the good compromise between oscillation and settling time. However, this response performance usually is unacceptable in the process industry due to concerns of stability and overshoot. Sluggish response (Line ①) is not acceptable either. The desired response is quick and smooth, with a small (maximum 5%) overshoot (Line ②).

9.1.2 Practical Considerations

Control performance can be affected by many factors such as noises and disturbances, nonlinearity and time variation in process dynamics, uncertainties in the final control elements, control path changes due to branching and merging of control signals. Poor PID tuning is not the primary cause of bad performance.

Because of the industrial process's inherent nonlinear and time-varying nature, there is no way to tell whether a given set of tuning parameters is the best. The best tuning today may become sub-optimal tomorrow or next week. Therefore, it is impossible, nor is it necessary, to reach the so-called optimal tunings. The tuning only

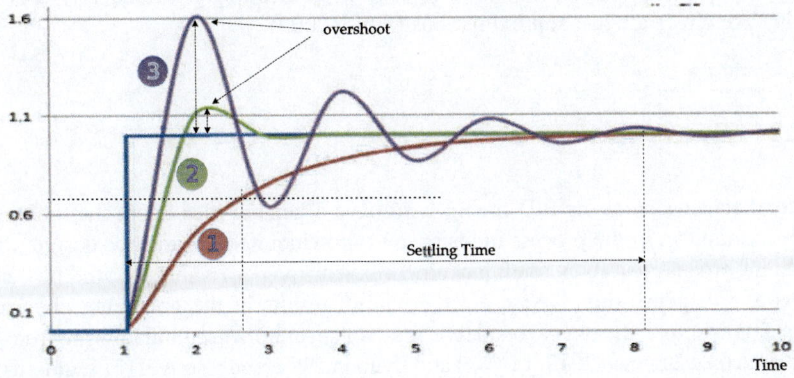

Fig. 9.2 PID closed-loop response

needs to provide decent and robust performance within a sufficiently large operating range. Any effort to achieve an improvement from 90% satisfaction to 100% is a waste of time.

9.2 Classical PID Controller Tuning Methods

Proper controller tuning is the quickest, easiest, and least expensive process control improvement to reduce process variability. As a result, most inexperienced engineers are often eager to jump to PID tunings when faced with a control performance issue.

The loop dynamics dictate the closed-loop transfer function instead of the process dynamics alone. All the data flow components contribute to the loop dynamics (see Sect. 3.3.2). The process and control valves are the dominant factors affecting the loop dynamics, while other components such as the analog/digital converters and transmitters may only have insignificant and relatively deterministic dynamics. Other factors such as branching and merging the data flow through selectors and splitters may significantly impact the loop dynamics.

The closed-loop transfer function is mainly determined by the process dynamics $G_p(s)$ and the controller dynamics $G_c(s)$ for a simple standalone PID loop. The desired closed-loop transfer function $G_{\mathrm{cl}}(s)$ is given by

$$G_{\mathrm{cl}}(s) = \frac{Y(s)}{Y_{\mathrm{sp}}(s)} = \frac{G_c(s)\,G_p(s)}{1 + G_c(s)\,G_p(s)} \tag{9.1}$$

and the desired response characteristic for this closed-loop transfer function (line ② in Fig. 9.2) is achieved by adjusting the two (for PI) or three parameters (for PID) in $G_c(s)$, which is the so-called PID tuning process.

Due to the complexity of the process dynamics and the added dynamics of all the components in the control loop, the loop dynamics are usually complex and non-deterministic. Thus, achieving the desired response performance with merely two or three tuning parameters in the PID algorithm is challenging. Consequently, all the tuning principles are just approximations (Smith 2009).

9.2.1 *Typical Tunings*

A good starting point for PID tuning is to have a general idea about the tunings' ballpark values. In the process industry, by following the best practice design, the different process variables often have a consistent range of values for the process gains and response time. These are the natural results of the governing physical laws. For example, the typical residence time for separators/accumulators are from 3 to 20 min (see Brannan 2018; Luyben and Lyuben 1997, and also API12J standards). The residence time is achieved through the vessel and flowline sizing, which dictates

Table 9.1 Typical values for process gains and controller tunings

Process type	Gain (%/%)	Response time (min)	Controller gain (%/%)	Integral (min)	Derivative (min)
Flow	1.0–5.0	0.25–1.0	0.3	0.5	0.0
Temperature	0.5–2.0	0.50–5.0	1.5	5.0	1.0
Pressure/liquid	1.0–5.0	0.25–1.0	0.5	0.5	0.0
Pressure/vapor	0.1–2.0	0.50–2.0	2.0	5.0	0.0
Level	1.0–5.0	5.0–30	1.0	10.0	0.0
Quality	1.0–5.0	0.5–30	0.3	40.0	0.0

the process gain and time constant (see Sect. 9.5.2). Table 9.1 lists the typical range of values for the process gains and response time for typical process variables (Blevins et al. 2003; Ford 2013; Svrcek et al. 2006).

The typical PID tuning parameters can be obtained in line with the typical process gain and time constants and are shown on the right side of Table 9.1. These values are on the conservative side and will generally result in a sluggish response. They can serve as the initial tunings to start a green-field plant if nothing is known about the loop and no test can be performed. They are by no means the final acceptable tunings (Brannan 2018; Ford 2013). The tunings must be improved with a proper tuning method later on.

These initial tuning parameters can also serve as an order-of-magnitude sanity check on existing PID tunings. For example, an average level control loop would expect an integral time in the order of minutes rather than seconds. If a level loop has an integral time of 10 s, it definitely raises a flag for further investigation.

Control systems all have their system default values for PID tunings. For example, in the DeltaV system, the system default for PID controller gain is 0.25 (%/%), and the integral is 10 s. However, these values are not appropriate for most applications and must be modified before putting the control loop online.

9.2.2 *Tuning by Trial and Error*

Tuning by trial-and-error is a "tune by feel" method and has been widely used in practice. Although quick, convenient, and sometimes effective, it has been proven time and again to be inefficient, inaccurate, and often dangerous. This method should be avoided whenever possible.

Below we give two examples showing where the trial-and-error method can be successful for one problem and ineffective and dangerous for another.

Example 9.1 Oscillating PID performance. A typical performance problem is oscillations in both process value and controller output while the setpoint is constant. This oscillation typically has a sinusoidal pattern with a fixed frequency or period. Figure 9.3 shows the historical trends of the process value and control output of a flow controller. The clear sinusoidal pattern suggests that the swing is due to the PID

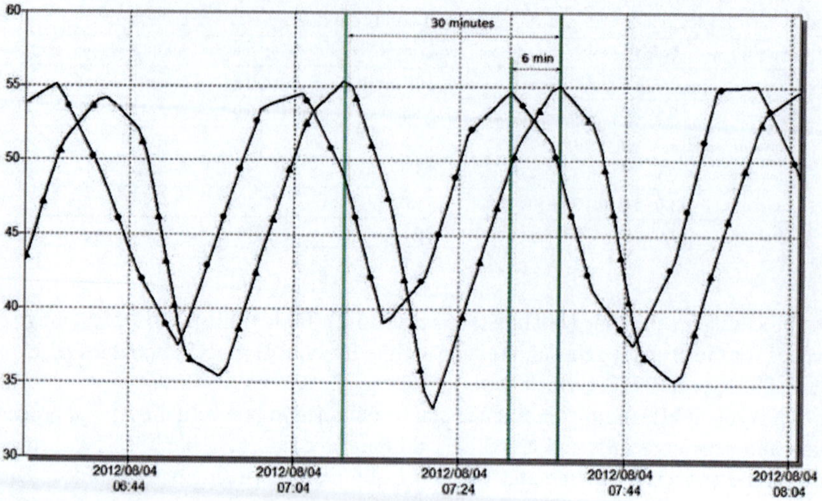

Fig. 9.3 Oscillating loop due to too strong integrating action

controller's excessive control action. Is it because the controller gain is too large or the integral action too strong?

One further indication is the phase shift. The integral term in the PID controller (Eq. 3.8) introduces a 90° phase shift between the process value and the controller output, while the proportional action does not cause a phase shift. A PI controller has a combination of both, and consequently, the phase shift is typically between 0 and 90 °C. Based on the phase shift (if pronounced), we can often determine whether the oscillation is caused mainly by too strong proportional action or too strong integral action.

In the example of Fig. 9.3, the process value and the controller output have been swinging with a 30 min cycle. Process value leads (due to closed-loop control) the control output by about 6 min, i.e., 1/5 of a cycle. Since a complete cycle has a phase shift of 360 °C, the 6 min phase shift between the process variable and controller output is equivalent to $360 \times \frac{6}{30} = 72°$, which suggests that the oscillation is primarily caused by too strong integral action.

Increasing the integral time from 180 to 1200 s (larger value means less strong integrating action) improved the control performance. The amplitude of the oscillation decreases almost immediately, as shown in Fig. 9.4.

The loop performance can be further improved by increasing the integral time or reducing the gain. However, many more trial-and-error rounds may be needed before reaching the final *satisfactory* tunings. It is also possible that the satisfactory performance may never be reached because the criterion for "satisfaction" is personal and subjective. Since the new tuning parameters are based purely on data patterns via trial-and-error, the confidence in the new tunings is low, and a temporary deviation from the "good" performance may trigger another round of trial-and-error tuning.

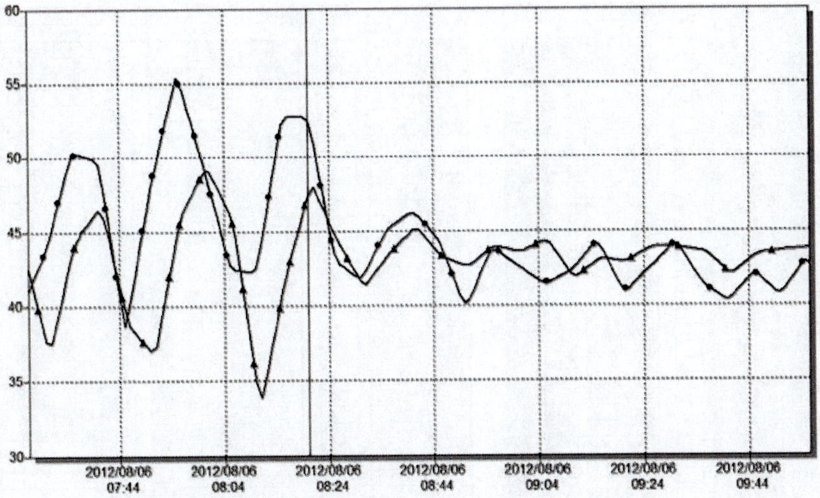

Fig. 9.4 Oscillating loop after tuning change

Now let us look at an extreme case of misuse or abuse of this trial-and-error method.

Example 9.2 Failure example of the trial-and-error tuning method. Figure 9.5 shows the tuning history of two control loops from a petrochemical plant, one pressure loop, and one temperature, with strong interactions between each other. Both are critical control loops on a high-pressure steam system in the utility plant.

The list shows the changes made to the two controllers' tuning parameters over one year:

1. Many tunings were tried, possibly by different engineers, with no success. Most of the tuning changes seem to be following the trial-and-error approach since there was no sound basis for any of the tuning parameters experimented. Little trust was given to any of them, and thus they were abandoned and replaced easily. "Easy come easy go!"
2. Some of the changes were very drastic without a clear rationale to support the decision. For example, on Feb 10, 2009, the integral time for loop `PC-811` was changed from 3800 s to 60 and the proportional band from 16.1 to 50.1.
3. On the other hand, some of the changes are so small that they should not have any noticeable impact on the control performance to worth the change. For example, the proportional band was changed from 101 to 96, then to 91, then to 96, 91, 81, 76, 91, and 81. This pattern of change indicates the lack of confidence in the new tuning parameters.

Unfortunately, the final tuning change on loop `TC-811` coincided with a unit trip that caused an expensive unscheduled shutdown.

PC811 PID Tunning Record

Proportional Band		Integral Time	
2008-02-11 9:09	101	2008-02-11 9:09	130
2008-05-24 8:32	101	2008-05-24 8:32	130
2008-05-24 8:32	101	2008-05-24 8:32	130
2008-05-24 16:19	101	2008-05-24 16:19	130
2008-05-24 17:49	101	2008-05-24 17:49	130
2008-05-29 12:53	101	2008-05-29 12:53	130
2008-05-29 12:53	101	2008-05-29 12:53	130
2008-05-29 19:09	101	2008-05-29 19:09	130
2008-07-24 10:10	96	2008-07-24 8:40	150
2008-07-24 12:00	91	2008-07-24 12:00	170
2008-07-24 16:30	96	2008-07-24 16:40	160
2008-07-24 16:40	91	2008-07-25 14:10	220
2008-07-25 14:20	81	2008-07-25 14:20	600
2008-07-25 14:40	76	2008-07-25 16:00	210
2008-07-25 15:30	91	2008-08-11 14:20	250
2008-08-11 14:20	81	2008-09-18 17:16	250
2008-09-18 17:16	81	2008-09-18 17:16	250
2008-09-18 17:16	81	2008-09-18 17:21	250
2008-09-18 17:21	81	2008-09-18 17:31	250
2008-09-18 17:31	81	2008-11-25 10:08	250
2008-11-25 10:08	81	2008-11-25 10:08	250
2008-11-25 10:08	81	2008-11-25 10:22	250
2008-11-25 10:22	81	2008-11-27 10:15	250
2008-11-27 10:15	81	2008-11-27 10:15	250
2008-11-27 10:15	81	2008-11-27 10:22	250
2008-11-27 10:22	81	2008-11-27 10:32	250
2008-11-27 10:32	81	2008-12-03 15:30	250
2008-12-03 15:30	81	2008-12-03 15:30	250
2008-12-03 15:30	81	2008-12-03 15:42	250
2008-12-03 15:42	81	2008-12-24 14:32	750
2008-12-24 14:42	71	2008-12-24 15:02	600
2008-12-24 16:12	41	2008-12-24 16:12	650
2009-01-08 16:13	26	2009-01-08 16:13	1800
2009-01-12 14:23	21	2009-01-12 14:13	2600
2009-01-13 11:53	19.1	2009-01-12 14:23	2800
2009-01-13 14:33	17.1	2009-01-13 11:53	3200
2009-01-13 16:13	16.1	2009-01-13 14:33	3800
2009-02-10 17:23	50.1	2009-02-10 17:23	60
2009-02-10 22:13	16.1	2009-02-10 22:13	3800
2009-02-10 22:33	261	2009-02-10 22:53	911
2009-02-10 22:53	41	2009-02-11 11:09	911
2009-02-11 11:09	41		

TC811 PID Tunning Record

Proportional Band		Integral Time		Derivative Time	
2008-02-11 9:34	71	2008-02-11 9:34	100	2008-02-11 9:34	10
2008-05-24 8:32	71	2008-05-24 8:32	100	2008-05-24 8:32	10
2008-05-24 8:32	71	2008-05-24 8:32	100	2008-05-24 8:32	10
2008-05-24 9:39	71	2008-05-24 9:39	100	2008-05-24 9:39	10
2008-05-24 17:19	71	2000-05-24 17:10	100	2008-05-24 17:19	10
2008-05-24 17:49	71	2008-05-24 17:49	100	2008-05-24 17:49	10
2008-05-29 12:53	71	2008-05-29 12:53	100	2008-05-29 12:53	10
2008-05-29 12:53	71	2008-05-29 12:53	100	2008-05-29 12:53	10
2008-05-29 19:09	71	2008-05-29 19:09	100	2008-05-29 19:09	10
2008-09-18 17:16	71	2008-09-18 17:16	100	2008-09-18 17:16	10
2008-09-18 17:16	71	2008-09-18 17:16	100	2008-09-18 17:16	10
2008-09-18 17:21	71	2008-09-18 17:21	100	2008-09-18 17:21	10
2008-09-18 17:31	71	2008-09-18 17:31	100	2008-09-18 17:31	10
2008-11-25 10:08	71	2008-11-25 10:08	100	2008-11-25 10:08	10
2008-11-25 10:08	71	2008-11-25 10:08	100	2008-11-25 10:08	10
2008-11-25 10:22	71	2008-11-25 10:22	100	2008-11-25 10:22	10
2008-11-27 10:15	71	2008-11-27 10:15	100	2008-11-27 10:15	10
2008-11-27 10:15	71	2008-11-27 10:15	100	2008-11-27 10:15	10
2008-11-27 10:22	71	2008-11-27 10:22	100	2008-11-27 10:22	10
2008-11-27 10:32	71	2008-11-27 10:32	100	2008-11-27 10:32	10
2008-12-03 15:30	71	2008-12-03 15:30	100	2008-12-03 15:30	10
2008-12-03 15:30	71	2008-12-03 15:30	100	2008-12-03 15:30	10
2008-12-03 15:42	71	2008-12-03 15:42	100	2008-12-03 15:42	10
2008-12-24 16:12	41	2008-12-24 16:12	500	2008-12-24 16:12	5
2009-01-05 15:12	21	2009-01-05 15:12	550	2009-02-11 11:34	5
2009-01-05 16:22	10.1	2009-01-09 10:13	1500		
2009-01-07 10:12	8.1	2009-01-09 14:23	2500		
2009-01-09 10:13	7.1	2009-01-12 11:13	2700		
2009-01-13 15:43	5.1	2009-01-13 15:43	3500		
2009-02-10 21:33	61	2009-02-11 11:34	3500		
2009-02-11 11:34	61				

Fig. 9.5 Failure example of trial-and-error tuning

Table 9.2 Loop dynamics and controller tunings

Loop dynamics	Control gain	Integral	Derivative
	K_c	T_i	T_d
Loop gain $K_p \nearrow$	$\searrow$		
Delay/lag ratio $\theta/\tau \nearrow$	$\searrow$	$\nearrow$	$\nearrow$
Delay $\theta \nearrow$	$\searrow$	$\nearrow$	$\nearrow$

Table 9.3 Controller tuning versus controller response

PID tuning	Rise time	Overshoot	Settling time	Offset	Stability
Gain $K_c \nearrow$	$\searrow$	$\nearrow$	–	$\searrow$	$\searrow$
Integral $T_i \nearrow$	$\searrow$	$\nearrow$	$\nearrow$	$\downarrow$	$\searrow$
Derivative $T_i \nearrow$	–	$\searrow$	$\searrow$	–	$\updownarrow$

The general qualitative indication between the loop dynamics and the controller tunings are given in Tables 9.2 and 9.3. More discussions along this line can be found in Smith (2009).

Table 9.4 Ziegler–Nichols tuning method

Parameter	Symbol	P-only controller	PI controller	PID controller
Gain	K_c	$K_u/2.0$	$K_u/2.2$	$K_u/1.7$
Integral	T_i		$P_u/1.2$	$P_u/2.0$
Derivative	T_d			$P_u/8.0$

9.2.3 Ziegler–Nichols Method

There are many empirical tuning methods in use, with variable levels of success. Most tuning methods developed in the early days had focused on fast closed-loop response, that is, on how to track the setpoint changes as fast as possible.

The Ziegler–Nichols method, or simply the Z–N method (Ziegler and Nichols 1942), is the best known empirical tuning method. It has had a significant influence on the practice of PID control. The Z–N method can be applied as follows: under P-only control, gradually increase the controller gain K_c until the control loop is in sustained oscillation. The corresponding controller gain K_u is called the *ultimate gain*, and the process oscillation period P_u is called the *ultimate period*. The tuning parameters are then calculated from the ultimate gain and period, as shown in Table 9.4.

Although popular and with a proven record in other industries, the Z–N method is rarely used in the process industry for many reasons:

- The controlled variables in the process industry are often interacting with each other. Putting the controlled variable to sustained oscillation could lead to unsafe operation to many loops and even an entire unit.
- The method is based on a quarter-wave-decay tuning objective and thus produces aggressive control tunings that often lead to process upsets.
- The significant overshoot in the controlled variable is typically not allowed because many controlled variables are already operated close to the operating limit, and overshoot could drive the controlled variable over the limit.

The Z–N method is limited to processes with relatively small process delays (comparing to process lag time) and integrating processes such as level control (Ziegler and Nichols 1942).

9.3 Model-Based Tuning

The trial-and-error method is simple and convenient because it does not require much knowledge about the underlining process dynamics. On the other hand, this lack of insight into the process dynamics is the exact reason that little confidence is placed in the tuning result.

Model-based tuning is based on a sufficient understanding of the cause and effect dynamic relationship between the manipulated variable (e.g., control valve) and the controlled variable. A mathematical model is used to represent this relationship, and this tuning method is thus called *model-based* controller tuning.

9.3.1 Model-Based Tuning Concept

The generic approach of model-based tuning can be simplified as

$$\text{Data} \xrightarrow{\text{Model Identification}} \text{Model} \xrightarrow{\text{Tuning Rules}} \text{Tunings.}$$

Compared with the trial-and-error or empirical tuning methods, the model-based approach has a sound theoretical basis. As a result, much higher trust and confidence can be placed in the resulting tuning parameters. It might require more effort and self-discipline to reach the tuning parameters, but overall it is more efficient and less time-consuming because it needs less re-work later.

The general model-based PID tuning involves three simple steps:

1. Identify the process model. There are many ways to find the process model, depending on the process type, model type, and personal preferences.
2. Calculate the tuning parameters. Based on the model parameters, apply a favorite tuning rule to find the controller parameters.
3. Apply the new tunings. Apply the new controller parameters online and make tuning adjustments as necessary.

A detailed tuning procedure in block diagrams is shown in Fig. 9.6.

9.3.2 Step Testing

Model-based tuning starts with a good process model. The model is typically derived from credible input/output response data obtained through *step tests* . One or more carefully planned step changes are applied to the manipulated variable and cause the controlled variable to respond. The dynamic response reveals the process dynamics for calculating the process model. A general procedure is recommended as follows:

1. Based on process knowledge and history data, obtain a guesstimate of the gain and time constant of the loop dynamics. Then decide on the step size and switch time of the step changes.
2. Explain to the operator the intent of the step test. Confirm with the operator that the proposed move size, move direction, and switch time are acceptable.
3. Put the control loop in MAN and let the process value stabilize. If loop output has been swinging, wait for the valve to come to its average value before opening the loop.

Fig. 9.6 Model-based PID tuning procedure

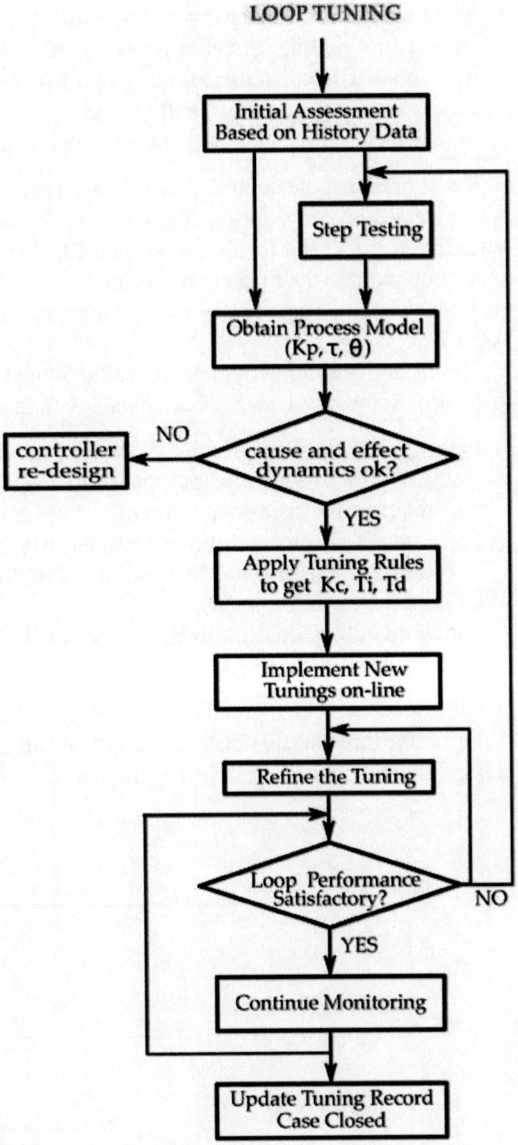

4. Make at least two step changes in opposite directions, with switch time about the same as the settling time. For noisy loops, make at least four-step changes (two up's and two down's), two moves with a switch time of one settling time, and two moves with about half the settling time.
5. Watch and record any valve problems such as stiction, deadband, and slippage.

Remember that the control calculation is based on the loop dynamics rather than the process dynamics, so the step changes should be applied from the controller output rather than the final control element. See Fig. 3.11 for the location of the control output and the final control element.

A single step change on the manipulated variable, and its response, is shown in Fig. 9.7.

The data quality directly determines the model quality and thus the control performance. Some disadvantages associated with this approach include the following:

1. Safety concerns: Running the process in an open-loop condition always risks losing control to a significant process upset during the test. Close monitoring is thus required for open-loop testing.
2. For a nonlinear process, the test results may be sensitive to the step change's direction and magnitude. The starting point, the step size, and the direction are all important considerations.
3. For an open-loop unstable system such as a level loop, it is possible to perform open-loop testing but must be doubly careful in the preparation and execution.

For a process with significant noise or disturbances, multiple step changes can minimize the influence of noisy measurement and drifting steady-state, as illustrated in Fig. 9.8. The modeling can be accomplished with dedicated identification tools.

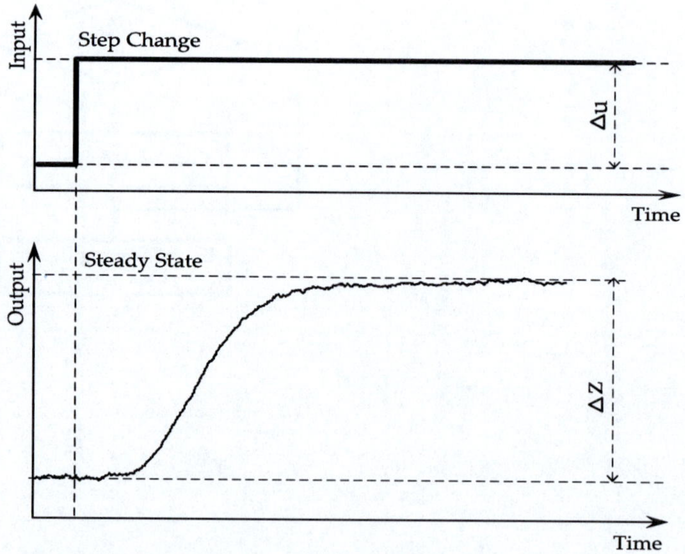

Fig. 9.7 Step response of a stable process

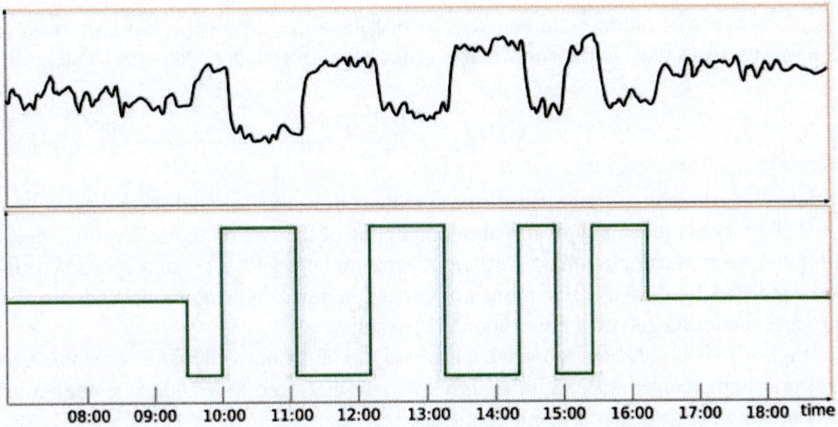

Fig. 9.8 Step testing with multiple step changes

The general guideline is that the step size should be large enough to be outside the noise band (typically at least five times larger than the noise band, or 1~2% of the measurement range). On the other hand, the step size should not be too large to drive the process to an unstable or undesirable operating region, e.g., nonlinear region, or trigger an alarm. The first step should be made in the direction of safer operation. For example, to test a reflux flow, step up the flow valve first.

9.3.3 Model Identification

A process model is a mathematical description of the cause and effect relationship between the process input and output. In other words, with the same process input $u(t)$, we would like to see that the model produces an output $\hat{z}(t)$ that is identical, or as close as possible, to the process output $z(t)$. See Fig. 9.9. The technology to extract the model from process data is *process identification* or *process modeling*.

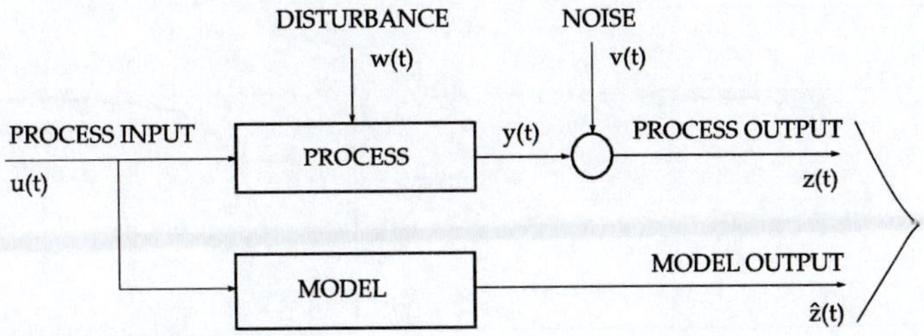

Fig. 9.9 Process and process model

Many types of models can be used to describe a given process. For loop tuning, the most widely used is the simple first-order-plus-time-delay (FOPTD) model:

$$G_p(s) = \frac{K_p}{\tau s + 1} e^{-\theta s} \tag{9.2}$$

where K_p is the process gain, τ the time constant, and θ the time-delay (or dead-time).

The process model is typically obtained in one of the two methods. Visual inspection and manual calculation are sufficient for a process with a clean response (well-behaved and low noise). For more complex dynamics with noisy data, dedicated process modeling software tools should be used.

Figure 9.10 shows how to derive a process model from the dynamics response to a single step change. Although the signal is not noise-free, its trend is clean enough to calculate the three parameters manually.

The process gain (normalized) is calculated as

$$\overline{K}_p = \frac{\Delta \text{CV}/\text{CVspan}}{\Delta \text{MV}/\text{MVspan}} = \frac{\text{CV}_2 - \text{CV}_1}{\text{MV}_2 - \text{MV}_1} \cdot \frac{\text{MVspan}}{\text{CVspan}}. \tag{9.3}$$

The time delay and time constant are obtained as shown in the time response. As shown in Sect. 3.3.2, the PID tuning is based on the loop gain, not the process gain. In other words, the MV in the gain calculation in Eq. 9.3 should be the controller output (OP), and the CV should be the controller PV. In other words, the step changes should be applied to the controller output (in manual mode), not at the valve. The rationale of this requirement is evident when performing step testing on a cascade control loop.

Suppose other components on the data flow path do not have significant dynamics. In that case, the difference between the process dynamics and the loop dynamics may be negligible, and either can be used for the gain calculation.

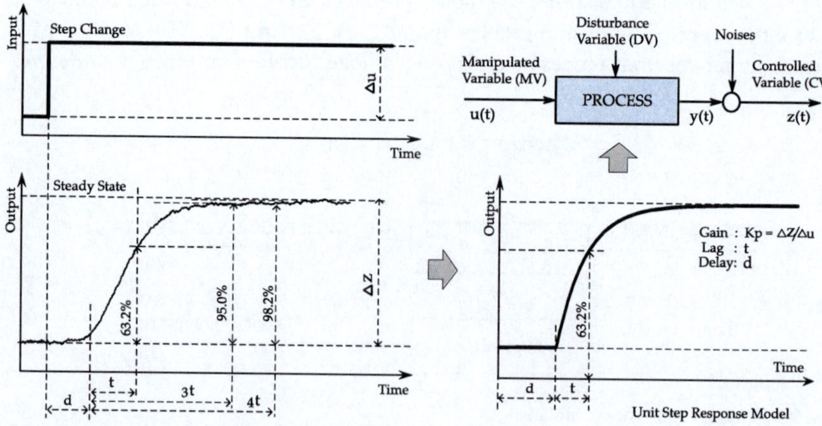

Fig. 9.10 Step response of a stable process

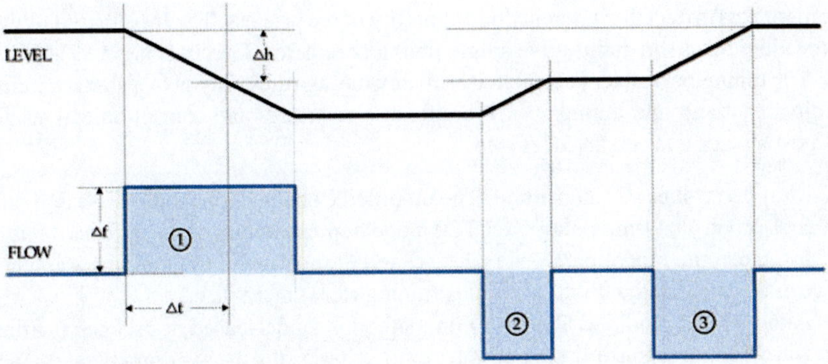

Fig. 9.11 Step response of a ramp process

Figure 9.11 illustrates the step response of a ramp process (e.g., level). The process gain (normalized) for a ramp process is calculated as follows:

$$\overline{K}_p = \frac{\Delta\mathrm{CVSlop}/\mathrm{CVspan}}{\Delta\mathrm{MVSlop}/\mathrm{MVspan}} = \frac{\left(\frac{\Delta\mathrm{CV}}{\Delta t}\right)_2 - \left(\frac{\Delta\mathrm{CV}}{\Delta t}\right)_1}{\mathrm{MV}_2 - \mathrm{MV}_1} \cdot \frac{\mathrm{MVspan}}{\mathrm{CVspan}}. \qquad (9.4)$$

When the above-discussed simple modeling methods are not feasible (e.g., for step test results from multiple step-moves), dedicated modeling tools should be used. The model identification step often needs to go back and forth several times before arriving at a satisfactory model. Process knowledge is critical in deciding if the model is satisfactory.

9.3.4 Tuning Parameter Calculation

The tuning parameters are calculated based on the process model. The calculation from model to tuning parameters is called the tuning rules and can be empirical or be based on sophisticated theoretical derivation. For convenience, the tuning rules are often simplified into a look-up table. Once the process model is known, the tuning formula is conveniently looked up from the tuning table.

There are many established tuning rules. Some rules may be better than others for different loops, but often, which rule to use is merely a personal preference. Among model-based tuning, the internal model control (IMC) (Garcia and Morari 1982; Garcia et al. 1989; Rivera et al. 1986)-based methods are becoming prevalent because of their sound theoretical basis and field proved track record. The IMC structure is a transparent framework for control design. It makes it easy to understand how the process characteristics (such as time delays and unstable transfer function components) affect the inherent controllability of the process. The IMC-based tuning procedure results in an intuitive single tuning parameter (Garcia et al. 1989).

The tuning parameters calculated with any tuning rules only give a starting point. Adjustment and fine-tunings based on process responses and control objectives are almost always a necessary next step:

1. Modified Cohen–Coon method: The modified Cohen–Coon method is based on a first-order-plus-time-delay (FOPTD) model representation of the process dynamics. Once the model parameters K_p, τ, and θ are known, the tuning parameters can be calculated with the formulas (tuning rules) in Table 9.5.
   Although the modified Cohen–Coon method is model-based, it uses the quarter-wave-decay response as the optimal tuning criteria, and the resulting performance is typically less stable than expected with excessive overshoots, thus not widely used in the process industry.
2. IMC-based tuning and λ-tuning rules: Internal model control (IMC) for PID tuning has been one of the most popular model-based tuning rules (Rivera et al. 1986). The IMC tuning rules have proven to produce robust performance when used in the control of common processes.
   It is beyond this book's scope to go into the detailed theory and derivations of IMC. We list the resulting tuning rules and show them in Tables 9.6 and 9.7 for easy reference (Bequette 2002; Smith 2009). These tuning rules are based on a first-order-plus-dead-time (FOPDT) model, sufficient for most processes. More sophisticated tuning rules should be followed for truly high-order or nonlinear dynamics, and many references are available (Bequette 2002; Chien and Freuhauf 1990; King 2011; Morari and Zafiriou 1989; Rivera et al. 1986).
   The λ value can be considered the closed-loop time constant and is a user-selected value. For this reason, this tuning method is also called λ-tuning. The value is typically set to be one to three times the process dead time θ. $\lambda = 3\tau$ can be used to achieve a very stable closed-loop response. Set the λ as low as τ for a faster response.
   Compared with Ziegler–Nichols and Cohen–Coon tuning rules, the IMC-based tuning rules have some clear advantages as follows:
   - The closed-loop response exhibits a first-order dynamic response, and the controlled variable has no or minimum overshoot. The process industry welcomes this type of response.
   - The tuning is robust and is less sensitive to changes in process dynamics than other tuning methods.
   - It is more intuitive since the user can specify the desired closed-loop time constant, which serves as the only tuning parameter required.

   The drawback is that the resulting controller tuning may produce a large integral value for slow process dynamics, resulting in a slow recovery from disturbances. The reason is that the controller's integral time is assumed to be the same as the process's time constant.

Table 9.5 Modified Cohen–Coon method for PID tuning

Parameter	Unit	PI controller	PID controller
Gain K_c	(%/%)	$\frac{0.45}{K_p} \cdot \left(\frac{\tau}{\theta} + 0.092\right)$	$\frac{0.67}{K_p} \cdot \left(\frac{\tau}{\theta} + 0.185\right)$
Integral T_i	(min)	$3.33\,\theta \cdot \left(\frac{\tau + 0.092\,\theta}{\tau + 2.22\,\theta}\right)$	$2.5\,\theta \cdot \left(\frac{\tau + 0.185\,\theta}{\tau + 0.611\,\theta}\right)$
Derivative T_d	(min)	–	$0.37\,\theta \cdot \left(\frac{\tau}{\tau + 0.185\,\theta}\right)$

Table 9.6 The IMC tuning rules for stable process

Parameter	Unit	PI controller	PID controller
with		$\lambda \geq \max(0.2, \tau, 1.7\,\theta)$	$\lambda \geq \max(0.2\,\tau, 1.7\theta)$
Gain K_c	(%/%)	$\frac{1}{K_p} \frac{\tau}{\lambda + \theta}$	$\frac{1}{K_p} \frac{\tau + \theta/2}{\lambda + \theta}$
Integral T_i	(min)	τ	$\tau + \frac{\theta}{2}$
Derivative T_d	(min)		$\frac{\tau\theta}{2\tau + \theta}$

Table 9.7 The IMC tuning rules for ramp process

Parameter	Unit	P-Only controller	PI controller	PI controller	PID controller
with			$\lambda > 0, \theta = 0$	$\lambda \geq \sqrt{10}\,\theta$	$\lambda \geq \sqrt{10}\,\theta$
Gain K_c	(%/%)	$\frac{1}{K_p} \frac{1}{\lambda + \theta}$	$\frac{1}{K_p} \frac{2}{\lambda}$	$\frac{1}{K_p} \frac{2\lambda + \theta}{(\lambda + \theta)^2}$	$\frac{1}{K_p} \frac{2\lambda + \theta}{(\lambda + \theta/2)^2}$
Integral T_i	(min)		2λ	$2\lambda + \theta$	$2\lambda + \theta$
Derivative T_d	(min)				$\frac{\lambda\theta + \theta^2/4}{2\lambda + \theta}$

9.3.5 *Tuning Implementation*

Once the tuning parameters are available, apply them to the PID controller with care. The recommended procedure is recommended as follows:

1. After a new tuning is put online, make one or two setpoint changes to verify that the control performance is adequate.
2. If the tuning changes are more than five times different from the existing tunings, especially in the more aggressive direction, stop and double-check the new tunings before implementation.
3. When making significant tuning changes, apply the changes in several steps, especially in the more aggressive direction. Leave enough time in between to observe the performance improvement.

4. When making changes to both gain and integral time, always change the one that will make the loop more stable (slower). For example, apply the gain change first if the controller gain needs to be changed from 1 to 0.3 and the integral time from 60 to 40 s.
5. For a nonlinear process, identify the highest gain and the largest dead-time and tune the control loop based on that worst-case scenario.
6. Unless in an emergency, never apply dramatic changes to a control loop without being able to test and monitor its performance for some time. For example, never apply a drastic tuning change to a loop at the last minute of the week and leave for the weekend.

9.3.6 PID Self-Tuning

PID *auto-tuning* (or *self-tuning*) has been extensively researched and is available in almost all major control systems. The so-called PID auto-tuner automatically identifies the process model online and calculates the tuning parameters to keep up with the changed loop dynamics. Typically the tuning procedure is *on-demand*, where the operator initiates the tuning procedure via a user interface.

The popular self-tuning method proposed by Åström and Hägglund (1984) uses a relay switch to create on/off changes in the manipulated variable to perturb the process. The input/output data from the artificial perturbation is then used to infer the process dynamics and calculate the PID tuning parameters.

Although the PID self-tuning function is available in most major control systems, its acceptance is not as well as anticipated. One of the primary reasons is that self-tuning is not fool-proof. It still requires a good knowledge of all the five control loop components, a sound judgment on the condition suitable to start the self-tuning, and proper validation of the resulting tuning parameters. Another reason is that the extra perturbations added are disturbances to the normal operation that the operator may not feel comfortable with. Blindly applying the auto-tuning function to all control loops inevitably leads to many failures and hurts the credibility of auto-tuners. If a controller cannot be tuned manually, it will not be tuned by the auto-tuner either.

Designing a fully automated and continuously running PID auto-tuning procedure involves many critical issues, especially with data quality and model quality, and thus remains an open challenge for research. A fully automated PID auto-tuner falls under the category of adaptive control (Visioli 2006).

9.3.7 Best Practice of PID Tuning

In a nutshell, model-based PID tuning is a simple three-step procedure:

1. Identify the process model.

2. Calculate tuning parameters.
3. Implement the new tuning.

The critical requirement for applying model-based tuning is the practitioner's self-discipline to adhere to the best practice work process. A checklist for model-based PID tuning is provided below for convenience:

1. Identify the process model:

   a. Record the original tuning parameters for later reference and book-keeping purposes.
   b. Decide if alternative tuning methods are applicable for the loop to be tuned. For example, flow loops may be tuned via the x-y plot of the history data (Sect. 9.4). Level loops may be tuned based on vessel sizing information (Sect. 9.5).
   c. Review history data and check if the normal operating data has sufficient information for modeling.
   d. If step testing is deemed necessary, conduct open-loop testing of the loop if possible. If it is too risky to operate the process in an open-loop, sometimes a closed-loop step testing can be performed instead. Closed-loop step testing requires that the identifiability conditions be satisfied for the data to be valid. Modeling identification in a closed-loop is more involved in both theory and practice. The interested readers should consult books on process identification, such as Ljung (1999), for more details.
   e. From the step test data, determine the process gain, time constant, and delay.

2. Calculate tuning parameters:

   a. Normalize the process gain to make it dimensionless, i.e., adjust the gain by input/output signal spans.
   b. Apply the tuning rules (e.g., IMC tuning) to obtain the controller gain (or proportional band) K_c, integral time T_i, and derivative time T_d.

3. Implement the new tuning:

   a. If the tuning is significantly different from the existing tuning, stop and double-check the data and calculation. Apply the changes in multiple steps to minimize risk and disturbance to operation.
   b. Test the loop to see how it responds to small setpoint changes in both directions. Also observe the controller's response to noise and disturbances. Make fine-tunings as necessary until satisfactory.
   c. If the new tuning is consistently worse than the original tuning, restore the original tunings, and re-validate the tuning procedure and calculations.

There are many empirical rules and best practices accumulated from practical experience. One of them is the *rule of five*'s, which is provided as follows:

1. For step testing, the changes in PV step response should be at least five times larger than the noise band.

2. The PV filter constant for any control variable should be no larger than 1/5 of the process's time constant (or the controller integral time T_i).
3. If the tuning change is more than five times different from the existing tunings, especially in the more aggressive direction, stop and double-check the new tunings before implementation.
4. Mass balance (inventory) control loops (e.g., level controls) should be tuned at least five times as slow as energy balance control loops (e.g., temperature control).
5. For a cascaded control loop or a nested control loop, the closed-loop time constant for the outer loop should be larger than five times that of the inner loop.
6. For PID controller tunings, the derivative action, if activated, should be no more than 1/5 of the integral action. That is, $T_i/T_d > 5$.
7. The positioner in a smart device (field bus) is a control loop by itself. Any control loop cascaded to the positioner should be at least five times slower.

9.4 Tuning Flow Control Loops with History Data

When the setpoint of a flow control loop has frequent setpoint changes, the gain for the flow loop can often be conveniently estimated using normal closed-loop operating data without needing step tests. One good example is the flow control loop in a level-to-flow cascade control. The outer level loop produces the setpoints for the inner flow loop with continuous changes. The PID tuning for flow control loops can thus be simplified. The general model-based tuning procedure discussed in Sect. 9.3 should be followed for difficult flow loops with delays and longer time constants.

9.4.1 Flow Loop Dynamics

Under particular circumstances, the flow control loop can be tuned with historical data without step testing. However, the effectiveness of this approach depends on the following assumptions and should be used with discretion:

1. Most flow control loops are so fast in response that the dead time is typically negligible. That is, we can assume the delay equals zero.
2. The time constant of a typical flow loop is about 0.25–0.5 min. If the time constant is much larger than this, then this method is not recommended (see Table 9.1 for typical process dynamics).
3. The process data meets the identifiability condition. For open-loop data, it is required that there be sufficient output changes to reveal the process dynamics. If using closed-loop data, there must be a sufficient number of setpoint changes, such as the inner loop of a cascade loop, to meet the model identifiability criteria.

9.4.2 *Flow Loop Tuning Procedure*

The recommended general procedure for flow loop tuning based on closed-loop PV-OP data is provided as follows:

1. Identify the process model parameters K_p, τ, and θ:
   a. Select the period of data that the controller has sufficient changes in setpoint SP (frequency and magnitude) and is free of saturation in the controller output OP.
   b. Draw a scattered plot of the controller PV versus OP.
   c. Visually check the plot for valve problems such as stiction, dead zone, friction.
   d. Estimate the normalized process gain:

$$\overline{K}_p = \frac{\Delta \text{PV}/[\text{PV Span}]}{\Delta \text{OP}/[\text{OP Span}]}. \tag{9.5}$$

   e. For normal flow controller, assume delay $\theta = 0$, and time constant $\tau = 0.25$ or 0.50 min. Note that if a PV filter is present, the filter value needs to be added to the selection of τ.

2. Calculate the controller tuning parameters K_c, T_i, and T_d:
   a. Decide on the desired closed-loop constant. Use $\lambda = 0.5$ min for the inner/nested control loop and $\lambda = 1.0$ min for the standalone loop or the primary loop in cascade control.
   b. Using tuning rules in Table 9.6 to calculate the tuning parameters.

3. Apply the new tuning parameters:
   a. Apply the new tuning parameters. If change is drastic, apply the change in multiple steps.
   b. Monitor and fine-tune the loop performance.

This "shortcut" approach can be demonstrated with a practical example.

Example 9.3 Flow loop tuning via PV-OP Plot. Consider the flow loop FC-303 in Fig. 9.12, the inner loop of the level-flow cascade control.

Figure 9.13 shows the setpoint SP, process value PV, and controller output OP for the flow control loop. The cycling is caused by the primary controller and provides constant setpoint changes to meet the requirement for using this shortcut approach.

Although the data shown in the trending plot (Fig. 9.13) is quite oscillatory and noisy, the PV-OP plot in Fig. 9.14 reveals that the valve-to-flow dynamics are pretty decent at the normal operating range (30–80%).

Because of the loop's fast dynamics, the time delay can be assumed to be zero, and the time constant is assumed to be known and small, say 0.25 min. What is remaining is just the process gain, which can be simply performed as the slope of the curve in the PV-OP plot. A relatively linear range, e.g., from 40 to 80% valve opening, is

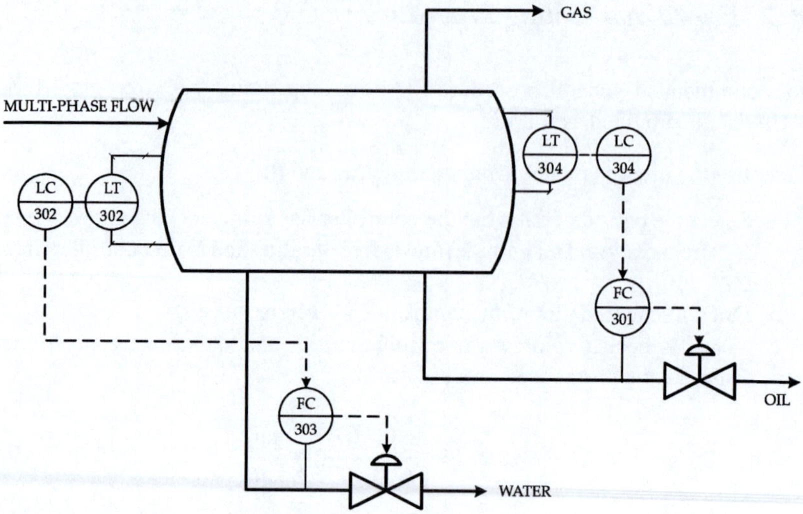

Fig. 9.12 A three-phase separator with level-flow cascade loop

selected to calculate the process gain. This selected range should fall into the normal operating window and be as wide as possible for better accuracy. The corresponding flow is between 900 and 3,350 m^3/d. The process gain in engineering units is then calculated as follows:

$$K_p = \frac{PV_2 - PV_1}{OP_2 - OP_1} = \frac{3350 - 900}{80 - 40} = 81.5\ m^3/d/\%. \tag{9.6}$$

Tuning calculation is based on normalized (dimensionless) process gain. Knowing that the PV range is 0–3,500 m^3/d, and the MV range is 0–100%, the normalized process gain is given as

$$\begin{aligned} \overline{K}_p &= \frac{(PV_2 - PV_1)/[PVSPAN]}{(OP_2 - OP_1)/[OPSPAN]} \\ &= \frac{(3350 - 900)/(3500 - 0)}{(80 - 40)/(100 - 0)} \\ &= 2.33\ (\%/\%). \end{aligned} \tag{9.7}$$

Now we have a flow process model with $\overline{K}_p = 2.33$ %/%, $\tau = 0.25$ min, and $\theta = 0$. Also assume the desired closed-loop response time is $\lambda = 0.5$ min (cascade loop), then from Table 9.6, we have

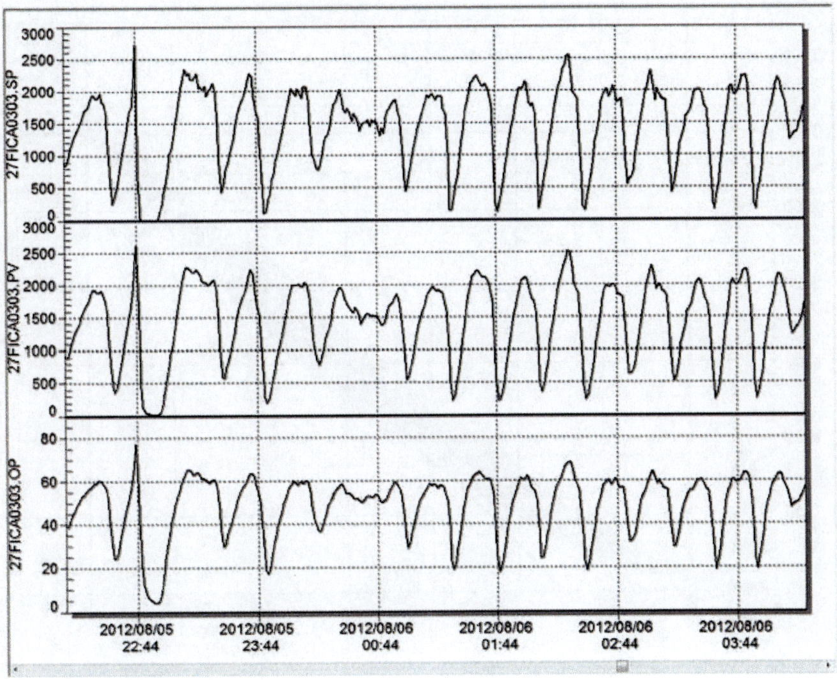

Fig. 9.13 Trending the flow controller data

$$\begin{cases} K_c = \dfrac{1}{\overline{K}_p} \dfrac{2\tau + \theta}{2\lambda} = \dfrac{1}{2.33} \dfrac{2 * 0.25 + 0}{2 * 0.5} = 0.21\ \%/\% \\ T_i = \tau + \theta/2 = 0.25 + 0/2 = 0.25\ \text{(min)} \\ T_d = 0\ \text{(min)}. \end{cases} \tag{9.8}$$

Example 9.4 Flow loop tuning. For loop FC-301, the time series plot is given in Fig. 9.15. Again, it is seen that the flow loop oscillates due to poor level control LC-0304. In addition to the setpoint changes driven by the level loop, the flow PV and OP signals are also noisier than FC-303 in themselves.

A PV-OP plot for the flow controller, shown in Fig. 9.16, indicates that the flow control valve does not perform well. There is excessive stick–slip with the valve dynamics, and service is needed. Knowing that the flow range is 2,200 m^3/d, as seen from Fig. 9.16, the valve might be oversized as well for its intended operation.

Nevertheless, to proceed with the tuning, let us pick two points within the normal operating range, say, at 20 and 60% valve openings. The corresponding flow rates can be approximately assumed to be 500 and 2,100 m^3/d, respectively. The normalized process gain can then be calculated as

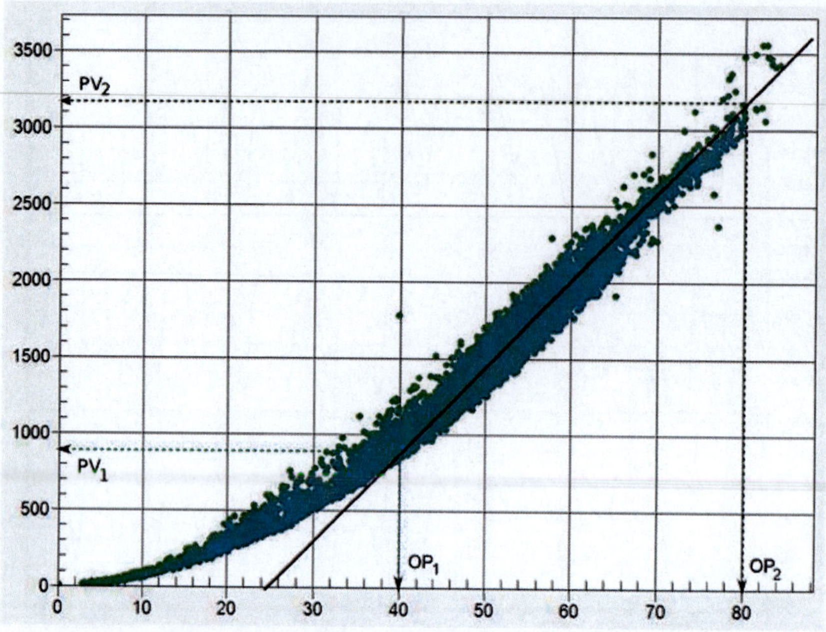

Fig. 9.14 PV-versus-OP plot for a flow loop

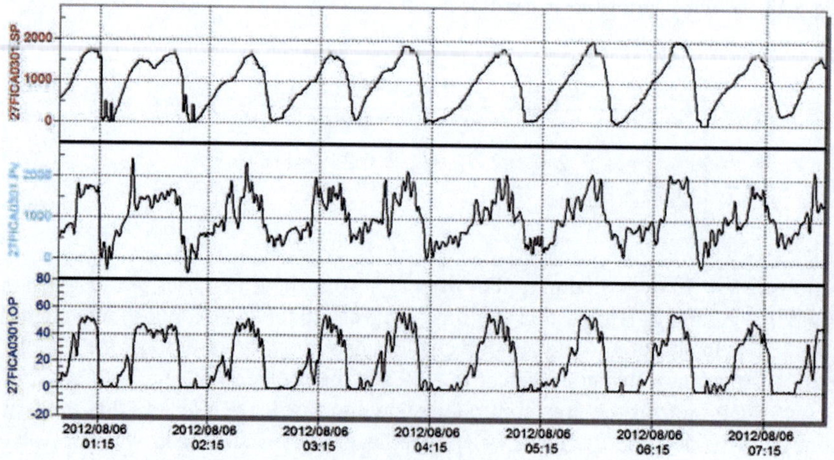

Fig. 9.15 Trending the flow controller data (incomplete)

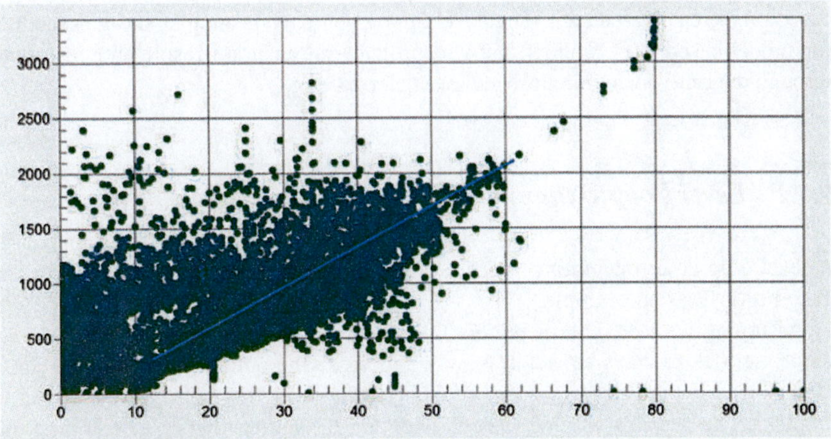

Fig. 9.16 X-Y plot for a flow loop

$$\overline{K}_p = \frac{(2100 - 500)/2200}{(60 - 20)/100} = 1.82\ \%/\%. \tag{9.9}$$

Since the valve dynamic is not ideal, let us assume the time constant to be 0.5 min and the delay to be 0. With a desired closed-loop time constant $\lambda = 1.0$ min, the controller tuning parameters are then given, from Table 9.6, as

$$\begin{cases} K_c = \dfrac{1}{\overline{K}_p} \dfrac{2\tau + \theta}{2\lambda} = \dfrac{1}{1.82} \dfrac{2 * 0.5 + 0}{2 * 1 + 0} = 0.27(\%/\%) \\ T_i = \tau + \dfrac{\theta}{2} = 0.5 + \dfrac{0}{2} = 0.5 \text{ (min)} \\ T_d = 0 \text{ (min)}. \end{cases} \tag{9.10}$$

The level control loops `LC-301` and `LC-304` cannot be tuned easily with history data. However, for many level control, if the vessel sizing information is available, there is also "shortcut" approaches to calculate the tuning parameters, as shown below.

9.5 Tuning Level Control Loops with Vessel Sizing Data

Level control is critical in maintaining the inventory in the process operation. Unfortunately, a vessel level loop is an inherently unstable process, and PID control parameter tuning is thus also tricky.

Luckily, the process gain can often be calculated based on the vessel sizing information and the flow rate. Consequently, the PID tuning parameters can be obtained based on the calculated gain. This level tuning approach is another shortcut method for model-based PID tunings that can produce an accurate result without going through the time-consuming procedure of step testing.

9.5.1 Level Loop Dynamics

A level loop is an integrating process rather than a self-regulating stable process. Take Fig. 9.17 as an example.

Assuming we start with a perfect mass balance between inlet and outlet flow, the vessel's level stays constant. Now we increase the outlet flow; the level starts to decrease. If the flow rate does not change, the level will fall until the vessel is emptied. The vessel's volume (vessel size) and the outlet flow rate determine the level's decrease rate. If we change the flow rate back to the original value, the level will stop decreasing and stay constant. However, it does not come back to where it was as the flow does since the vessel's inventory has decreased.

A pure level loop has a transfer function of

$$G(s) = \frac{K_p}{s} \tag{9.11}$$

that is, only one parameter, the process gain, is in the model. For a pure integrating process, the normalized process gain is calculated as

$$\overline{K_p} = \frac{\Delta \text{CVSlop}/\text{CVspan}}{\Delta \text{MVSlop}/\text{MVspan}} = \frac{\left(\frac{\Delta \text{CV}}{\Delta T}\right)_2 - \left(\frac{\Delta \text{CV}}{\Delta T}\right)_1}{\text{MV}_2 - \text{MV}_1} \cdot \frac{\text{MVspan}}{\text{CVspan}}. \tag{9.12}$$

For the level process in Fig. 9.17, we can calculate the gain based on the two data segments ① and ②. The flow change from segment ① to ② is 0–100% of the flow

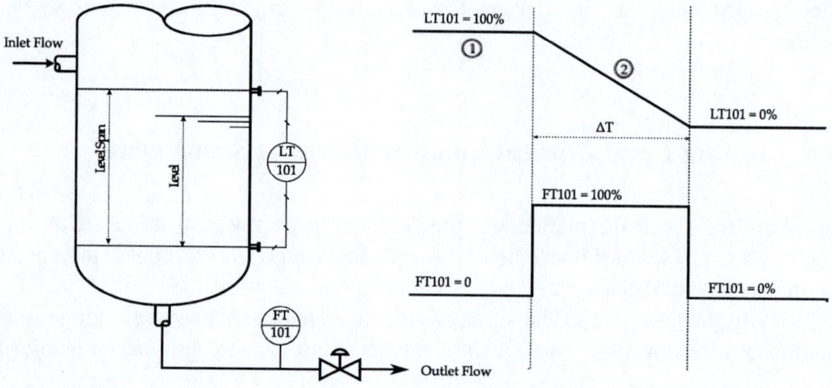

Fig. 9.17 Level in a vessel

range, while level change is from 100 to 0% over a period of ΔT. Assuming the level is H, and the flow is F, the normalized process gain is then given by

$$\overline{K}_p = \frac{\frac{0-H}{\Delta T} - 0}{F - 0} \cdot \frac{F-0}{H-0} = \frac{\frac{0-100\%}{\Delta T} - 0}{100\% - 0} \cdot \frac{100\% - 0}{100\% - 0} = -\frac{1}{\Delta T}. \tag{9.13}$$

This ΔT is the duration it takes to empty the vessel from 100 to 0% with the maximum outlet flow (100%). This duration is also called the *residence time*, denoted as T_r. The residence time is given as the control volume (0–100% level) divided by the maximum volumetric flow rate (at 100% flow). The control volume can be calculated with the vessel sizing information such as shape, height, and diameter. The maximum flow rate can typically be found from the flow meter datasheet or valve datasheet.

9.5.2 *Level Control Loop Tuning Procedure*

Once the process gain K_p is known, the level controller tuning can follow the generic tuning rules in Table 9.7. A general procedure for level loop tuning based on vessel sizing information is summarized as follows:

1. Calculate the vessel residence time based on vessel sizing information:
   a. Calculate the control volume, V, of the vessel.
   b. Find or calculate the maximum flow rate (flow range), F.
   c. Calculate the residence time:

$$\text{Residence Time } T_r = \frac{\text{Control Volume}}{\text{Maximum Flow Rate}} = \frac{V}{F}. \tag{9.14}$$

2. Compute the normalized process gain, which is the reciprocal of the residence time (see Eq. 9.12):

$$\text{Process Gain } \overline{K}_p = \frac{1}{\text{Residence Time}} = \frac{1}{T_r}. \tag{9.15}$$

3. Calculate the controller tuning parameters based on the tuning rule in Table 9.7:
   a. Calculate the recommended closed-loop time constant $\lambda = T_r = 1/\overline{K}_p$, or choose a different λ value of choice.
   b. Calculate controller tunings $K_c = \frac{1}{\overline{K}_p} \cdot \frac{1}{\lambda}$. Note that $K_c \equiv 2$ if $\lambda = 2\,T_r$ is used.
   c. Calculate the controller integral time $T_i = 2\,\lambda$.
4. Apply the new tunings online and fine-tune the integral time until satisfaction.

Example 9.5 Level loop tuning with sizing information. Consider the level control loop in Fig. 9.17. Assume that the vessel has an inner diameter of 3.0 m. The height of the vessel is 11.5 m, with a tap-to-tap distance of 10.0 m.

To calculate the tuning parameters for the level controller LC-101, let us first calculate the residence time. The vessel is a standard vertical cylindrical vessel. The control volume, i.e., the volume between the two level-measuring taps, is given by

$$V = A \cdot H = \pi \cdot \left(\frac{D}{2}\right)^2 \cdot H = \pi \cdot \left(\frac{3}{2}\right)^2 \cdot 10 = 70.7\ \text{m}^3. \tag{9.16}$$

The maximum flow rate is given as a 2500 kg/min mass flow, with a density of 960 kg/m^3. Therefore, the maximum volumetric flow rate is 2500.0/960 = 2.6 m^3/min. The residence time is then given by

$$T_r = \frac{V}{F} = \frac{70.7\ \text{m}^3}{2.60\ \text{m}^3/\ \text{min}} = 27.2\ \text{min}. \tag{9.17}$$

The process gain is thus

$$\overline{K}_p = \frac{1}{T_r} = \frac{1}{27.2} = 0.0367. \tag{9.18}$$

If we would like the closed-loop response to having a time constant the same as the residence time, then set $\lambda = T_r = 27.2$. From Table 9.7, the controller tunings are given as

$$\begin{cases} K_c = \frac{1}{\overline{K}_p} \cdot \frac{1}{\lambda} = 2\ (\%/\%) \\ T_i = 2\lambda = 2 \times 27.2 = 54.4\ (\text{min}) \\ T_d = 0\ (\text{min}). \end{cases} \tag{9.19}$$

For less tight control, the λ value can be set to $\lambda = 2$, or $T_r = 54.4$ min. The resulting controller tunings would be $K_c = 1.0$ and $T_i = 108.8$ min.

In this example, the vessel volume is a linear function of the level (height). In case the relationship is not linear, this method provides only an average value for the process gain. Adjustment of the result based on actual conditions should be applied. Below is another example illustrating the case of nonlinear gain.

Example 9.6 Vessel with nonlinear gain. Consider the level loop tuning for the three-phase separator in Fig. 9.12. There are two level/flow cascade control loops on the separator, one loop for the total oil level (LC304/FC301) and the other is the interface level control (LC302/FC303). The vessel sizing information is supplied in the vessel manufacturer's datasheet. The relevant information for the loop level tuning is extracted from the datasheet and shown in Fig. 9.18. The flow ranges for the two flow controllers are 0–2,200 m^3/d for `FC-301` and 0–3,500 m^3/d for `FC-303`.

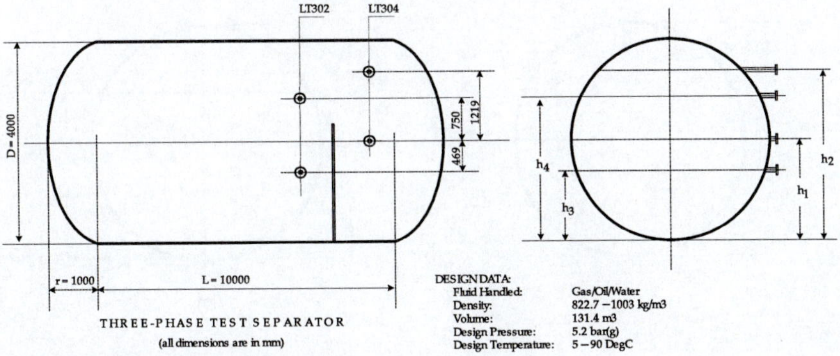

Fig. 9.18 Vessel sizing information for level tuning

Following the loop tuning procedure, let us first calculate the vessel's residence time, which requires calculating each vessel's control volume. By definition, the control volume is the volume in the vessel between the two level taps. For `LC-302`, the control volume is between height h_1 and h_2; and for `LC-304`, between height h_3 and h_4. Fig. 9.19 shows the cross-section areas corresponding to the four level transmitter taps. The blue/dashed horizontal line designates the liquid level. The level transmitter range for `LC-304` is between the level lines in Fig. 9.19a and b, while the level transmitter range for `LC-302` is between the level lines in Figs. 9.19c and d.

The cross-section areas are calculated as follows:

1. Control volume for level loop `LC-304`: From Fig. 9.19a, the liquid area is half the circle, and thus the vertical cross-section area is given by

$$A_1 = \left[\frac{\pi}{4} D^2\right]/2 = \left[\frac{\pi}{4} 4^2\right]/2 = 6.28 \text{ m}^2. \tag{9.20}$$

From Fig. 9.19b, the vertical cross-section area up to height $h_2 = 3.22$ m can be calculated as

$$\alpha_2 = \arcsin\left(\frac{2\,h_2}{D} - 1\right) = \arcsin\left(\frac{2 \times 3.22}{4} - 1\right)$$
$$= \arcsin(0.61) = 0.66 \tag{9.21}$$

$$A_2 = \frac{1}{8}\,(\pi/ + 2\,\alpha/_2)\, D^2 + \frac{1}{4}\, D\,(2h_2 - D)\, \cos(\alpha)$$
$$= \frac{1}{8}\,(\pi/ + 2 \times 0.66) \times 4^2 + \frac{1}{4} \times 4 \times (2 \times 3.22 - 4) \times \cos(0.66)$$
$$= 10.83 \text{ m}^2. \tag{9.22}$$

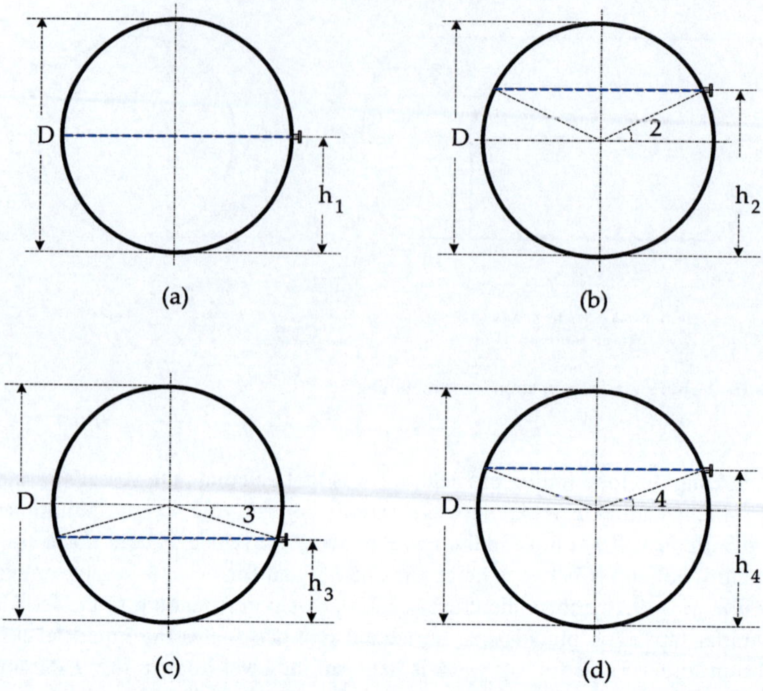

Fig. 9.19 Vessel cross-section area calculation

The area corresponding to the level transmitter range is thus

$$A = A_2 - A_1 = 10.83 - 6.28 = 4.55 \text{ m}^2 \tag{9.23}$$

and the control volume (the middle cylindrical part without the elliptical ends) is thus given by

$$V = A \cdot L = 4.55 \times 10 = 45.5 \text{ m}^3. \tag{9.24}$$

Assume that the two elliptical bulges at the two ends contribute 4.5 m^3 to the control volume without detailed calculation. That makes the total control volume to be about 50.0 m^3.

2. Control volume for level loop `LC-302`: The control volume for the `LC-302` controller can be calculated similarly. From Fig. 9.19b, the vertical cross-section area up to level transmitter tap $h_3 = 1.53$ m can be calculated as

$$\alpha_3 = \arcsin\left(1 - \frac{2\,h_3}{D}\right) = \arcsin\left(1 - \frac{2 \times 1.53}{4}\right) = 0.24$$

$$A_3 = \frac{1}{8}\,(\pi - 2\,\alpha_3)D^2 - \frac{1}{4}D\,(D - 2\,h_3)\,\cos(\alpha_3)$$

$$= \frac{1}{8}(\pi - 2 \times 0.237) \times 4^2 - \frac{1}{4} \times 4 \times (4 - 2 \times 1.53) \times \cos(0.24)$$
$$= 4.42\ \mathrm{m}^2.$$

The calculation of the area up to height h_4 in Fig. 9.19d is similar to that in Fig. 9.19b, with $h_4 = 2.75$ m

$$\alpha_4 = \arcsin\left(\frac{2\,h_4}{D} - 1\right) = \arcsin\left(\frac{2 * 2.75}{4} - 1\right) = 0.384$$
$$A_4 = \frac{1}{8}(\pi + 2\,\alpha_4)\,D^2 + \frac{1}{4}\,D\,(2\,h_4 - D)\,\cos(\alpha_4)$$
$$= \frac{1}{8}(\pi + 2 \times 0.384) \times 4^2 + \frac{1}{4} \times 4 \times (2 \times 2.75 - 4) \times \cos(0.384)$$
$$= 9.21\ \mathrm{m}^2.$$

The area between the level transmitter taps is then given by

$$A = A_4 - A_3 = 9.21 - 4.42 = 4.79\ \mathrm{m}^2. \tag{9.25}$$

The control volume of the vessel without counting the bulges on the two ends is given as

$$V = A \cdot L = 4.79 \times 10 = 47.9\ \mathrm{m}^3. \tag{9.26}$$

The volume of the two bulges can be calculated separately but is more involved. However, without loss of much accuracy, assume it accounts for 5.1 m ^{3} to make the total control volume 53.0 m^3.

In summary, the level controller tuning, whether by the generic step testing method or the vessel sizing method, has significant challenges due to the nonlinearity.

Now follow the level tuning rules (Table 9.7) and calculate the two level controllers' tuning parameters. The results are shown in Table 9.8. For comparison, the tunings for Level `LC-301` are also included. To demonstrate the ease and flexibility on achieving different closed-loop performance, we choose $\lambda = 1.5T_r$ instead of $\lambda = T_r$ for calculating the tuning parameters.

9.5.3 *Characterization of Level Loop Dynamics*

Depending on the vessel's shape, the level is often a very nonlinear function of the flow rate. This nonlinearity poses a significant challenge to PID control. We now show another example to demonstrate how nonlinear the level process dynamic can become. Fig. 9.20 shows a conical vessel that requires level control. The volume is

Table 9.8 Level tuning parameters

	Symbol	Unit	LC-101	LC-302	LC-304
Control volume	V	m^3	70.7	53	50
Max flow range	F	m^3/d	3,744	3,500	2,200
Residence time	$T_r = V/F$	min	27.2	21.8	32.7
Process gain	$\overline{K}_p = 1/T_r$	%/%	0.037	0.046	0.031
Closed-loop	$\lambda = 1.5\,T_r$	min	40.8	32.7	49.0
Controller gain	$K_c = \frac{1}{\overline{K}_p}\frac{2}{\lambda}$	%/%	1.33	1.33	1.33
Integral time	$T_i = 2\,\lambda$	min	81.6	65.4	98.0
Derivative time	$T_d = 0$	min	–	–	–

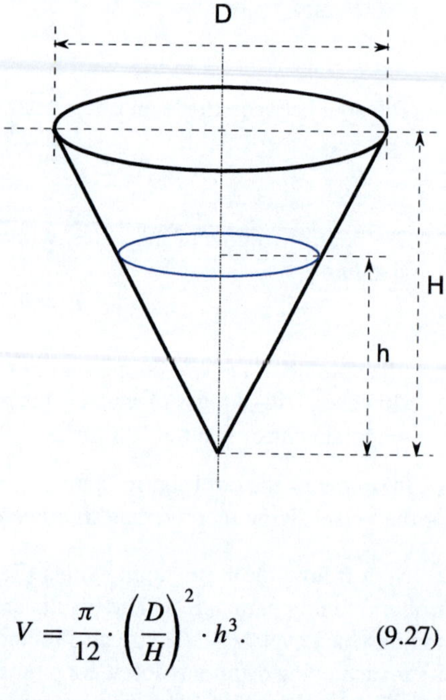

Fig. 9.20 Nonlinearity of conic shaped vessel

given by

$$\text{volume up to level } h: \quad V = \frac{\pi}{12}\cdot\left(\frac{D}{H}\right)^2\cdot h^3. \tag{9.27}$$

The relationship between the volume and the level is a cubic function and is highly nonlinear. For example, assuming $D = 10$ m and $H = 5$ m, a level change of one meter (10%) at the lowest liquid level leads to a volume change of

$$\Delta V = \frac{\pi}{12}\left(\frac{10}{5}\right)^2 \times 1^3 - 0 = \frac{1}{3}\pi\ (\text{m}^3) \tag{9.28}$$

while a level change of one meter (10%) at the highest level (from 9 to 10 m) will cause a volume change of

$$\Delta V = \frac{\pi}{12}\left(\frac{10}{5}\right)^2 \times 10^3 - \frac{\pi}{12}\left(\frac{10}{5}\right)^2 \times 9^3 = \frac{271}{3}\pi\ (\mathrm{m}^3). \tag{9.29}$$

Also, for comparison, for the same vessel, the volume at 50% level (5 m) is

$$V_{50\%} = \frac{\pi}{12}\left(\frac{10}{5}\right)^2 \times 5^3 = \frac{125}{3}\pi\ (\mathrm{m}^3) \tag{9.30}$$

while at level 100%, the volume is

$$V_{100\%} = \frac{\pi}{12}\left(\frac{10}{5}\right)^2 \times 10^3 = \frac{1000}{3}\pi\ (\mathrm{m}^3) \tag{9.31}$$

which is eight times more liquid than at 50% level.

Between Eqs. 9.28 and 9.29, there is a difference of 271 times! It is almost impossible for a PID controller with constant tuning parameters to cope with such a wide range of dynamic changes.

In practice, level characterization (transformation) can be applied to help PID control. A new DCS tag can be created to calculate the vessel volume from the vessel height h with Eq. 9.27. A volume controller replaces the level controller to control the volume instead of the level. The calculated volume serves as the process value (PV) for the PID controller. Since the volume and flow relationship is linear, this volume PID control should be much more robust across the whole vessel level range. After all, we are interested more in the volume than the height for inventory control.

9.6 Summary

A control loop consists of five essential components (Fig. 2.2). The controller performance is determined by the loop dynamics rather than process dynamics alone. As long as the loop dynamics stay the same, the PID tuning should have no reason to change. The possibility that the PID tuning causes the loop problem is surprisingly lower than commonly believed. Most of the time, the poor performance is caused by other reasons such as issues with the control valve or the data flow in the control loop. For this reason, a systematic troubleshooting procedure (see Chap. 8) is the recommended first step to rule out other problems before starting PID tuning. Jumping too quickly to PID tuning often proves to be counter-productive. Nevertheless, tuning a PID control loop is an essential skill that a competent process control engineer must possess.

PID tuning is still a mixed bag of art and technology and requires knowledge, experience, and patience. Controller tuning starts with a clear expectation of the targeted performance to be achieved (Sect. 9.1). The best PID performance in automatic control, such as the "quarter-wave decay," may not necessarily be acceptable in the process industry.

Among the many different loop tuning methods, trial-and-error and empirical methods (Sect. 9.2) are quick and simple to start. However, it typically takes longer to complete since there is low confidence in the tuning parameters and thus no "endpoint" to call a stop. Model-based PID tuning methods (Sect. 9.3) are based on an explicit process model, representing a direct insight into the process dynamics, and thus offer much higher confidence on the tuning parameters produced. As a result, even though the model-based method may sound more time-consuming than other approaches, the overall effort is typically less since there is less repeated effort.

There are some "shortcut" methods for tuning flow (Sect. 9.4) and level (Sect. 9.5) control loops. They can produce sufficiently good performance while saving time. However, a good understanding of their principle and limitations should be in order.

References

Åström K, Hägglund T (1984) Automatic tuning of simple regulators with specifications on phase and amplitude margins. Automatica 20(5):645–651

Bequette BW (2002) Process control—modeling, design, and simulation. Prentice Hall

Blevins TL, McMillan GK, Wojsznis WK, Brown MW (2003) Advanced control unleashed—plant performance management for optimum benefit. ISA

Brannan C (2018) Rules of thumb for chemical engineers—a manual of quick, accurate solutions to everyday process engineering problems, 6th edn. Elsevier

Chien IL, Freuhauf PS (1990) Consider IMC tuning to improve controller performance. Chem Eng Prog 86:33–41

Ford J (2013) Initial settings for PID controllers. Control Eng. https://www.controleng.com/articles/initial-settings-for-pid-controllers

Garcia CE, Morari M (1982) Internal model control—1: a unifying review and some new results. Ind Eng Chem Process Des Dev 21:308–323

Garcia CE, MPrett D, Morari M, (1989) Model predictive control: theory and practice—a survey. Automatica 25(3):335–348

King M (2011) Process control—a practical approach. Wiley

Ljung L (1999) System identification: theory for the user, 2nd edn. Prentice-Hall, Englewood Cliffs, NJ

Luyben WL, Lyuben ML (1997) Essentials of process control. McGraw-Hill

Morari M, Zafiriou E (1989) Robust process control. Prentice Hall

Rivera DE, Skogestad S, Morari M (1986) Internal model control 4: PID controller design. Ind Eng Chem Process Des Dev 25:252–265

Smith CL (2009) Practical process control—tuning and troubleshooting. Wiley

Svrcek WY, Mahoney DP, Young BR (2006) A real-time approach to process control, 2nd edn. Wiley

Visioli A (2006) Practical PID control. Advances in industrial control. Springer, London

Ziegler JG, Nichols NB (1942) Optimum settings for automatic controllers. J Dyn Syst Meas Control-Trans ASME 64:759–768

Part IV
Process Control Typicals

With the essential knowledge in Part II and core skills and methodologies in Part III, this book's final part devotes to some typical control applications. The goal is to demonstrate how the knowledge and skills are applied to practical applications. The specific applications selected include the following:

1. Common control-related calculations (Chap. 10): Many online calculations are needed to support control solutions, from signal processing, measurement compensation, to inferential property estimation. Online calculations, especially inferential property estimation, are very cost-effective ways of enhancing and improving control performance. Online calculations are also essential for real-time monitoring applications such as data validation, performance monitoring, abnormal situation management, and operator advisory systems.
2. Control of typical equipment (Chap. 11): Pumps and compressors are critical and prevalent equipment in the process industry, and their optimal control and reliable protection are essential for safe operation and efficient production. Traditionally, equipment control is regarded as the specialty of equipment manufacturers or third-party specialists. However, equipment control is not more mysterious than a distillation column or chemical reactor as part of the process flow. The standard process analysis and control design skills and methodology apply.
3. Plantwide and unit-level control (Chap. 12): Most process control books covered distillation columns or chemical reactors to showcase the process analysis and control design. However, this book purposely selects several less discussed yet pervasive processes to demonstrate the thinking and analytical processes for designing decent control solutions. The purpose is not to provide control solutions as go-by templates, which is helpful but inefficient since even the typical processes are countless. This chapter attempts to demonstrate a more practical approach, which is to learn the general thinking process and structured analytical skills that can be applied to any process control problem.

All the selected applications are typical in practical settings, and their proper implementation or control has a significant impact on HSE or production profitability.

Each application demonstrates one significant aspect of process analysis, control design, or control improvement. The selected applications are modestly complex so that each can be covered in its entirety in one section to show the complete thinking and analysis process, at the same time are not too trivial. As the discussions are based on the knowledge and skills in Part I–Part III, cross-references to the relevant sections are provided wherever appropriate. It is recommended to check back on these sections as needed.

Designing, troubleshooting, and improving a process control solution is a mix of art and technology. We strive for simplification and standardization to improve reliability and sustainability. Therefore, the solutions should be as simple as practically possible. On the other hand, we must also pay attention to the distinct characteristics of each process to avoid blind copy and paste. The standardization should be on the philosophy and methodology, not on the solution.

There is no standard solution in process control, and the best solution is always the one that is simple and fits the purpose. As a result, the final solutions still depend to a large extent on the practitioner's knowledge, skill, experience, and personal preference for complicated control problems such as plant-level or unit-level control.

Chapter 10
Control Typicals: Common Calculations

Online calculations are essential to real-time control. This chapter presents some commonly encountered control-related calculations, such as signal transformation (Sect. 10.1), data quality assessment (Sect. 10.2), measurement compensation (Sects. 10.3–10.5), and internal reflux (Sect. 10.6). They are critical components of process control but are often misunderstood or incorrectly implemented. For consistency and ease of understanding, each section follows a standard format, starting with application background, theory and design, implementation and operation, and finally, some application examples.

10.1 Signal Transmission and Transformation

Signal processing is an essential and standard component in the control data flow and may include input processing, calculation processing, output processing, and alarm processing (Yokogawa 2004). In a closed control loop consisting of multiple function blocks, the signal transmission and processing can be a source of confusion and configuration errors, negatively impacting the control design and performance. A basic understanding of the calculations involved is required.

10.1.1 Application Background

In a typical PID loop, the process variable (PV) originates from the sensors in the field as a measurement signal, transmitted from the sensor, through the transmitter, to the control system, and back to the actuator and valve in the field.

Along this transmission path, signal scaling and conversion are performed at multiple places, some being simple range scaling to deal with the change of engineering unit, while others may involve more complex signal transformation such as square-rooting and signal characterization.

The control design is based on the loop dynamics, including everything on the loop path, from the controller output, through the process, and back to the controller PV

S. S. Niu and D. Xiao, *Process Control*, Advances in Industrial Control,
https://doi.org/10.1007/978-3-030-97067-3_10

(see Sect. 3.3.2). The signal transmission and transformation may have a significant influence on control design and performance. For instance, a PV filter on the process value adds additional dynamics to the loop. The filter's impact may be insignificant for most loops but must be considered for fast control loops such as flow control.

10.1.2 Theory and Design

1. Signal transmission: As shown in Fig. 10.1, the signals from the sensors in the field are typically converted to a 4–20 mA electric current signal for transmitting to DCS (e.g., as the PV for a PID controller). Similarly, the PID controller output is converted to a 4–20 mA electric current signal before transmitting from DCS to the field device. This 4–20 mA electric current signal is further converted to a 3–15 psi pressure signal to operate a pneumatic valve. With 4–20 current mA signal or 3–15 psi pressure signal, all values must be converted from the real-zero to the *live-zero* (see Sect. 3.3.1) and back to the real-zero.
   One example is a differential pressure (DP)-based (e.g., orifice) flow meter. A differential pressure signal (e.g., 0–10 kPa) is produced by the sensor. This signal is converted to 4–20 mA at the transmitter for sending to DCS. On the receiving side, which typically is the analog input (AI) module in DCS or PLC, the 4–20 mA signal is converted back to a digital flow signal (e.g., 0–40 kg/s). Square-root transformation is done along the way, either in the transmitter or the AI function block.
   The latest Fieldbus technology allows the digital control signal to extend directly to the field equipment before converting to an analog signal.
2. Signal transformation: A signal goes through many transformations when transmitted from a sensor in the field to the controller in the DCS. The transformation can be linear or nonlinear. Many times, both linear and nonlinear transformations are performed simultaneously.

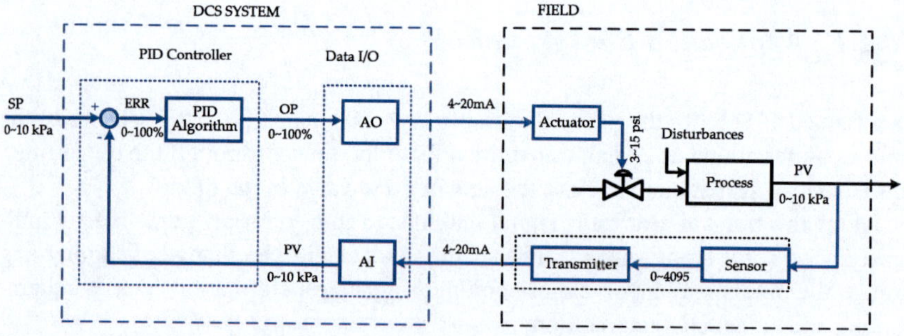

Fig. 10.1 Transformation of control signals in a PID control loop

A linear transformation changes the signal range but keeps the same *percentage-in-range* (PIR) value; However, a nonlinear transformation changes the PIR but may retain the signal range. Some common transformations are discussed as follows:

a. Signal scaling: Signal scaling is a linear transformation. It adjusts the signal range but retains the percentage-in-range (PIR). For example, converting a level signal of 0–5 m to a percentage value of 0–100% is a linear transformation.
   Assume the incoming signal is x with range x_{min} to x_{max}. The signal is to be converted to y with a range from y_{min} to y_{max}. The conversion is performed as

$$\text{PIR} = \frac{y - y_{min}}{y_{max} - y_{min}} = \frac{x - x_{min}}{x_{max} - x_{min}} \tag{10.1}$$

   which gives

$$y = \frac{x - x_{min}}{x_{max} - x_{min}} \cdot (y_{max} - y_{min}) + y_{min}. \tag{10.2}$$

   If both x_{min} and y_{min} are zeros, then Eq. 10.2 is simplified into

$$y = \frac{x}{x_{max}} \cdot y_{max}. \tag{10.3}$$

b. Signal characterization: Signal characterization is a nonlinear transformation that changes the signal characteristics. It typically retains the range of the signal but changes the PIR.
   Most modern DCS has built-in support for signal transformation in many of its basic control function blocks such as the transmitters, analog in (AI) and analog out (AO) blocks (see Sect. 4.2.1 for DCS function blocks). Transformation can be performed with a mathematical formula or via a lookup table (characterization table). Although a simple mathematical equation is preferred, an analytic formula cannot always represent the characteristics; a lookup table can be used. Most transmitters and DCS I/O blocks support signal characterization.
   Assume the signal transformation (both linear and nonlinear) is defined by five mapping points as in the following table:

x (input)	y (output)
x_{max}	y_{max}
x_4	y_4
x_3	y_3
x_2	y_2
x_{min}	y_{min}

When a new measurement value x is received, the table is searched for an exact match with x in the first column. If found, the corresponding y value in the second column is the result. Most of the time, there would be no exact match to the valve x in the table, but rather in between two values. Let us say, for an input value $x_2 < x < x_3$, a *linear interpolation* using the two neighboring points is performed to produce the corresponding y value as

$$y = \frac{x - x_2}{x_3 - x_2} \cdot (y_3 - y_2) + y_2, \quad \text{or} \quad y = \frac{x - x_3}{x_2 - x_3} \cdot (y_2 - y_3) + y_3$$

which is the same operation as in Eq. 10.2.

The advantage of the lookup table is that it is extremely flexible to represent any transformation as long as the relationship between the two values is monotonic. For example, a valve with linear characteristics can be converted to fast open via a nonlinear transformation.

c. Square-rooting: Square-Rooting is a nonlinear signal conversion that is commonly applied to a differential pressure type of flow measurement. Square-rooting is based on

$$\frac{y - y_{min}}{y_{max} - y_{min}} = \sqrt{\frac{x - x_{min}}{x_{max} - x_{min}}}$$

which leads to

$$y = \sqrt{\frac{x - x_{min}}{x_{max} - x_{min}}} \cdot (y_{max} - y_{min}) + y_{min}. \tag{10.4}$$

If both x_{min} and y_{min} are zeros, then Eq. 10.4 can be simplified as

$$y = \sqrt{\frac{x}{x_{max}}} \cdot y_{max}. \tag{10.5}$$

A square-rooting transformation changes the PIR of the signal. If the signal is converted to a different engineering unit simultaneously, the range of the signal will be changed as well. That is, signal characterization and signal scaling are performed at the same time.

d. Signal filtering: A signal filter, most commonly a first-order exponential filter, is equivalent to a nonlinear signal characterization. While keeping the range intact, it modifies the PIR of the signal as the signal goes through the filter. A first-order filter is given as

$$y_f(k) = (1 - \alpha)\, y_f(k-1) + \alpha\, x(k) \tag{10.6}$$

where $x(k)$ is the raw signal and $y_f(k)$ is the filtered signal; $y_f(k-1)$ is the filtered signal saved from the previous execution (see Eq. 5.2).

The filter constant α is related to the filter time constant τ by $\alpha = e^{-\tau/T_s}$, where T_s is the sampling frequency. For example, assuming $T_s = 1$ min, a 2-second filter is equivalent to $\alpha = e^{-2/60} = 0.967$, while a 10 second filter is equivalent to $\alpha = 0.846$.

On a modern control system, filtering is provided as a standard configuration option. Adding a filter to a PV is just a matter of activating this function by specifying a non-zero filter constant. However, beware that a PV filter introduces extra lag to the loop dynamics and potentially affects the controller tuning. Generally speaking, the filter constant should be at least five times smaller than the time constant of the loop dynamics.

e. Signal compensation: The signal generated by the sensor often does not reflect the actual physical value, e.g., due to different operating conditions. It often requires compensation with other information to arrive at the correct measurement value. Signal compensation, such as flow compensation, adjusts the signal's PIR but retains the measurement range. See Sects. 10.3 and 10.4 for examples on flow and temperature compensation.

10.1.3 Implementation and Operation

The signal scaling and transformation are essential for signal processing. Support for this operation is provided in multiple devices or control modules such as the transmitters in the field, the analog in (AI), or the analog out (AO) function blocks in DCS. The decision is often on where to perform the signal processing.

Normally, signal scaling should be performed closest to the signal source and preferably with the built-in functions. However, more complex signal transformation shall be performed in a location best for the application if multiple options are available. For example, the square-rooting transformation from DP to flow shall typically be performed in the flow transmitter. One exception is compressor anti-surge control, where the transformation shall be done in DCS so that both the DP and flow signals are accessible from DCS for purposes of display and troubleshooting.

Signal transformation via table look up shall be used with discretion. For a complex relationship that requires a large table to define, it is recommended that an analytic equation be obtained via curve fitting and be used in place of the large lookup table.

All complex signal transformations shall be appropriately documented with detailed narratives. Due to the lack of documentation, maintenance work on back-end calculation often relies on reverse engineering and guesswork in practical applications, which is highly inefficient and error-prone.

10.1.4 Application Examples

Online calculations are ubiquitous. Here are a few examples illustrating the basics.

Signal Scaling

Assume a level sensor is calibrated to measure a level from 0 to 2.0 m. The level signal in % is then given, following Eq. 10.2, by

$$\mathrm{LI} = \frac{\mathrm{LT} - 0}{2.0 - 0} \cdot (100\% - 0) + 0 = 50\% \cdot \mathrm{LT}.$$

For example, an LT = 1.2 m of the level at the sensor will produce LI = 50% × 1.2 = 60% of the level.

In reality, however, the level signal may be transmitted as an analog electric current signal of IT = 4–20 mA from the field to the DCS, and then converted to a digital level indicator value of LI = 0–100%. The conversion is performed as follows: from LT = 0–2 m to IT = 4–20 mA in the transmitter:

$$\mathrm{IT} = \frac{\mathrm{LT} - 0}{2 - 0} \cdot (20 - 4) + 4 = 8 \cdot \mathrm{LT} + 4$$

and then from IT = 4–20 mA to LI = 0–100% in DCS:

$$\mathrm{LI} = \frac{\mathrm{IT} - 4}{20 - 4} \times (100 - 0) + 0 = \frac{\mathrm{IT} - 4}{16} \times 100\%.$$

A 1.2 m of level will produce an electric current signal of $8 \times 1.2 + 4 = 13.6$ mA, and then to a level indication of $(13.6 - 4)/16 \times 100 = 60\%$.

Square-Rooting for Flow Measurement

A differential pressure (DP)-based flow meter, such as an orifice or a Venturi tube, directly measures the pressure difference across the device. The pressure signal requires square-rooting to produce a flow signal. The flow FT and ΔP are related by

$$\begin{aligned} \mathrm{FT} &= \sqrt{\frac{\Delta P - \Delta P_{\mathrm{min}}}{\Delta P_{\mathrm{max}} - \Delta P_{\mathrm{min}}}} \cdot (\mathrm{FT}_{\mathrm{max}} - \mathrm{FT}_{\mathrm{min}}) + \mathrm{FT}_{\mathrm{min}} \\ &= \sqrt{\frac{\Delta P}{\Delta P_{\mathrm{max}}}} \cdot \mathrm{FT}_{\mathrm{max}}. \end{aligned} \tag{10.7}$$

Equation 10.7 follows Eq. 10.5 with both $\Delta P_{\min}$ and $\text{FT}_{\min}$ being zero. For example, assume the differential pressure (DP) measurement range is $0 \sim 50$ kPa under a given condition. The flow meter is calibrated to produce a flow signal of 0–20,000 kg/h corresponding to the $0 \sim 50$ kPa range at this given condition. The signal transformation can be performed following Eq. 10.7 as

$$\text{FT} = \sqrt{\frac{\Delta P}{50}} \times 20000.$$

That is, a ΔP signal of 12.5 kPa indicates a flow of $\sqrt{12.5/50} \times 20,000 = 10,000$ kg/h.

Internally, the ΔP signal is first converted to a 4–20 mA electric current signal in the flow transmitter for transmission in the field. The analog signal is converted to a digital signal in DCS before transforming into a flow signal. The transformation can be performed in two ways:

1. Square-rooting in the transmitter: Linear scaling with square-rooting from ΔP = 0–50 kPa to IT = 4–20 mA in the transmitter, followed by linear scaling from IT = 4–20 mA to FT = 0–20,000 kg/hr in the AI block:

$$\text{IT} = \sqrt{\frac{\Delta P - 0}{50 - 0}} \cdot (20 - 4) + 4 = \sqrt{\frac{\Delta P}{50}} \cdot 16 + 4$$
$$\text{FT} = \frac{\text{IT} - 4}{20 - 4} \cdot (20000 - 0) + 0 = \frac{\text{IT} - 4}{16} \times 20000.$$

   The 12.5 kPa of ΔP signal will produce an electric current signal of $\sqrt{12.5/50.0} \times 16 + 4 = 12$ mA for transmission, and then a value of $(12 - 4)/16 \times 20{,}000 = 10{,}000$ kg/hr as flow indication.
2. Square-rooting in DCS: Linear scaling from ΔP = 0–50 kPa to IT = 4–20 mA in the transmitter, followed by linear scaling with square-rooting in the AI block:

$$\text{IT} = \frac{\Delta P - 0}{50 - 0} \times (20 - 4) + 4 = \frac{\Delta P}{50} \times 16 + 4$$
$$\text{FT} = \sqrt{\frac{\text{IT} - 4}{20 - 4}} \times (20000 - 0) + 0 = \sqrt{\frac{\text{IT} - 4}{16}} \times 20000.$$

   That is, the 12.5 kPa ΔP signal produces an electric current signal of $12.5/50 \times 16 + 4 = 8$ mA, and then a flow signal of $\sqrt{(8 - 4)/16} \times 20{,}000 = 10{,}000$ kg/hr. One advantage of the first approach is that the square-rooting of the sensor signal in the transmitter results in a larger electric current (12 mA vs. 8 mA in this example) for transmission. This analog electric current tends to have a higher signal-to-noise ratio and thus offers higher accuracy during transmission. Although the square-rooting can be performed in multiple locations along the signal transmission path, it shall only be performed once!

Signal Characterization

In this example, a crude oil pipeline travels over mountainous terrain to reach the storage tanks in the tank farm at the coastal port. The maximum flow rate is close to a million barrels a day.

A feedback + feedforward control scheme is implemented to control the arriving pressure at the storage tank. Flow rate is the feedforward variable, and the change in flow rate will cause pressure friction loss along the pipeline. The pressure friction loss in the crude oil pipeline is given in Table 10.1 as a function of the flow rate.

This pressure–flow relationship was implemented as a lookup table in Foxboro DCS using a CHARC (for signal characterizer) function block. This approach works as expected.

Alternatively, a simple mathematical formula can be used to approximate the relationship defined in this table. Using a formula instead of a large lookup table allows a cleaner implementation and faster execution in DCS. A trivial curve-fitting

Table 10.1 Pressure friction loss versus flow rate

Flow rate	Frictional losses
F	P
(bpd × 10^6)	(kPa)
0	0.00
0.0525	30.74
0.1050	105.02
0.1575	215.59
0.2100	359.24
0.2625	533.92
0.3150	738.13
0.3675	970.71
0.4200	1230.72
0.4725	1517.36
0.5250	1829.97
0.5775	2167.94
0.6300	2530.74
0.6825	2917.93
0.7350	3329.07
0.7875	3763.80
0.8400	4221.79
0.8925	4712.64
0.9450	5228.14
0.9975	5769.36
1.0500	6335.11

exercise reveals that a simple second-order polynomial function between the pressure and flow can reasonably approximate the lookup table:[1]

$$\text{[pressure loss]} \approx 858.9 * \text{[flow rate]} + 4959.4 * \text{[flow rate]}^2. \tag{10.8}$$

10.1.5 Special Considerations

In practical applications, there are more scenarios involving single transformations and processing than discussed here. Many are typically *hidden* from process control engineers, but it is beneficial to know their existence and the potential impacts.

Raw Counts

The word *raw count* is often encountered when dealing with more primitive PID loops such as those implemented in a remote terminal unit (RTU) (see Sect. 3.1.9). This raw count involves another step of signal scaling during signal transmission from the field to DCS. This step converts the signal from analog to digital by the so-called A/D converter and is part of the AI block's functionality. It is an essential step but is usually not interested by process control engineers.

With a 12-bit A/D converter, the range of the digital signal (raw counts) is between 0 and 4095 (= $2^{12} - 1$). Therefore, the electric current signal is transformed into the digital signal of the range 0–4095 as the A/D converter output. This digital signal is typically shifted by 2 to 4 bits by different vendors for improved resolution. So the range may become 0–16383 (14 bits), 0–32767 (15 bits), or even 0–65535 (16 bits). This *raw count* signal is then scaled to the correct engineering range and becomes the output of the Analog Input (AI) block. A detailed illustration of the signal scaling and transformation is provided in Table 10.2.

However, for historical reasons, the raw count range varies between PLC/DCS vendors and rarely matches the ranges in Table 10.2. For example, Allen Bradley PLC uses 6241–31206 for 4–20 mA mapping. Following Eq. 10.2, the conversion is given by

$$R = \frac{I - 4}{20 - 4} \times (31206 - 6241) + 6241 \tag{10.9}$$

$$I = \frac{R - 6241}{31206 - 6241} \times (20 - 4) + 4. \tag{10.10}$$

The maximum raw count value of 32,767 corresponds to a current signal of 21 mA, which is defined as the high range (not compliant with NAMUR NE43), and the live-zero of 4 mA in current corresponds to 6241 in the raw count.

---

[1] With MATLAB/Octave: `x=[ones(size(F)),F,F.^2]\P;` where `F` is the flow rate and `P` is the pressure loss.

Table 10.2 Example of flow signal scaling and transformation

	Transmitter		A/D raw count			DPI		FI	
	kPa	PIR%	mA	12-bit	15-bit	kPa	PIR%	m^3/h	PIR%
Span		105%	22.0	4,095	32,767				
Burnout (BAD)			21.0	3,909	31,278				
Over range		102.5%	20.5	3,816	30,533	10.31		40.62	
Range high	10.0	100%	**20.0**	3,723	29,788	10.00	100%	40.0	100%
75% range	7.5	75%	16.0	2,978	23,831	7.50	75%	34.6	86.6%
50% range	5.0	50%	12.0	2,234	17,873	5.00	50%	28.3	70.7%
25% range	2.5	25%	8.0	1,489	11,915	2.50	25%	17.9	50.0%
Range low	0.0	0	**4.0**	745	5,958	0.00	0	0.0	0
Under range		−1.25%	3.8	707	5,660	−0.13			
Burnout (BAD)			3.6	670	5,362	−0.25			
Zero			0.0	0	0				

Pneumatic Signals

Most control valves are driven by instrument air and controlled with pneumatic signals (i.e., pressure). The controller output, typically 0–100%, is transmitted to the field as a 4–20 mA analog signal. This electric current signal of 4–20 mA is converted to a 3–15 psi pneumatic pressure signal through an I/P transducer to drive the control valve. This I/P conversion is another signal scaling step involving two signals that both have *live-zeros*.

The 4–20 mA signal transmission has long been adopted as the standard for signal transmission. With the *live-zero* set to 4 mA, the over-range and under-range reading is 3.8 mA and 20.5 mA (i.e., −1.25% and 103.125%), while the burn-out out-of-range values are 3.6 mA and 21.0 mA (−2.5% and 106.25%), respectively, per the widely accepted standards of NAMUR NE 43 and API 551.

10.2 Data Quality and Bad Data Handling

Data quality is a measure of the condition of data based on factors such as accuracy, completeness, consistency, reliability, and timeliness. Measuring data quality can help identify data errors that need to be resolved and assess whether the data is adequate for the intended purpose.

Low data quality is one of the critical factors that prevent many advanced applications from online 24/7 operations. Many complexities in applications are catered to deal with reliability issues in real-time data.

10.2.1 Application Background

Nothing lasts forever. Measurement can fail and will fail; it is just a matter of probability. Bad process values (`BAD PV`) are one of the most common issues in practice. How to gracefully respond to bad process values is a crucial consideration in process control design. Below are some examples:

1. In a simple level-flow-valve cascade control loop, both the process variables (`PV`) for the level controller and the flow controller can fail, so do the control valve positioner and network communication. How should the control loop detect the failure and respond? In a much more complex control solution, such as compressor control in Fig. 4.2, what is the mitigating action if one of the process values becomes unavailable or erroneous?
2. Many model predictive control (MPC) applications are implemented in a standalone machine. The MPC application communicates with DCS to receive the process values, calculates the setpoint for the PID controllers, and sends them back to DCS. How should MPC monitor the network communication's health? How should MPC and DCS respond if the communication is lost or becomes suspicious?
3. Online analyzers are becoming prevalent, especially in petrochemical and chemical plants. However, their availability and reliability are still not sufficiently good to support online real-time control. How to assess the quality of data received from the analyzer to decide whether the closed-loop control should stay in control or temporarily go off-line?

These questions must be considered, and mitigation actions must be included in the control design.

10.2.2 Theory and Design

Measurement failure refers to either loss of data or data with erroneous value. The control system may react unpredictably to measurement failures and is thus a potentially serious hazard. There have been no standard solutions for bad data handling, but some general guidelines or approaches exist:

1. Reactive approach: Once a bad measurement is detected, the control scheme should safely and gracefully downgrade itself to a less sophisticated operating mode, and at the same time, alarm the operations personnel to intervene. This degraded operating condition is called *crippled mode operation*.
2. Proactive approach: For mission-critical process variables, redundant measurement is commonly implemented to reduce the probability of failure. Real-time monitoring and fault detection are also an active research area.

3. Intelligent handling: In a modern plant, the instrumentation level has improved to such an extent that redundant information, both *spatial* and *causal*, is available to cross-validate each other and provide quality indications and early warnings.

One key area to improve is reliably detecting and indicating the quality of measurement. One change in recent years was the generation and transmission of data quality as part of the process data. For example, with the open platform communication (OPC[2]) protocol, all the data from the field has a data quality attribute with one of the following three statuses:

1. *Good*: The tag value is trustworthy.
2. *Uncertain*: There is a non-critical abnormality associated with the tag value.
3. *Bad*: There is a critical abnormality associated with the tag value.

See the OPC data access specification from the OPC Foundation for a detailed definition of data quality.

A general recommendation is that only *good* data is acceptable for real-time and automated decision-making such as closed-loop control. *Uncertain* data cannot be trusted for automated decision-making but may still have reference values and thus can be used for information purposes such as monitoring and display.

The OPC quality indication, however, is related chiefly to device-level measurement issues. More advanced data quality indicators suitable for process control are still not widely available. For example, there are no standard and reliable approaches in detecting common data anomalies and generating credible data quality indicators. Below are some common data anomalies:

1. Out-of-range value: The measurement value is out of the measurement range.
2. Frozen value (flat line): A dead sensor often produces a constant (frozen) value, especially from online analyzers.
3. Spikes and outliers: A sudden and significant change in value.
4. Runaway value: A measurement value that is within range but gradually running away from its correct value.

10.2.3 Implementation and Operation

A careful evaluation of the potential causes and consequences of bad measurements is needed. This exercise is part of the hazard and operability (HAZOP) review. *Bad measurement* and *poor measurement* are the two prevalent abnormal scenarios that have to be addressed from the beginning.

The field implementation infrastructure (DCS/PLC) has provided various means for handling abnormal situations. However, many of the decisions are dependent on the operating objectives and must be dealt with case by case. Here are a few practical options available for handling bad data issues:

[2] The abbreviation OPC has changed from the original *OLE for Process Control* to *Open Platform Communication*.

1. Middle-of-three voting: The middle-of-three configuration provides higher availability and is thus favored in practical applications. The voting mechanism ensures that the correct signal is always selected even if one of three input signals fails. Challenge remains when two or all the three input signals are not available or erroneous.
2. Measurement comparison and validation: For specific measurements such as online analyzers, validation algorithms must detect analyzer malfunction such as flat-lining and spikes.
   When redundant measurements are available, for example, one for control and one for safeguarding, the two measurements can be constantly compared, and an alarm is generated if the difference exceeds a predetermined value. The comparison should consider the inherent noise in the signals and apply a realistic threshold to avoid nuisance alarms.
3. Dual transmitters: All meters have an effective range. For example, a flow meter has a typical turndown ratio of 3:1 for DP-based measurement or 10:1 for other types. The measurement reading below this turndown ratio will either be unreliable (e.g., orifice meter) or unavailable at all (e.g., vortex meter).
   For measurement with a large span, dual flow meters can be used to improve the accuracy. See Fig. 10.2 for a typical arrangement. The high-range transmitter loses accuracy when it measures below its turndown limit (typically 10% of the full range). Below the turndown limit, the source of the measurement value switches to the low-range transmitter.
   To achieve a smooth transition between the two signals, which typically takes place between 80% and 100% of the range of the low- range transmitter, and to maximize the measurement accuracy over the entire operating range, the following calculation can be used, where k defines the transition factor:

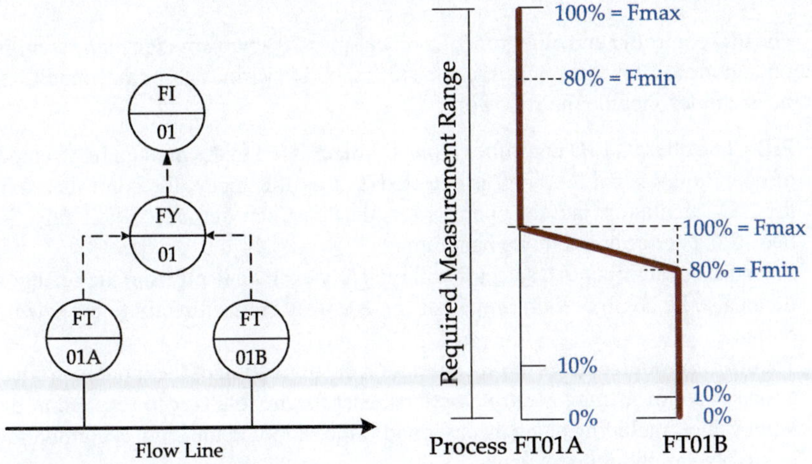

Fig. 10.2 Dual-range measurement

$$k = \max\left[\min\left(\frac{F_{TX2} - F_{min}}{F_{max} - F_{min}}, 1\right), 0\right]$$

$$F = (1 - k) \cdot F_{TX1} + k \cdot F_{TX2}$$

where

k: Transition range factor.
F_{TX1}: Measurement from low range transmitter.
F_{TX2}: Measurement from high range transmitter.
F_{min}: Minimum of the transition range (typically 80% of F_{TX1} range).
F_{max}: Maximum of the transition range (typically 100% of F_{TX1} range).

When the measurement F_{TX2} is below the minimum transition range (80% of low-range transmitter), $k = 0$, the final measure is determined only by the low-range transmitter. When the actual flow is higher than the maximum transition range (100% of the low-range transmitter), $k = 1$, then the high-range transmitter provides the measurement. When the actual measurement value is within the transition range, the measurement value is a weighted average of the two measurements, determined by the above equation ($0 \leq k \leq 1$).
If flow compensation needs to be performed, it should be applied to the individual flows rather than the combined flow unless the calibration conditions are identical for both measurements.
4. Soft sensors: Certain process variables such as product quality (compositions, cut-points, vapor pressures, and dew points) are inherently difficult to measure, and online analyzers are notorious for low availability and reliability. Soft sensor technology (see Sect. 5.3) has become popular where more reliable process measurements are used to infer the process values for supplementing the online measurement. The soft sensor technology can also be used to cross-validate some online measurements.

The PID controller and other function blocks in DCS have provided many configuration options to allow the control scheme to respond to various abnormal conditions. Some examples include the following:

1. PID Controller: A PID controller typically *sheds* itself to the manual (`MAN`) mode of operation when a `BAD PV` is detected. In the `MAN` mode, the controller skips the PID calculation, and the operator sets the controller output (`OP`) directly. See Sect. 3.6.2 for crippled mode handling.
2. Selector: A bad input to a selector will typically exclude itself from the selection. It can also be configured to propagate the `BAD PV` status directly to the selector output.
3. Totalizer: A totalizer does the accumulation or summation of the input to produce a total value over time. Multiple options exist for the totalizer to respond to bad input values, including using the last good value, skipping the input accumulation, or simply stopping the totalizer.

10.3 Flow Compensation and Conversion

Flow is one of the most fundamental measurements. Due to the diversity of measuring principles and devices, the raw measurement signal may need various signal processing to provide an accurate estimate of the flow rate at conditions different from the conditions at which the flow meter was calibrated.

10.3.1 Application Background

Flow is not directly measurable. Instead, it is always inferred from other directly measurable values such as differential pressure (DP), time, and frequency. Based on the measurement principle, there are many different flow meter types. Correspondingly, there are many ways of inferring the flow from the raw measurements.

Flow meters typically consist of a primary element (e.g., orifice plate, and vortex meter), one or more secondary devices such as pressure and temperature transmitters, and sometimes tertiary devices such as a flow computer. See Fig. 10.3 for a depiction.

All measurements have errors. Some may be inherent to the measuring principle, while others may be due to the environment, such as heat expansion. For example, a differential pressure (DP)-based flow meter such as an orifice plate produces a DP signal. The DP signal is then converted to a flow signal with the assumed fluid property and flow-line condition. If the flow-line condition changes, the meter will produce the correct DP value, but the flow value derived from the DP may no longer be accurate since the correlation between the two has changed under the new condition. Flow compensation is thus required with a modified correlation equation.

If the flow-line temperature changes significantly, the orifice's heat expansion may cause the orifice to produce an inaccurate DP value. A wrong DP leads to an incorrect flow value. For precise measurement, temperature compensation for the flow is also required. However, this error is relatively insignificant and is not commonly a concern.

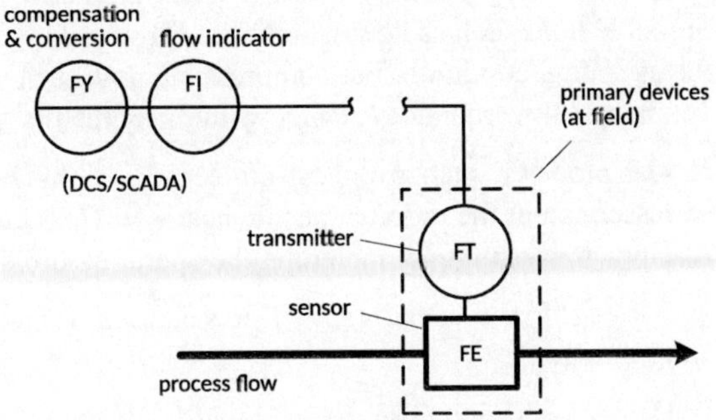

Fig. 10.3 Flow meter components

10.3.2 Theory and Design

A perfect flow meter would be reliable, accurate, inexpensive, with a high turndown ratio, requiring little calibration/maintenance, and can apply to any fluid at any condition. In real life, of course, this perfect flow meter does not exist. The best flow meter is always the one that is fit-for-purpose.

Type of Flow Meters

Based on the measurement principles, flow meters can be broadly divided into linear and square-root types:

1. Linear type meters: The so-called *linear type* flow meters are typically based on velocity or density, with which the flow is a linear function:
    a. Velocity-based: Both vortex meter and ultrasonic meter are velocity-based. Volumetric flow rate is the product of the flow velocity v and cross-section area of the pipe A, i.e., $F = v\,A$. By measuring the flow velocity, the flow rate can be readily calculated.
    b. Density-based: The Coriolis flow meter directly measures the density of the fluid. The mass flow can then be calculated.
2. Square-root-type meters: Square-root-type flow meters such as those based on orifice plate, Venturi tube, or flow nozzle measure the pressure drop (ΔP) across the device. A fixed relationship between the pressure drop and flow rate can be established at a controlled condition (calibration condition). Consequently, the flow rate is calculated from the actual measured ΔP value based on this relationship. The DP-based flow meter is called the *square-root type* since the flow rate is proportional to the squared root of ΔP.

Type of Flow Values

There are two types of flow measurements: *mass* flow and *volumetric* flow. For incompressible fluid such as liquid, both mass flow and volume flow are not influenced by the operating condition. For compressible flow such as gas and steam, however, the volumetric flow rate does change with many factors, primarily:

1. Gas property, such as the gas molecular weight (M), compressibility Z, and to a lesser extent the ratio of specific heat $\kappa = C_p/C_v$.
2. Operating condition, including the operating pressure P and temperature T.

For clarity and consistency of presentation, let us define the following types of *flow conditions*:

1. Design condition: The *design condition* is the expected normal operating condition specified during process design.

2. Calibration condition: The *calibration condition* is the reference condition under which the flow meter is calibrated. Typically, the calibration is chosen to be the same as the design condition of the process.
3. Reference condition: The *reference condition* is a pre-specified pressure and temperature condition. The reference condition is often selected to be the same as the design condition and the calibration condition.
4. Flowing condition, line condition, or actual condition: This condition refers to the actual pressure and temperature that the fluid is flowing under.
5. Standard condition: The definition of the *standard condition* varies by industry and country. The standard condition for the oil and gas industry is typically $P = 1$ atm and $T = 15°C$.
6. Normal condition: The definition of the *normal condition* varies by industry and country. A typical normal condition is $P = 1$ atm and $T = 0°C$.

For clarity, let us also define the different types of *flow values*:

1. Indicated flow (FI): The *indicated flow* refers to the flow that is directly read from the flow transmitter. Sometimes, this is also called the *raw flow*. For a DP-based flow meter, the indicated flow is not the correct flow unless the flowing condition is the same as the calibration condition.
2. Actual flow: The *actual flow* refers to the true process flow, compensated to the flowing condition if necessary. The actual flow is also called the *compensated* flow or *true* flow. This flow is only meaningful when the associated flowing condition is mentioned.
3. Equivalent flow: The *equivalent flow* refers to the same actual flow converted to a different baseline condition such as *standard condition* or *normal condition*. The baseline condition may be chosen as the design condition or the *standard condition*.

Flow Compensation

A DP-based flow meter can produce correct results only if the fluid density is the same as that under the calibration condition. Otherwise, the indicated flow is *inaccurate* and needs to be *compensated* or *corrected* with the density to arrive at the correct flow measurement. On the other hand, velocity or density-based linear flow meters directly produce the correct volumetric or mass flow, and thus no compensation is needed.

Assuming that the fluid density is ρ_a under actual condition (line condition), ρ_r under reference/calibration condition, and ρ_s under standard condition, the compensation formulas are given in Table 10.3.

Since density is a function of pressure P, temperature T, and molecular weight M, in case the fluid density is unavailable or remains constant, it is often acceptable to perform with the available measurements (P, T, and M). The accuracy of the compensation is reduced accordingly. That is, in the order of preference,

Table 10.3 Flow compensation and conversion with density

Meter type	Raw	Flow type	Indicated	Compensated	Standardized
Orificc	ΔP	Mass	$\mathrm{FI} = C\sqrt{\Delta P\,\rho_r}$	$F_m = \mathrm{FI}\cdot\sqrt{\frac{\rho_a}{\rho_r}}$	$F_s = \mathrm{FI}\cdot\sqrt{\frac{\rho_a}{\rho_r}}\cdot\frac{1}{\rho_s}$
Orifice	ΔP	Volume	$\mathrm{FI} = C\sqrt{\frac{\Delta P}{\rho_r}}$	$F_v = \mathrm{FI}\cdot\sqrt{\frac{\rho_r}{\rho_a}}$	$F_s = \mathrm{FI}\cdot\frac{\sqrt{\rho_r\,\rho_a}}{\rho_s}$
Vortex	v	Volume	$\mathrm{FI} = v\cdot A$	$F_v = \mathrm{FI}$	$F_s = \mathrm{FI}\cdot\frac{\rho_a}{\rho_s}$
Ultrasonic	v	Volume	$\mathrm{FI} = v\cdot A$	$F_v = \mathrm{FI}$	$F_s = \mathrm{FI}\cdot\frac{\rho_a}{\rho_s}$
Coriolis	ω	Mass	$\mathrm{FI} = f(\omega)$	$F_m = \mathrm{FI}$	$F_s = \mathrm{FI}\cdot\frac{1}{\rho_s}$

Table 10.4 Flow compensation/conversion with pressure/temperature

Meter	Flow	Indicated	Compensated	Standardized
Orifice	Mass	$\mathrm{FI} = C\sqrt{\Delta P\,\rho_r}$	$F_m = \mathrm{FI}\cdot\sqrt{\frac{P_a}{P_r}\frac{T_r}{T_a}}$	$F_s = \mathrm{FI}\cdot\sqrt{\frac{P_a}{P_r}\cdot\frac{T_r}{T_a}}\cdot\frac{Z_s\,R\,T_s}{P_s\,M_s}$
Orifice	Volume	$\mathrm{FI} = C\sqrt{\frac{\Delta P}{\rho_r}}$	$F_v = \mathrm{FI}\cdot\sqrt{\frac{P_r}{P_a}\frac{T_a}{T_r}}$	$F_s = \mathrm{FI}\cdot\sqrt{\frac{P_r}{P_s}\frac{P_a}{P_s}\frac{T_s}{T_r}\frac{T_s}{T_a}}$
Vortex	Volume	$\mathrm{FI} = F_v$	$F_v = \mathrm{FI}$	$F_s = \mathrm{FI}\cdot\frac{P_a}{P_s}\frac{T_s}{T_a}$
Ultrasonic	Volume	$\mathrm{FI} = F_v$	$F_v = \mathrm{FI}$	$F_s = \mathrm{FI}\cdot\frac{P_a}{P_s}\frac{T_s}{T_a}$
Coriolis	Mass	$\mathrm{FI} = F_m$	$F_m = \mathrm{FI}$	$F_s = \mathrm{FI}\cdot\frac{1}{\rho_s}$

$$\frac{\rho_1}{\rho_2} \quad\rightarrow\quad \frac{P_1}{P_2}\cdot\frac{T_2}{T_1} \quad\rightarrow\quad \frac{P_1}{P_2} \quad\rightarrow\quad \frac{T_2}{T_1}$$

The pressure and temperature are commonly available. The compensation formulas with pressure and temperature are given in Table 10.4, with the density ρ replaced by P and T.

Flow Conversion

Flow conversion is a technique to convert the flow from one condition to another. Flow conversion assumes that the flow measurement is accurate and correct, and the same flow value is to be represented in a different engineering unit. For example, the flow rate can be converted from *actual* flow rate (m^3/h) to *standard* flow rate sm^3/h or *normal* flow rate nm^3/h.

Certain flow meters such as vortex and ultrasonic meters directly produce volumetric flow, which is the true actual flow, then there is no need to compensate. However, the flow can be converted to other desired conditions, such as standard flow or normal flow. Coriolis meters directly produce mass flow, and typically no need to convert unless volumetric flow is expected.

Compensation versus Conversion

A flow transmitter provides a raw flow (or indicated flow) under actual conditions. This raw actual flow is compensated to become the *compensated flow under actual conditions* and then can be converted to *compensated flow under standardized conditions*. Alternatively, the *raw flow under actual condition* can be converted to standardized flow first to get the *raw flow under standard condition* (uncompensated) and then compensated to the *compensated flow under standard condition*. Either way, it arrives at the same final *compensated standardized flow*.

This conversion procedure is illustrated in Table 10.5. It is essential to know what type of flow representation is needed.

For DP-based *volumetric* flow measurement, assuming the indicated flow is represented by FI, the formulas for performing the compensation and conversion are summarized in Table 10.8.

For example, if the orifice is calibrated to produce a flow value of `FI` at the calibration (reference) condition, to compensate the flow to the actual condition, we simply multiply the indicated flow `FI` by $\sqrt{\rho_r/\rho_a}$. Sometimes, the flow meter is calibrated to provide standardized actual flow directly. To obtain the compensated

Table 10.5 Flow compensation verses flow conversion

Raw Actual Flow	⇒ `compensation` ⇒	Compensated Actual Flow
⇓ `conversion` ⇓		⇓ `conversion` ⇓
Raw Standard Flow	⇒ `compensation` ⇒	Compensated Standard Flow

Table 10.6 Linear volumetric flow meter

	Raw Flow	⇒ `compensation` ⇒	Compensated Flow
Actual Condition (kg/s)	$\mathrm{FI} \triangleq F_v$	$\times 1$	$= \mathrm{FI}$
⇓ `conversion` ⇓	$\times \dfrac{\rho_a}{\rho_s}$		$\times \dfrac{\rho_a}{\rho_s}$
Standard Condition (sm3/h)	$= \mathrm{FI} \cdot \dfrac{\rho_a}{\rho_s}$	$\times 1$	$= \mathrm{FI} \cdot \dfrac{\rho_a}{\rho_s}$
	Raw Flow	⇒ `compensation` ⇒	Compensated Flow

Table 10.7 Linear mass flow meter

	Raw Flow	⇒ `compensation` ⇒	Compensated Flow
Actual Condition (m3/h)	$\text{FI} \triangleq F_m$	$\times 1$	$= \text{FI}$
⇓ `conversion` ⇓	$\times \frac{1}{\rho_s}$		$\times \frac{1}{\rho_s}$
Standard Condition (sm3/h)	$= \text{FI} \cdot \frac{1}{\rho_s}$	$\times 1$	$= \text{FI} \cdot \frac{1}{\rho_s}$
	Raw Flow	⇒ `compensation` ⇒	Compensated Flow

Table 10.8 DP-based volumetric flow meter

	Raw Flow	⇒ `compensation` ⇒	Compensated Flow
Actual Condition (m3/h)	$\text{FI} \triangleq C\sqrt{\frac{\Delta P}{\rho_r}}$	$\times \sqrt{\frac{\rho_r}{\rho_a}}$	$= \text{FI} \cdot \sqrt{\frac{\rho_r}{\rho_a}}$
⇓ `conversion` ⇓	$\times \frac{\rho_a}{\rho_s}$		$\times \frac{\rho_a}{\rho_s}$
Standard Condition (sm3/h)	$= \text{FI} \cdot \frac{\rho_a}{\rho_s}$	$\times \sqrt{\frac{\rho_r}{\rho_a}}$	$= \text{FI} \cdot \frac{\sqrt{\rho_r \rho_a}}{\rho_s}$
	Raw Flow	⇒ `compensation` ⇒	Compensated Flow

actual volume flow, we need to multiply the flow by $\sqrt{\rho_r/\rho_a}$ (compensation), and then divide it by ρ_a/ρ_s. Similarly, the formulas for performing the compensation and conversion for DP-based *mass* flow measurement are summarized in Table 10.9.

10.3.3 Implementation and Operation

There are many different ways of performing flow compensation, depending on what is available online.

Dynamic and Static Compensation/Conversion

Flow compensation can be either static, dynamic, or quasi-dynamic depending on the number of relevant parameters. Since the density of the flow is given by $\rho = \frac{PM}{ZRT}$, the compensation is done either by density ρ, or by one or more of the individual parameters in the formula, including pressure (P), temperature (T), molecular weight (M), and compressibility (Z), if the density is not available:

1. Static compensation: Suppose it is known that the density at the actual operating condition is different from that at the calibration condition but remains relatively constant. In that case, a fixed compensation term can be used in place of ρ_a in the above formulas in Tables 10.7, 10.8 and 10.9. Similarly, if the molecular weight is known and is different at the flowing line condition, then static compensation on molecular weight can be performed.
2. Dynamic compensation: Since the pressure and temperature at flow conditions are typically different from that at calibration conditions, online compensation can be performed with the actual pressure and temperature measurements. See Tables 10.3 and 10.4.
3. Quasi-dynamic compensation: Some of the properties are known and relatively constant while others vary with time, then mixed compensation can be applied. For example, the gas's molecular weight and compressibility are different from that of the calibration gas but do not vary much with time, but the pressure and temperature vary significantly. The flow compensation can then be performed with static compensation on molecular weight and compressibility, and dynamic compensation on pressure and temperature.

10.3.4 Application Examples

To correctly apply compensation and conversion, start with the type of flow meter used, the type of flow (mass vs. volume); it is calibrated to measure under what condition it operates, and for what process condition to display the value. Table 10.10 provides a lookup table for finding the right formula for compensation and conversion. Several examples are provided below to demonstrate the procedure.

Table 10.9 DP-based mass flow meter

	Raw Flow	⇒ `compensation` ⇒	Compensated Flow
Actual Condition (kg/s)	$\text{FI} \triangleq C\sqrt{\Delta P \rho}$	$\times \sqrt{\frac{\rho_a}{\rho_r}}$	$= \text{FI}\sqrt{\frac{\rho_a}{\rho_r}}$
⇓ `conversion` ⇓	$\times \frac{1}{\rho_s}$		$\times \frac{1}{\rho_s}$
Standard Condition (sm3/h)	$= \text{FI} \cdot \frac{1}{\rho_s}$	$\times \sqrt{\frac{\rho_a}{\rho_r}}$	$= \text{FI} \cdot \sqrt{\frac{\rho_a}{\rho_r}} \frac{1}{\rho_s}$
	Raw Flow	⇒ `compensation` ⇒	Compensated Flow

Table 10.10 Compensation method lookup table

	For volumetric flow	For mass flow
Linear type meter	See Table 10.6	See Table 10.7
Square-root-type meter	See Table 10.8	See Table 10.9

Pressure and Temperature Compensation on Flow

The most common application of flow compensation is with pressure and temperature on the gas flow. Assuming a DP-based flow meter is calibrated to the design condition of 2,200 kPa(g) for pressure and 60 DegC for temperature. With the actual flow-line condition at 2,400 kPa(g) and 65 oC, the indicated flow rate is 89,000 m^3/h. Since the actual pressure and temperature condition is different from the calibration condition, we know that the indicated flow of 89,000 m^3/h is not the correct value.

To convert from indicated flow to compensated flow, we can start from Table 10.10. Since the compensation is on volumetric flow with a square-root type of meter, the compensation formula is in Table 10.8. However, since density is unavailable, the online compensation is commonly simplified to pressure and temperature compensation only:

$$\rho = \frac{PM}{ZRT}, \quad \frac{\rho}{\rho_s} \approx \frac{P}{P_s}\frac{T_s}{T}. \tag{10.11}$$

To obtain the compensated actual flow, the following formula from Table 10.8 can be used:

$$\begin{aligned}
\mathtt{FI}_{\text{comp}} &= \mathtt{FI} \cdot \sqrt{\frac{\rho_r}{\rho_a}} \\
&\approx \mathtt{FI} \cdot \sqrt{\frac{P_r}{P_a} \cdot \frac{T_a}{T_r}} \\
&= 89,000 \times \sqrt{\frac{2200 + 101.35}{2400 + 101.35} \times \frac{65 + 273.15}{60 + 273.15}} \\
&= 86,006\,\text{m}^3/\text{hr}
\end{aligned} \tag{10.12}$$

where 101.35 and 273.15 are added to the actual pressure and temperature, respectively, to arrive at the absolute pressure and temperature.

Converting the *indicated flow* to *compensated standard flow* is a two-step calculation, as shown in Table 10.8:

$$\begin{aligned}
\mathtt{FI}_{\text{comp}} &= \text{FI} \cdot \sqrt{\frac{\rho_r}{\rho_s}\frac{\rho_a}{\rho_s}} \\
&\approx \mathtt{FI} \cdot \sqrt{\frac{P_r}{P_s}\frac{P_a}{P_s}}\sqrt{\frac{T_s}{T_r}\frac{T_s}{T_a}}
\end{aligned}$$

$$= 89,000 \times \sqrt{\frac{2200 + 101.35}{101.35}\frac{2400 + 101.35}{101.35}}\sqrt{\frac{15 + 273.15}{60 + 273.15}\frac{15 + 273.15}{65 + 273.15}}$$
$$= 1,808,793\text{sm}^3/\text{hr} \tag{10.13}$$

where the standard condition is defined as $P_s = 101.3$ kPa and $T_s = 15°C$.

Vortex Meter Flow

A vortex meter produces the correct actual volumetric flow at the flow-line condition with no need for compensation. In practice, the flow may typically need to be converted to some common base condition for comparison and understanding. For example,

1. Mass flow: There is no distinction between actual mass flow and standard mass flow, so there is no conversion necessary between the two mass flows. However, converting the indicated volumetric flow to a mass flow requires the gas density ρ. In addition to the pressure and temperature, the gas molecular weight M and compressibility Z are also required:

$$F_m = \text{FI}/\rho.$$

2. Standard volume flow: Table 10.10 suggests that the formula in Table 10.6 be used to convert the actual volume flow to standard flow with a linear flow meter:

$$F_{v,s} = F_{v,a}\frac{\rho_a}{\rho_s} \approx F_{v,a}\frac{P_a}{P_s}\frac{T_s}{T_a}. \tag{10.14}$$

For standard condition $P_s = 101.3$ kPa and $T_s = 15°C$, sometimes the following simplified formula is given:

$$F_{v,s} = 2.844\, F_{v,a} \cdot \frac{P_a}{T_a} \quad \text{since} \quad \frac{T_s}{P_s} = \frac{273.15 + 15}{101.3} = 2.844.$$

In addition, if the molecular weight and compressibility are known, static compensation with molecular weight and compressibility can be applied.

Compensated Flow to Raw Flow

In some applications, such as compressor anti-surge control (details can be found in Sect. 11.3), the required flow measurement is the raw volumetric flow at the actual condition.

Assume the orifice-based flow meter is calibrated to provide mass flow measurement. The design condition of the pressure is 2200 kpa(g) and temperature of 60 °C.

The fluid is natural gas with a molecular weight of 21.5 kg/kmol, compressibility 0.96, and the ratio of specific heat of 1.26.

For square-root-type flow meter calibrated for mass flow measurement, Table 10.10 suggests that the formula in Table 10.9 be used for compensation and conversion. In actual operation, the flow meter gives a flow reading of 24 kg/s, with a pressure measurement of 2,100 kpa(g) and a temperature of 58 °C:

1. This 24 kg/s is the indicated flow (raw flow) at the actual condition.
2. To get the compensated mass flow (true flow) at the actual condition, the formula given in Table 10.9 is as follows:

$$\mathrm{FI}_{m,a} = \mathrm{FI}\sqrt{\frac{\rho_a}{\rho_r}} \tag{10.15}$$

   where ρ_r is the design density, and ρ_a is the actual density. Since the molecular weight is not known under the actual condition, the compensation is reduced, based on $\rho = \frac{PM}{ZRT}$, to the following:

$$\begin{aligned}\mathrm{FI}_{m,a} &= \mathrm{FI}\sqrt{\frac{\rho_a}{\rho_r}}\\ &\approx \mathrm{FI}\sqrt{\frac{P_a}{P_r}\frac{T_r}{T_a}}\\ &= 24\times\sqrt{\frac{2100+101.325}{2200+101.325}\frac{60+273.15}{58+273.15}}\\ &= 23.5\,\mathrm{kg/s}.\end{aligned}$$

   The indicated flow from the meter is 24 kg/s, while true compensated flow at actual condition is 23.5 kg/s.
3. If the compensated volumetric flow at standard condition (pressure = 101.325 kPa and temperature = 15 °C) is required, then again from Table 10.9, the following compensation formula can be used:

$$\begin{aligned}\mathrm{FI}_{m,a} &= \mathrm{FI}\sqrt{\frac{\rho_a}{\rho_r}\frac{1}{\rho_s}}\\ &\approx \mathrm{FI}\sqrt{\frac{P_a}{P_r}\frac{T_r}{T_a}\Big/\frac{P_s\,M_s}{Z_s\,R\,T_s}}\\ &= 24\times\sqrt{\frac{2100+101.325}{2200+101.325}\frac{60+273.15}{58+273.15}\Big/\frac{101.325\times 21.5}{0.96\times 8.314\times(15+273.15)}}\\ &= 24.855\,\mathrm{sm^3/s}.\end{aligned}$$

   The compensated volumetric flow rate at standard condition is thus 24.855 m^3/s.

4. Now assume that the flow meter is calibrated to produce a standard volumetric flow of 24.855 sm^3/s; that is, the indicated value at the flow meter is the true volumetric flow at standard condition. Anti-surge control requires that the raw value at actual condition be provided. Therefore, the indicated flow must be converted to raw flow at actual condition.
   The same Table 10.9 is used but in reverse order. The raw value at the actual condition is given by

$$\mathrm{FI} = F_{v,s} \Big/ \sqrt{\frac{\rho_a}{\rho_r}\frac{1}{\rho_s}}.$$

   It is easy to verify that a standard flow at 24.855 sm^3/s is 24 kg/s at actual condition.

10.4 Temperature Compensation with Pressure

Gas and steam temperature are heavily correlated with their pressure. For better control, the temperature measurement often needs to be compensated with the pressure variation to improve the accuracy.

10.4.1 Application Background

Gas and steam temperature are a function of their pressure. Temperature measurement is calibrated to be accurate for a certain pressure. In practice, the pressure often fluctuates and could lead to significant temperature measurement errors. Temperature compensation with pressure is often required. For instance, in a distillation column, the temperature inside the column is a critical controlled variable because the temperature is used to infer purity. Due to the correlation with pressure, the column temperature will also increase if the column pressure increases due to inadequate pressure control. This temperature increase due to pressure change will incorrectly indicate that the composition has changed, although it is the same composition. Thus, for more accurate temperature control, which leads to better composition control, the temperature is often compensated with pressure to eliminate the pressure influence.

10.4.2 Theory and Design

The objective is to derive a pressure-compensated temperature that correlates better with composition.

Pressure compensation of the vapor temperature is based on an equation to remove the correlation between pressure and temperature. That is, if the temperature increase is caused by pressure, this will be automatically excluded. If a composition change causes an increase in both pressure and temperature, then the part of the temperature change caused by pressure change will be nullified by the calculation.

Temperature–Pressure Relationship

The basis for temperature compensation is the relationship between temperature and pressure described by vapor pressure equations. Most of them are semi-empirical equations derived from the integration of the Clausius–Clapeyron equation. These equations are for pure components, and the coefficients are obtained by experimental procedures (Towler and Sinnott 2012).

One of the best known equations is Antoine's equation given by

$$\log P = A - \frac{B}{T + C}, \quad \text{or} \quad T = \frac{B}{A - \log P} - C \tag{10.16}$$

where P is the vapor pressure (absolute) in mmHg, T is the temperature in °C, and log is the base-10 logarithm function.

Antoine's equation uniquely relates the pressure and temperature for a pure component in vapor–liquid equilibrium. The base-10 logarithm of the pressure is a linear function of the inverse of the temperature. If there is a pressure swing, the temperature will change immediately according to the ideal gas law. The vapor–liquid equilibrium will shift, and the pressure and temperature will eventually re-balance as the purity changes.

The Antoine constants A, B, and C are readily available from chemical engineering data books or the Internet for pure compounds. For example, for the following compounds, the constants are given in Table 10.11 (see physical properties in Appendix C of Towler and Sinnott 2012).

Table 10.11 Antoine Coefficients for some compounds (data from http://ddbonline.ddbst.com and Towler and Sinnott (2012))

	P in mmHg, T in °C			P in Pascal, T in Kelvin			Valid T Range in°C	
	A	B	C	A	B	C	T_{min}	T_{max}
Methane	6.612	259.63	265.99	8.737	259.63	−7.16	−180	−153
Ethane	6.803	656.40	255.99	8.928	656.40	−17.16	−143	−74
Propane	6.830	813.20	247.99	8.955	247.99	−25.16	−109	−24
n-Butane	6.809	935.86	238.73	9.058	935.86	−34.42	−78	17
iso-Butane	6.748	882.80	240.00	8.873	882.80	−33.15	−86	7
Pentane	6.876	1,075.78	233.21	9.001	1,075.78	−39.94	−50	58
Steam	8.140	1,810.94	244.49	10.265	1,810.94	−28.67	99	374
Water	8.071	1,730.63	233.43	10.196	1,730.63	−39.72	1	100

Even though SI units are preferred, the coefficients are customarily given in mmHg for pressure and °C for temperature. To convert from P in mmHg and T in °C to P in Pascal and T in Kelvin, use the following equations (the converted results are shown in Table 10.11):

$$A + \log \frac{101325}{760} = A + 2.124903 \mapsto A$$
$$B \mapsto B$$
$$C - 273.15 \mapsto C.$$

A linear approximation approach may be used for compounds whose exact composition is unknown or changes with time. Knowledge of linear regression (or data fitting) is required. Since Antoine's equation (Eq. 10.16) is nonlinear between pressure P and temperature T, it needs to be linearized first to use data-based regression. Rearranging Eq. 10.16 gives the equation that is linear in parameters:

$$\begin{aligned} \log P &= A + \frac{AC - B}{T} - C\frac{\log P}{T} \\ &= K_1 + K_2\frac{1}{T} + K_3\frac{\log P}{T}. \end{aligned} \tag{10.17}$$

A simplified form of the Antoine equation is the August equation:

$$\log P = A - \frac{B}{T}, \quad \text{or} \quad T = \frac{B}{A - \log P}. \tag{10.18}$$

Temperature Compensation with Pressure

There are two common methods for performing pressure compensation on temperature measurement. The rigorous approach is based on accurate knowledge of the composition of the fluid. For complex compounds, this requirement may not be realistic, especially if the composition changes online. An approximate data-based empirical compensation may be applied.

The pressure-compensated temperature (PCT) is the measured temperature, minus the part caused by pressure change. That is,

$$T_c = T - \Delta T \tag{10.19}$$

where T_c is the compensated temperature and T is the indicated temperature. ΔT is the temperature change caused by pressure deviation from baseline pressure P_b and should be removed from the indicated temperature.

The correction term ΔT can be calculated in several ways, depending on the equation adopted for the temperature–pressure relationship:

1. Direct compensation based on Antoine's equation:

$$
\begin{aligned}
T_c &= T - \left[\left(\frac{B}{A - \log P} - C\right) - \left(\frac{B}{A - \log P_b} - C\right)\right] \\
&= T - \left(\frac{B}{A - \log P} - \frac{B}{A - \log P_b}\right). \qquad (10.20)
\end{aligned}
$$

2. Compensation based on Antoine's equation via local linearization:

$$
\begin{aligned}
T_c &= T - \frac{dT}{dP}(P - P_b) \\
&= T - \frac{B}{\ln(10)\,P\,(A - \log P)^2}(P - P_b) \qquad (10.21)
\end{aligned}
$$

where log is the base-10 logarithm and ln is the base-e natural logarithm. The temperature gradient dT/dP is calculated from Antoine's equation (Eq. 10.16):

$$
\frac{dT}{dP} = \frac{d\left(\frac{B}{A - \log P} - C\right)}{dP} = \frac{B}{\ln(10)\,P\,(A - \log P)^2}. \qquad (10.22)
$$

This method works only within the base operating point's close vicinity due to the assumption of a linear relationship between pressure and temperature, which is technically not true. The range of validity of the compensation should be specified in actual implementation, and the calculation result should be clamped accordingly. If the pressure moves outside this validity range, the clamped value should be used, and the data quality should be marked as *uncertain*.

10.4.3 Application Example

Consider a debutanizer column as an example. The bubble points of the vapor inside the column are produced via dynamic simulation, and the temperatures/pressures are shown in the first two columns in Table 10.12 (Stephan 2013).

Calculation of Antoine's Coefficients

Let us calculate Antoine's coefficients based on the bubbling point data. First, convert the pressure from kg/cm^2 (gauge) to mmHg (absolute) by multiplying the pressure

Table 10.12 Bubble points: pressure and temperature in a debutanizer

Temperature	Pressure	↦	Pressure	Estimated		Compensated		
T	P		P	$\hat{T}^{(1)}$	$\hat{T}^{(2)}$	$\hat{T}^{(3)}$	$\hat{T}^{(4)}$	$\hat{T}^{(5)}$
°C	kg/cm^2(g)		mmHg(a)	°C				
90.00	4.83		4312.8	89.99	90.73	99.963	100.663	98.770
91.25	4.98		4423.1	91.25	91.78	99.965	100.492	98.984
92.50	5.13		4533.4	92.47	92.84	99.987	100.372	99.198
93.75	5.29		4651.1	93.76	93.94	99.950	100.210	99.343
95.00	5.45		4768.8	95.03	95.05	99.935	100.098	99.488
96.25	5.61		4886.5	96.27	96.17	99.940	100.030	99.634
97.50	5.77		5004.2	97.50	97.28	99.966	100.050	99.779
98.75	5.94		5129.2	98.77	98.46	99.936	99.945	99.855
100.00	**6.10**		**5246.9**	**99.96**	**99.57**	**100.000**	**100.000**	**100.000**
101.25	6.28		5379.3	101.27	100.82	99.937	99.948	100.007
102.50	6.45		5504.4	102.49	101.99	99.967	100.007	100.083
103.75	6.63		5636.8	103.76	103.24	99.945	100.034	100.090
105.00	6.81		5769.2	105.02	104.49	99.943	100.098	100.097
106.25	6.99		5901.6	106.25	105.74	99.959	100.197	100.104
107.50	7.17		6034.0	107.47	106.99	99.994	100.330	100.111
108.75	7.36		6173.7	108.73	108.32	99.981	100.433	100.049
110.00	7.55		6313.5	109.97	109.65	99.986	100.569	99.987
111.25	7.75		6460.6	111.26	111.05	99.945	100.680	99.856
112.50	7.94		6600.3	112.47	112.38	99.986	100.878	99.794
113.75	8.14		6747.5	113.73	113.79	99.982	101.049	99.663
115.00	8.35		6901.9	115.03	115.27	99.934	101.198	99.463
116.25	8.55		7049.0	116.24	116.68	99.966	101.427	99.332
117.50	8.76		7203.5	117.51	118.17	99.954	101.632	99.132
118.75	8.97		7358.0	118.75	119.67	99.960	101.864	98.931
120.00	9.19		7519.8	120.04	121.25	99.925	102.075	98.662

values with 735.559 and adding the atmospheric pressure of 760 mmHg. The result is in the third column of Table 10.12.

Using the linearized Antoine equation (Eq. 10.17), with the pressure and temperature data in Table 10.12, the coefficients K_1, K_2, K_3 can be obtained by solving 25 simultaneous equations.[3] The result is given as $K_1 = 6.349$, $K_2 = 546.630, and K_3 = -217.585$. Antoine's coefficients are then given by Eq. 10.17 as $A = K_1 = 6.349$, $C = -K_3 = 217.585$, $B = AC - K_2 = 6.349 \times 217.585 - 546.630 = 834.730$. The values are close to that of n-Butane in Table 10.11. The estimated temperatures using Eq. 10.16 are shown as $\hat{T}^{(1)}$ in Table 10.12, very close to the true temperatures in Column 1. In other words, Antoine's equation with the

[3] With MATLAB/Octave:

```
K=[ones(size(T)),1./T,log10(P)./T]\log10(P); A=K(1); C=-K(3); B=A*C-K(2);.
```

above coefficients is an excellent approximation. This close match is not surprising since the dynamic simulation is likely based on the same equation.

Now let us try the simplified August equation in Eq. 10.18. Solving for parameters A and B with the data in Table 10.12, we have $A = 4.594$, $B = 87.036$.[4] The estimated temperature with August's equation is given as $\hat{T}^{(2)}$ in Table 10.12. The simplified August equation produces a close approximation as well.

Temperature Compensation

Up to now, the effort has been on obtaining and validating Antoine's coefficients from pressure and temperature data. The data in Table 10.11 is for the vapor in the debutanizer with a fixed composition. The different pressure/temperature combinations in the table indicate the same composition. The temperature variation is caused entirely by the pressure change. Therefore, the ΔT term in Eq. 10.19 accounts for the temperature change caused by pressure change. By subtracting ΔT from the actual temperature, we obtain the pressure-compensated temperature, which correlates to the composition and is expected to be a constant value since the composition is constant in this example.

To use Antoine's equation for temperature compensation, we have several approaches as explained in Sect. 10.4.2. Assuming that the base pressure is $P_b = 6.10$ kg/cm^2(g) (5246.9 mmHg(a)) indicated by the line in bold text in the table, with Eq. 10.20, the compensated temperature T_c is calculated, and the results are listed as $\hat{T}^{(3)}$ in Table 10.12. As expected, the compensated temperature is nearly constant at 100°C, since the composition is fixed.

The simplified equation in Eq. 10.21 can also be used, but the results are accurate only at the vicinity of the base pressure due to the local linearization. See $\hat{T}^{(4)}$ in Table 10.12 for the estimated results and Fig. 10.4 for an illustration. This method is based on piece-wise linearization. The slope dT/dP around $T = 100$ as calculated by Eq. 10.22 is 0.0099988.

The methods mentioned above are all based on Antoine's equation. They assume that the Antoine coefficients A, B, and C are explicitly available either by table lookup (for pure compounds) or via data analysis with Eq. 10.20 or 10.21. In practice, there is a more direct data-based linearization method based on the following:

$$T_c = T - K\,(P - P_b) \tag{10.23}$$

where K is the slope of the T-P curve in Fig. 10.4. The slope can be calculated by solving the following equation:

$$T = K\,P + K_0. \tag{10.24}$$

---

[4] With MATLAB/Octave: `x=[ones(size(T)),-1./T] \log10(P); A=x(1); B=x(2);`.

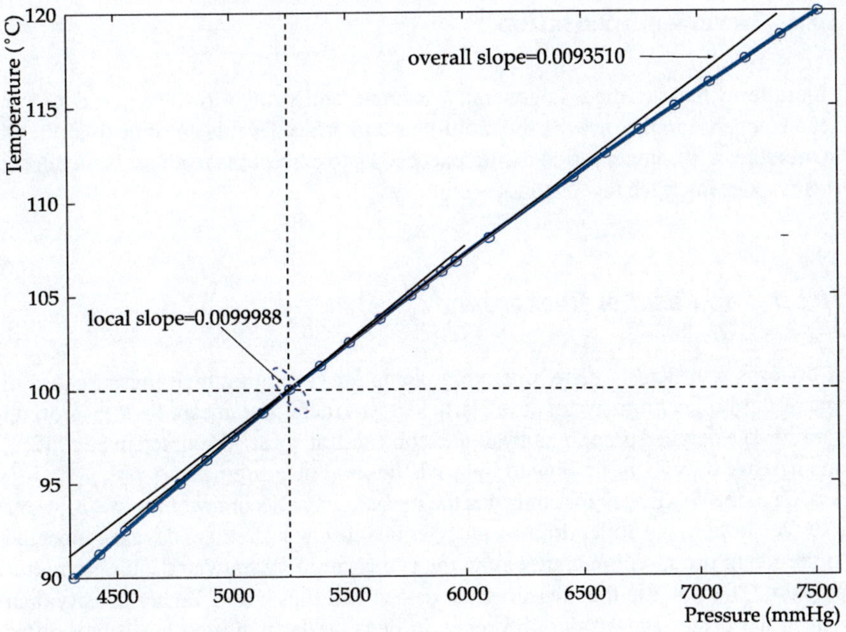

Fig. 10.4 Nonlinearity in pressure compensation

For the data in Table 10.12, the K value is given by $K = 0.009351$.[5] K is the overall T-P slope in Fig. 10.4 based on all the data, which is different from the local slope of 0.0099988 calculated above. The compensated temperatures with Eq. 10.23, assuming a base pressure of 5246.9 mmHg(a), are given as $\hat{T}^{(5)}$ in Table 10.12, which is also close to the expected value of 100 °C.

10.4.4 Special Considerations

The example in the previous section assumes that the composition is constant, and the temperature variation is solely caused by pressure variation. In practice, if actual operation data is used, either historical data or step test data, the most likely scenario is that the temperature change results from both composition change and pressure variation. The challenge is how to separate the two. The recommended approach is to select a period where the composition is relatively constant, but sufficient pressure variations are present.

[5] With MATLAB/Octave: `x=[P,ones(size(T))] \T; K=x(1); K0=x(2);`.

10.5 Level Compensation

Liquid level measurement is generally accurate and straightforward and does not need compensation. However, the multi-phase interface level is notoriously difficult to measure, and compensation is often necessary to correct the reading. Boiler drum steam/water interface level is one example.

10.5.1 Application Background

A boiler is prevalent in many operating plants for generating high-pressure steam, where the boiler drum water level is the most crucial parameter to measure and control. The detailed process analysis and control strategy are discussed in Sect. 12.2, and here we only focus on how to improve the level measurement.

One of the challenges for control is the measurement accuracy of the steam/water interface level in the boiler drum. Any external level measurement device connected to the steam drum will operate below the actual drum steam/water saturation temperature. The water in the measurement device will thus have a higher density than that in the drum. Due to the difference in density, the indicated level may differ significantly from the drum's actual level.

10.5.2 Theory and Design

The boiler level is indicated either by direct reading from a *gauge glass* or through a *remote level indicator* of various types (ASME 2019). Here, we will explain the working principles of density compensation for the two most widely used methods:

1. Gauge glass: Gauge glass is a direct reading method. A gauge glass is more commonly called a *gage glass*, and is a mandatory requirement by ASME code (ASME 2019) for any boiler.
2. DP-based level measurement: This indirect measurement is deduced from a differential pressure measurement.

Level measurement based on other principles such as displacers, conductivity, and radar is also common but will not be covered here.

Gauge Glass for Steam/Water Interface Level

As shown in Fig. 10.5, a gauge glass is for direct visual reading of the level in the drum.

Because the density of the water inside and outside the drum is different, the indicated drum water level is always lower than its true level. The indicated level

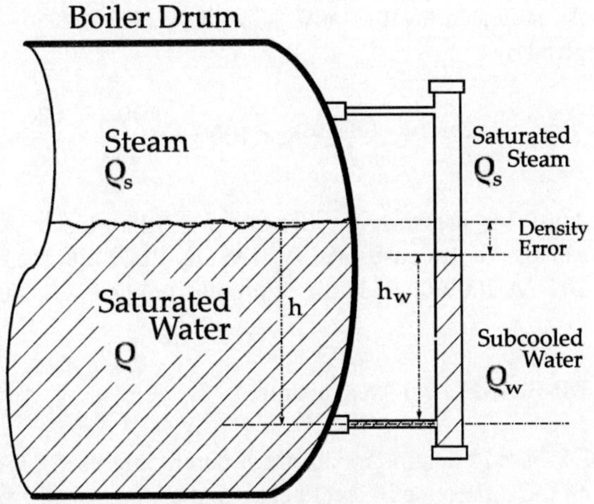

Fig. 10.5 Boiler drum level indicator: gauge glass

is h_w while the true level is h. H is assumed to be the height between the level measurement taps, and ρ_s the steam density. The difference $h - h_w$ is the error caused by the difference in density and is commonly called the *density error*. For accurate level reading and control, the density error must be corrected.

The correction is performed by inferring the true level h from the indicated level h_w. The static pressure is related to static height via

$$P = \rho\, g\, h \tag{10.25}$$

where ρ is the density, h is the height, and g is the gravitational acceleration constant. Based on pressure balance at the steam/water interface level in the gauge, we have

$$\rho\, g\, h = \rho_w\, g\, h_w + \rho_s\, g\, (h - h_w) \tag{10.26}$$

solving for h gives

$$h = \frac{\rho_w - \rho_s}{\rho - \rho_s}\, h_w. \tag{10.27}$$

Here, $\frac{\rho_w - \rho_s}{\rho - \rho_s}$ is the compensation factor. It is always greater than 1.0 because $\rho \leq \rho_w$. Note that if the water inside the drum is the same cold water as in the gauge glass, i.e., $\rho_w = \rho$, then the indicated level is the same as the true level, $h = h_w$.

At the design pressure, the density of water and steam can be looked up from steam tables. For example, at atmospheric pressure, the densities of the water in the gauge glass (at 35 °C) and inside the drum (at 100 °C) are $\rho_w = 994\, kg/m^3$ and $\rho = 958\, kg/m^3$, respectively, and the density of the saturated steam is $\rho_s = 0.598\, kg/m^3$ (at 1 atm). The compensated level is then given by

$$h = \frac{994 - 0.6}{958 - 0.6}\, h_w = 1.038\, h_w. \tag{10.28}$$

An indicated level of 50% is actually 51.9% inside the drum. The density error is given by

$$\frac{h - h_w}{h} = \frac{\rho_w - \rho}{\rho_w - \rho_s} \times 100\% = \frac{994 - 958}{994 - 0.6} \times 100\% = 3.6\%. \tag{10.29}$$

At 40 bar pressure, the density ρ is 798.4 kg/m^3 for water, and 20.10 kg/m^3 for steam. We can easily verify with Eq. 10.29 that the density error can be as high as 20%. A 100% level in the drum may only show as 80% inside the glass gauge!

DP-Based Level Measurement

DP-based transmitters are most commonly used for drum level control. It is based on the difference in level between the water inside the drum and the external static column (Fig. 10.6). The level difference generates a pressure difference. This column is also called the *reference leg* (or *wet leg*), which is filled with sub-cooled water from the drum up to the condenser pot.

The drum level can be *inferred* from an online DP measurement. A higher DP indicates a lower water drum level, while a higher DP indicates a lower drum level.

Assume that the water in the reference leg has height H, and the steam density is ρ_s. The pressure on the drum side, LP, and the pressure on the reference leg side, HP, are given by

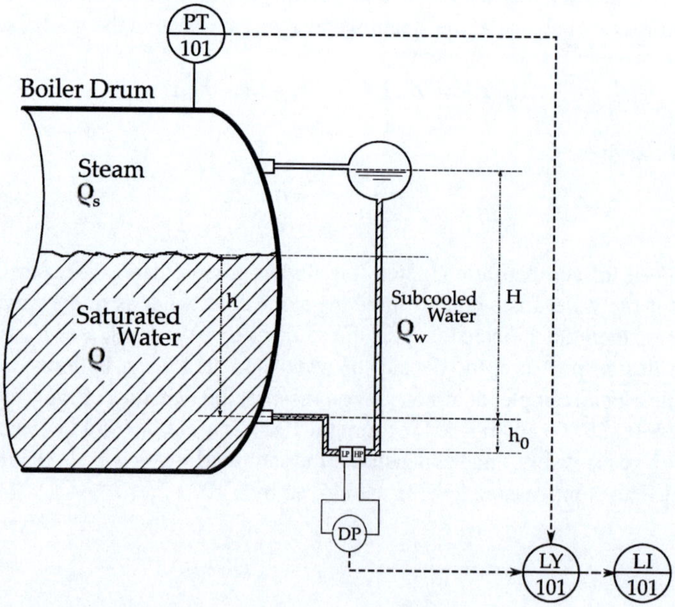

Fig. 10.6 Boiler drum level measurement: DP-based transmitter

$$\text{LP} = \rho\, g\, h + \rho_s\, g\, (H - h) + \rho_0\, h_0 \tag{10.30}$$

$$\text{HP} = \rho_w\, H + \rho_0\, h_0 \tag{10.31}$$

the pressure difference DP is then given by

$$\begin{aligned}\Delta P &= \text{HP} - \text{LP}\\ &= (\rho_w\, H + \rho_0\, h_0) - (\rho\, g\, h + \rho_s\, g\, (H - h) + \rho_0\, h_0)\\ &= (\rho_w - \rho_s)\, g\, H - (\rho - \rho_s)\, g\, h.\end{aligned} \tag{10.32}$$

When the drum is filled with the same liquid (e.g., cold water) as in the reference leg (i.e., $h = H$, $\rho = \rho_w$), the differential pressure ΔP should be zero.

From Eq. 10.32, the water level h is easily derived from ΔP as

$$h = \frac{(\rho_w - \rho_s)\, g\, H - \Delta P}{(\rho - \rho_s)\, g} \tag{10.33}$$

$$= \frac{\rho_w - \rho_s}{\rho - \rho_s} H - \frac{\Delta P}{(\rho - \rho_s)\, g}. \tag{10.34}$$

10.5.3 *Implementation and Operation*

The water and steam density is a function of the pressure inside the Drum. Therefore, density compensation is mainly pressure compensation. The densities of the steam ρ_s and water ρ are a function of the drum pressure. They are assumed to be known at design time and remain unchanged after calibration.

However, the drum pressure changes significantly during operation, especially during startup, shutdown, and major upsets. Real-time online compensation based on pressure changes becomes mandatory. Typically, the drum pressure measurement is provided, and an online calculation of the compensation factor is needed. See Fig. 10.6, where the pressure reading `PT-101` is used to look up the water and steam's actual density. The calculation is performed in the calculation block `LY-101`.

The water and steam density can be looked up from the *steam tables* widely available in reference books or the internet. Table 10.13 shows some sample values for the density of water and steam under several different pressures.

10.6 Internal Reflux Versus External Reflux

Reflux flow is a crucial manipulated variable in distillation column control and directly affects the product quality. However, it is also challenging to use reflux to achieve the desired control performance because of the strong correlation between energy and mass.

Table 10.13 Density of water and steam at different pressures

Drum pressure	Temperature	Water density ρ_w	Steam density ρ_s
bar(a)	°C	kg/m^3	kg/m^3
1.000	120.2	958.6	0.59
5.000	151.8	915.3	2.67
10.000	179.9	887.1	5.15
15.000	198.3	866.6	7.60
20.000	212.4	849.8	10.05
25.000	224.0	835.1	12.52
30.000	233.9	821.9	15.01
40.000	250.4	798.4	20.10

Internal reflux is a combination of mass transfer and heat transfer and often proves to be a better handle than external reflux for column control.

10.6.1 Application Background

A distillation column is a standard unit operation involving multi-phase, multi-stage, and counter-current mass and heat transfer. For many columns, the reflux is used to control the top and upper tray temperature. The temperature control loop serves as the inner loop for a cascade control with top product composition control. See Fig. 5.17 for a typical fractionator column, where the reflux is responsible for the column overhead control.

Control with reflux can be challenging due to the strong interaction between vapor temperature and pressure in the column, determined by the heat duty change brought up by the reflux. Operators often break the cascade loop and directly manipulate the reflux flow, leading to inconsistent performance.

10.6.2 Theory and Design

Reflux provides down-flowing liquid throughout the rectification section and contacts the up-flowing vapor to achieve stage-by-stage equilibrium heat and mass transfer. The heat exchange and mass transfer are complicated and interact with each other. For this reason, control with external reflux is not accurate.

Heat duty is better than temperature as a measure of heat content. For sub-cooled reflux flow, the heat duty is defined as

$$Q = F_m \, C_p \, \Delta T \tag{10.35}$$

where Q (in Watt) is the heat duty or the total heat transferred, F_m (in kg/s) is the mass flow rate for the reflux flow, C_p is the heat capacity of the reflux fluid going through temperature change (at constant pressure), and ΔT is the temperature change in the reflux after entering the column.

As reflux enters the column, the mass/heat transfer and phase changes are complicated, and it is impossible to be described by flow or temperature alone. The reflux is typically sub-cooled at the condenser before entering the column. The degree of sub-cooling varies with the cooling medium, ambient air, or rainstorms and is a significant disturbance. *Internal reflux* is a derived variable that combines mass transfer and heat transfer and often provides a better indicator of the top product's purity than *external reflux* (Shinskey 1984).

The internal reflux is calculated as

$$R_i = R_e \left[1 + \frac{C_p}{\Delta H}(T_o - T_r) \right] \tag{10.36}$$

where

- R_i internal reflux,
- R_e external reflux,
- C_p heat capacity of liquid reflux,
- ΔH heat of condensation of overhead vapor,
- T_o overhead temperature,
- T_r reflux temperature, and

the heat of condensation $\Delta H \approx$ heat of vaporization of reflux. Overhead temperature T_o is used to approximate the temperature inside the column. The reflux temperature T_r is assumed to be available.

The internal reflux can then be used as a derived or inferred variable to control the overhead temperature or top product draw quality, depending on the control strategy. The internal reflux is a nonlinear function of flow and temperature and can be deemed a static inferential property (Sect. 5.3.1).

10.6.3 Implementation and Operation

The temperature change $\Delta T = T_o - T_r$ in Eq. 10.36 is assumed to be the difference in the reflux temperature before and after it enters the column. In practice, the temperature at the best location may not be measured, but the overhead temperature is almost always available. So using the overhead temperature as T_0 is a close approximation.

The heat capacity C_p and heat of vaporization H can be easily looked up based on the reflux fluid property. The values depend on the degree of sub-cooling of the reflux flow. As long as the reflux stays sub-cooled in one phase, it is acceptable to use a constant value for both. The exact value is not critical since feedback control can eliminate the error introduced by the C_p value inaccuracy.

10.7 Summary

The chapter presents several typical calculations that are commonly encountered in process control. Most of the calculations in this chapter can be regarded as special cases of static inferential properties discussed in Sect. 5.3:

1. Signal transformation and conditioning (Sect. 10.1): Signal transformation is ubiquitous in process control but often hidden in applications. Following the signal transmission path in the control loops, many types of signal transformations or processing occur. It is essential to know how these transformations can impact control performance.
2. Data quality and bad data handling (Sect. 10.2): Reliable measurement is a necessary condition for good control. It is crucial but challenging to monitor the data quality in real time and assess its usability for closed-loop control. For example, if a measurement signal is deemed bad or cannot be trusted, the control loop must take appropriate mitigation measures.
3. Flow compensation and conversion (Sect. 10.3): Flow is not directly measurable and is always deduced from other measurable quantities such as pressure or velocity based on a known correlation. The correlation is established under a specific operating condition. Many times, compensation must be performed to correct or improve the measurements due to the change in operating conditions. Flow compensation is a requirement and technique that is widely used but often incorrectly implemented.
4. Temperature compensation with pressure (Sect. 10.4): Vapor or gas temperature changes with pressure. In many applications, the measured temperature must be compensated with pressure to arrive at the correct temperature value.
5. Level compensation with density (Sect. 10.5): Steam/water interface level is heavily influenced by the relative density of the steam and water. Inaccurate level measurement is one of the key reasons for poor control in boiler drum level. Online compensation of the level measurement is essential, especially for high-pressure steam-generating boilers.
6. Simple inferential properties such as internal reflux (Sect. 10.6): Similar to pressure-compensated temperature, in an actual application, many process variables correlate and interact with each other, and a derived variable that “links” or “lumps” the variables together can provide cleaner and improved process dynamic responses.

References

ASME (2019) BPV complete code – 2019, ASME Boiler and pressure vessel code. American Society of Mechanical Engineers

Shinskey FG (1984) Distillation control: for productivity and energy conservation. McGraw-Hill

Stephan D (2013) How to compensate pressure changes in temperature control. Process-Worldwide.com. https://www.process-worldwide.com/how-to-compensate-pressure-changes-in-temperature-control-a-394278/

Towler G, Sinnott R (2012) Chemical engineering design - principles, practice and economics of plant and process design, 2nd edn. Butterworth-Heinemann

Yokogawa (2004) CS1000/CS3000 reference – function block details. Yokogawa, 11th edn

Chapter 11
Control Typicals: Equipment Control

There are all kinds of equipment in an operating plant, with pumps and compressors being the most common and essential. This chapter chooses ejectors (Sect. 11.1), centrifugal pumps (Sect. 11.2), and centrifugal compressors (Sect. 11.3) as typical equipment examples for detailed analysis and control design. Equipment is part of process flow, but equipment control has traditionally been the specialty of equipment manufacturers. With sufficient engineering insight into the working principle, the same structured process analysis (Chap. 6) and control design (Chap. 7) skills and methodology can be followed to produce a sound control solution, as demonstrated with the selected equipment in this chapter.

11.1 Ejector and Educer

Ejectors and educers have been widely used for the pumping and compressing of gas and liquids, respectively. They have a simple geometry, have no moving parts, and do not require mechanical or electric drivers; therefore, they are very compact and cost-effective in construction. They are also very reliable and efficient in operation (Berkeley 1958; Keenan et al. 1950; Liao 2008).

Ejector and educer are selected as typical equipment in this chapter because they are probably the simplest and most reliable liquid pumping and gas compressing equipment. However, their design and operation are backed by very sound theories. They serve as an excellent introduction to the pump and compressor discussions in the sections that follow. Ejectors and educers work on the same principle, so we will use a gas ejector as an example for our discussions in this section.

S. S. Niu and D. Xiao, *Process Control*, Advances in Industrial Control,
https://doi.org/10.1007/978-3-030-97067-3_11

11.1.1 Process Description and Analysis

In E&P operations, a gas ejector is commonly found in a glycol dehydration process (Campbell 2004) to compress the waste gas to a higher pressure to serve a useful purpose. High-pressure motive gas that might otherwise be throttled for use in an evaporator or heat exchanger can be used to actuate the ejector. For example, high-pressure natural gas at 6 bar or higher as motive gas can bring the pressure of the process gas at atmospheric pressure to one bar or higher.

Operating Requirements

A gas ejector can serve one or more of the following purposes:

1. Transport and compress a weight of process gas from the suction pressure to the exit pressure.
2. Throttle the high-pressure gas to the desired pressure.
3. Blend two gases to the desired proportion.

The operating requirements are typically specified on the ejector capacity, suction pressure, and discharge pressure, depending on the operating objectives.

Process Flow Analysis

A gas ejector consists of an actuating nozzle, a suction chamber, and a diffuser. The high-pressure *motive gas* (activating gas) is expanded from its initial pressure to a pressure equal to that of the secondary fluid. See Fig. 11.1 for typical construction.

The process gas enters the suction side and becomes entrained in the motive gas. The high-pressure motive gas elevates the pressure of the process gas and exits the ejector mixed as the discharge gas. Inside the ejector, the motive gas induces a region of low pressure and high-velocity gas flow and causes the process gas to

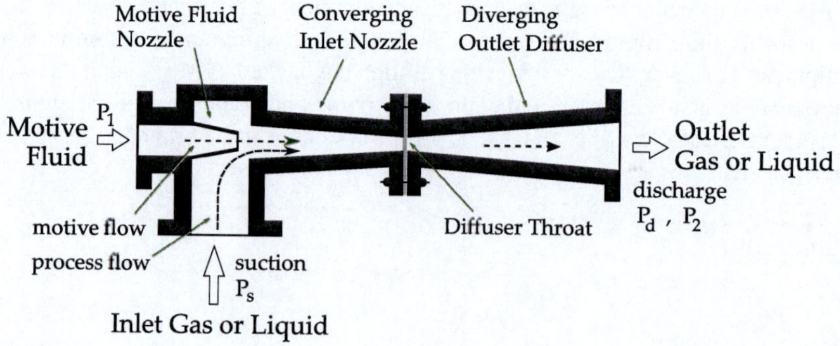

Fig. 11.1 A gas ejector

become entrained and mixed with the motive gas. During the mixing process, the actuating fluid is retarded, and the secondary fluid is accelerated. As the mixture enters the diffuser, it is compressed to the exit pressure by rapid deceleration.

Ejectors can be used singly or in series to create a wide range of vacuum conditions.

Supply and Demand Analysis

As usual, the supply and demand analysis needs to include the entire process surrounding the ejector operation, from the source of supply to the final destination of the demand. In a simple ejector operation, there are two inlet flow streams and one exit flow stream:

1. The process gas flow comes from an operating unit such as a glycol dehydration unit, a vacuum deaerator, and a flash cooler, so the operation is typically supply-driven by the process gas production.
2. The motive gas compresses the process gas and is thus demand-driven by the process gas.
3. The exit gas goes to another ejector or a downstream operation unit and is supply-driven. The downstream operation must have full capacity to consume the exit gas and absorb its fluctuations.

A recycle flow path is sometimes provided to facilitate capacity control.

The *potential* measurement instruments and valves are appropriately placed in line with this supply and demand model. See Fig. 11.2 for the potential valves and measurements and their locations (see guidelines in Fig. 6.2). A recycle line from discharge to process gas inlet is sometimes used to help with the turndown operation.

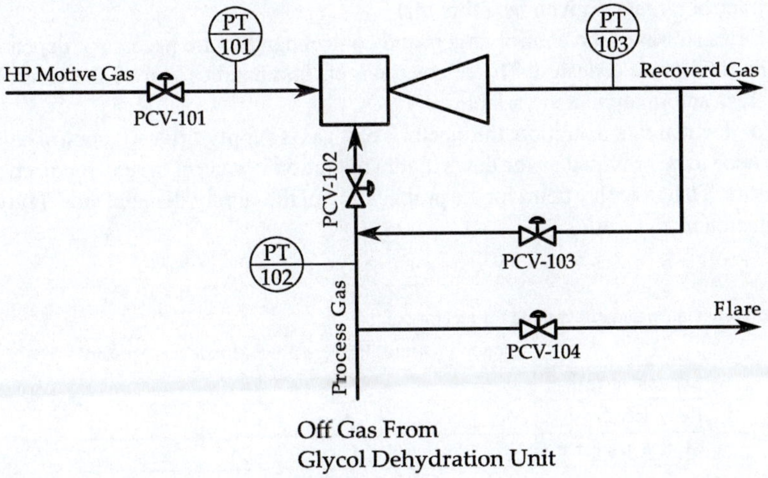

Fig. 11.2 Instrumentation for an ejector

The relationship between the flow streams is given by

$$[\text{exit flow}] = [\text{motive gas flow}] + [\text{process flow}] + [\text{recycle flow}]. \quad (11.1)$$

Not all valves are needed for all applications. Which one to be in the final control design is a joint decision by process engineers and process control engineers in early process design.

Cause and Effect Analysis

The ejector capacity is usually the key control target and is thus a controlled variable. The motive gas pressure may need to be controlled depending on the supply pressure and whether the ejector operates at a fixed or variable capacity. The suction pressure may be a protective control target, depending on the process configuration. The discharge pressure affects the ejector stability and may need to be a control target.

The control handles for ejector control include one or more of the following:

1. The suction valve `PCV-102`: The suction valve is primarily for capacity control.
2. The recycle valve `PCV-103`: The recycle valve adjusts the recycle flow from discharge to suction side to help capacity control.
3. The flare valve `PCV-104` prevents excessively high suction pressure.
4. For variable capacity ejectors, the pressure of the motive gas is often used as one additional handle for process control, and a throttling valve `PCV-101` is thus provided.

There are three pressures of interest: suction pressure of the process gas `PT-102`, inlet pressure of the motive gas `PT-101`, and the discharge pressure of the mixed exit gas `PT-103`. The full cause and effect table with all the potential control handles and control targets is given by Table 11.1.

Only a subset of the control targets and control handles are necessary, depending on the final process design. The selected targets and handles will be the controlled variables and manipulated variables.

For downstream operation, the ejector's exit gas is supply-driven; a control handle may need to be provided by the downstream operation to control the ejector discharge pressure. That is a relay point for the propagation of the supply/demand model further downstream.

Table 11.1 Cause and effect table for ejector control

	Suction pressure	Process flow	Motive gas pressure
Suction valve ↑	↗	↗	–
Recycle valve ↑	↗	↘	–
Motive gas valve ↑	↘	↗	↗

Dynamic Response Analysis

A key concept of ejector operation is *critical flow* and *critical condition* (Brannan 2018). When a flowing fluid at a given pressure and temperature passes through an ejector (or any constriction such as an orifice or nozzle in a pipe) to a lower pressure environment, the velocity of the fluid increases. The flow rate increases as the upstream pressure increases or the downstream pressure decreases. However, when the pressure ratio between the upstream pressure and downstream exceeds a specific limit, the velocity of the fluid would reach sonic speed, and the flow becomes the so-called *critical flow* or *choked flow*, which indicates that the *velocity is choked or limited.*

The volumetric flow is the product of gas velocity and cross-sectional area A; thus, the volume flow becomes constant once the flow is *choked* at sonic speed:

$$F_v = A \cdot v \tag{11.2}$$

where F_v is the volumetric flow, A is the cross-section area of exit, and v is the flow velocity.

The flow at the critical condition is called the *critical flow*. Assuming single-phase ideal gas behavior, the pressure ratio at which the flow becomes choked is given by

$$\frac{P_2}{P_1} = \left(\frac{2}{\kappa+1}\right)^{\frac{\kappa}{\kappa-1}} \tag{11.3}$$

where P_1 is the upstream (motive gas) pressure and P_2 is the downstream (exit gas) pressure; $\kappa = C_p/C_v$ is the ratio of heat capacity, with C_p and C_v as the heat capacity at constant pressure and constant volume, respectively. For most gases, the value of κ ranges from 1.09 to 1.41. For natural gas with $\kappa \approx 1.3$, a pressure ratio of about 0.5 is required for the gas to reach critical flow (Campbell 2004).

Although the volumetric flow is constant under critical conditions, the choked mass flow is not. It is a function of the gas density that varies with pressure and temperature. The choked mass flow rate is given by Perry and Green (1984):

$$F_m = C_d\, A \sqrt{\kappa\, \rho_1\, P_1 \left(\frac{2}{\kappa+1}\right)^{\frac{\kappa+1}{\kappa-1}}} \tag{11.4}$$

where C_d is the discharge coefficient, ρ_1 is the gas density at absolute pressure P_1 and absolute temperature T_1 upstream. For an ideal gas, $\rho = \frac{P\,M}{Z\,R\,T}$. Note that the mass flow rate is primarily dependent on the cross-section area A of the nozzle throat and the upstream pressure P_1, and is weakly dependent on the upstream temperature T_1. The flow rate does not depend on the downstream pressure at all.

For completeness, the volumetric flow rate can be derived from Eq. 11.4 as

$$F_v = \frac{F_m}{\rho_1} = C_d\, A \sqrt{\kappa\, \frac{P_1}{\rho_1} \left(\frac{2}{\kappa+1}\right)^{\frac{\kappa+1}{\kappa-1}}} = C_d\, A \sqrt{\kappa\, \frac{Z\,R\,T}{M} \left(\frac{2}{\kappa+1}\right)^{\frac{\kappa+1}{\kappa-1}}} \tag{11.5}$$

where it is seen that the choked volumetric flow is independent of the upstream or downstream pressures. However, the choked velocity is a function of the gas property and the temperature.

An ejector can operate in two different modes because of choked flow which are fixed capacity and variable capacity:

1. A fixed capacity ejector typically runs at the sonic speed with the critical flow. The pressure of the motive gas does not affect the capacity as long as the pressure is above the minimum pressure required for stable operation.
2. Variable capacity ejectors typically operate at a subsonic speed with a relatively lower compression ratio than fixed capacity ejectors. For variable capacity ejectors, the pressure of the motive gas does affect the capacity. That is, the ejector's capacity is a function of both the pressure of the motive gas and pressure at the suction inlet.

Note that the capacity here refers to the exit gas flow, which combines the process gas flow and motive gas flow (Eq. 11.1). The exact rate of process gas flow is not easy to calculate. Also, note that the fixed capacity refers to the volumetric flow. The mass flow changes with the motive gas pressure even after reaching the critical flow condition (see Eq. 11.4).

Depending on the pressure ratio, the ejector can be running in different operating conditions:

1. Stable operation: For most ejectors, the pressure of the motive gas must be higher than a threshold to sustain the stable operation. The value of this minimum pressure is a direct function of the discharge pressure and also depends on whether the pressure is increasing from unstable operation region to stable region or is decreasing from stable region to unstable region.[1]
2. Nonlinear behavior: Before the pressure ratio reaches the critical flow threshold, a higher motive gas pressure will increase the discharge flow. After reaching the critical pressure ratio, a further increase in the motive pressure may decrease the process flow due to the choking effect of the excess gas in the diffuser throat, because the exit gas flow is the total of motive gas flow and process flow (Berkeley 1958). The motive gas pressure setpoint should be carefully selected.
3. Discharge pressure: Ejectors are sensitive to discharge pressure. The ejector may become unstable if the discharge pressure exceeds a high-pressure limit, and the relationship between process flow and suction pressure may change.

[1] The former is called "pickup" pressure, while the latter is called "break" pressure. The "pickup" pressure is always higher than the "break" pressure. Operation between the "pickup" pressure and the "break" pressure can be either stable or unstable depending on which direction the pressure is changing (Berkeley 1958).

11.1.2 Control Strategy

The control objective serves the operating objective, and the control strategy is subject to the process design and operating constraints. The available motive gas pressure, and downstream pressure, and capacity are all factors to consider.

Control Structure

The control requirements include both *normal regulatory control* and *abnormal protective control*; thus, a layered control design is recommended:

1. Normal regulatory control: Capacity control is the primary control objective in normal operation. Capacity control is achieved through suction pressure control as the capacity change is reflected in pressure change.
   The requirements for suction pressure control are determined by upstream operation. Some process operations may require that a constant suction pressure be maintained, while most others do not.
2. Abnormal protective control: Depending on the operating mode (fixed capacity versus variable capacity) and process configuration, protective control against high and low pressures may be needed (see control schematics in Fig. 11.2):
   a. Low-pressure protection: If the process unit that produces the off-gas is fully rated for vacuum operation, low-pressure protection is not needed, and the recycle valve `PCV-103` is not required. The suction side can float with the pressure of the process unit, even if it draws a vacuum.
   b. High-pressure protection: In case of process upset or ejector shutdown, the suction pressure may go excessively high. High pressure protection should be provided. `PCV-104` is provided to send the excess gas to flare.
   c. Motive gas pressure: If the ejector operates in fixed capacity mode, low-pressure protection on the motive gas inlet may be required to ensure that the pressure ratio of the ejector is above the critical flow threshold. In the case of a variable capacity ejector, the pressure of the motive gas needs to be controlled for stable operation of the ejector.

11.1.3 Functional Design

Figure 11.3 is a typical control scheme for fixed capacity ejectors where the pressure of the motive gas is controlled by `PC-101` at the desired pressure to maintain stable ejector operation, and the suction valve is used to control the suction pressure via `PC-102`. If a high turndown ratio is required, the discharge recycle line can be added. The suction pressure control `PC-103` can manipulate the recycle valve to maintain

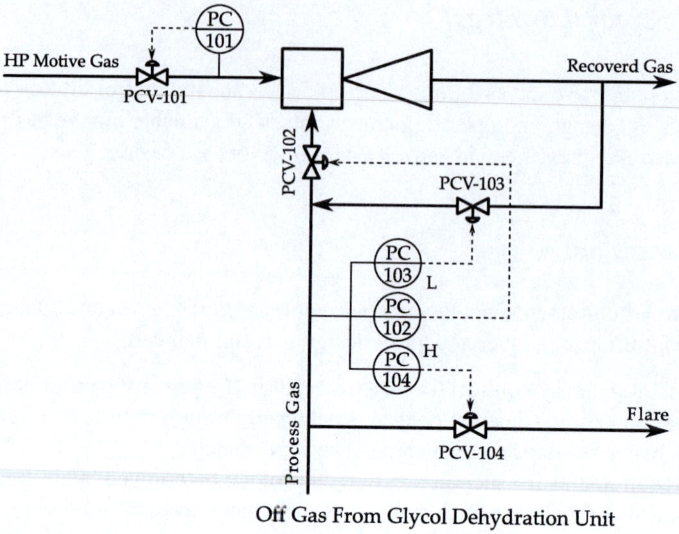

Fig. 11.3 Control design for an ejector: fixed capacity

the desired turndown ratio. The high-pressure protective controller PC-104 provides high-pressure protection.

Figure 11.4 is a control scheme for variable capacity ejectors. In comparison with Fig. 11.3, the recycle line is removed. The pressure PC-101 of the motive gas is used under cascade control to achieve the desired turndown ratio since recycling is inefficient. Note that a minimum pressure (via low setpoint limit or minimum valve opening) should be imposed on the motive gas to ensure that the ejector operates in the stable region.

If the suction side is fully rated for vacuum operation, the control scheme can be simplified by eliminating the suction pressure controller PC-103. The high-pressure protective controller PC-104 is still needed to flare the excess gas if the suction pressure does go too high. See Fig. 11.5.

11.1.4 Implementation and Operation

All controllers are either standalone PID controllers or standard complex PID loops. The configuration of the final control scheme can follow the checklist in Table 3.9 and is omitted here.

Multiple ejectors can be used in series to boost the pressure. The process and control design can be analyzed in the same way.

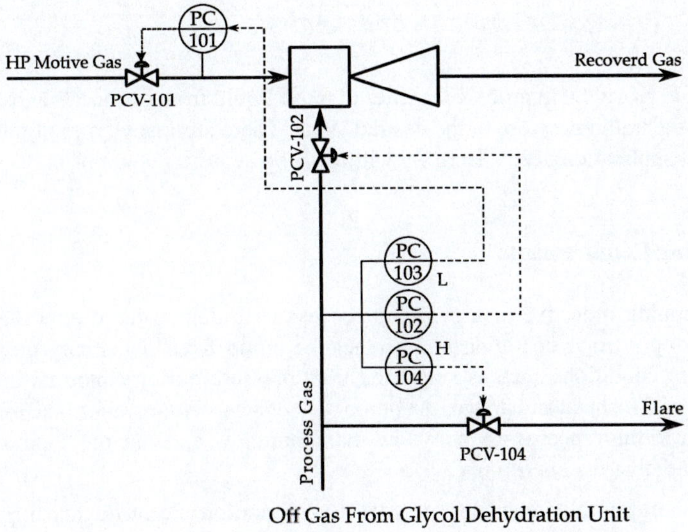

Fig. 11.4 Control design for an ejector: variable capacity

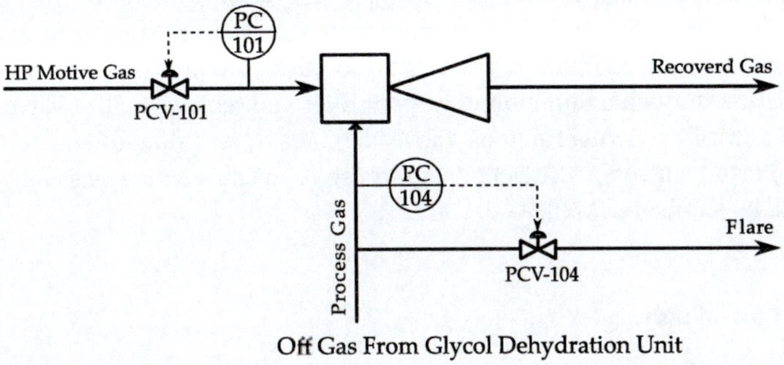

Fig. 11.5 Control design for an ejector: fully rated for vacuum

11.2 Centrifugal Pump

Pumps are extremely widely used. A 2001 report by the United States Department of Energy claims that "... pumping systems account for nearly 20% of the world's electrical energy demand and range from 25 to 50% of the energy usage in certain industrial plant operations" (DOE 2001). Therefore, the reliable and optimal operation of pumps is of great economic value.

11.2.1 Process Description and Analysis

Pumps are prevalent in process facilities to move liquid from one location to another or raise the liquid pressure to the desired value. There are many types of pumps for different application needs. Here we will focus on centrifugal pumps only.

Operating Requirements

The operating objective of a pumping process is to deliver the desired *head* and *flow* during normal operation and protect the pump from potentially unsafe and inefficient conditions such as surge and high pressure during abnormal operating operations. The head and flow of the pump define the *operating point*. The minimum speed, maximum speed, minimum flow, maximum flow, and maximum motor power typically define the *operating envelope*.

1. Minimum continuous stable flow (MCSF):[2] Operation of centrifugal pumps below their minimum flow requirements is the primary cause of premature pump failure. Hydraulic instability occurs at low flows and can cause surge, cavitation, and excessive vibration in the pump. The type and extent of damage to the pump depend on how long the low flow persists and the magnitude of the generated forces and vibrations. The surge phenomenon is a common threat to all centrifugal and axial rotating equipment. The typical solution is to provide automatic bypass circulation through a minimum flow controller and recycle control valve.
2. Operation of centrifugal pumps above their maximum flow tolerance (end-of-curve) results in very inefficient operation and can cause vibrations and typically should be avoided as well.

Process Flow Analysis

Figure 11.6 shows a typical pumping process. The center of the process is a centrifugal pump, where the mechanical energy from the driver (e.g., a motor) is injected into the liquid to increase its potential and kinetic energy. The potential energy is the increase in the pump head, while the kinetic energy is reflected in the change of the pump flow.

Pumps are generally motor-driven and operate at either fixed speed or variable speed. Based on operating needs, multiple pumps can operate in parallel or serial. Therefore, depending on the type of pump and the operating requirements, the piping and instrumentation diagrams (P&ID) can be very different and complex.

---

[2] There are three types of minimum flows for a centrifugal pump: minimum continuous thermal flow (MCTF), intermittent minimum flow, and minimum continuous stable flow (MCSF). By minimum flow, we typically refer to the minimum continuous stable flow (MCSF). MCSF is a value that can range from roughly 10% to 80% of the best efficiency point flow depending on pump size and type, operating speed, impeller suction geometry, liquid density, and other factors.

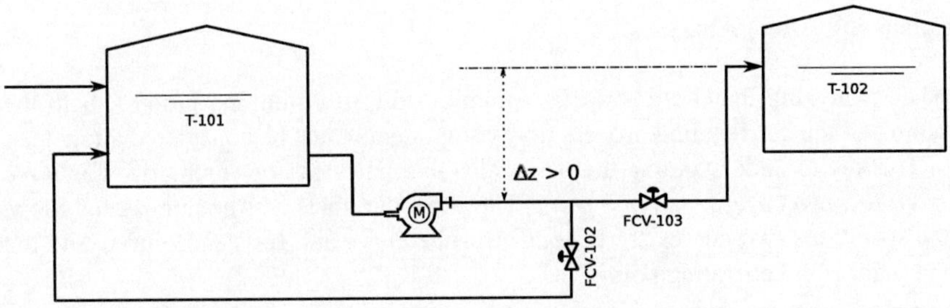

Fig. 11.6 A typical pumping process, with difference in elevation

Supply and Demand Analysis

A pumping process can be as simple as a single pump with one stream coming in and going out. It can also be as complicated as multiple pumps operating in parallel or serial with multiple streams merging and branching out. The supply and demand model can also vary from supply-driven, demand-driven, to mixed.

Figure 11.7 is a simple process flow configured as a supply-driven operation with a loop-back. The same process flow can easily be configured as a demand-driven operation by relocating the measurements and control valves.

The loop-back is a recycle flow from pump discharge to the tank to ensure that the pump flow is above the minimum flow requirement. Obviously,

$$[\text{pump flow}] = [\text{process flow}] + [\text{recycle flow}]. \tag{11.6}$$

The recycle flow remains closed when the pump flow is above the minimum flow limit. In case the process flow falls below the minimum flow limit, e.g., during startup or turndown operation, the recycle valve opens, and the recycle flow supplements the process flow to bring the pump flow above the minimum flow limit. With a properly sized recycle line and valve, the pump operation can achieve 100% turndown, i.e., full recycle and zero process flow.

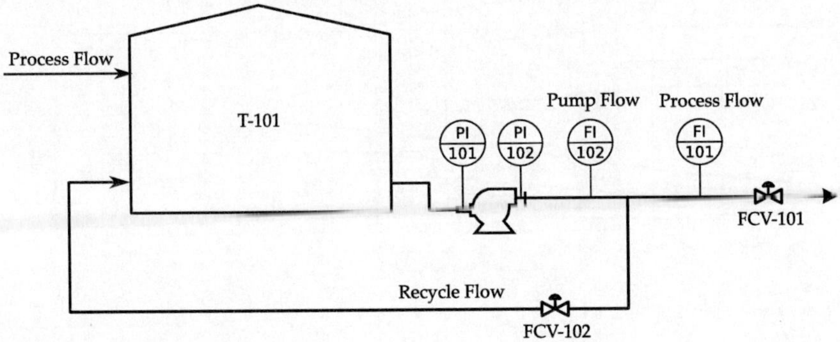

Fig. 11.7 Flows in pumping process

Cause and Effect Analysis

When analyzing the cause and effect relationship of a pumping process, both the pump and the surrounding process flow components must be considered altogether.

The key to understanding the pump characteristics can be summarized as *three variables*, *two curves*, and *one point*. The three variables are the pump head, flow, and speed; the two curves are the performance curve and resistance curve, and the one point is the operating point:

1. Head: The pump head is the liquid force measured by the height of elevation, typically by meters or inches of the liquid.
2. Flow: Pump flow is the liquid volume flow through the pump. The pump flow rate is also called the pump capacity.
3. Speed: The pump speed is the pump shaft speed measured in revolutions per minute (rpm).

The relationship between the head, flow, and speed defines the pump operation. The relationship can be graphically represented by the so-called *pump performance curves* or simply the *pump curves*. An example of the pump curve is given in Fig. 11.8.

The head–flow relationship at a constant speed is represented by one head–flow pump curve. If the pump speed can vary, then there would be a range of parallel head–flow pump curves. As pump speed changes, the operating point moves up and down, following roughly a parabolic path. This trajectory is called the *system resistance curve*.

The pump curves do not change once the pump is built. However, the resistance curve can move or be moved by changing the process flow components such as a throttling valve position or recycle flow. These adjustable components serve as the control handles to influence the operating conditions:

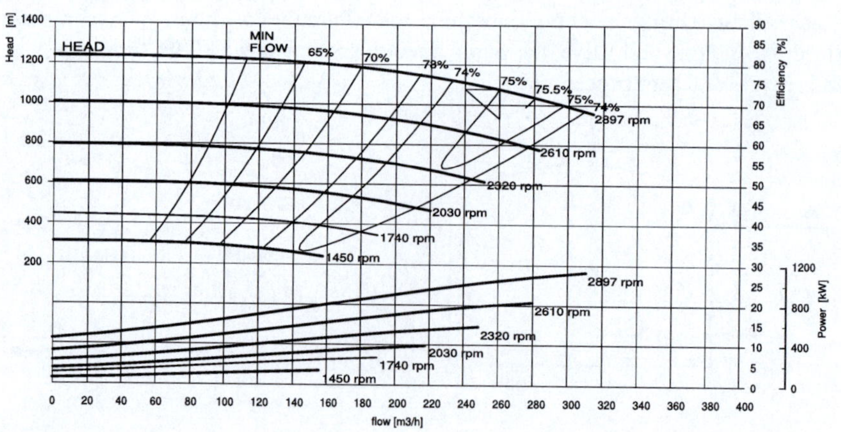

Fig. 11.8 A typical pump curve

Table 11.2 Cause and effect table for a centrifugal pump

	Pump head	Pump flow	Process flow	Recycle flow
Pump speed ↑	↗	↗	↗	–
Discharge valve ↑	↘	↗	↗	–
Recycle valve ↑	↘	–	↘	↗

1. Discharge valve: For fixed-speed pumping, a discharge control valve is typically used to adjust the operating point.
2. Recycle valve: The recycle control valve is typically reserved to maintain the minimum pump flow.

The pump performance curve determines where the operating point *can* be, while the system resistance determines where the operating point *should* be; the former is determined by the mechanical characteristic of the pump while the latter is desired by process operation. Thus, the operating point is always at the intersection of the pump performance curve and the system resistance curve. By changing the pump speed and system resistance, the operating point can be moved to anywhere on the performance map (subject to constraints).

The controlled variable is typically the pump flow (pumping capacity). The manipulated variable is the pump speed in a variable-speed pump or discharge throttling valve for a fixed-speed pump. The cause and effect analysis of a pumping process can be simplified to four control targets and three control handles. Depending on the application, all the control handles are not necessary.

Dynamic Response Analysis

The power, head, flow, and speed are related by the so-called pump *affinity laws*, which are more popularly known as the *fan laws*. The fan laws state the following:

1. Change in pump flow is directly proportional to change in velocity or speed.
2. Change in pump head is directly proportional to the square of the change in velocity or speed.
3. Change in Power is directly proportional to the cube of the change in velocity or speed.

In equation format,

$$\frac{W_2}{W_1} = \left(\frac{N_2}{N_1}\right)^3, \quad \frac{H_2}{H_1} = \left(\frac{N_2}{N_1}\right)^2, \quad \frac{F_{v,2}}{F_{v,1}} = \frac{N_2}{N_1} \tag{11.7}$$

where W is the *brake horsepower* (BHP), N is the speed, H is the pump head, and F_v is the volume flow rate. Subscripts "$_1$" and "$_2$" denote two operating conditions.

The affinity laws imply that if the speed of a pump is increased by 10% (i.e., $N_1/N_2 = 1.10$), then the changes in flow, head, and power are given as follows:

1. The volumetric flow increases by 10%: $F_{v,2}/F_{v,1} = 1.10$.
2. The head increases by 21%: $H_2/H_1 = 1.1^2 = 1.21$.
3. The power consumption increases by 33%: $W_2/W_1 = 1.1^3 = 1.33$.

The *best efficiency points* (BEP) are the flow rates at which the pump unit's combined hydraulic losses are the lowest. Moving the operating point along the system resistance curve does not cause drastic changes in pump efficiency. On the other hand, moving the operating point along the performance curve can cause significant changes in pump efficiency. See the pump efficiency curves in Fig. 11.8.

Capacity control is usually achieved through pump speed or via a discharge valve; see Table 11.3. In specific applications, the capacity control may be through discharge pressure control or direct flow control. One decisive factor is the dynamic characteristics. For example, a flat performance curve results in a low process gain for pressure change and makes the capacity control a challenge. It is typically a joint decision between process engineering, rotating equipment, and process control to determine which control targets to use the controlled variables and which control handles to use as manipulated variables.

For completeness, the performance curves from pump manufacturers typically include the following information, as shown in Fig. 11.8. They are all important considerations for abnormal protective control:

- Head versus mass or volume flow.
- Efficiency versus mass or volume flow.
- Brake horsepower (BHP) versus mass or volume flow. Brake power is the energy needed to pump the liquid.
- Best efficiency point (BEP).
- Minimum flow limit.
- Choked flow limit.
- Operating limits: Minimum speed and maximum speed.
- Operating limits: Minimum flow (surge line) and maximum flow (end of curve line).

11.2.2 Control Strategy

Centrifugal pumps are widely used in many industries and process operations. The operating objectives and requirements can significantly vary. It is impossible to cover all the application scenarios. As part of process control, pump control must be considered within the overall control strategy. The critical connection with the plant-wide control is the supply and demand balancing, which may affect the supply-driven or

Table 11.3 Cause and effect table for pump capacity control

	Discharge pressure	Pump flow
Pump speed ↑	↗	↗
Discharge valve position ↑	↘	↗

demand-driven control. The best approach is to follow the systematic methodology discussed in Chap. 7 to arrive at the best fit-for-purpose design. Here, we will only present some typical control designs.

Control Objectives

The main objective of pump control is to provide the desired throughput (capacity) during normal operation, and keep the pump operation within the operating envelope during abnormal operating conditions.

Control Structure

The control design follows the layered approach of *normal regulatory control* and *abnormal protective control*. Equation 11.8 provides a list of the full scope of pump control and protection. Not all features are required for all applications:

- Pump Control
  - Regulatory Control
    - Capacity Control
      - Fixed Speed Pump
      - Variable Speed Pump
    - Load Balancing
      - Pumps in Parallel
      - Pumps in Series
  - Protective Control
    - Envelope Protection
      - Minimum Flow (Surge)
      - Maximum Flow (End of Curve)
      - High Speed
      - Low Speed
    - Constraint Handling
      - Discharge Pressure
      - Motor Power
  - Sequential Control
    - Startup Sequence
      - Cold Startup
      - Hot Swapping
    - Shutdown Sequence
      - Emergency Shutdown
      - Planned Shutdown
  - Instrumented Safeguarding

(11.8)

1. Normal regulatory control: Normal regulatory control is typically about quality and capacity. There is no quality specification for pumping operation, so the only normal regulatory control requirement is pump capacity control.
   For fixed-speed pumps, capacity control is typically achieved by discharge throttling with a control valve. Throttling the discharge value increases the discharge pressure and causes the operating point to move to the left along the pump curve.

That is, higher pressure leads to lower flow and vice versa.
However, valve throttling results in permanent pressure losses. The more efficient way is to use a variable-speed pump with which the pump capacity is adjusted by varying the pump speed. Reducing the pump speed will cause both the head and flow to decrease.
The recycle flow can also be used to adjust the pump capacity, although it is very inefficient.

2. Abnormal protective control: The objectives of protective controls are to protect both the pump and the process. Protective controls for the process depend on the overall process flow, and thus are application-specific. Typical pump protection includes minimum flow (surge) control and maximum flow (end-of-curve) control:

    a. Minimum flow: Centrifugal pump suffers from the damaging phenomenon of *surge* when the flow throughput is below the minimum flow requirement. The minimum flow is typically maintained by recirculating a portion of the process flow back to the suction side to prevent surge via a simple PID controller. For a fixed-speed pump, the minimum flow limit is a fixed value, and the controller has a fixed setpoint. For a variable-speed pump, however, the minimum flow value is different at different speeds, and thus a more elaborated control scheme is needed for efficient control.
    b. Maximum flow: The maximum flow protection is also called the *end-of-curve protection*. The pump is designed to operate in a specific range of back pressures. In real applications, the discharge pressure may be different during startup and at different operating conditions. If the back pressure falls too low, *i.e.*, the system resistance is too little, it could drive the pumps to run at the end of the performance curve or beyond and results in very inefficient operation and even potential damage to the pump. See Fig. 11.8. End-of-curve control is typically achieved by reducing the pump speed or throttling a control valve on the discharge.

3. Sequential Control: Pump startup is a routine operation and sometimes involves risks in itself. An automated startup sequence is desired to ensure a consistently safe and smooth startup of the pump.

11.2.3 Functional Design

The functional design of process control varies with process design. This section provides several typical designs to illustrate the general design principle.

Capacity Control

Pump capacity control is to maintain the operating point at the desired flow and head. The pump discharge flow control usually serves as the capacity control. From the cause and effect table (Table 11.2), there are two common approaches:

1. For variable-speed pumping, manipulating the pump speed to control the pump capacity (flow).
2. For fixed-speed pumping, throttling the discharge valve to change the pump flow.

Figure 11.9 is a typical control design for a supply-driven fixed-speed pumping process. `FC-101` is the process flow, and `FC-102` is the pump flow. The control target for the flow is usually supplied by the tank level control for supply-driven operation. If the operation is demand-driven, which is less common, the flow set-point may be supplied manually by the operator or automatically by a downstream controller.

The performance curve in Fig. 11.8 correlates the pump head, flow, and speed. If two are specified, the third is determined. A higher tank level demands a higher discharge flow, and a high flow will result in a lower head (and thus the discharge pressure).

For a variable-speed pump, the pump speed is the manipulated variable for capacity control. Figure 11.10 shows a typical control design. With a low tank level, the level controller demands a lower flow rate. The flow controller will decrease the pump speed to achieve a lower pump flow (also a lower head). If the required flow rate is below the pump's minimum flow limit, the recycle valve will open to increase the recycle flow, helping reduce the process flow further while maintaining the pump flow above the limit. The process flow, recycle flow, and pump flow are related by Eq. 11.6 and illustrated in Fig. 11.11a.

From the pump performance curves in Fig. 11.8, the minimum flow limit decreases with speed. The amount of recycle flow also decreases. This change is illustrated in Fig. 11.11b. Since recycling is a very inefficient operation, the recycling flow should be reduced to the minimum possible.

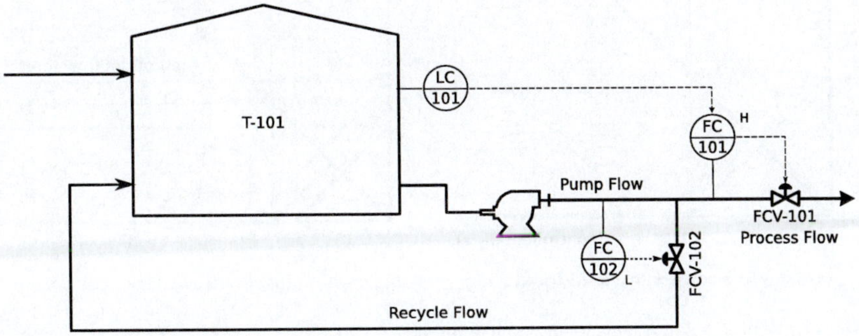

Fig. 11.9 Capacity control for fixed-speed pumping

Minimum Flow Control

One of the leading causes of pump damage is running the pump below the minimum flow limit. Maintaining the pump flow (not the process flow! see Fig. 11.11) above the minimum flow limit is mandatory for most large-capacity centrifugal pumps. The standard approach is to recirculate a portion (or all if necessary) of the discharge flow back to the suction side. See the flow controller FC-102 in both Figs. 11.9 and 11.10. The minimum flow controller FC-102 is a standalone protective controller discussed in Sect. 4.1.3.

The challenge for minimum flow control, however, is in determining the flow control setpoint. There is one single performance curve for a fixed-speed pump, and thus one fixed minimum flow limit value, as shown by the point marked as MIN

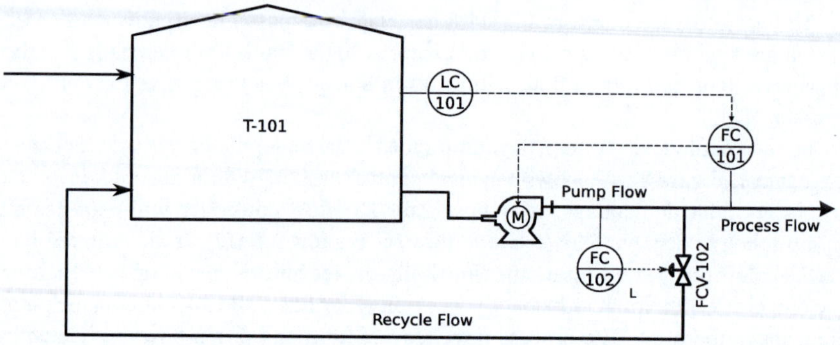

Fig. 11.10 Capacity control for variable-speed pumping

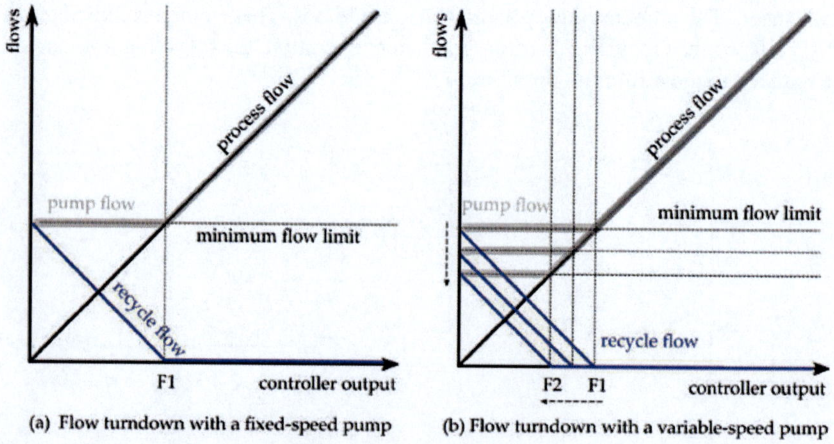

Fig. 11.11 Process flow, recycle flow, and pump flow

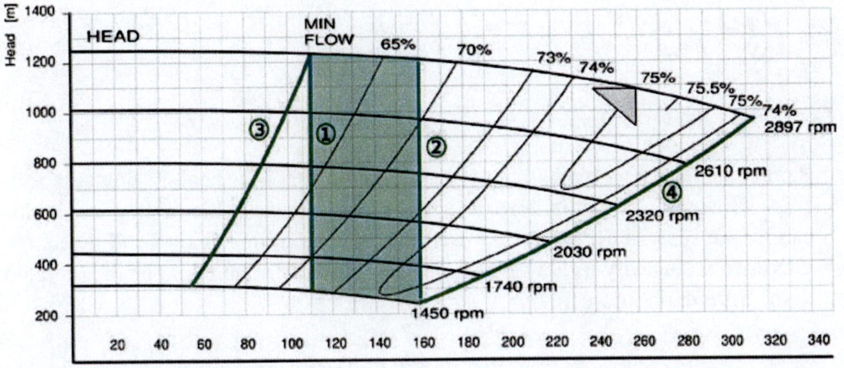

Fig. 11.12 Head–flow pump curve

FLOW in Fig. 11.12. This minimum flow value serves as the setpoint for the flow controller.

However, for a variable-speed pump, there is a performance curve for every speed, and the minimum flow limit is a function of the pump speed. The parabolic line ③ in Fig. 11.12 indicates the minimum flow values at different speeds. Suppose a fixed flow setpoint is used, indicated by the vertical dashed line ①. In that case, the area between line ① and ③ will not be reachable by the pump operation, because if the pump needs to operate at the flow rate lower than indicated by line ①, then recycle starts from line ① instead of ③. See Fig. 11.11b for the two scenarios where the recycle flow F1>F2. This unnecessarily high setpoint is a waste of energy since recycling is very inefficient for capacity control.

The approach to address this inefficiency is to use a variable flow control setpoint as indicated by line ③. There are multiple ways of achieving this:

1. Variable setpoints for the flow controller can be used to maintain the minimum recycle flow. It is easy to derive a function that relates the flow rate with the pump head or speed from the performance curve. The pump affinity laws state that the pump head change is a quadratic function of the change in the flow rate. That is,

$$\frac{H}{H_0} = \left(\frac{Fv}{F_{v,0}}\right)^2 \quad \mapsto \quad H = \frac{H_0}{F_{v,0}^2} \cdot F_v^2 = K \cdot F_v^2. \tag{11.9}$$

   In other words, the parabolic resistance curves in the head (H) versus flow (F_v) coordinate in Fig. 11.13a become a straight -line in the head (H) versus squared flow (F_v^2) coordinate in Fig. 11.13b.

   The line that links all the minimum flow points at the different pump speeds is called the *surge line*. A straight line approximation to the surge line is given by fitting Eq. (11.9) to the surge points. Picking one of the surge points (for example, 1220 m, 114 m^3/hr) as (H_0, $F_{v,0}$), then from Eq. (11.9), the minimum flow setpoint is derived as

$$H = \frac{H_0}{F_{v,0}^2} \cdot F_v^2 \quad \mapsto \quad F_v = \frac{F_{v,0}}{\sqrt{H_0}} \sqrt{H}$$
$$\approx \frac{114}{\sqrt{1220}} \sqrt{H} = 3.26\sqrt{H} \quad (\mathrm{m^3/hr}). \tag{11.10}$$

Pump manufacturers typically provide the pump curves as head-to-flow relationships. However, the pump head H is not directly measurable in practical applications, and Eq. 11.10 cannot be used online. The pump suction and discharge pressures are usually used in place of the pump head. The pump head is related to the differential pressure by

$$P_d - P_s = \rho\, g\, H \qquad (\mathrm{Pa} = \mathrm{kg/m^3} \cdot \mathrm{m/s^2} \cdot \mathrm{m}) \tag{11.11}$$

where P_s and P_d (in Pascal) are the suction and discharge pressure, respectively; H (in meters) is the pump head; ρ (in kg/m^3) is the density of the liquid being pumped. Assuming that the density of the liquid is 960 kg/m^3, then the setpoint for the minimum flow controller given by Eq. (11.10) becomes

$$F_v = \frac{F_{v,0}}{\sqrt{H_0}} \sqrt{H} = \frac{F_{v,0}}{\sqrt{H_0}} \sqrt{\frac{P_d - P_s}{\rho\, g}} = \frac{F_{v,0}}{\sqrt{H_0\, \rho\, g}} \sqrt{P_d - P_s} \tag{11.12}$$
$$\approx \frac{114}{\sqrt{1220 \times 960 \times 9.8}} \sqrt{P_d - P_s} = 0.0336\sqrt{P_d - P_s} \quad (\mathrm{m^3/hr}). \tag{11.13}$$

The pump speed can also be used to relate to the minimum flow limit. However, the online speed measurement is not as fast and reliable as the head measurement and is not recommended for real-time control.

2. Minimum flow control via a surge indicator. Another approach is to define an intuitive *surge indicator*. On the surge reference line, this surge indicator has a constant value regardless of pump speed. For any other points on the pump curves, there is a corresponding surge indicator that indicates its location relative to the surge reference line.
One definition of this indicator is the ratio of the slopes between the operating point and the surge reference line (see Fig. 11.13b). The slope in the Head–Flow coordinate is given by

$$S = H/F_v^2. \tag{11.14}$$

Assume that the slope of the surge reference line is $S_0 = H_0/F_{v,0}^2$, and the slope of the current operating point is $S = H/F_v^2$, then the ratio of slope between the surge reference line and the operating point is given by

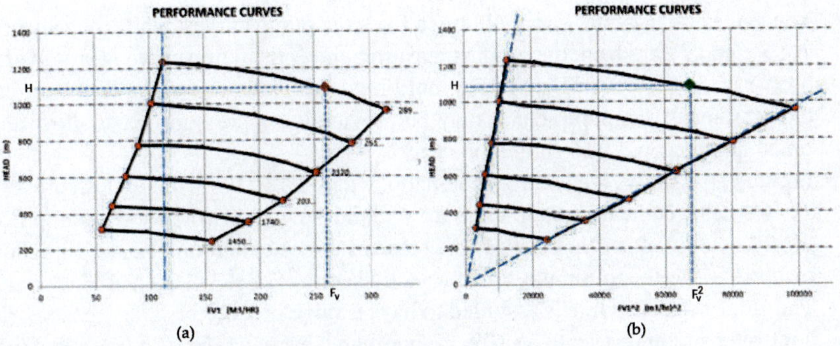

Fig. 11.13 Head–flow and head–flow-squared curve

$$K_S = \sqrt{\frac{S_0}{S}} = \sqrt{\frac{H_0/F_{v,0}^2}{H/F_v^2}} = \frac{\sqrt{H_0}}{F_{v,0}}\frac{F_v}{\sqrt{H}} \tag{11.15}$$

$$= \frac{\sqrt{H_0}}{F_{v,0}}\frac{F_v}{\sqrt{\dfrac{P_d - P_s}{\rho g}}} = \frac{\sqrt{\rho g H_0}}{F_{v,0}}\frac{F_v}{\sqrt{P_d - P_s}}. \tag{11.16}$$

Obviously, Eq. 11.16 is a different form of Eq. 11.12. The absolute value of K_S indicates the current operating point:

$$K_s \begin{cases} = 1.0 & \text{on the surge reference line} \\ > 1.0 & \text{on the right of the surge reference line} \\ < 1.0 & \text{on the left of the surge reference line.} \end{cases} \tag{11.17}$$

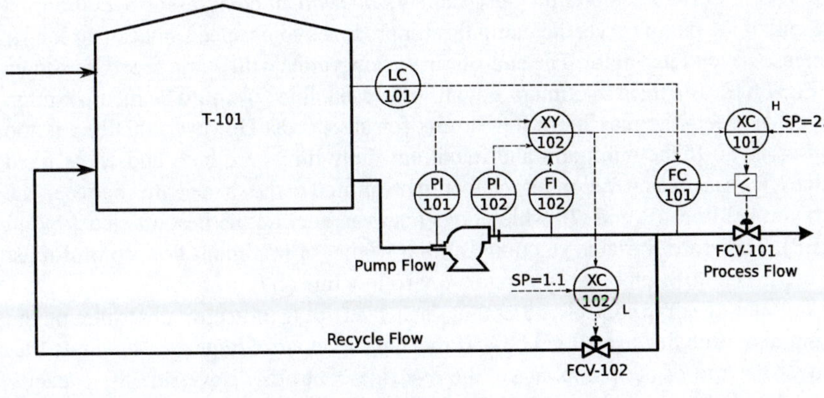

Fig. 11.14 Protective flow control for centrifugal pump

The values of H_0 and $F_{v,0}$ are obtained from the pump curves, while the values of P_d, P_s, and F_v are from the online measurement. If the pump curve is provided as head-versus-flow, then Eq. 11.15 should be used. However, in the less common circumstances, the pump curve may be provided as a pressure-versus-flow relationship, then Eq. 11.16 should be used.

Figure 11.14 shows the control schematic, where `PI-101` is the suction pressure P_s, `PI-102` is the discharge pressure P_d, and `FI-102` provides the pump flow F_v. The calculation block `XY-102` performs the calculation in Eq. 11.16, whose result is fed to the minimum flow protective controller `XC-102` as PV.

The surge reference line is assumed to have a value of 1.0. For safety reasons, the controller setpoint is set to be 10% higher than the surge reference line value, i.e., 1.10. The controller output is sent down to recycle valve `FCV-102`.

If the flow is measured with a DP-based flow meter such as an orifice plate, the differential pressure ΔP across the orifice is available as $F_v = C_v \sqrt{\Delta P/\rho}$ where C_v is the flow constant. The above formula (11.16) can be further simplified into

$$K_S = \frac{\sqrt{H_0}}{F_{v,0}} \frac{F_v}{\sqrt{H}} = \frac{\sqrt{H_0}}{F_{v,0}} \frac{C_v \sqrt{\dfrac{\Delta P}{\rho}}}{\sqrt{\dfrac{P_d - P_s}{\rho g}}}$$

$$= C_v \frac{\sqrt{g H_0}}{F_{v,0}} \sqrt{\frac{\Delta P}{P_d - P_s}} = C \sqrt{\frac{\Delta P}{P_d - P_s}}. \tag{11.18}$$

We will come back to this formula when we discuss the compressor anti-surge control in the next section.

End-of-Curve Control

The end of curve for a pump is also called *stonewall* in compressor operation. At the end of the pump curve, the pump flow approaches sonic speed and can no longer increase beyond this point. The end-of-curve flow varies with pump speed, as shown in Fig. 11.12. If a fixed maximum flow limit is used, line ② would be the maximum flow that meets the maximum flow limits for all speeds. However, this line is too conservative. If the minimum and maximum flow limits are both chosen as fixed values, the feasible operation region would be limited to the shaded area between the two vertical lines ① and ②, which does not even enclose the best efficiency point (BEP). Therefore, variable setpoints similar to that in minimum flow control must be used. The flow setpoints approximately follow line ④.

The end-of-curve controller `XC-101` is an overriding protective controller on the pump discharge flow; see Fig. 11.14. If the capacity control requires a flow rate that is over the end of the pump curve, the override control will override the capacity control and limit the flow to below the maximum flow limit.

The surge indicator calculated by Eqs. 11.16 and 11.18 produces a value of 1.0 for the surge line. The same equation would produce a value of approximately $\sqrt{10}$ at the end of the curve, consistent with the $\sqrt{10}$:1 turndown ratio of pressure measurement.

We will use the end-of-curve control to show an alternative procedure to calculate the surge indicator. Different engineering units for pressure and flow measurements are used on purpose in this example to demonstrate the unit conversion required in the calculation.

Example 11.1 Surge indicator calculation. Figure 11.14 is the pumping process schematic, and the pump curve is given in Fig. 11.12. The pressure measurements `PI-101` and `PI-102` are provided in kPa(g). The flow measurements `FI-101` and `FI-102` are provided as mass flow in kg/s. Now we use a simplified form Eq. 11.16 as

$$K_S = C\,\frac{F_m}{\sqrt{P_d - P_s}} \quad\mapsto\quad C = \frac{\sqrt{P_d - P_s}}{F_m} \quad \text{(with } K_S = 1) \tag{11.19}$$

where C is a constant to be determined. We have assumed that the surge points have a surge indicator value K_S of 1.0; we can readily determine the constant value C from the pump curve.

Let us choose a few surge points from the pump curve to calculate the C value. The pump curve is given as head (meter) versus volume flow (m^3/hr). The head and flow data must be converted to kPa(g) and kg/s to be the same as the online measurements. Conversion from head to differential pressure can follow Eq. 11.11, while mass flow and volume flow are related via

$$F_m = \rho\, F_v \qquad (\text{kg/s} = \text{kg/m}^3 \cdot \text{m}^3/s). \tag{11.20}$$

Assuming the liquid density is 960 kg/m^3, the raw data for the surge points are read from the pump curves. The converted data with the same engineering units as online measurements are listed in Table 11.4. The calculated values for the constant C with Eq. 11.19 are listed in the last column.

For safer operation, let us take the most conservative value of C from the last column, 3.42. The surge indicator is then given by

$$K_S = 3.42\,\frac{F_v}{\sqrt{P_d - P_s}}. \tag{11.21}$$

Table 11.4 Surge points from pump curves

	Pump head (m)	Volume flow (m^3/hr)	$P_d - P_s$ (kPa)	F_m (kg/s)	C from Eq. (11.19) ($\sqrt{\text{kPa}}$/kg/s)
Surge point 1	1220	114	11477.8	31.0	3.45
Surge point 2	1000	104	9408.0	28.3	3.43
Surge point 3	780	90	7338.2	24.5	3.50
Surge point 4	600	79	5644.8	21.5	3.49
Surge point 5	440	68	4139.5	18.5	3.48
Surge point 6	300	57	2822.4	15.5	3.42

Table 11.5 Pump operating range by surge indicator

	Pump head (m)	Volume flow (m^3/h)	$P_d - P_s$ (kPa)	F_m (kg/s)	K_S from Eq. (11.21) (–)
Surge point 1	1220	114	11477.8	31.0	0.99
Surge point 2	1000	104	9408.0	28.3	1.00
Surge point 3	780	90	7338.2	24.5	0.98
Surge point 4	600	79	5644.8	21.5	0.98
Surge point 5	440	68	4139.5	18.5	0.98
Surge point 6	300	57	2822.4	15.5	1.00
Choke point 1	950	315	8937.6	85.8	3.10
Choke point 2	770	285	7244.2	77.6	3.12
Choke point 3	605	252	5691.8	68.6	3.11
Choke point 4	470	220	4421.8	59.9	3.08
Choke point 5	340	190	3198.7	51.7	3.13
Choke point 6	240	158	2257.9	43.0	3.10

Equation 11.21 is the formula for the online calculation block `XY-102` in Fig. 11.14. We can verify the surge points and choke (end-of-curve) points, as shown in Table 11.5.

The end-of-curve control is a standard protective controller with the PV calculated from Eq. 11.21 and a setpoint set to 3.10. See `XC-101` in Fig. 11.14.

In summary, by limiting the surge indicator value K_S to between 1.0 and 3.1, the pump effectively operates between the minimum flow limit defined by the surge reference line and the maximum flow limit defined by the end-of-curve points. For safety reasons, a margin of approximately 10% should be given to both setpoints, resulting in SP = 1.1 for the minimum flow controller and SP = 2.8 for the end-of-curve controller.

11.2.4 Implementation and Operation

The actual implementation depends on the functional design and can vary from application to application. Here, a few key considerations are presented and discussed.

PID Configuration

The minimum recycle flow control `FC-102` is a simple protective flow control (see Sect. 4.1.3 for protective control), with the setpoint provided by Eq. 11.12. The setpoint calculation is typically implemented in a standard `CALC` function block in DCS. To the minimum flow controller, the setpoint is provided by another program, and thus the normal controller mode should be `RCAS` or equivalent (see Sect. 3.2.1 for `RCAS` mode). Controllers `FC-101` and `XC-101` may override each other, and anti-

Table 11.6 Loop configuration details for pump control

Configuration	LC-101	FC-101	XC-101	XC-102
Description	Tank level	Pump discharge	Minimum flow	End-of-curve flow
Equation type	I-PD	PI-D	I-PD	I-PD
Normal mode	AUTO	CAS	AUTO	AUTO
Action direction	DIRECT	REVERSE	REVERSE	REVERSE
Anti-reset windup	n/a	YES	n/a	YES
PV-tracking	Disabled	Enabled	Disabled	Disabled
Output scaling	0% = fully closed, 100% = fully open			

reset windup needs to be correctly configured (see Sect. 4.3.3 for anti-reset windup) (Table 11.6).

Multi-pump Operation

Multiple pumps can run in parallel or series to deliver a higher pressure or flow. Typical pump configurations and control include the following:

1. Pumps in series: Booster pump + main pump. Pumps in series share the same flow, but the pump head (thus pressure) is extended. The startup of serially operating pumps is challenging.
2. Pumps in parallel: Pumps running in parallel extend the overall capacity. One of the considerations is load balancing, i.e., how to distribute the capacity among the pumps. Common approaches include split-range control and fan-out control. Switching the pumps on and off creates a significant disturbance to operation and is a challenge for control design.

Due to the length limit, we will not detail the control complexity of multi-pump operations.

Sequential Control

Pump startup is a routine operation. However, frequent startup and shutdown of the pump are harmful to the pump due to the higher electric current required to start the pump, which leads to motor heating. The electric current required during startup can be ten times higher than keeping the pump at a normal speed. Since motor heating is proportional to the current squared ($I^2 R$), the heat generated during startup may be 100 times higher than normal. Typical startup scenarios include the following:

1. Single pump: An operator manually starts and stops the pump based on operational requirements.

2. Duty/standby: Standby pump automatically starts upon failure of duty pump or electrical/vibration trip of the duty pump. Pump switching may cause a temporary disturbance to the flow control.
3. Normal auto-start: With two pumps, the standby pump starts automatically based on a process variable that provides the switching point corresponding to a specific operating condition. For example, the process variable could be a low flow or low-pressure switch point.
4. Special auto-start: With three or more pumps operating in parallel, the standby pump can be selected via a selector switch. A duty/standby switch may be required to designate the standby pump.

The startup can be *cold* or *hot*. One particular challenge for a cold startup is the lack of initial system resistance. From the pump curves, it is clear that the pump's power and pressure are related to pump resistance. If the discharge pressure (thus pump head) is low, the operating point may be over the end of curve; then the pump flow would be very high, and so is the power required. One extreme example is deep water disposal (DWD), where the pump discharge connects via a long pipeline to the deep disposal wells. The pipeline, which can be several kilometers long, is initially empty and has almost zero resistance. During a cold start, the discharge pressure remains very low until the pipeline is filled with liquid.

With a fixed-speed pump, once the pump starts, the operating point rapidly moves to the far right on the performance curve due to the low system resistance (low head). The large flow rate quickly overwhelms the motor and trips the motor by power or vibration.

Pump startup is a complex sequence. The trajectory of the operating point during startup is a crucial consideration for process and process control design. Adequate resistance should be provided by design. The pump response during startup is briefly discussed as follows:

1. Pump starts with the downstream block valve closed and the recycle valve fully open.
2. Once the pump is started, the fluid quickly fills the recycle line and enters full recycle mode. Depending on the size of the recycle piping and valve, the operating point will typically be on the far right side of the performance curve.
3. As more process flow enters the suction side, the discharge pressure (thus the head) starts to build up.
4. For a warm startup or a pump joining an established header, the block valve would be opened as soon as the discharge pressure is higher than the downstream line pressure, and thus the transition is bumpless.
5. The downstream pipeline may be a long empty pipeline with virtually zero pressure. During a cold startup, if the block valve (an on/off valve!) is suddenly opened, and the pump discharge is faced with virtually zero resistance, the operating point would abruptly move beyond the end of curve and trips the pump.

It is not uncommon to see careless process engineering and process control designs that have overlooked the system resistance requirement and find themselves in a situation that the pump simply cannot be started.

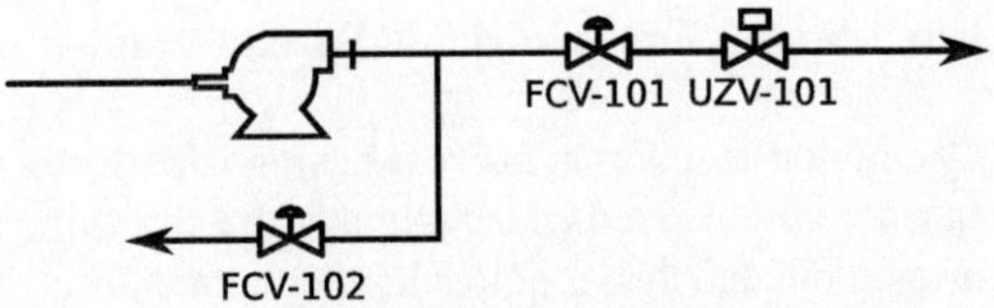

Fig. 11.15 Pump startup with a throttling valve

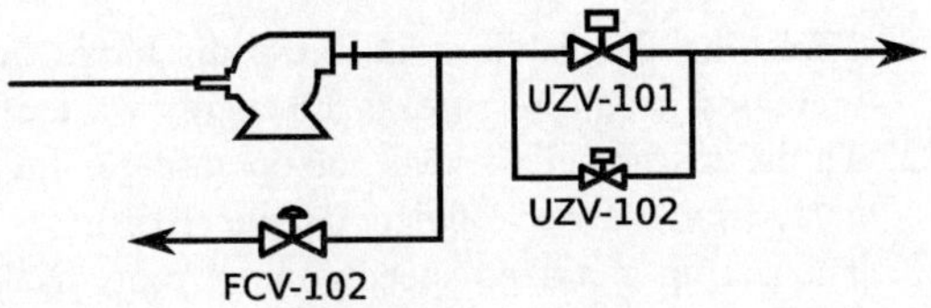

Fig. 11.16 Pump startup with double block valves

The process solution, and with proper accompanying control design and startup sequence, is to create the required system resistance artificially during cold startup. Two popular designs include the following:

1. Discharge throttling valve: The discharge control valve for capacity control can be used to create the initial system resistance during startup. Once the pump reaches full recycle mode, the valve can be manually and slowly opened to build up the back pressure in the downstream pipeline. See the piping configuration at the top in Fig. 11.15.
2. For a variable-speed pumping process, there may not be a discharge valve installed. Even if available, the control valve may not be rated for the type of high pressure drop across the valve in high-pressure pumping applications. For this scenario, a working example is to utilize two block valves in parallel, one large and one small. At the initial phase of the startup, the large valve is closed, and the small valve is open. The small valve provides enough resistance to keep the pump running and *prime* the pipeline. Once the back pressure in the pipeline reaches a specific value, the large valve is opened. See the configuration in Fig. 11.16. The startup sequence is responsible for the opening and closing of the valves.

11.2.5 Practical Considerations

Many practical considerations are dictated by process design and operational needs. However, these process decisions often have an enormous impact on the process control strategy.

Fixed-Speed Versus Variable-Speed Pumping

One major decision in process design is between using a variable-speed or a fixed-speed pump. A fixed-speed pump with a throttling discharge valve is lower in capital investment and also requires lower maintenance. However, there are also significant drawbacks as follows:

1. Throttling the valve reduces the discharge flow and increases the head. The increased head (pressure) is the energy wasted/lost across the throttling valve.
2. Throttling the valve moves the operating point along the performance curve and away from the best efficiency point (BEP).
3. The startup of a fixed-speed pump is more challenging. Typically, a much larger motor is needed to address the initial power spike.

Variable-speed operating is typically provided by either a motor with variable frequency drive (VFD) or a variable-speed gearbox (VSGB) between the pump and a fixed-speed motor. Although the initial capital investment (CAPEX) is higher, variable-speed drives are more energy-efficient, and thus the operating cost (OPEX) is lower.

The startup of a variable-speed pump is also much more manageable. The pump speed can be quickly ramped up to the minimum stable speed and then slowly be brought to the normal operating point. This startup procedure allows the operating point not to go too far to the right on the performance curve, so less initial motor power is required to allow the use of a smaller motor.

Suction Throttling Versus Discharge Throttling

When pumping incompressible (liquid) flow, discharge throttling is always preferred. The main reason is that suction throttling could potentially cause low pressure and increase the risk of cavitation.

The formation of bubbles causes cavitation as the liquid flashes under low pressure (below the vapor pressure) and can damage a pump. Therefore, it is imperative to operate the pump only in its operating region where the *net positive suction head required* (NPSHR) is satisfied. In addition to proper design, low suction pressure protective control and safeguarding can be implemented to avoid low suction pressure if needed.

11.3 Centrifugal Compressor

A gas compressor is a vital equipment in many operating plants. The gas compression system has the following unique properties (Bloch 2006; Hanlon 2001):

1. Large capital investment: A gas compressor can easily cost millions of dollars to purchase and install.

2. Critical to operation: The gas compression system is typically an indispensable part of the process operation. Tripping or shutting down the compressor often results in a significant interruption to production, which may directly lead to production loss or deferment.
3. High operating cost: Gas compression also constitutes a considerable part of the operating cost. It is common to see compressors rated at over ten megawatts of power consumption in the oil and gas industry.[3]
4. Maintenance intensive: A compressor is a piece of maintenance-intensive equipment and requires routine and costly maintenance work.

For these reasons, a mandatory requirement is that a reliable compressor control system be implemented to protect the capital investment and maintain efficient operation.

A centrifugal compressor is similar to a centrifugal pump in many aspects. Many of the discussions on pumps in Sect. 11.2 also apply to compressors. However, a significant difference between a pump and a compressor is that the former works on incompressible liquid and the latter on compressible gas. Most of the complexities of compressor control originate from the compressibility of the gas.

Chapter 6 has used a compression process for process analysis. This section focuses more on the compressor itself with the overall process flow in mind, as shown in Fig. 6.1.

11.3.1 *Process Description and Analysis*

A compressor is a piece of rotating equipment that compresses the gas from a lower pressure to a higher pressure and moves it from one location to another for further processing, storage, or transportation.

A compressor is part of the overall gas compression process. Process analysis should be carried out on the entire process flow rather than the compressor alone. Since process configurations vary from operation to operation, it is impossible to analyze all possible variations. Instead, the generic analysis methodology provided in Chap. 6 can be followed to perform the analysis. Here in this section, we only focus on the process flow scheme around the compressor.

Operating Requirements

The operating objectives of a gas compressor include the following:

1. During normal operation, maintain the desired pressure and flow (operating point) at their target values. The compressor is part of the process flow and serves the

[3] A 10 MW compressor consumes more than 80 million kWh of power (10.0 MW × 1000 kw/MW × 24 h/day × 365 day/yr = 87,600,000 kWh), or over 8 million dollars per year (assuming $0.1/kwh for electricity).

overall process operation. As a result, the operating requirements from the process side determine the operating point.

2. During abnormal operating conditions such as process upsets or severe turndown operation, keep the operating point within the safe operating region (operating envelope).
3. During the transition between different operating conditions, including startup and shutdown, ensure that the compressor quickly and smoothly reaches the new steady-state.

A compressor must operate within many constraints and limits imposed by both the compressor and the process to ensure safe and efficient operation. The typical constraints that are inherent to the compressor include the following:

1. Surge limit: When the compressor flow falls below a specific low limit, an unstable operating condition may occur where the flow rapidly reverses direction. This flow reversal may cause violent vibrations that can potentially damage the compressor and associated piping and fittings. This phenomenon is called compressor *surge* and must be prevented.
2. Stonewall: On the other hand, when the compressor flow exceeds a specific high limit, the compressor may reach a condition called *stonewall* or *choke* condition, and the flow can no longer increase. The impact of a choke on the compressor operation is not as severe as that of a surge. However, the compressor efficiency drops significantly at the stonewall, and thus compressor operation should stay away from the stonewall if possible.
3. Maximum and minimum speed: Another two inherent constraints with the compressor are the maximum and minimum rotating speed. Beyond the speed limits, the compressor operation may become less stable and thus should be prevented.

The process side can impose additional operating constraints. They vary from operation to operation subject to the process design. Common operating constraints from the process side include the following:

1. High limits on discharge pressure and temperature: Compression increases the gas pressure, and at the same time, also increases the gas temperature. As the process flow components such as compressor bearing, sealing, process piping, and vessels all have pressure and temperature ratings, keeping the temperature and pressure within the safe operating limits is imperative. For example, for centrifugal compressors in process service, the maximum actual discharge temperature shall generally not exceed 180 °C.
2. Motor power: For motor-driven compressors, the motor is another expensive piece of equipment to protect. A motor has a rated power limit that the compressor operation should not exceed. There is usually a spike in the motor load during startup, which can be several times higher than regular demand.
3. Gearbox torque: A variable-speed gearbox (VSGB) has a limit on the torque rating. Compressor operation should monitor and limit the torque it generates on the gearbox.

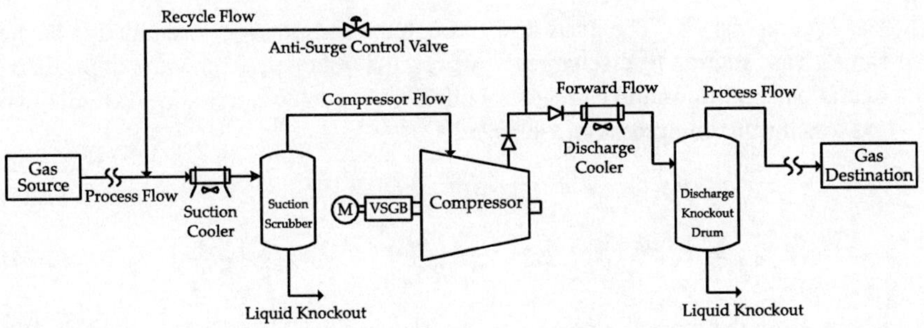

Fig. 11.17 A single-stage gas compression unit

The limits and constraints mentioned above define the operating envelope. They reduce the feasible operating region, and some may even change over time.

Process Flow Analysis

A typical gas compression system consists of the gas source, a suction scrubber, the compressor, a discharge cooler, a discharge knockout drum, the recycle line and valve, and the gas destination. Figure 11.17 shows the essential components in a simple single-stage variable-speed gas compression circuit.

Three types of flows are critical to our analysis for the compressor: the process flow, compressor flow, and the recycle flow (see Fig. 11.17). The process flow, also called the forward flow, is the process gas to be moved forward to downstream operation. The recycle flow is typically not flowing and is needed only for surge prevention (see later). The three flows are related to each other by

$$[\text{compressor flow}] = [\text{process flow}] + [\text{recycle flow}]. \qquad (11.22)$$

Compressor flow is usually measured at the suction side using an orifice plate or a Venturi tube but can also be measured at the discharge side.

A compressor can operate at a fixed speed or variable speed. A compressor driven by a fixed-speed motor is a fixed-speed compressor. A compressor driven by a variable-speed motor, or a fixed-speed motor with a variable-speed gearbox, becomes a variable-speed compressor.

Supply and Demand Analysis

The material balance in the compression process follows the general principles below:

1. Gas is compressible. The mass flow through the compressor is assumed to be the same[4] from suction to discharge. However, the volumetric flow rate depends on operation conditions and can be very different due to compression. Assuming no gas loss during compression, we have

$$F_{m,d} = F_{m,s} \tag{11.23}$$

$$F_{v,d} = F_{v,s}\,\frac{\rho_d}{\rho_s}, \qquad \rho_d = \frac{P_d\,M_d}{Z_d\,R\,T_d}, \qquad \rho_s = \frac{P_s\,M_s}{Z_s\,R\,T_s} \tag{11.24}$$

where F_m is the mass flow and F_v is the volume flow. ρ is the gas density. The subscripts "$_s$" and "$_d$" indicate the suction side or the discharge side, respectively. P is the operating pressure, and T is the operating temperature. M is molecular weight, and R is the gas constant.
2. The imbalance in supply and demand causes the gas accumulation to change, reflected in the pressure. The pressure–flow profile[5] is thus the focus for supply and demand analysis.

The compression process's energy balance is about converting the motor power to compressor impellers' mechanical energy and then imparting it to the process gas as potential energy (head or pressure) and kinetic energy (gas flow). Some energy is lost as heat, causing temperature increases in the gas and the casing/bearing. The temperature ratio and pressure ratio are linked via

$$\frac{T_d}{T_s} = \left(\frac{P_d}{P_s}\right)^{\frac{n-1}{n}} \tag{11.25}$$

where n is the polytropic exponent that will be discussed later.

The compressor is part of the supply and demand propagation. The capacity of the compressor can be set by either the supply or the demand. The compressor has to take whatever gas flow comes in, compress it, and send it further downstream with the supply-driven operation. The pressure–flow profile (Sect. 6.3) would require that suction pressure be the controlled variable (with discharge pressure as an override). The supply/demand propagation goes from upstream to downstream. On the other hand, if the downstream user determines how much gas the compressor should supply, this is a demand-driven operation. The controlled variable would be the discharge pressure, and the supply and demand relationship would propagate backward from downstream to upstream.

The supply and demand model may become very complicated for a complex compressor system involving multiple compressor trains, gas suppliers, and consumers. The supply and demand analysis can follow the guidelines in Sect. 6.3. A real-world example is provided in Chap. 6.

---

[4] There can be liquid knockout or side streams that cause the mass flows to differ at suction and discharge.

[5] In liquid operation, the material balance is reflected by the level-flow profile, while for gas processing, the material balance is reflected by the pressure–flow profile.

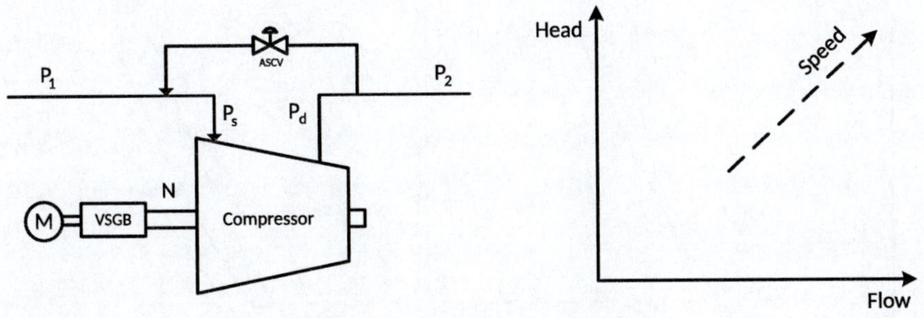

Fig. 11.18 The three variables that define compressor dynamics

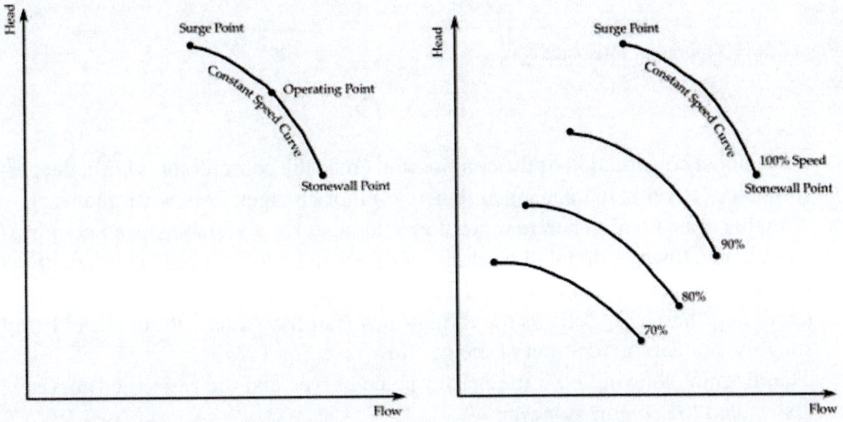

Fig. 11.19 Compressor performance curves (Left: fixed speed; Right: variable speed)

Cause and Effect Analysis

Similar to pump operation, the cause and effect relationship of compressor operation can also be simplified and summarized as three variables, two lines, and one point:

1. Three variables: The gas flow through the compressor, the head (or pressure increase) that the compressor generates, and the compressor speed are the three variables that fully define the compressor operation. See Fig. 11.18 for an illustration.
2. Two curves: The relationship between the three variables is defined by the so-called performance curve(s). A fixed-speed compressor has one performance curve, while a variable- speed compressor can have a series of curves within the operating range. See Fig. 11.19.
   At a constant speed, the compressor head is inversely proportional to the compressor flow. The more the gas is compressed, the less the head it will generate, and vice versa. The exact relationship is complicated and is determined by the

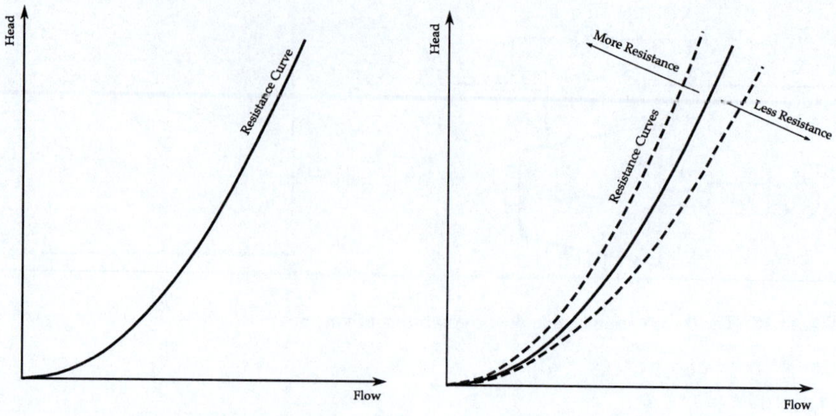

Fig. 11.20 Compressor resistance curve(s)

mechanical construction of the compressor. Once the compressor is built, the performance curves stay unchanged (barring minor changes from wear and tear). A higher speed will generate more flow and head for a variable-speed compressor, while a lower speed will produce less flow and a lower head. This head/flow versus speed relationship is called the system resistance curve(s). The resistance curve approximately follows the affinity law (fan law) and with the head being roughly a quadratic function of the gas flow (see Fig. 11.20).
The diagram showing both the performance curves and the resistance curves is also called the *compressor map*.

3. One point: The operating point of the compressor must fall on both the performance curve and the system resistance curve. The operating point is always at the intersection of the performance curve and the resistance curve. The compressor map in Fig. 11.21 shows the possible locations of the operating points.
Unlike the performance curve, which is fixed after the compressor is constructed, the system resistance can be changed dynamically by many factors such as valve position, recycle, supply pressure, and discharge back pressure, and thus the system resistance curve can move right or left on the compressor map as the system resistance changes.

With a fixed-speed compressor (Fig. 11.21 left), the operating point always moves along the performance curve through system resistance changes. For a variable-speed compressor, the operating point can be moved continuously to any location within a region by changing the system resistance and compressor speed.

Example 11.2 A two-speed hairdryer. Figure 11.22 shows a two-speed hair dryer with a concentrator to increase the air pressure.

A hairdryer is essentially a low-pressure air compressor. With the choice of two operating speeds, and with or without the concentrator, there are four operating points, as shown in Table 11.7.

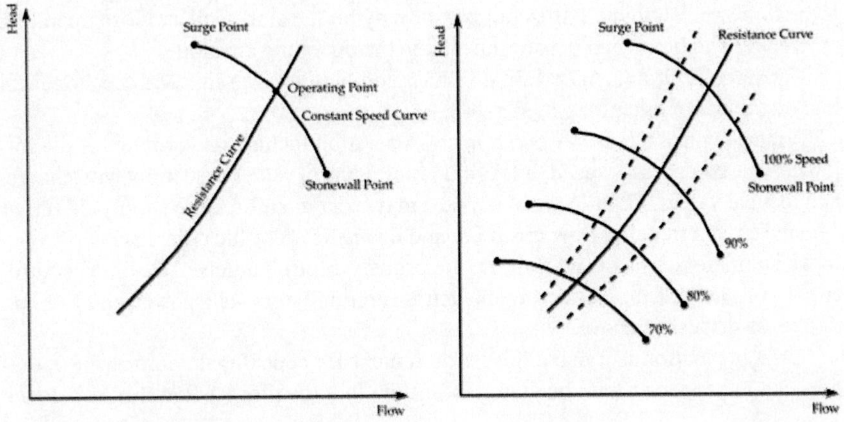

Fig. 11.21 Compressor operating points

Fig. 11.22 A two-speed hair dryer with an add-on concentrator

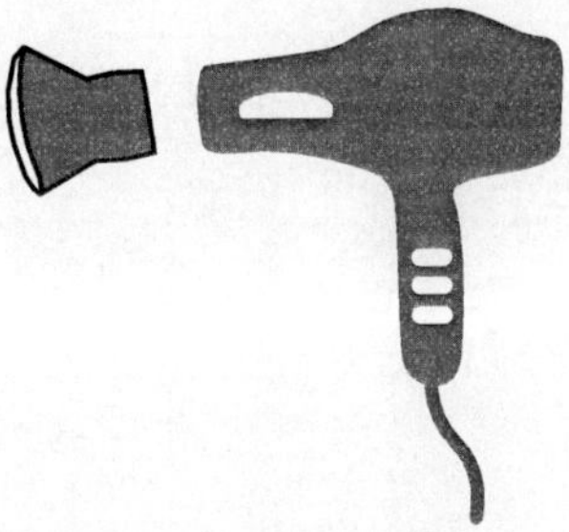

The performance curves of the hair dryer can be illustrated as in Fig. 11.23.

Based on the cause and effect relationship, at the low speed and without the concentrator, the operating point is at A. Switching to the higher speed will move the operating point up along the system resistance to point B. Adding the concentrator increases the system resistance but does not change the motor speed, so the operating point moves along the performance from A to C or B to D. The concentrator functions as a discharge valve.

There are design limits and operating constraints that restrict the operating point to a specific region. To the left on the compressor map is the surge line. When the flow goes too low, the compressor can enter the surge cycle and lose stable operation.

Table 11.7 Operating points of a two-speed hair dryer

	Concentrator	
	OFF	ON
Low speed	Point A	Point C
High speed	Point B	Point D

If the flow rate is too high, the compressor may hit the stonewall and lose operating efficiency. The limits and constraints define the operating envelope.

The surge limit and stonewall are illustrated by the surge line and the stonewall line on the compressor map in Fig. 11.24.

Typical control handles for compressor operation include one or more of the following: the compressor speed, inlet guide vane, control valves at suction or discharge, and recycle valves. Other control handles may exist to enforce the supply-driven or demand-driven material flow upstream and downstream of the compressor.

The compressor head and flow are the primary control targets. Additional control targets may include protective targets such as surge and stonewall, power, and gearbox torque, as discussed above.

Surge prevention is a mandatory requirement for centrifugal compressor operation. Surge happens when the compressor flow (not the process flow!) runs below a

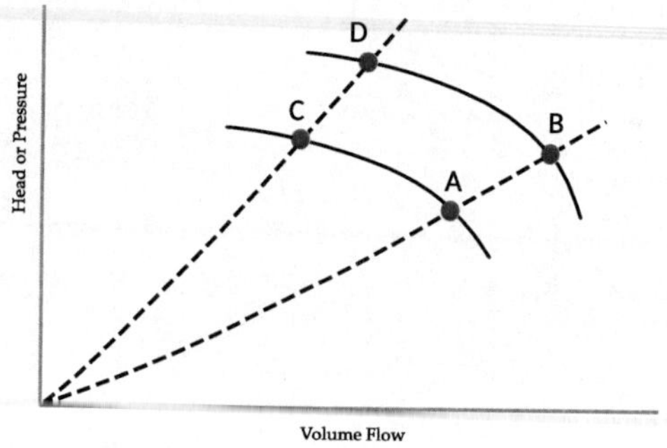

Fig. 11.23 Performance curves of a two-speed hair dryer

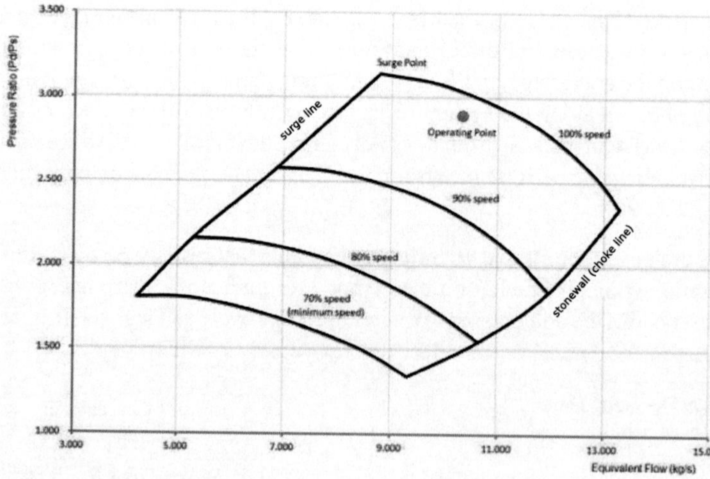

Fig. 11.24 Compressor operating envelope

low limit. The challenge, however, is that this minimum flow limit is not constant. It is different at different compressor speeds and varies with gas property (molecular weight and compressibility) and operating conditions (pressure and temperature).

In practice, an *inferential property* called *surge indicator* is usually defined as an online variable to indicate how far the operating point is from the surge limit (see Sect. 5.3 for inferential property estimation). Surge indicator is inferred from fast and reliable measurements such as flow, pressure, and optionally temperature. A general formula is as follows:

$$K_s = f(P_s, P_d, F_v, T_s, T_d) \tag{11.26}$$

where P_s is the suction pressure, and P_d the discharge pressure. F_v is the raw volumetric flow (Sect. 10.3) at either the suction or discharge side. T_s and T_d are the temperatures at the suction and discharge sides, respectively. There are many ways to define the surge indicator, and there are also different names for surge indicator. A more detailed discussion will be provided later in the control design section.

The surge indicator is the most critical protective control variable for a centrifugal compressor. Other protective targets such as motor power or gearbox torque can be controlled variables as well.

The potential manipulated and controlled variables, and the cause and effect relationship between them, are summarized in Table 11.8.

Not all the control handles are available for a gas compression system. The speed, suction valve, and discharge valve are mutually exclusive. For a fixed-speed compressor, the typical configuration is with the suction valve and the recycle valve. For a variable-speed compressor, the typical manipulated variables are the compressor speed and the recycle valve.

The compressor speed or suction valve as a manipulated variable is primarily for maintaining the compressor capacity. The recycle valve, also called the *anti-surge control valve* (ASCV), is mainly for surge prevention (and choke prevention if needed).

Degree-of-Freedom Analysis

Table 11.8 shows that there are more targets than handles for compressor control. Per the degree-of-freedom requirement, the controlled variables and manipulated variables must be fully paired up and exhaust all the degrees of freedom (DOF) available (Sect. 6.5).

Table 11.8 Cause and effect relationship for compressor operation

	Head	Compr flow	Proc flow	Surge margin	Power	Dschg press
Speed ↑	↗	↗	↗	–	↗	↗
Suction valve ↑	↘	↗	↗	↗	↗	↘
Discharge valve ↑	↘	↗	↗	↗	↗	↘
Recycle valve ↑	↘	↗	–	↗	–	↘

The available handle is typically the suction valve, discharge valve, or compressor speed for normal regulatory control. Usually, only one of these control handles is installed. It is the manipulated variables responsible for maintaining the compressor head and flow (the capacity).

The compressor performance curves (Fig. 11.19) show that the pressure and flow are completely correlated at a given speed, so the compressor head and flow are treated as one controlled variable and consume one DOF. Typically, the polytropic head is selected as a controlled variable to indirectly control the compressor flow (capacity).

The practical challenge is that the polytropic head is not directly measurable. However, since the polytropic head is a function of the pressure ratio between discharge and suction pressure, the pressure is typically chosen as the controlled variable instead. Either the suction pressure or the discharge pressure will serve as the capacity control variable depending on whether the operation is supply-driven or demand-driven.

The recycle flow is provided primarily for surge protection. Therefore, choosing the recycle valve as the manipulated variable and the surge indicator as the controlled variable is an easy decision.

Now we still have two remaining challenges with the process variable selection and pairing:

1. There are still more control targets to be paired up with control handles. The targets include the compressor power limit, discharge pressure, gearbox torque, and maybe more.
2. There are strong interactions between the manipulated variables and controlled variables and remaining control targets.

Based on the guidelines in Sect. 2.2 and Chap. 7, complex control loops such as split-range control and overriding control can be considered to meet the DOF requirements. The pairing of the process variables will be further explored in the next section's control strategy discussion.

Dynamic Response Analysis

Typically, step tests are recommended to learn more about the dynamic characteristics of each causal relationship in the cause and effect table (Table 11.8). Centrifugal compressor has been a piece of well-studied equipment, and some general understandings of the compressor dynamic characteristics include the following:

1. Compressor dynamics are extremely fast. Among the cause and effect relationships in Table 11.8, the dynamic response between the recycle valve and the surge indicator is typically in sub-seconds. The dynamics between the compressor speed and capacity are orders of magnitude slower.
2. The dynamic response characteristics between all the control handles and control targets are assumed to be linear, time-invariant, and self-stabilizing. For example,

by increasing the compressor speed, the operating point will move to and settle at a new location by itself provided that the operating points are within the operating envelope.

3. For a multi-stage compressor, opening the recycle valve of one stage can inadvertently interfere with the operation of other stages.

A compressor is a self-regulating process. That means if the system resistance changes, the operating point automatically moves to a new location and settles there by itself. This behavior is reflected by the negative slope between the head and the flow in the performance curve and the positive slope of the system resistance curves (see Fig. 11.19). As long as the operating point stays inside the operating envelope, the head and the flow will automatically find a new balance point, and thus the compressor operation is relatively stable.

The surge phenomenon occurs when the operating point moves too far to the left on the performance curve. To the left of the surge point (i.e., below the minimum flow limit), the head–flow slope turns positive and thus becomes self-accelerating and unstable. This phenomenon is like pushing a cart up a hill (see Fig. 11.25). The energy from the motor is like the force pushing the cart, while the system resistance is like the gravitational force pulling the cart down. Once the cart is over the top, the two forces work in the same direction, and the cart quickly falls over the top.

Therefore, the surge point's location (the top of the hill) at each compressor speed is the most critical information for compressor control.

The compressor is typically designed and built based on off-line simulation and calculation data, per the client's specifications. The *predicted* (estimated) performance data is provided in advance for front-end engineering of the overall process design (including process control). After the compressor is constructed, surge tests are usually conducted to confirm that the actual dynamic responses (e.g., performance curves) match the predicted performance. The test is typically performed

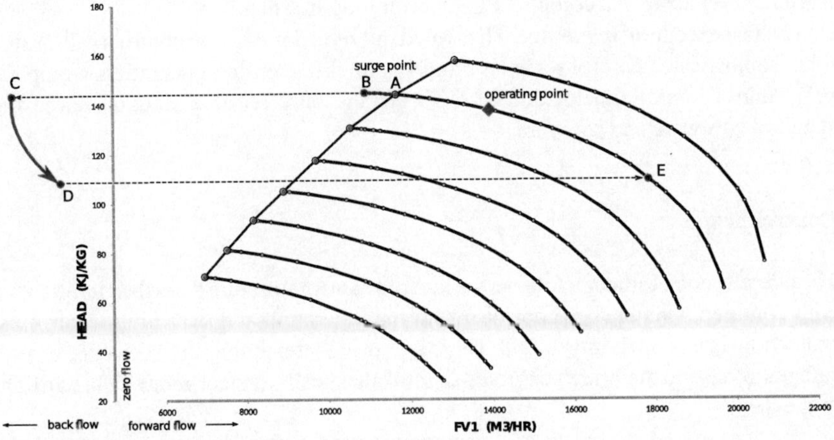

Fig. 11.25 Surge and reverse flow

with common gases such as nitrogen (N_2) or fuel gas because the actual process gas is not available yet.

After the compressor is installed at the site and before startup, a formal surge test is usually required to confirm the compressor dynamics response with the actual process gas.

11.3.2 Control Strategy

The process analysis paves the way for control design. Chapter 7 provides some guidelines for process control design at both plant level and individual loop level.

Control Objectives

Control objectives serve operating objectives. The generic objectives of compressor control are maintaining the operating point at the desired value during normal operation and keeping the operating point inside the operating envelope during abnormal process conditions. Protecting the compressor is vital; therefore, integrated control and safeguarding is also a common requirement.

The compressor dynamic response, particularly the surge phenomenon, is extremely fast. Compressor surge causes the reversal of compressor flow, which happens in tens of milliseconds to hundreds of milliseconds. The control solution must execute at a sub-second frequency, typically as fast as the DCS system can support. In olden days, dedicated hardware and software were utilized to support fast execution. As technology advances, many DCS systems can now execute at 100ms scan time or faster. Consequently, many compressor control applications are now implemented in DCS using the standard function blocks. Compressor control is no more mysterious than a complex PID control loop in Chap. 4.

The fast execution requirement has ruled out model predictive control (MPC) and other sophisticated control schemes requiring intensive online calculation. Complex PID control loops implemented in DCS are the only feasible solution based on standard hardware and software.

Control Scope

The overall control strategy follows the layered control structure described in Chap. 7, with continuous regulatory control for normal operation and overriding control as protection during operating upsets. There are strong interactions between control and safeguarding, and therefore integrated control also needs to incorporate safeguarding in the design.

The startup and shutdown of a gas compression system is an operational challenge, and automated startup and shutdown procedures greatly help operations. The

- Compressor Operation
  - Regulatory Control
    - Capacity Control
    - Load Balancing Control
  - Protective Control
    - High/Low Speed Protection
    - Surge/Choke Prevention
    - Discharge Pressure/Temperature Protection
    - Motor/Gearbox Power/Torque Protection
  - Sequential Control
    - Startup/Shutdown Sequence
    - Transition of Operating Mode
  - Instrumented Safeguarding
    - Instrument Failures
    - Compressor Surge
  - Online Monitoring
    - Key Performance Indicators
    - Live Operating Envelope

Fig. 11.26 Operating requirements of gas compression process

automated startup sequence is part of sequential control and is typically included in the integrated compressor control design.

Figure 11.26 provides a list of the components in a fully integrated compressor solution.

It is beyond this book's scope to provide an in-depth discussion of all the components in the control strategy. A discussion on the essential components is provided as follows:

1. Normal regulatory control: Normal regulatory control is typically about quality and capacity. There is typically no requirement on product quality for gas compression; thus, the normal regulatory control is mainly about capacity control.
   For capacity control, a clear understanding of the supply/demand model is required for placing the appropriate pressure/flow controllers at suitable locations to ensure proper supply/demand propagation. For example, the control configuration for a supply-driven operation of a variable-speed compressor is shown in Fig. 11.27. The blue dashed line indicates the direction of supply/demand propagation. The two relay points for the propagation are the suction pressure and the discharge pressure.
   The downstream consumer of the gas is assumed to have sufficient capacity to absorb the fluctuation in supply. Otherwise, another swing stream needs to be provided as swing control.
   If the operation is demand-driven, the control configuration would be as shown in Fig. 11.28. In this case, the supply side needs to adjust the supplied gas rate in response to the demand change. Otherwise, a separate swing stream must be provided.

   Suppose the supply-driven operation may become demand-driven, either temporarily or permanently. In this case, override control must be provided in the

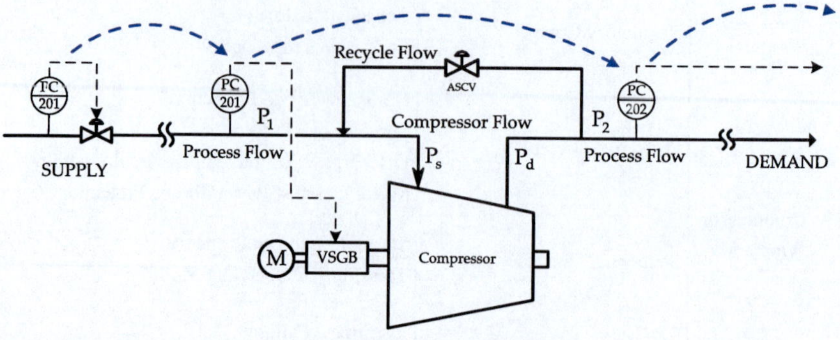

Fig. 11.27 Supply-driven control of a variable-speed compressor

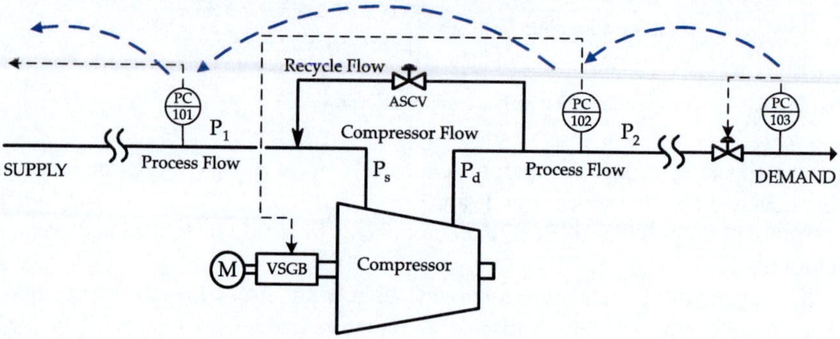

Fig. 11.28 Demand-driven control of a variable-speed compressor

design. Figure 11.29 shows an example of supply-driven control with demand override. The dashed line indicates the propagation during regular supply-driven operation, while the dotted line illustrates propagation when the supply-driven operation reverses to demand-driven. In terms of control configuration, Fig. 11.29 is a combination of Figs. 11.27 and 11.28. The automatic reversal of the supply/demand model is via the control selector function block and the proper configuration of the normal regulatory control setpoints and the protective control setpoints.

2. Overriding protective control: The most crucial protective control for a compressor is the anti-surge control. The focus of this section is anti-surge control only, as other protective controls listed in Fig. 11.26 are relatively straightforward. Compressor surge is a dangerous condition to compressor operation and must be prevented. Low process flow (see Fig. 11.17) causes low compressor flow and subsequently leads to compressor surge. The most effective approach to prevent the compressor surge is to open the recycle valve and recirculate part of the compressed gas back to the suction side. As the compressor flow is the sum of the

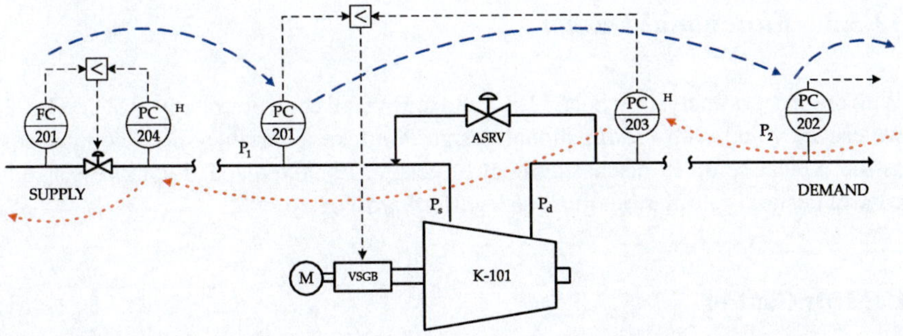

Fig. 11.29 Supply-driven control of a variable-speed compressor, with demand override

process flow and recycle flow, the compressor does not know, neither does it care, whether the gas is from the process or recycle.

Reducing system resistance can increase the surge margin as well. For example, opening the suction or discharge valve (if available) moves the operating point to the right side of the performance curve.

3. Instrumented safeguarding: If both capacity and protective controls fail, the safety instrument system (SIS) shuts down the compressor operation. The safeguarding design should be included in the overall control design because the anti-surge tripping must be consistent with the anti-surge control. The integrated control design has the following requirements on the safeguarding function:

   a. The same calculation for the surge indicator must be used.
   b. The pressure and flow measurement must be from the same locations. They are from independent sensors and transmitters than those used for anti-surge control to increase reliability by avoiding common source failures.

   Emergency shutdown of the compressor has its own risk, so the trip logic needs to be carefully calibrated to safeguard the compressor effectively, and at the same time, minimize nuisance shutdowns. For example, an isolated surge of short duration (e.g., less than 3 s) is not a serious concern, but a frequent or sustained surge is very hazardous.

4. Sequential Control: Compressor startup is a complicated task. There are many checks and balances in an automated startup sequence. Multiple steps in the startup sequence require coordination with the process control scheme. For example, the startup sequence needs to quickly bring the operating point to a stable location inside the operating envelope. The anti-surge control will then be activated. After the compressor reaches normal operating conditions, the capacity controller can be put into automatic operation.

11.3.3 Functional Design

With the process analysis in Sect. 11.3.1 and the overall control strategy in Sect. 11.3.2, we can move on with the functional design. We take a variable-speed compressor as the typical setup in discussing control design. For fixed-speed compressor, the control design is similar but more straightforward.

Capacity Control

The capacity control around the compressor is provided in Figs. 11.27, 11.28, and 11.29. However, the actual gas compression process is much more complicated than a simple single-train operation. Example 6.1 is a compression process of medium complexity. A more complex process may have multiple compressor trains, and each may have multiple stages, with multiple streams in and multiple streams out. See Fig. 12.29 for an example.

The overall capacity control design can follow the generic supply and demand analysis methodology in Sect. 6.3. The capacity control propagation may span across multiple operating units or even beyond the project boundary. As mentioned in Sect. 7.1.2, capacity control is a consideration at the plant operation level. A holistic view of the supply and demand model must be kept in mind, while a process overview diagram (POD) is a handy tool to keep this *big picture* at sight.

Surge Indicator Calculation

A healthy surge margin needs to be maintained in compressor operation to prevent the compressor from going into surge. Fast and reliable calculation of the surge indicator is thus a requirement. There have been many efforts and solutions for the surge indicator calculation (White 1972; Staroselsky and Ladin 1979). A good summary is provided in Mirsky et al. (2012).

In a nutshell, the surge point is the extreme operating point beyond which the compressor would enter a surge cycle. When the compressor flow falls below a minimum flow limit, the flow can reverse direction in less than 200 ms seconds and goes back and forth. It is like running up a hill. Right over the top is a cliff. Once at the top, there is no time to stop from going over the edge. So it is crucial to stay away from the top, but be able to get near the top.

The line connecting all the surge points is called the surge line, which may have a highly irregular shape due to noises and irregularities in data (Fig. 11.24). For online calculation, a simple mathematical formula is desired to approximate the surge. This simple formula represents a smooth line going through some or most of the surge points. This line is called the *surge reference line* or *surge limit line* (see Fig. 11.30). The selection and definition of the surge reference line is the primary difference between the different compressor control vendors' technologies.

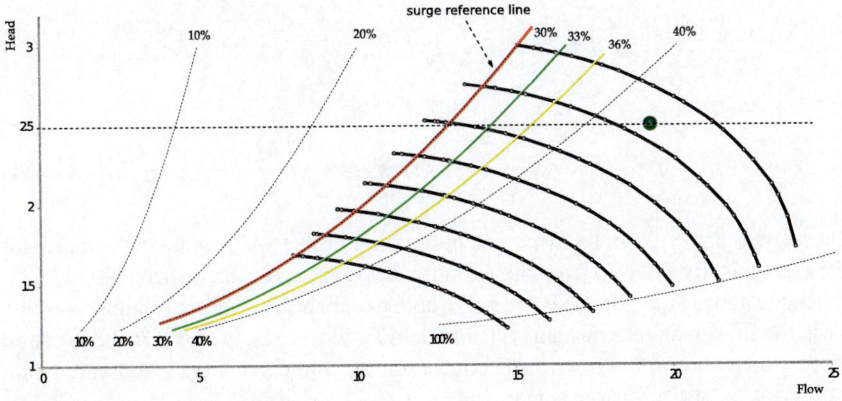

Fig. 11.30 Surge reference line

Compressor affinity law states that the head is a quadratic function of the flow:

$$\frac{H_{p,1}}{H_{p,2}} \approx \left(\frac{F_{v,1}}{F_{v,2}}\right)^2, \quad \text{or} \quad \frac{H_p}{H_{p,0}} \approx \left(\frac{F_v}{F_{v,0}}\right)^2 \tag{11.27}$$

where H_p and $H_{p,0}$ are the polytropic head of two arbitrary points along a system resistance line; F_v and $F_{v,0}$ are the corresponding volumetric flows. In the head-versus-flow coordinate, the fan law formula represents a parabolic line. That is, the system resistance line is parabolic in theory.

If $H_{p,0}$ and $F_{v,0}$ are assumed to be a known point on the surge reference line, then any point on the surge reference line can be represented by the following equation:

$$H_p = K \cdot F_v^2. \tag{11.28}$$

Again, the surge reference line does not necessarily go through all the surge points due to measurement noises and some non-conformity to theory. With a conservative design, the surge reference line should be drawn in such a way that all the surge points are either on or to the left of the line.

In practical application, the polytropic head is not directly measurable, nor is the compressor flow. Therefore, they are not available for online calculation of the surge indicator. Other reliable and online measurements must be used instead. From the definition of polytropic head and volumetric flow (Eqs. 11.29 and 11.31),

$$H_p = \frac{n}{n-1}\frac{Z_s R T_s}{M}\left[\left(\frac{P_d}{P_s}\right)^{\frac{n-1}{n}} - 1\right] = \frac{1}{\sigma}\frac{P_s}{\rho_s}\left[\left(\frac{P_d}{P_s}\right)^{\sigma} - 1\right] \tag{11.29}$$

$$\text{with} \quad n = \ln\left(\frac{P_d}{P_s}\right) \Big/ \ln\left(\frac{P_d}{P_s}\frac{T_s}{T_d}\right), \quad \rho = \frac{P\,M}{Z\,R,T}, \quad \sigma = \frac{n-1}{n} \tag{11.30}$$

$$F_v = C\sqrt{\frac{\Delta P_s}{\rho_s}} \quad \text{or} \quad F_v = C\sqrt{\frac{\Delta P_d}{\rho_d}} \tag{11.31}$$

the polytropic head can be expressed in pressure ratio P_d/P_s, while the compressor flow is typically inferred from the pressure drop ΔP across an orifice plate.

In the above equations, n is the polytropic exponent, Z is the gas compressibility, R is the universal gas constant, ρ is the density of the gas, and ΔP is the pressure drop across the orifice flow meter. Subscript "$_s$" indicates the suction side while subscript "$_d$" the discharge side.

With Eqs. 11.29 and 11.31, the compressor affinity laws can be reformulated as

$$\frac{\frac{n}{n-1}\frac{P_s}{\rho_s}\left[\left(\frac{P_d}{P_s}\right)^{\frac{n-1}{n}} - 1\right]}{H_{p,0}} \approx \left(\frac{C\sqrt{\frac{\Delta P_s}{\rho_s}}}{F_{v,0}}\right)^2 = \frac{C^2\frac{\Delta P_s}{\rho_s}}{F_{v,0}}. \tag{11.32}$$

Lumping all the constant values $H_{p,0}$, $F_{s,0}$, and C together into a single constant K, and reformulating the equation, we have

$$\frac{1}{\sigma}\left[\left(\frac{P_d}{P_s}\right)^{\sigma} - 1\right] \approx K \cdot \frac{\Delta P_s}{P_s}. \tag{11.33}$$

Most surge indicator calculation formulas are based on this relationship, all with a certain degree of simplification or approximation.

Head versus flow is known as an *invariant coordinate*, meaning that the head–flow relationship is determined by the compressor's mechanical construction and does not vary with gas property or operating condition. It is proven that the compression ratio P_d/P_s versus $\Delta P/P_s$ is an alternative coordinate that is *almost* invariant and suitable for online applications since it is based on available online measurements only (Batson 1996; Nägeli et al. 1973; White 1972).

Anti-surge technologies based on Eq. 11.33 require online temperature measurements of T_s and T_d, because polytropic exponent n is needed (see Eq. 11.30) (Kurz et al. 2016; Mirsky et al. 2012). Since temperature is a slower and less reliable measurement than pressure and flow, it is not ideal to be included in online calculations. Equation 11.33 can be simplified to eliminate the need for temperature:

$$\frac{1}{\sigma}\left[\left(\frac{P_d}{P_s}\right)^{\sigma} - 1\right] \approx C_1 + C_2\frac{P_d}{P_s}. \tag{11.34}$$

Some early anti-surge control utilizes two differential pressures, one across the orifice plate (ΔP), the other between the compressor discharge and suction ($P_d - P_s$). It is evident that if Eq. 11.34 is simplified further by assuming $\sigma = 1$, then

$$\frac{1}{\sigma}\left[\left(\frac{P_d}{P_s}\right)^{\sigma} - 1\right] \approx C_1 + C_2\frac{P_d}{P_s} \approx P_d - P_s. \tag{11.35}$$

The typical approach is to define a more intuitive variable to indicate the operating point's location relative to the surge reference line for online anti-surge control. When the operating point is on the reference line, the value is set to a fixed value called *surge reference value*. When the operating point is at another location on the map, its value is compared with the surge reference value. The relative value indicates how far the point is from the surge limit.

The parabolic surge reference line becomes a straight line if redrawn in the coordinate of H_p versus F_v^2. The surge parameter can be defined as the ratio between the slope at the operating point and the slope of the surge reference line:

$$S = \frac{\text{Slope of Surge Point}}{\text{Slope of Operating Point}} \begin{cases} = 1.0, & \text{at surge point} \\ > 1.0, & \text{at right of surge point} \\ < 1.0, & \text{at left of surge point} \end{cases}$$

$$= \frac{H_{p,0}}{F_{v,0}^2} \bigg/ \frac{H_p}{F_v^2} = \frac{\dfrac{\frac{1}{\sigma}\left[(P_{d,0}/P_{s,0})^{\sigma} - 1\right]}{\Delta P_0/P_{s,0}}}{\dfrac{\frac{1}{\sigma}\left[(P_d/P_s)^{\sigma} - 1\right]}{\Delta P/P_s}} = C\,\frac{\Delta P/P_s}{\dfrac{1}{\sigma}\left[\left(\dfrac{P_d}{P_s}\right)^{\sigma} - 1\right]} \tag{11.36}$$

where S is the surge indicator.[6] The surge reference line has a surge value of $S \equiv 1$. A value of $S > 1$ indicates that the operating point is on the right-hand side of the surge reference line, while $S < 1$ is on the left-hand side (Kurz et al. 2016; Mirsky et al. 2012).

The surge value can also be based on the so-called equivalent flow, with a definition derived from Eqs. 11.33 and 11.34:

$$S = \sqrt{\frac{\Delta P/P_s}{\dfrac{1}{\sigma}\left[\left(\dfrac{P_d}{P_s}\right)^{\sigma} - 1\right]}} \approx \frac{\text{FTE}}{\sqrt{C_1 + C_2\dfrac{P_d}{P_s}}} \approx \frac{\text{FT}}{\sqrt{C_1 P_s + C_2 P_d}} \tag{11.37}$$

$$\text{FT} = F_{m,\text{range}} \cdot \sqrt{\frac{\Delta P}{\Delta P_{\text{range}}}} \tag{11.38}$$

$$\text{FTE} = F_{m,\text{range}} \cdot \sqrt{\frac{\Delta P}{\Delta P_{\text{range}}} \cdot \frac{P_{s,\text{range}}}{P_s}} \tag{11.39}$$

[6] When the polytropic exponent n approaches ∞, value of σ would approach 1. Equation 11.36 is reduced to $S \approx C\frac{\Delta P}{P_d - P_s}$. This is the same as Eq. 11.18 for pump.

where FT is the raw actual flow (uncompensated actual flow as discussed in Sect. 10.3.4), and FTE is the equivalent flow. It can be proven that FTE is linearly proportional to $\sqrt{\Delta P/P_s}$; therefore, the compression ratio P_d/P_s versus equivalent flow is another choice of invariant coordinates. If the surge indicator S is set to 30% for the surge reference line, then the stonewall line would have a value close to 100%. The operating region lies between approximately 30% and 100%.[7]

A 10% safety margin is typically provided for control. So a surge control line is typically chosen to be about 10% higher than the minimum flow rate. The surge control line is the setpoint for the anti-surge controller. Above this setpoint, the anti-surge controller drives the recycle valve to the fully closed position. Below this setpoint, the anti-surge aggressively open the recycle valve to prevent surge. Equation 11.36 is modified to become

$$\mathrm{DEV} = S - 1 - B1 \begin{cases} = 0.0, & \text{on control line} \\ > 0.0, & \text{above control setpoint} \\ < 0.0, & \text{below control setpoint} \end{cases} \tag{11.40}$$

where $B1$ is the 10% safety margin for control. The objective of the anti-surge controller is to maintain the DEV value above zero, corresponding to an operating point to the right of the surge control line.

11.3.4 Implementation and Operation

The actual compressor control implementation can have many variations. This section only provides a few typical configurations to give a flavor of the complexity.

Capacity Control

Capacity control is closely related to the supply and demand model and varies from design to design. Figure 7.5 in Sect. 7.2.2 has already provided an excellent example of capacity control design. Figures 11.27, 11.28, and 11.29 provide several typical implementations. Therefore, no further discussions will be provided here.

[7] A compressor behaves similar to an orifice in terms of restriction to flow. An orifice plate has a turndown ratio of about 10:1 in DP. Below 10% of the DP measurement range, the flow may not be stable enough to produce credible readings. Since the flow is a square-root function of DP, this 10:1 DP turndown corresponds to a $\sqrt{10} : 1 \approx 3 : 1$ turndown for flow. In other words, for flow measurement with an orifice plate, the effective range of reading is approximately between 30% and 100%.

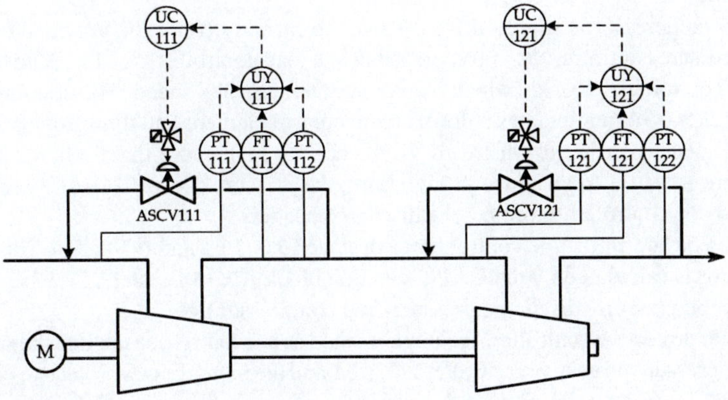

Fig. 11.31 Basic anti-surge control

Anti-surge Control

The anti-surge control is a protective PID control that monitors the surge indicator. Whenever the surge indicator is below its setpoint, the anti-surge controller quickly opens the recycle valve to bring the compressor above the minimum flow limit. Figure 11.31 shows a basic anti-surge control configuration. The surge indicator is calculated with Eq. 11.26 from the suction pressure, the discharge pressure, and the compressor flow. The surge indicator is the process value (PV) for the anti-surge controller. The controller setpoint is the desired surge indicator, while the controller output is sent down to the anti-surge recycle valve. The compressor flow can be measured at either the discharge side (`FT-111`) or the suction side (`FT-121`).

Other Protective Controls

Other protective controls such as high discharge pressure protection, high motor power protection, and high gearbox torque protection are all overriding protective control. They share the existing final control elements such as the compressor speed and recycle valve, override the capacity control or other protective controls as needed. For example, high discharge pressure protective controller can override the anti-surge controller through a high selector to open the anti-surge recycle valve if it demands a larger opening than required by anti-surge control (Fig. 11.32).

Integrated Control

Figure 11.33 shows a typical compressor control implementation.

The compressor is supply-driven. Thus, the capacity control is provided by suction pressure controller PC-102, cascaded to a load controller XC-101, and then to speed controller SC-101, which manipulates the gearbox speed. The pressure controller PC-004 manipulates a downstream control element and thus propagates the supply change further downstream. High capacity may cause the discharge header pressure PC-003 to go above its high limit; the pressure control PC-003 serves as an override controller to protect the discharge header.

There is one anti-surge control for each stage (UC-111 and UC-121). The surge indicator is calculated by the CALC function block (UY-111 and UY-121) based on suction pressure, discharge pressure, and compressor flow.

Other protective controllers are implemented as needed. These include the suction header pressure protective controller PC-001 and the motor power protective control EC-101.

For optimal performance, a split-range arrangement can be implemented so that the capacity control uses the compressor speed as the primary control handle, supplemented by a recycle valve in case a severe turndown operation is needed. Due to the length limit, we will skip the detailed discussion on this split-range arrangement.

11.3.5 Practical Considerations

Integrated compressor control is one of the most complicated baselayer control schemes. In addition to the intricate control design, there are also numerous practical considerations to be addressed. Below are some examples:

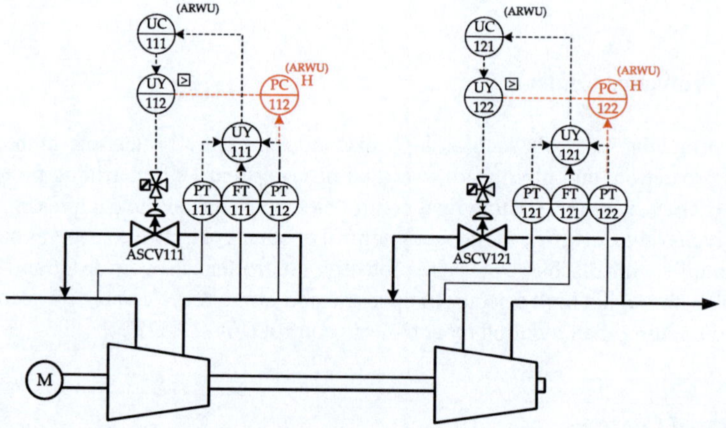

Fig. 11.32 Anti-surge control with high discharge pressure override

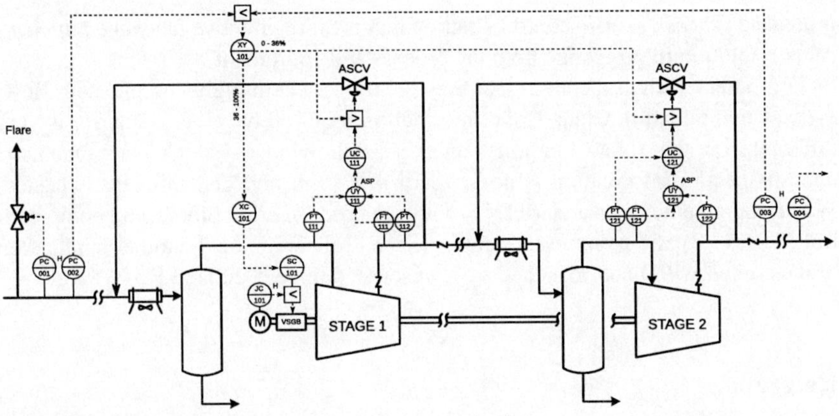

Fig. 11.33 A typical integrated compressor control scheme

1. The fast compressor dynamics require high-speed calculation and control action. This requirement poses some unique challenges on all the components in the control loop, especially on the control valves.
2. PID configurations are quite involved because of the extensive use of override controls, fan-out controls, and cascade controls. Anti-reset windup, bumpless transfer, and back initialization are essential considerations.
3. Startup and shutdown. Starting the compressor is a complicated procedure. The compressor goes through multiple stages of the transition process, from zero flow and zero pressure ratio, through full recycling, quickly increasing the discharge pressure, pushing for the forward flow, and finally reaching the steady-state. The startup sequence is responsible for bringing the compressor to the normal operating condition safely and quickly. Process control is part of the startup sequence.

11.4 Summary

Equipment control such as pump and compressor control is traditionally regarded as the specialty of equipment manufacturers or their contractors, assuming that equipment is too complex and critical to leave their control to somebody else. However, one of the issues with the control solutions from equipment manufacturers or third parties is that their primary objective is to protect their equipment rather than the overall process control. They do not have the required process knowledge and control expertise to produce an optimal integrated control strategy aiming at overall optimality in control.

Process control depends on a good understanding of the process dynamics, equipment characteristics, and operating requirements. For this reason, process control engineers with a holistic view of both the process and the equipment can typically produce a *process-centric* control solution that is more effective than an *equipment-centric* solution in protecting both the process and equipment.

Equipment such as a compressor may seem overwhelmingly complicated. However, as mentioned in Chap. 1, the information required by process control is very different from that for an equipment engineer. Following a systematic approach, the control-specific information is not as hard to extract and generalize as expected. When implemented in standard DCS with standard function blocks, the pump control (Sect. 11.2) and compressor control (Sect. 11.3) are merely another application of the complex PID control schemes, as discussed in Sect. 2.2 and Fig. 4.1.

References

Batson BW (1996) Invariant coordinate systems for compressor control. In: Proceedings of the international gas turbine and Aeroengine congress & exhibition. The American Society of Mechanical Engineers, Birmingham, UK

Berkeley D (1958) Ejectors have a wide range of uses. Petroleum Refiner December

Bloch H (2006) A practical guide to compressor technology, 2nd edn. Wiley Interscience, New York

Brannan C (2018) Rules of thumb for chemical engineers - a manual of quick, accurate solutions to everyday process engineering problems, 6th edn. Elsevier, Amsterdam

Campbell J (2004) Gas conditioning and processing - the equipment modules, vol 2, 8th edn. John M, Campbell and Company

DOE (2001) Pump life cycle costs: a guide to LCC analysis for pumping systems. Technical Report DOE-GO-102001-1190, Hydraulic Institute, Euro-Pump, and the US Department of Energy's Office of Industrial Technologies (OIT)

Hanlon PC (2001) Compressor handbook, 2nd edn. McGraw-Hill, New York

Keenan JH, Neumann EP, Lustwerk F (1950) An investigation of ejector design by analysis and experiment. ASME J Appl Mech 72(17):299–309

Kurz R, White RC, Brun K, Winkelmann B (2016) Surge control and dynamic behavior for centrifugal gas compressors. In: Proceedings of Asia Turbomachinery and Pump Symposium, Turbomachinery Laboratory, Singapore

Liao C (2008) Gas ejector modeling for design and analysis. PhD thesis, Texas A&M University

Mirsky S, Jacobson W, Tiscornia D, McWhirter J, Zaghloul M (2012) Development and design of anti-surge and performance control systems for centrifugal comrpessors. In: Proceedings of the Forty-Second Turbomachinery Symposium, Houston, Texas

Nägeli JP, Spechtenhauser A, Aicher W (1973) Turbomachinary in base load natural gas liquidification plants. In: Proceedings of the 12th World Gas Conference, Nice, France

Perry R, Green D (1984) Perry's chemical engineers' handbook, 6th edn. McGraw-Hill, New York

Staroselsky N, Ladin L (1979) Improved surge control for centrifugal compressors. Chem Eng 175–184

White MH (1972) Surge control for centrifugal compressors. Chem Eng 54–62

Chapter 12
Control Typicals: Plant-Wide Control and Unit Control

The number of processes in operation or under design is innumerable. Therefore, it is not practical to provide standard control designs for every typical process. The more practical approach is to learn the general thinking process and structured analytical skills that can be applied to any process control problem. This chapter presents several commonly encountered but not much-discussed processes to demonstrate the process analysis and control design skills and methodology. The selected processes are all plant-level or unit-level operations with manageable complexity, from tank blanketing (Sect. 12.1), boiler drum level (Sect. 12.2), well test separator (Sect. 12.3), to firewater control (Sect. 12.3). An acetylene hydrogenation process is presented to show the work process for unit-level control design (Sect. 12.5), and an E&P surface production plant is used to illustrate the overall control strategy of a plant-wide control solution (Sect. 12.6). The discussions focus not on the specific solutions but rather on the thinking processes and design procedure.

12.1 Liquid Tank Blanketing Control

Gas blanketing control of a liquid tank is a common and straightforward application. However, this application is chosen as a typical control problem to show that even a simple control problem can have different control designs, from good, bad, to ugly, with long-lasting consequences.

The ugly designs may work to some extent and would not impact the operation as severely as tripping the plant operation. They are just less efficient or less reliable. For this very reason, plant operations may not take the poor designs as severe problems and choose to live with the poor performance. The bad control designs are often carried over to the next project through careless copy and paste. Due to the massive number of tanks in operation and the many incidents associated with tank operation, the proper control design is critical.

S. S. Niu and D. Xiao, *Process Control*, Advances in Industrial Control,
https://doi.org/10.1007/978-3-030-97067-3_12

12.1.1 *Process Description and Analysis*

A tank is typically used for separation or storage. For example, there are many liquid tanks of different sizes and shapes in the oil and gas processing facilities, such as refineries, petrochemical plants, and terminals. Some tanks are for the separation or storage of oil rich in hydrocarbon, while others are for produced water that has entrained hydrocarbons.

Operating Requirements

Tank operation is typically a low-pressure operation (near atmospheric pressure). The critical operating requirement for tanks with hydrocarbon content is to prevent oxygen or air from entering the tank. The hydrocarbon content in the liquid must be adequately isolated from oxygen or air to avoid potential fire and explosion. Because the tank operation is seemingly straightforward, the importance of proper control is often underestimated. Due to negligence in design and operation, many incidents are linked to the collapse or explosions of liquid tanks.

One common approach to protecting the tank is to blanket the liquid in the tank with oxygen-free gas. *Tank blanketing*, also referred to as *tank padding*, applies gas to the empty space in a storage container. The purpose is to keep oxygen levels in and around the product at a low level to reduce oxidation. The oxygen-free blanketing gas can be inert nitrogen gas or processed natural gas from the same facility. As the blanketing gas has its cost, the tank is expected to be protected with minimal gas consumption (Brannan 2018; Kinsley 2001).

Process Flow Analysis

Figure 12.1 shows a free water knockout tank (FWKO) for produced water from production separators. There are entrained hydrocarbons in the produced water that may flash out and fill the tank with flammable hydrocarbon vapor. No contact with oxygen is allowed.

Supply and Demand Analysis

There are three gas streams for the tank operation: the inlet blanketing gas, the flashed gas from the liquid, and the outlet venting gas to flare. They are related to each other by the following material balance equation:

$$\text{[venting gas]} = \text{[blanketing gas]} + \text{[flash gas]}. \tag{12.1}$$

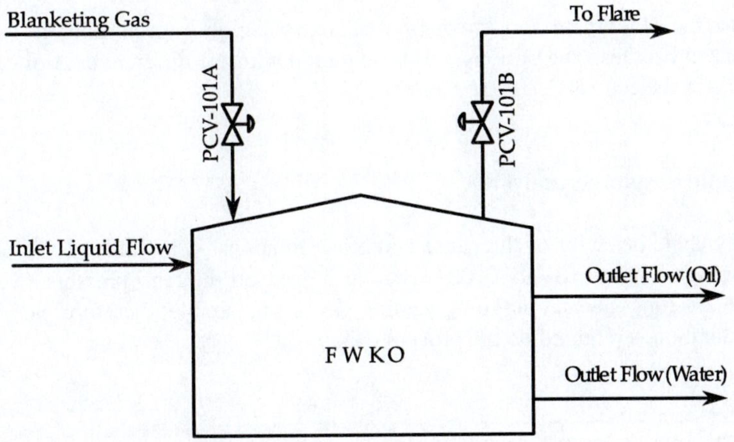

Fig. 12.1 A free water knockout drum with gas blanketing

The operation is *tank-driven* (see supply and demand analysis in Sect. 6.3). To the tank, the blanketing gas stream is demand-driven, while the venting gas stream is supply-driven. This tank-driven operation requires that both the blanketing gas and venting be able to *swing*. Two independent control valves meet the blanketing operation requirements.

The inlet blanketing gas flow and outlet venting gas flow are two independent control handles serving as two manipulated variables for control. The tank pressure is the control target chosen as the controlled variable.

Cause and Effect Analysis

The available control handles are the inlet blanketing gas and outlet flaring gas. Increasing the blanketing gas intake will increase the tank pressure. Increasing the venting gas flow to flare will decrease the tank pressure. This relationship can be summarized by the simple cause and effect table (CET) shown in Table 12.1.

Degree-of-Freedom Analysis

There are two control handles and one control target, implying that one extra degree of freedom (DOF) must be consumed. As discussed in Sect. 6.5, there are two ways to

Table 12.1 Cause and effect table for gas blanketing control

	Tank pressure
Blanketing gas flow ↑	↗
Venting gas flow ↑	↘

exhaust the DOF: increasing the number of control targets or decreasing the number of control handles. The options will be explored with the different control designs later on in the control design section.

Dynamic Response Analysis

The dynamic behavior of the cause and effect relationship in Table 12.1 is all fast and straightforward. Besides, the dynamic impact on the tank pressure from the inlet blanketing gas and outgoing venting gas is comparable; therefore, no special consideration is required on the process dynamics.

12.1.2 Control Strategy

The control objective for the liquid tank operation is to maintain the tank pressure within the desired range of 0–1.0 kPa for three reasons:

1. Keep the tank above atmospheric pressure to prevent oxygen ingress.
2. Avoid vacuum in the tank to prevent buckling collapse of the tank.
3. Minimize the consumption of the blanketing gas.

This application has one control target and two control handles; thus, as suggested in Table 2.7, a complex PID control scheme is sufficient. There is no need to explore more advanced control schemes.

12.1.3 Functional Design

The process is simple, and the control objective is clear; however, the detailed functional design can vary from good, bad, to ugly.

The "Ugly" Design

A naïve control design uses two independent pressure controllers to maintain the tank pressure, as shown in Fig. 12.2. One controller manipulates the inlet blanketing gas, while the other controls the flaring gas rate. Both controllers are assigned with the same setpoint of 0.5 kPa(g).

This design will not work as desired! The pressure measurements for the two pressure controllers are highly correlated, so they are essentially the same variable. As a result, this design has one controlled (tank pressure `PC-101A` $\approx$ `PC-101B`) and two manipulated variables (`PCV-101A` and `PCV-101B`).

As explained in Sect. 8.3.4 on troubleshooting, there is no unique solution for this under-determined control problem. The two controllers will continuously fight each other until one is driven to saturation because when the valve is saturated, it is no longer a DOF. The worst-case scenario is that the blanketing gas control valves saturate to the fully open position, and excessive blanketing gas is sent to flare. This scenario can happen because a PID controller does not know or care about the absolute opening of the valve. It only cares about the incremental change to be added to the valve opening to reduce the control error (see Sect. 3.1.6). The two controllers can maintain the 0.5 bar(g) setpoint with infinite combinations of valve openings, including one fully open. This design is similar to the troubleshooting example of Fig. 8.5 discussed in Sect. 8.3.4.

This design may have achieved the goal of maintaining the tank pressure but is far from optimal due to the potential excessive blanketing gas consumption.

The Bad Design

Split-range control is a common approach to reducing the number of independent control handles by *linking* the two valves in series (see Table 2.7). For this tank pressure control example, as shown in Fig. 12.3, a split-range controller can improve the control performance over the design shown in Fig. 12.2.

A single pressure controller `PC-101` manipulates two control valves, `PCV-101A` and `PCV-101B`. The control with `PCV-101A` is `reverse` action, while the one with `PCV-101B` is `direct` action.

Since the dynamics from the two valves to the tank pressure are comparable, the split point is set to the default 50%. When the tank pressure is below 0.5 kPa(g), the pressure controller first reduces the venting gas flow until the valve `PCV-101B` is fully closed. If the pressure is still below the setpoint, the controller will introduce

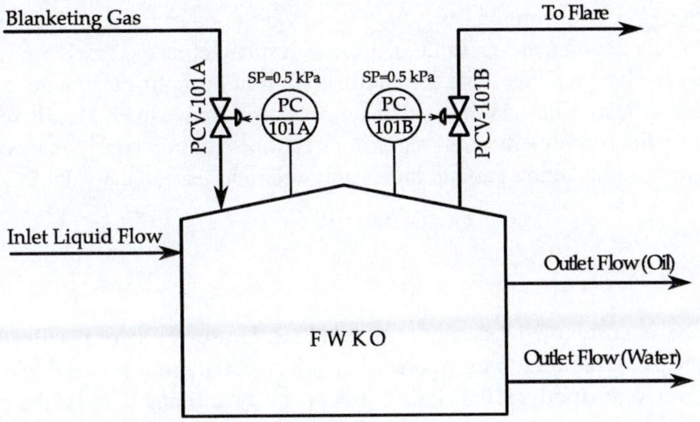

Fig. 12.2 Tank gas blanketing control: a naive design

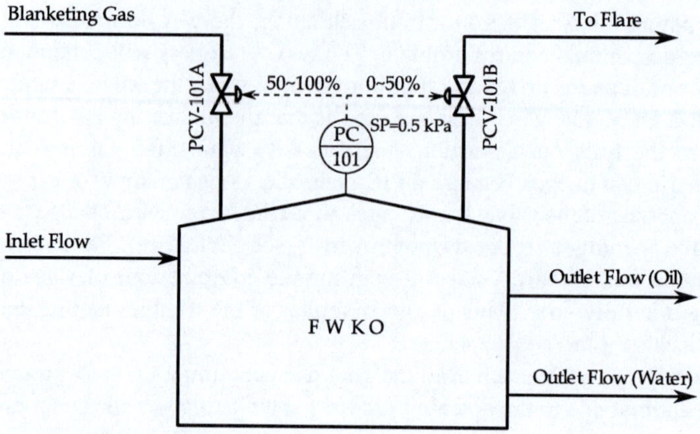

Fig. 12.3 Tank gas blanketing control: a improved design

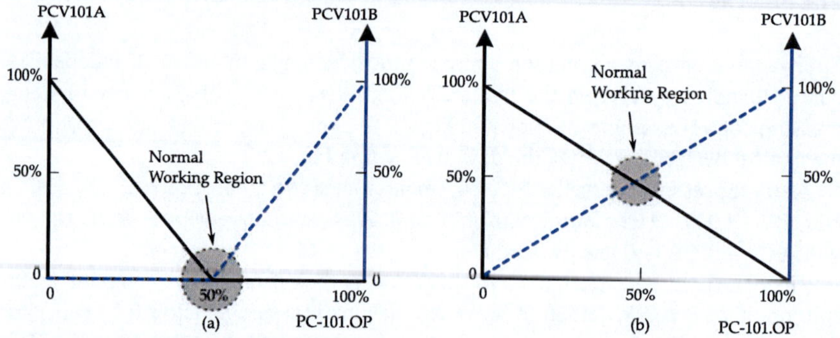

Fig. 12.4 Tank gas blanketing control: split point

blanketing gas by opening valve `PCV-101A`. On the other hand, if the pressure is higher than its setpoint, the controller will close the blanketing gas first before opening the venting valve. See Fig. 12.4a for an illustration of the split-range pattern.

A careless design may result in a split-range control shown in Fig. 12.4b, where the controller adjusts both valves in opposite directions simultaneously, unnecessarily consuming the blanketing gas and increasing wear and tear on the valves.

The Good Design

Figure 12.3 is an improved design over Fig. 12.2. However, there is still room for improvement. According to the operating requirement, there is no need to maintain the tank pressure strictly at 0.5 kPa(g). It is perfectly acceptable to let the pressure fluctuate without taking any control action, as long as the pressure fluctuation is within an acceptable range.

Along this line, a further improved design uses two pressure controllers with different setpoints, i.e., the dual pressure controller approach (Sect. 2.2.2). As shown in Table 2.7, dual-controller control increases the number of control targets to consume the extra DOF, contrary to the split-range control scheme just discussed above.

The process values for the two pressure controllers come from the same pressure transmitter.[1] See Fig. 12.5 for an illustration of the controller configuration and note the difference with the naïve design in Fig. 12.2.

With the dual-controller control design, there are three operating scenarios:

1. If the tank pressure `PT-101` is higher than the setpoint of controller `PC-101B` (=0.8 kPa), the flaring valve will be opened to let out the excess gas and the inlet valve for the blanket gas remains closed.
2. If the pressure is lower than the setpoint of `PC-101A` (=0.2 kPa), then the inlet blanketing gas flow valve is opened to let in more gas, while the flaring valve is fully closed.
3. When the pressure is between 0.2 and 0.8 kPa, which is expected to be the case most of the time, both valves will be fully closed.

Compared with the split-range control scheme where the pressure controller acts as the normal regulatory controller, the dual controllers perform as two independent protective controllers, one against low pressure, the other high pressure. At steady-state, at least one valve is fully closed. When the pressure is between 0.2 and 0.8 kPa(g), both valves are fully closed. Therefore, the gas consumption and venting are both minimized.

See Sect. 4.1.6 for the pros and cons between split-range control and dual-pressure control.

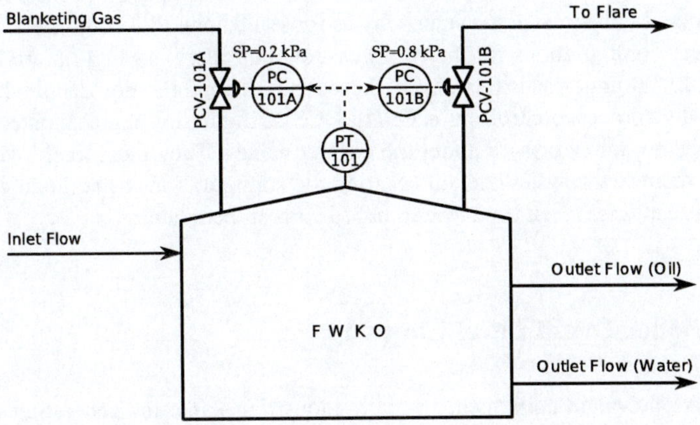

Fig. 12.5 Tank gas blanketing control: a good design

[1] If two independent pressure meters are used, the measurement may have discrepancies that potentially affect the two pressure controllers' setpoint spacing. Besides, the use of two meters also increases the cost.

Table 12.2 Loop configuration details

Parameter	Unit	PC-101A	PC-101B
Algorithm type		PI	PI
Equation type		I-PD	I-PD
Typical PV value	kPa(g)	0.5	0.5
Setpoint	kPa(g)	0.2	0.8
Setpoint limit	kPa	0.2–0.3	0.6–0.8
Initial tuning: gain	%/%	2.0	2.0
Initial tuning: integral	Min	5.0	5.0
Initial tuning: derivative	Min	0.0	0.0
Direction of action		REVERSE	DIRECT
Fastest scan rate/sequence		Default	Default
Output scaling		0% = Fully closed, 100% = Fully open	
Fail-safe position		Fail-close	
PV filtering	Min	TBD	TBD
PV tracking		No	No
Anti-reset windup		N/A	N/A
Bumpless transfer		N/A	N/A

12.1.4 *Implementation and Operation*

The implementation and operation are straightforward because the two pressure controllers (PC-101A and PC-101B) are independent simple PID controllers. The PID tunings can be treated the same way as for standalone PID controllers.

The key configurations for the two pressure controllers in Fig. 12.5 are listed in Table 12.2. As suggested in Sect. 7.4.4, these types of configuration details should be provided by a process control engineer. The DCS configuration engineer or technician often does not know or care about the correct values. They most likely will leave many of them to the system default. Setting the configurations right the first time is much more efficient than troubleshooting to correct them later.

12.2 Boiler Drum Level Control

Boilers are prevalent equipment in many industrial plants for generating process steam. Boilers are also central to thermal power generation. Boiler drum level control is a vital requirement for operational safety and efficiency.

The boiler drum level control is chosen as a typical process to demonstrate the non-minimum phase process behavior and the challenges it causes for process control.

For this application, we will again follow the work process in Figs. 6.14 and 7.6, but with the focus on dynamic response analysis and functional design.

12.2.1 Process Description and Analysis

The boiler drum is where water and steam are separated. A typical steam boiler application is shown in Fig. 12.6.

Operating Requirements

The steam must be superheated to avoid water carryover to the downstream consumer, so the operating objective of the boiler drum is the proper separation of steam and water. Maintaining the proper steam/water interface level is critical—if the level becomes too low, the boiler can run dry, resulting in mechanical damage to the drum and boiler water tubes. If the level becomes too high, water can be carried over into the superheater and steam pipework, possibly damaging downstream equipment. Therefore, for the steam-generating process, the critical requirement is to maintain the required steam quality and ensure that the drum's water level is within a safe range.

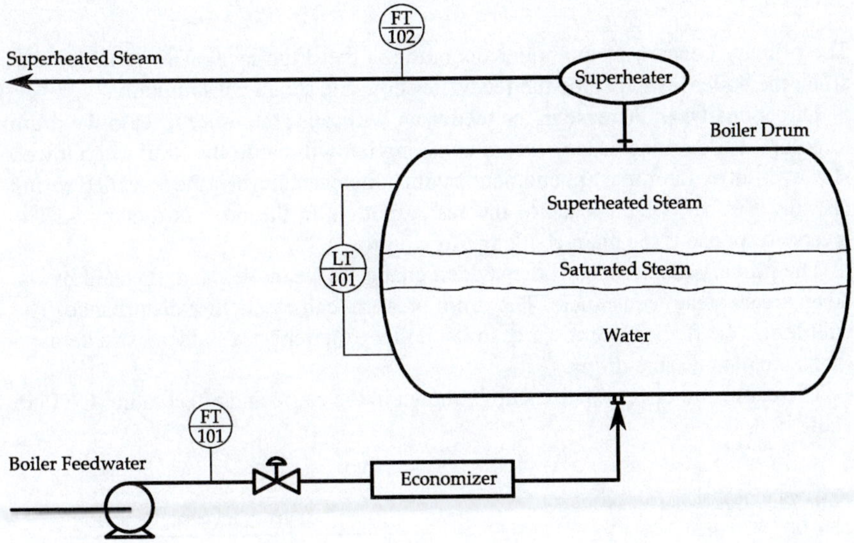

Fig. 12.6 Boiler drum level process

Process Flow Analysis

The process flow of the steam-generating process with a boiler is relatively straightforward. Boiler feedwater is pumped to the boiler, where it is heated into superheated steam and feeds the downstream steam consumers. The streams of concern include the inlet feedwater and the outlet steam flow. There are requirements on pressure and the temperature inside the drum, but that is a separate topic and is not included in the discussions here.

Supply and Demand Analysis

The mass balance in the boiler is between the boiler feedwater and the generated steam. Feedwater supply is typically assumed to be unlimited.

Since the level in the drum is an unstable integrating process, the mass balance between the feedwater and generated steam is reflected in the drum level. Any imbalance will cause the drum level to ramp up or down continuously. The drum level is thus a critical variable to control.

As a steam generator, the operation is demand-driven, the steam need from downstream dictates the production rate. The boiler feedwater is assumed to be unlimited and can serve as the swing stream with unlimited capacity.

Cause and Effect Analysis

The primary concern of the boiler operation is the drum level. The variables that affect the boiler level include the feedwater flow and steam consumption.

Directionally, an increase in the feedwater will cause the water level in the drum to go up, while an increase in steam consumption will cause the level to go lower. However, an extraordinary phenomenon with boiler water level is the so-called *shrink and swell* effect, which leads to inverse responses in the process dynamics. The inverse response is the main challenge for control.

The primary disturbance is the sudden change in steam demand, dictated by the downstream steam consumers. The drum pressure can be another disturbance. The bubble size is affected by pressure, so if a boiler experiences a sudden extra demand for steam, its pressure drops.

The cause and effect relationship is shown in the cause and effect table (CET) in Table 12.3.

Table 12.3 Cause and effect table for boiler drum level control

	Boiler drum level	Remark
Feedwater flow rate ↑	↗	Inverse response
Steam flow rate ↑	↘	Inverse response
Drum pressure ↑	↘	

Degree-of-Freedom Analysis

Naturally, the drum level is the controlled variable in the level control scheme. The boiler feedwater is the main handle for the level control and is the manipulated variable.

The steam flow rate is a key process variable (KPV) but is determined by downstream operation. The control handle for the steam flow is not available to the drum level control scheme. However, the steam flow is measured online. Based on Table 2.7, the option for including the steam flow in the control design is to treat it as a disturbance variable, and the choices for control include feedforward control and ratio control.

Dynamic Response Analysis

The dynamic behavior of the level response to the changes in feedwater flow and steam flow is exceptional. The boiler drum level is an integrating process. Any mismatch between the feedwater supply and steam consumption will cause a continuous ramping in the drum level. To further complicate matters, the boiler drum level is notorious for its inverse response. The inverse response poses a unique challenge for control, and thus the complexity in control design:

1. From feedwater flow to the drum level: When the drum level is low, more feedwater should be added to increase the level. However, because the feedwater is cooler than the water/steam in the drum, the cooler water will cause some steam to condense, which in turn causes the volume of water/steam to decrease. See Fig. 12.7 for an illustration. As a result, adding more feedwater tends to decrease the drum level instead of increasing it initially.
   Once the temperature is restored, the drum level will start to increase. This inverse response may confuse the controller with feedback control and cause it to take the wrong actions.

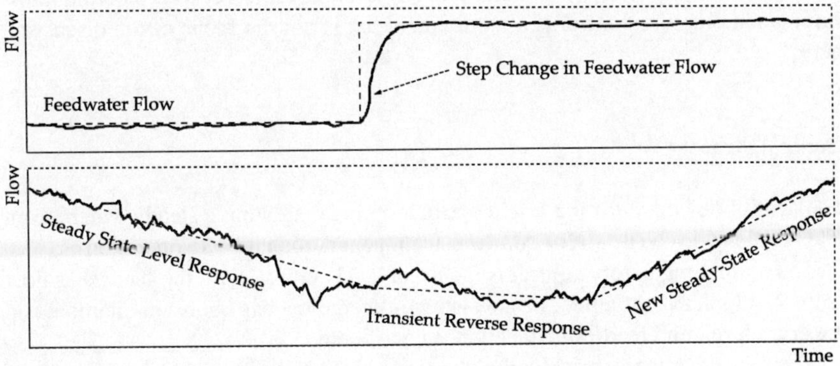

Fig. 12.7 Inverse response of boiler drum level

2. From steam flow to the drum level: A sudden steam load increase will cause a drop in the steam drum pressure because the firing rate cannot increase fast enough to match the steam production rate at the new demand level. The decrease in pressure causes a fraction of the saturated water in the boiler to vaporize immediately and increase the steam drum level. Once boiler pressure is restored, the steam bubbles contract, and the measured water level drops suddenly. This behavior is an inverse response to the load change.
   As the net steam draw rate has gone up, the net water flow to the boiler is expected to increase because the total mass of water inside the boiler is falling. The false response in the level may cause a feedback control to respond in the wrong direction, at least initially.

Inverse response (or non-minimum phase response) is illustrated in Fig. 6.11d and is one of the challenges for feedback control. The difficult dynamics lead to different control designs.

12.2.2 *Control Design*

The challenge in the drum level control design is not in the complexity of the process flow scheme but rather related to the nonlinear dynamic behavior between the feedwater and the drum level and the unreliable measurement of the drum level.

Control Scope

The boiler operation is a relatively isolated process. As explained above, the interaction with the upstream process flow is via the boiler feedwater, and the interaction with the downstream process flow is the steam flow rate. The boiler feedwater is assumed to be unlimited, and the steam flow is treated as a critical disturbance variable.

Within the boiler operating unit, the drum level may interact with the drum pressure. Good pressure control is a requirement but is not the focus of the discussion here.

Control Strategy

The control objective for the level control loop is to maintain a steady water/steam interface level in the drum for efficient steam generation and safe operation.

The normal regulatory control is the drum level control, with the feedwater flow as the manipulated variable. The downstream steam flow has a substantial impact on the drum level, and feedforward action is considered.

There are no process variables that need to be protected with protective controllers.

Control Algorithm

The challenge for boiler drum level is dealing with the nonlinear dynamic response and the difficulties in level measurements. The control technology selection is limited to DCS-based PID control schemes, typically with three options, from simple to more complex: single-element, two-element, and three-element control:

1. Single-element control: The single-element control strategy is a simple standalone PID-level control, as shown in Fig. 12.8.
   The drum level is the controller PV. The level controller directly manipulates the feedwater valve, which is at the pump discharge.
   One or more boiler feedwater pumps push water through feedwater control valves into the boiler drum. In the case of multiple pumps in parallel, split-range control, fan-out control, or selective control (Sect. 2.2) can be considered to manipulate the pumps. Variable-speed boiler feed pumps are sometimes used, in which case the pump speed will be the manipulated variable instead of the flow control valve (see Sect. 11.2 on pump control).
   The simple feedback control design described above is called *single-element control* because it uses only one feedback element—the drum level measurement.
2. Two-element control: The *two-element control* is a standard-level-flow-valve cascade control and offers some incremental improvements over the single-element control. The inner loop is a flow controller which stabilizes the pump discharge flow, and the outer loop is the level controller manipulating the flow control setpoint. See Fig. 12.9 for the control schematic.

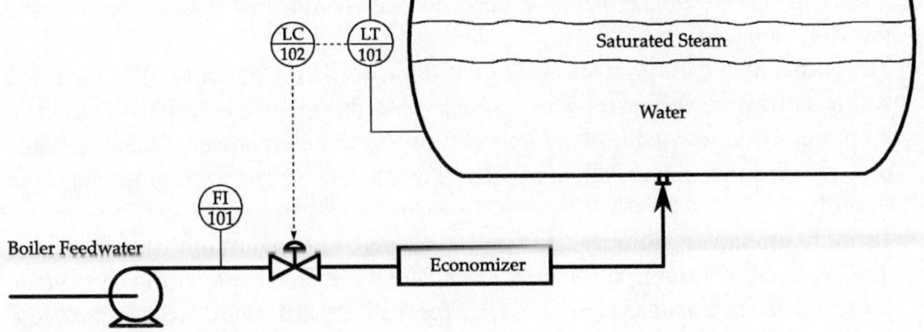

Fig. 12.8 One-element control for boiler drum level

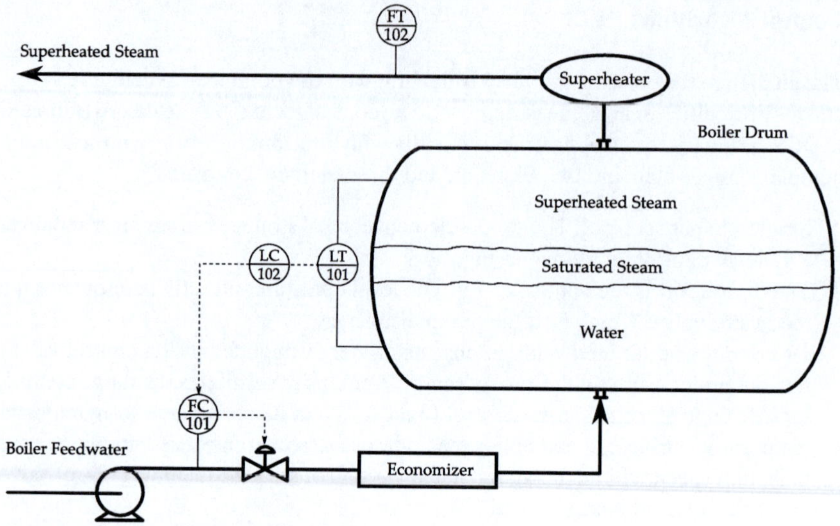

Fig. 12.9 Two-element control for boiler drum level

For large capacity boilers, there may be two or three feed pumps. The pumps will be switched on or off to follow the variation in feedwater needs. As a result, the overall dynamics of the pump discharge valve to the discharge flow can be complex and slow. Introducing a flow control as the inner control loop for level control can result in a noticeable improvement in the level control performance. See Sect. 2.2.4 for the advantages of cascade control over simple control.

3. Three-element control: Like the feed flow, the steam flow changes can also cause significant deviations in the drum level and may even trip the boiler. Changes in steam flow will directly affect the drum level before the controller starts to adjust the feedwater flow. Due to the slow and challenging level dynamics, the feedback control performance of single-element and two-element controls is often not satisfactory in responding to steam flow disturbance. In addition, the drum level's inverse response to feedwater flow and steam flow makes control even more challenging.
The difficulty lies in that the PID controller needs a high controller gain and strong derivative action for fast response. On the other hand, the PID control loop requires a low gain, long integral time, and no derivative to achieve stable performance. It is thus challenging to strike a balance between performance and stability.
To more effectively compensate for the influence of steam flow variations, the three-element control is commonly used. The three-element control is a cascade control with feedforward (see Sect. 4.1.2 for feedforward control). Steam flow rate is a measurable disturbance as a feedforward variable to the single-element or two-element control scheme. Changes in the steam flow rate will almost immediately be counteracted by similar changes in the feedwater flow rate (Fig. 12.10).

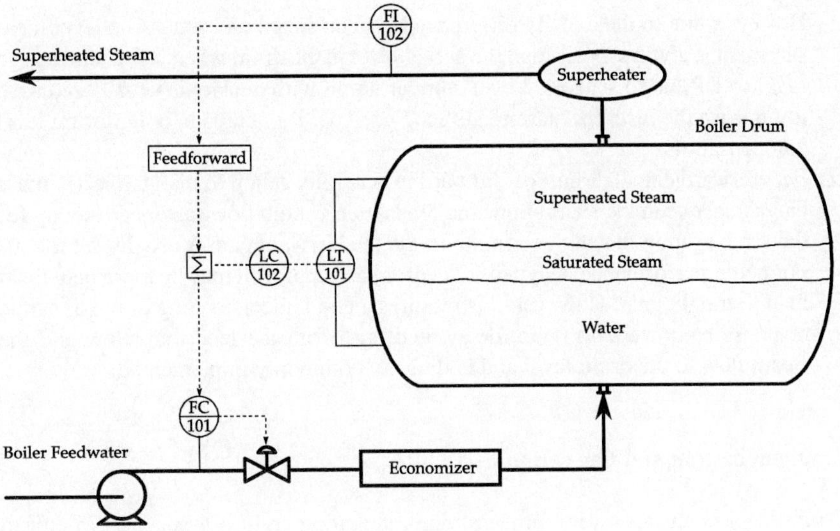

Fig. 12.10 Three-element control for boiler drum level

Although three-element drum level control is typically superior in performance to single- or two-element control, the feedforward element is typically turned off at low boiler loads. The reason is that steam flow measurement can be inaccurate at low rates of steam flow. Once the boiler load is high enough for steam flow to be measured accurately, the feedforward must be activated bumplessly (via the incremental form).

Functional Design

The overall functional design of the level control scheme for the boiler drum is straightforward. However, there are several things to watch out for, including the level compensation and feedforward calculation:

1. Level compensation: The drum level is challenging to measure accurately due to the turbulent condition with the water/steam interface and the density difference between the water and steam phases. Both are inherent to the process.
   Typical level measurement is by gauge glass for direct visual indication. Remote level sensors based on differential pressure, displacers, conductivity, and radar are typically used for closed-loop control. The DP-based level meter is the most commonly used for control.

However, due to the density difference between steam and water, the level measurement is always lower than the actual level in the drum when using either gage glass or DP-based sensors. Level compensation with density is usually required to improve the level measurement. See Sect. 10.5 for details in boiler drum level compensation.

2. Feedforward calculation: Feedforward is generally set up to maintain a 1:1 mass balance between the steam flow and feedwater. If both flow meters are set up for the same span in the engineering unit, the feedforward gain is usually set to 1.0. For better performance, a dynamic feedforward approach may be more beneficial than a straight gain. However, this requires a good understanding of the dynamic response characteristic (gain, delay, and lag) from the feedwater flow and the steam flow to the drum level and is thus not commonly implemented.

Implementation and Operation

Drum level is an integrating process, and integrating control loops are difficult to tune. The control loop can quickly become unstable if the controller's integral time is too short (i.e., high integral gain). The process-imposed requirement for a long integral time makes the loop slow to recover from disturbances to the drum level.

Following the tuning recommendation for cascade control loops, the inner flow loop should be tuned first, then the level loop. The flow loop tuning is straightforward, but the level tuning can be tricky. Model-based tuning based on a proper step test is recommended. The Lambda tuning method (Table 9.7) for controllers usually provides stable control loops:

$$K_c = \frac{1}{K_p}\frac{2\lambda + \theta}{(\lambda + \theta)^2}. \tag{12.2}$$

12.2.3 *Practical Considerations*

A practical challenge is the credibility of the level measurement. Depending on the technology used, the interface level measurement between the steam and water is usually inaccurate and inconsistent due to the density difference between steam and water. Level compensation with density is typically required, as discussed in Sect. 10.5. After all, we can't control what we can't measure.

For boiler drum level control, some improvements in process design or operating procedure can simplify the response dynamics to facilitate process control. For example, preheating the boiler feedwater and adding drum baffling can reduce the inverse response behavior.

12.3 Well Test Separator Control

Well testing is an essential and routine operation in upstream production to acquire production data on individual wells. The collected data is used to monitor the production rate for better production management and improve the understanding of the hydrocarbon properties of the underground reservoir where the hydrocarbons are trapped. Well testing is chosen as a typical process in this chapter for the following reasons:

1. There are some unique control challenges related to the inherent limitations in measurement technology. Process control can help to alleviate the limitation.
2. Well test sequence is a typical sequential control. Sequential control is a gray area that falls under the scope of process engineering, operation, instrumentation, and process control. This application shows the connection and interaction between sequential control and standard process control.
3. This application also demonstrates that the most effective improvement to control solutions is often from improvement in the process design or advancement in the instrumentation technology. A controller cannot do the impossible. The process has to be designed to accommodate its control function. Without the process doing its part, the controller cannot adequately do its job either.

The classical approach for well tests is through a *well test separator* where the oil, gas, and water are separated and individually measured.

12.3.1 Process Description and Analysis

In upstream production, a three-phase test separator is standard equipment that measures the individual phase production of oil, gas, and water. However, the control and operation of test separators have many practical issues, and there is much room for improvement.

Operating Requirements

An upstream exploration and production (E&P) company may have thousands of production wells in operation and dozens or hundreds of test separators for well testing. Each test separator is responsible for dozens of wells with various production capacities. The test separator can test only one well at a time, and each well test may last up to 24 h.

Two operating requirements are of significant importance to us. One is the scheduling and executing of well test in rotation. Well testing has traditionally been a labor-intensive manual operation involving multiple dedicated teams. Automated well testing has been a strong demand and is making progress as part of improved sequential control. The other requirement is to ensure accurate and reliable measurements of the individual phase flows, which is not as trivial as commonly believed. The two requirements dictate that the test separator operation needs to

1. emulate the normal production condition. The well production rate is affected by the wellhead pressure. When the well is diverted from the production header to the test header, the wellhead pressure should be maintained as close as possible to the wellhead pressure during normal production so that the same production rate is maintained while in test and normal production;
2. separate the incoming multi-phase flow into individual phases (gas, oil, and water). The phase separation is affected by the pressure, oil level, and interface level. Adequate control of the pressure and levels in the separator is required;
3. measure the individual phase flow rates. The separated oil, gas, and water are individually measured and recorded. The production rate can vary from well to well and over the well's life cycle.

Process Flow Analysis

A typical production system from wellhead to test separator is shown in Figs. 12.11 or 12.12.

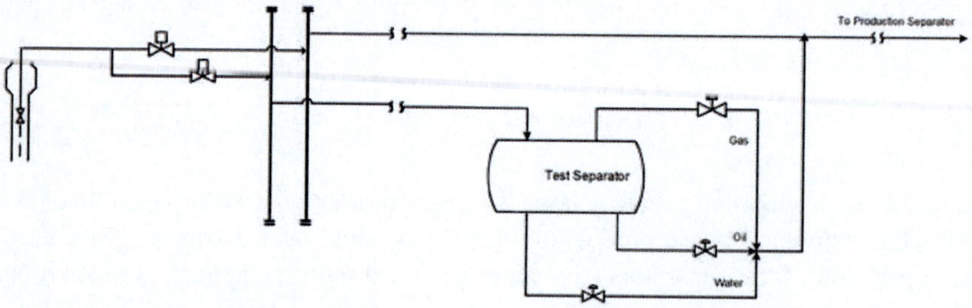

Fig. 12.11 From wells to test separator

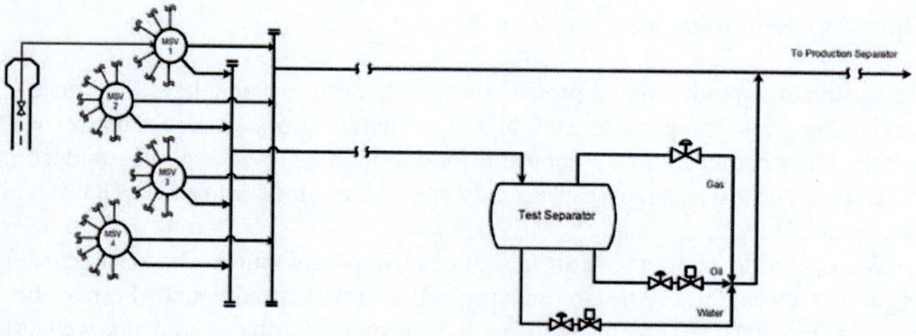

Fig. 12.12 From wells to test separator, with multi-port selection valve (MSV)

The produced multi-phase fluid from a well is sent to a production manifold. The fluids from multiple wells commingle at the manifold and flow under pressure to one or more production separators (bulk separators) in the downstream facility (processing station). The multi-phase fluid is separated into gas, oil, and water at the production separator for further processing.

The multi-phase fluid enters the separator as a turbulent flow. The separator provides sufficient residence time for the oil and emulsion to form an oil layer ("oil pad") at the top. The free water settles to the bottom. The oil weir maintains the oil level, and the controller maintains the water/oil interface level.

Although the total production rates of oil, gas, and water from multiple wells are measured, each well's production rate is typically unavailable online. For production monitoring and planning, it is crucial to know the production rate at each well.

A test separator is used for measuring the individual well production, one well at a time. The produced well fluid from a selected well is regularly diverted to a test separator, typically monthly, where the fluid is separated into gas, oil, and water. The flow rate of each phase is individually measured and totalized to calculate the production rate. The diversion of the fluid from the production separator to the test separator is typically automated, either via a range of isolation valves (Fig. 12.11), or more commonly, a multi-port selector valve (MSV) (see Fig. 12.12).

Figure 12.13 is an illustration of a conventional single-weir test separator. The oil is skimmed over the weir and flows to the oil chamber downstream of the weir. The oil level is controlled by a level controller that operates the oil outlet flow. The produced water flows from a nozzle located upstream of the oil weir. An interface level controller senses the oil/water interface and manipulates the outlet water flow to maintain the desired interface level. The gas flows horizontally and then out through

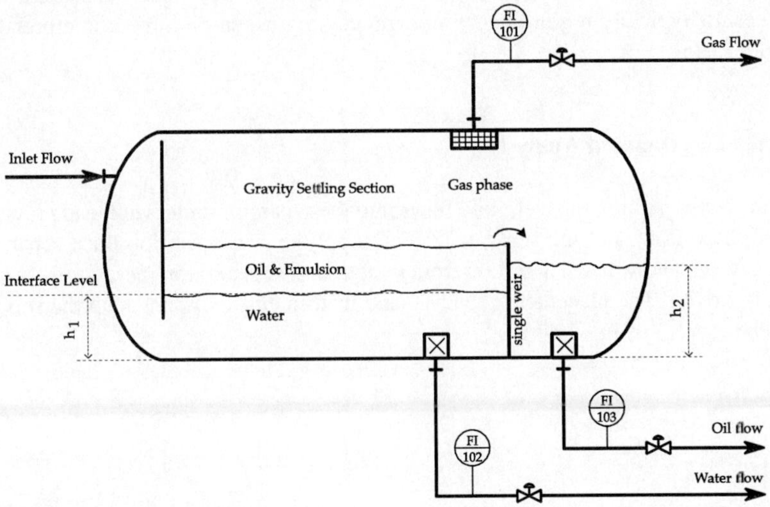

Fig. 12.13 A three-phase test separator

a mist extractor. A pressure controller is usually provided to keep the pressure inside the vessel. In some designs, the vessel pressure is let to float with the downstream operation.

Operating Limits and Constraints

Even though the test separator may be running round-the-clock, individual wells are tested only in turns, and thus the test data is very infrequent, typically once every few weeks. Besides, the measurements with the test separator are very unreliable as the results are affected by many factors. The key constraints that affect the operation include the following:

1. Inconsistent well test procedure: A well test has many steps that are prone to human operating errors. For example, wellhead pressure condition, well and separator line-up, separator purging, and reporting.
2. Inaccurate flow measurements: The test separator is designed to test multiple wells, of which the production rate may vary drastically. For the same well, as the well production depletes over time, the flow rates may also change substantially.[2] The production flow is measured with online flow meters. Flow meters have effective measurement ranges, typically between 10 and 100% of the full range (typical turndown ratio of 10:1). Below the turndown limit, the measurement would become less accurate (e.g., orifice-based) or even unavailable (e.g., vortex meters). Since the flow meter is sized to cover the highest production rate, the same flow meter may be significantly too large for low-producing wells, and thus no accurate measurement may be obtained.

With the current well testing practice, the well test data is notoriously unreliable. The test results typically require heavy post-processing based on human interpretation and cleansing.

Supply and Demand Analysis

For the test separator, the well fluid flows into the separator under wellhead pressure. Each phase flows out the separator under supply-driven operation once separated into gas, oil, and water. Therefore, from inlet to outlet, separator operation is strictly supply-driven. The placement of valves and instruments complies with the supply-driven model. See Fig. 12.13.

[2] For example, a low gas–oil ratio (GOR) oil well may gradually deplete to become a high GOR well after some years of production.

Cause and Effect Analysis

The controlled variables for the test separator include the pressure inside the vessel, the oil level, and the oil/water interface level. The manipulated variables include the flow control valve for the outlet oil, the flow control valve for the outlet water, and the pressure control valve for the gas. The cause and effect relationship is straightforward for the test separator (Table 12.4).

Dynamic Response Analysis

Both the level and pressure inside the vessel behave as an integrating process. Most of the time, test separator control is treated as a standard level control application. However, a practical challenge is in the control valve. Due to wear and tear, the flow control valve may leak even when fully closed (also called *passing*). When testing low production wells, even if the level control output is 0%, there may still be a non-negligible amount of flow through the control valve. This flow is below the low cutoff threshold of the flow meter and thus, cannot be accounted for. These abnormal valve characteristics as part of the loop dynamics affect both flow measurement and level control.

Another factor that affects the process dynamics is the wellhead pressure. When a well is diverted from a production separator to a test separator, the wellhead pressure may become different from its normal production pressure due to the different flow line-up, which affects the flow rate, water cut, gas–oil ratio, and other parameters. The test result will thus not reflect the true well behavior. The wellhead pressure should be maintained as close as possible between the test and normal operating modes to achieve more accurate production flow data.

12.3.2 *Control Strategy*

The main control objectives for the well testing are to achieve accurate and consistent test results. The goals are achieved by

1. automating the well testing procedure and ensuring accurate production flow measurement;
2. maintaining the level within the desired range by manipulating the outlet flow while at the same time ensuring that the outlet flow does not fall below the flow meter turndown limit when testing low producing wells.

Table 12.4 Cause and effect table for test separator

	Vessel pressure	Oil level	Interface level
Outlet oil flow ↑			↘
Outlet water flow ↑		↘	
Outlet gas flow ↑	↘		

Well Test Sequence

The well tests need to follow a well-defined procedure to produce consistent and accurate test results. Well test used to be a manual process but is being automated in many operations. However, the generic test sequence is the same whether the test is done manually or automatically.

The design of the well test sequence is a multi-team activity involving different specialties. Well test sequence falls under sequential control and is closely integrated with the standard process control requirements. A typical well test will go through the following five stages:

1. Purge: Purging is intended to clear the test line of the previously tested fluid and fill it with the fluid from the well to be tested.
2. Forced dump: Dump refers to the action of opening the outlet valve and letting go of the fluid accumulated in the separator. Before the well test starts, the separator level needs to be brought to a pre-defined value via a forced dump action. This level reflects the initial inventory in the tank for flow totalization.
3. Well test: The flow data for all the phases from the metering devices are collected and totalized during the well test.
4. Forced dump: After the well test, the separator level shall be brought back to the same level as before the well test starts, via a forced dump action. The same level before and after the test will ensure that the test results will not be affected by inventory changes in the test separator.
5. Reporting: The test results shall be calculated, reported, and archived as needed.

We will not go into more details than the above general procedures. However, the separator control interacts with the test procedure and requires a good understanding of the context of the control requirements.

Test Separator Control

The separator level needs to be controlled in a relatively tight range for proper separation of the multi-phase fluid. The fluctuation in the liquid phase in the inlet flow, from well to well due to different capacities, is reflected in the outlet liquid flow. The typical test separator control includes the following functionalities, and a basic control scheme is illustrated in Fig. 12.14:

1. Maintain the test separator pressure so that the well production in test mode closely mimics that in normal production mode.
2. Provide proper oil level control while ensuring that the flow rate is within the effective range of the flow meter for accurate oil production measurement.
3. Provide adequate water level (i.e., oil/water interface level) control while ensuring that the outlet water flow is within the effective range of the flow meter for proper water production measurement.

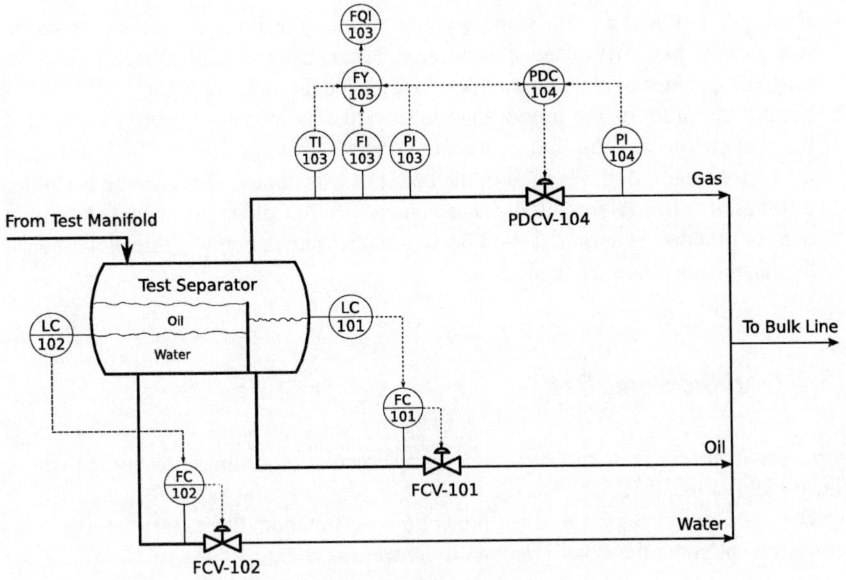

Fig. 12.14 Basic control scheme for a test separator

The following measurement and control are part of the well test separator operation:

1. Pressure control to maintain a minimum differential pressure across the valve to ensure the gas flows forward. The pressure is controlled by manipulating the pressure control valve at the gas outlet.
2. Gas production measurement with a gas flow meter. The raw flow is typically compensated with pressure and temperature for improved measurement accuracy. The compensated gas flow is generally converted to standard volumetric flow (see Sect. 10.3 for flow compensation and conversion).
3. Oil level control and oil flow measurement. Standard level-to-flow cascade control is used to maintain the oil/condensate level. A flow meter is installed to provide flow measurement.
4. Water level control and water flow measurement. Standard level-to-flow cascade control is used to maintain the interface level. A flow meter is installed to provide water measurement.

In practical applications, the extensive range of variations in the production flows from well to well may demand a very high turndown ratio of the flow meters beyond what the current technology can economically offer. Similarly, a control valve also has an effective range or turndown limit, beyond which the control will become less accurate.

Under these circumstances, one or more of the following enhancements can be considered to improve measurement accuracy and ensure control stability:

1. Dual gas flow meters: For improved pressure and differential pressure control, two parallel gas flow meters (even lines) can be used in tandem to increase the range of gas measurement. See Sect. 10.2 for a dual meter approach.
2. Semi-dump control for liquid level: The outlet flow is temporarily stopped if the production rate goes below the flow meter's turndown limit. Once sufficient inventory is built in the separator, the outlet flow resumes. This scheme is similar to on/off control (also called *dump control*), but is different in that the dump control function is only activated when the flow rate is below a threshold, hence the name *semi-dump* control.

12.3.3 Functional Design

A process control schematic showing an implementation of the semi-dump logic is illustrated in Fig. 12.15.

The semi-dump logic provides the following functionalities to make the flow measurement cover the flow range of all the wells:

1. For high-producing wells, the level is controlled like a standard level-flow-valve cascade control loop.
2. For low-producing wells, if the outlet flow drops below the minimum flow limit, the outlet flow line is temporarily closed to allow the liquid inventory in the separator to build up. Once the level is high enough, the level-flow-valve cascade loop is re-activated, and the outlet flow should be above the minimum flow limit again.
3. Transition: The transition between the high and low producers is automatic and bumpless.

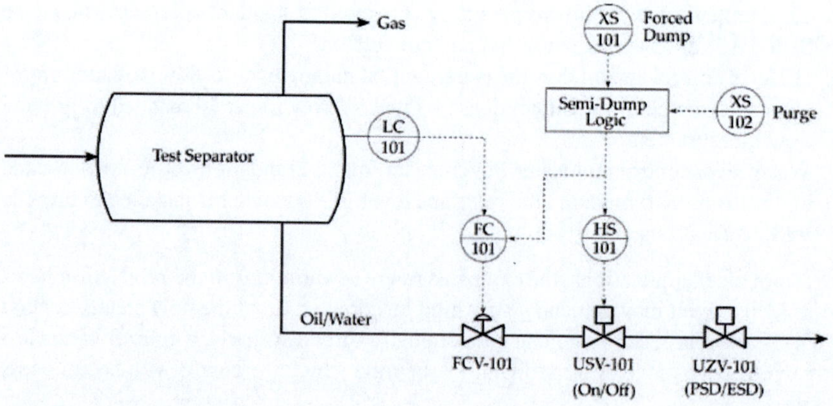

Fig. 12.15 Semi-dump control test separator level

When testing low production wells, the level control starts with the standard cascade control. Under level control, the outlet flow will decrease and be clapped at the minimum flow rate `Fmin`. The level would then start to fall. Once the level is below the level setpoint by a pre-defined margin, the semi-dump logic would kick in and switch off the on/off valve USV-101. With the outlet flow stopped, the level will then slowly rebuild.

Once the level is above a specific limit (e.g., 10% higher than the setpoint), the on/off valve is opened again, and cascade control resumes with outlet flow above the turndown limit. Figure 12.16 shows the possible level/flow interaction during semi-dump control.

Semi-dump control is a unique control scheme designed for test separators. It is not a universally recognized term. Nevertheless, the practice of semi-dump control has been very successful and is thus provided here as a creative example of PID control.

Integration Between Test Sequence and Control

The well test sequence and the semi-dump logic interact with each other and thus need to be considered together:

1. Purge: Purging is to clear the existing fluid in the pipeline and replace it with the fluid of the well to be tested. Depending on the production rate, the purge time varies from well to well. The test sequence needs to set a fixed duration or define a triggering event to end the purging.
   During the purge stage, the level control is kept in cascade control. Since the purpose of the purge step is to prepare the pipeline only, there are three scenarios:
   a. There is sufficient purging fluid to replace all the liquid in the separator vessel. The final level is at the setpoint under level control.
   b. The separator starts at a low level, and there is not enough purging flow to fill the separator to the setpoint. So the well test starts with the separator level that is lower than the setpoint.

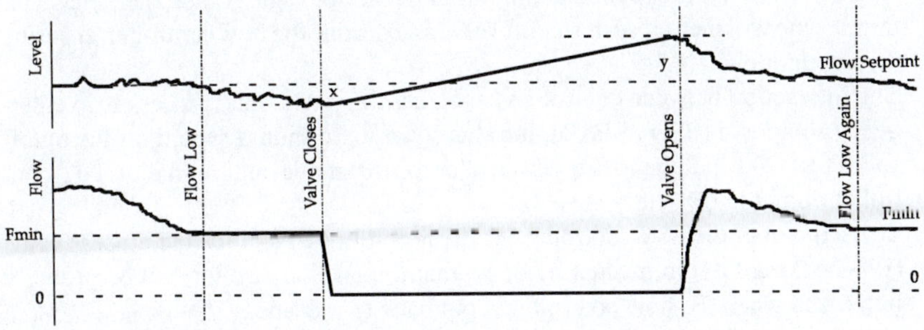

Fig. 12.16 Level and flow during semi-dump control

c. The purging flow is so low that the separator level control is in semi-dump mode with the on/off valve closed. At the end of the purge phase, the separator level is above the setpoint (see Fig. 12.16) but below the dumping limit (e.g., 10% above the setpoint).

The well test logic needs to provide appropriate considerations for all the above scenarios.

2. Forced dump: A forced-dump step is necessary to ensure that the separator level is at the level control setpoint. Forced dump stops automatically when the level is at the setpoint. The criterion for stopping the forced dump is based on the separator level and outlet flow.
3. Well test: During the well test step, the level control automatically transitions between standard cascade control mode and semi-dump control mode.
4. Forced dump: Again, the semi-dump logic is designed to automatically close the valve once the separator level reaches its setpoint. The forced dump ensures that the well test starts and stops with the same liquid level in the separator. The logic to reliably start and end the forced dump is an important consideration.
5. Reporting: The totalization of the different phases of flows stops now, and the production rate is calculated. Presenting the results to the operator and recording the results to an electronic report is part of the test sequence.

12.3.4 Implementation and Operation

Semi-dump control is an ad hoc approach to compensate for inaccurate flow measurement caused by low flow rate. The side-effect is that the interface level is artificially disturbed and may inadvertently affect the oil/water separation as well as the gas/liquid separation:

1. The flow controller `FC101` should have a low setpoint limit configured to be the same as the low flow limit `Fmin`. The level controller output will saturate at this low limit. If the outlet flow remains low-limited, the separator level will start to fall (*out of control*). When the level falls below the deadband (`LC101.PV< LC101.SP-x%`), the second condition in the above logic is then met; the semi-dump logic will then close the on/off valve and disable the flow controller, ensuring zero outlet flow.
2. The interaction between control and safeguarding. When the process is in emergency shutdown (PSD or ESD), the shutdown logic should switch off the on/off valve `USV101`, put the `FC101` controller to `MAN` mode, and set the `FC101` controller output to 0% (fully closed).
3. The setpoint of the flow controller FC101 is configured to track the process value (PV-Tracking ON) in manual mode so that the level controller will be properly initialized when the flow controller is put back to cascade.

Compared with a straightforward on/off dump control, the semi-dump control scheme provides a smoother and more stable control on both the level and flow variables.

12.3.5 Practical Considerations

To conclude this discussion on separator control, let us back off one step and think about the separation process for improvement.

First and foremost, an inexpensive and reliable multi-phase flow meter (MPFM) at every wellhead is the ultimate solution. MPFM provides continuous measurement of the well flows, and can consequently open the door for many real-time control and optimization opportunities. However, in the foreseeable future, an MPFM at every wellhead is still not realistic due to affordability reasons. Test separator will continue to be the primary choice for well testing.

However, some new separator designs, such as a *double-weir* separator, can mechanically ensure the oil and interface level; therefore, no tight level control is needed. For example, the so-called *bucket and weir* separator (Campbell 2004) is illustrated in Fig. 12.17. The height of the oil weir controls the liquid level in the vessel. The difference in the height of the oil and water weirs controls the oil pad's thickness based on specific gravity differences.

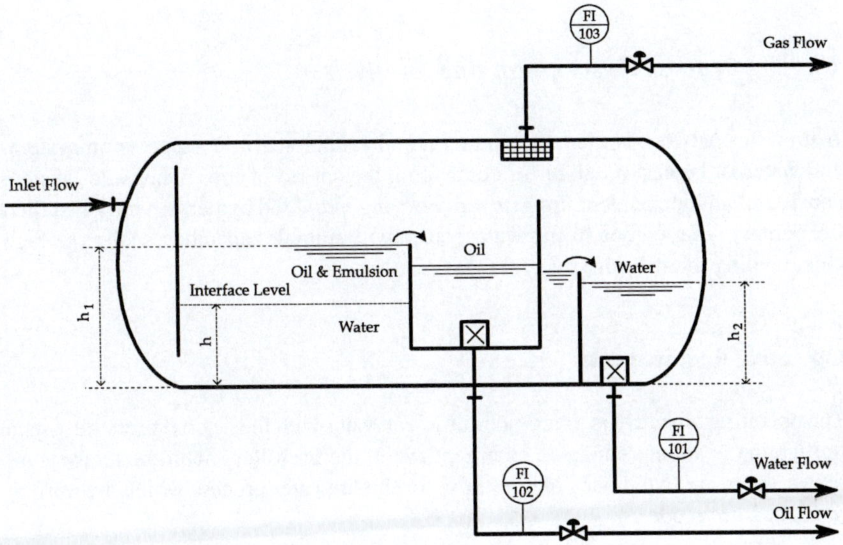

Fig. 12.17 A three-phase test separator, with double weirs

It is easy to verify that

$$(h_1 - h)\,\rho_1 = (h_2 - h)\,\rho_2 \quad \mapsto \quad h = \frac{h_2\,\rho_2 - h_1\,\rho_1}{\rho_2 - \rho_1} \tag{12.3}$$

where h is the interface level, h_1 is the height of the oil weir, and h_2 is the water weir height. ρ_1 and ρ_2 are the oil and water density, respectively.

With this design, the interface level is mechanically guaranteed, and thus the water/oil interface level controller `LC-102` in Fig. 12.14 is no longer needed. Both the oil and water flow over the weirs to the buckets of their own, where simple level control is needed for inventory control purposes only, and the level control performance does not impact the separation.

This application demonstrates that process design has a considerably higher impact on controllability than control design, whether classical or advanced (Luyben et al. 1998).

12.4 Firewater Control

Adequate firewater supply is essential to the operation of any operating plant. However, the proper design of the firewater control system was not receiving the attention it deserves.

12.4.1 *Process Description and Analysis*

A firewater network is the most common and economical fire suppression system, and water is the best medium for controlling the spread of fire. A firewater system consists of an independent fire *grid main* or *ring main* fed by permanently installed fire pumps taking suction from a water source with suitable capacities, such as storage tank, cooling tower basin, river, and sea.

Operating Requirements

The operating objective is to provide sufficient water with the desired pressure. Since firefighting is an unpredictable emergent event, the firewater should be ready at all times. There are two modes of operation for the firewater process which are normal mode and fire mode:

1. Normal mode: When there is no demand for firewater, the ring pressure should be maintained within a given range, e.g., 800–850 kPa. This pressure serves two purposes; one is to keep the system ready for emergency fire demand. The other

is for automatically detecting a fire demand and triggering the transition to fire mode.

2. Fire mode: In case of an increase in firewater demand, for example, from the opening of a fire hydrant, the ring pressure should be quickly brought up to firefighting pressure, e.g., 1150 kPa(g), and this pressure should be maintained. The actual pressure requirement is dependent on the height the firewater needs to be delivered and thus differs case by case. The pressure of the ring main should be protected against going overly high, e.g., over 1200 kPa(g), to potentially damage the piping and pump.

Process Flow Analysis

A firewater system usually consists of three essential parts, the firewater storage tank, fire pumps, and the firewater distribution system (Fig. 12.18):

1. The firewater is typically drawn from a large-capacity water source such as storage tanks, cooling tower basins, rivers, and the sea. The capacity should be adequate to meet peak demand for at least 4–6 hours (Vervalin 1981).
2. Centrifugal pump has been the standard choice for fire pumps because of its compactness, reliability, easy maintenance, hydraulic characteristics, and variety of available drivers. An outstanding feature of a centrifugal pump is the inverse relation between discharge pressure and flow rate at a constant speed, whereby flow rate is reduced when the discharge pressure (or head) increases. See Sect. 11.2 for detailed discussions on pump control.
   The fire pumps are a group of pumps consisting of at least a jockey pump and the main pump. The jockey pump runs on and off frequently to maintain the distribution system's correct pressure during normal operation. In case of a fire, the main pump automatically starts and provides the firewater for fighting the fire. Typically, there are at least two main pumps, one is electric motor-driven, and the other is driven by a fully independent power source such as a diesel engine. The latter serves as a backup. Each jockey pump typically also has a backup.
3. The distribution system is generally designed with a ring-main system. The ring main is a primary loop permanently connected to the pumps so that there are two routes for water to flow in case one side is blocked. Fire hydrants are distributed along the ring main to deliver the high-pressure firewater.

Supply and Demand Analysis

The firewater process is a typical demand-driven operation. During normal mode, no firewater is needed; the jockey pump automatically starts and stops to compensate for the slow and gradual loss of water in the ring main due to leakage and evaporation.

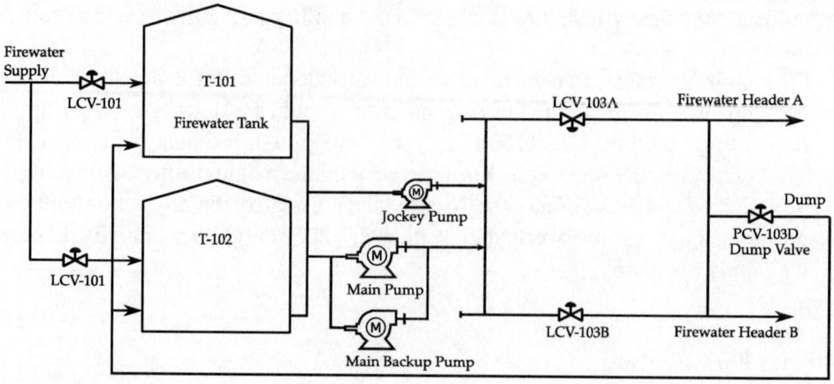

Fig. 12.18 Process flow scheme for a firewater process

In fire mode, the firefighting need determines the water demand. The main pump automatically starts and supplies the water needed. The dump line sends excessive water back to the storage tanks.

Cause and Effect Analysis

The control targets include the ring main pressure and the pump discharge pressure. The available control handles include the ring main control valves and the dump valve. The most critical control handle, however, is the on/off switch of the pumps. Firewater demand causes the ring main pressure to decrease, which in turn causes the pump discharge header pressure to decrease. In the normal mode (idle mode), this sudden drop in pressure triggers the start of the main pump to increase the pressure to the firefighting pressure quickly. Once in the fire mode, the ring main pressure is regulated by the pressure controllers. A decrease in the ring pressure causes the pressure valve to open and the dump valve to close. The cause and effect table is thus given as in Table 12.5.

Dynamics Response Analysis

The centrifugal fire pump has an important characteristic for the capacity control of the firewater. The discharge pressure (or head) and the flow have an inverse relation-

Table 12.5 Cause and effect table for firewater process

	Firewater flow	Ring pressure	Discharge pressure
Ring main pressure valve ↑	↗	↗	↘
Dump valve ↑	↘	↘	↘
Pump starts ↑	↗	↗	↗

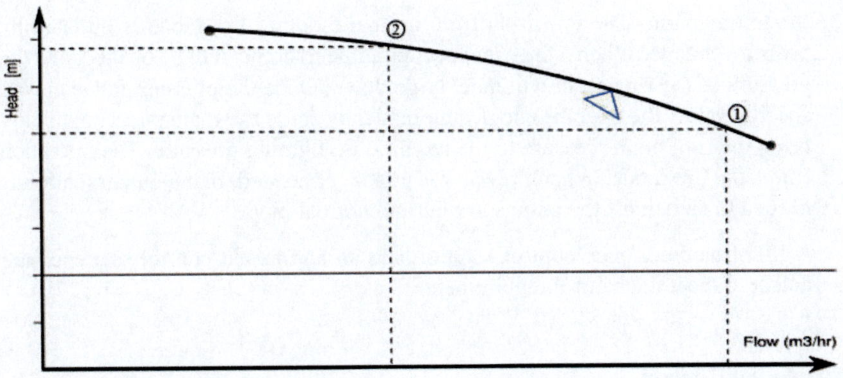

Fig. 12.19 Pump curve in a firewater process

ship, as shown in Fig. 12.19. Throttling the ring main valves will reduce the flow but increase the pump discharge pressure. Opening the valves will increase the flow and cause the pressure to go low. The pump pressure and flow are thus self-regulating, within a specific operating range.

These characteristics also enable the pump to serve as a dynamic swing control to respond to the fluctuation in firewater demand during fire mode. The centrifugal pump for firewater pumping typically has a relatively flat performance curve, offering a decent range for flow variations. On the other hand, this flat curve also implies that the main pump has a narrow range for operating pressure variations. As a result, the pump capacity needs to be carefully considered to meet the operating pressure requirements.

12.4.2 Control Strategy

The control requirement for the firewater process is relatively simple. The challenge is in the integration with the operating logic and sequence.

The control objective is to ensure that an adequate supply of firewater is always available, which is achieved with the following control design:

1. Normal regulatory control: The normal regulatory control is via the jockey pumps' on/off switching to maintain a normal operating pressure. In fire mode, the regulatory control is the pressure controller for the ring main.
2. Protective control: A high-pressure protective controller protects the ring's main pressure from going excessively high. In case the ring pressure goes above the high limit, the protective control opens the dump valve and sends the excess water back to the storage tanks.

3. Mode transition: The transition from normal mode to fire mode is initiated by pressure change. When a fire event occurs, some hydrants will be opened, and the pressure of the ring main will quickly go down. Once the pressure falls below a specific value, the operating logic automatically starts the main pump to quickly bring the ring main pressure to the required firefighting pressure. The operation enters the fire mode. When firewater is no longer needed, manual intervention is needed to switch off the pumps to return to normal mode.

A PID-based complex control solution and an automated control sequence are sufficient to meet the control requirements.

Sequential Control

The critical design consideration of the firewater process is the mode transition. Sequence and logic are required to monitor the ring pressure, initiate the mode transition, and change the operation modes. The sequence is summarized as follows:

1. Normal mode: During normal mode, the main pump is switched off. The pressure controllers remain in `MAN` modes. The jockey pump automatically starts and stops to maintain the ring pressure between 800 and 850 kPa(g). The ring main pressure control valve should stay at the low mechanical lock position (typically 10%) to keep the water flowing.
2. Transition to fire mode: If the firefighters open the hydrants, the ring pressure will drop. When the pressure drops below 750 kPa(g), the main pump starts automatically to increase the pressure. If the pressure continues to fall after a short time (e.g., 20 s) due to a failed start of the main pump, the backup pump should start.
3. Fire mode: The two pressure controllers, `PC-103A` and `PC-103D`, should now be switched to automatic (`AUTO`) mode. `PC-103A` is responsible for maintaining the firewater pressure at 1,150 kPa(a). If the supply from the pump is higher than the demand and causes the pressure to go higher than 1,200 kPa(g), the dump controller `PC-103D` opens the dump valve and sends some water back to the storage tanks.
4. Transition to normal mode: When there are no more firewater needs, the main pumps must be switched off manually, and both pressure controllers to manual (`MAN`) mode. The ring pressure will slowly fall back to the range of 800–850 kPa(g). The firewater system is in normal standby mode again.

12.4.3 Functional Design

Figure 12.20 is the control overview diagram (COD) for a typical control design. An actual design may be more complex depending on specific requirements.

The pressures of the two firewater headers are individually measured with `PI-103A` and `PI-103B`. The lower of the two pressure values is fed to the pressure

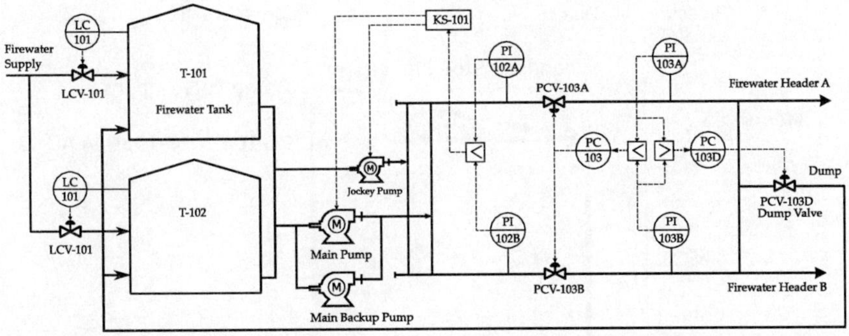

Fig. 12.20 Firewater control scheme

controller PC-103 as the process value (PV) and ensures the firewater pressures of both headers meet the operating requirement. The output of the pressure controller PC-103 is fanned out to the two ring pressure valves, PCV-103A and PCV-103B. Fan-out control is used because the two pressure measurements are highly correlated and can only be counted as one controlled variable.

The higher value of the two pressure measurements is fed to the pressure protective controller PC-103D. If the pressure goes excessively high, the protective controller PC-103D simply increases the dump valve (PCV-103D) to let more water go back to the storage tank. The excessive water can be caused by reduced firewater demand that is below the pump turndown limit.

The operating sequence is illustrated in Fig. 12.21. The main pump starts when the lower of the two pressure measurements at the pump discharge fall below 750 kPa(g).

During normal mode, the firewater is maintained between 800 and 850 kPa with on/off control through the start and stop of the jockey pumps. The drop in pressure transition triggers the transition to the fire mode through the startup of the main pump. Once the main pump is started, the pressure controller PC-103 is switched to AUTO and maintains the firewater pressure at 1,150 kPa. The operator manually initiates the transition back to normal mode after confirming that firewater is no longer needed.

12.4.4 Implementations and Operation

Some practical considerations for the firewater control design are briefly explained here:

1. Dump valve: Fail-open or fail-close? In case of loss of instrument air, a fail-open dump valve goes to its fully open position and releases the pressure in the ring main. This action protects the piping from potential damage due to high pressure; however, the ring main loses the required water pressure for fire fighting. On the other hand, a fail-close dump valve will maintain the ring main pressure, but the

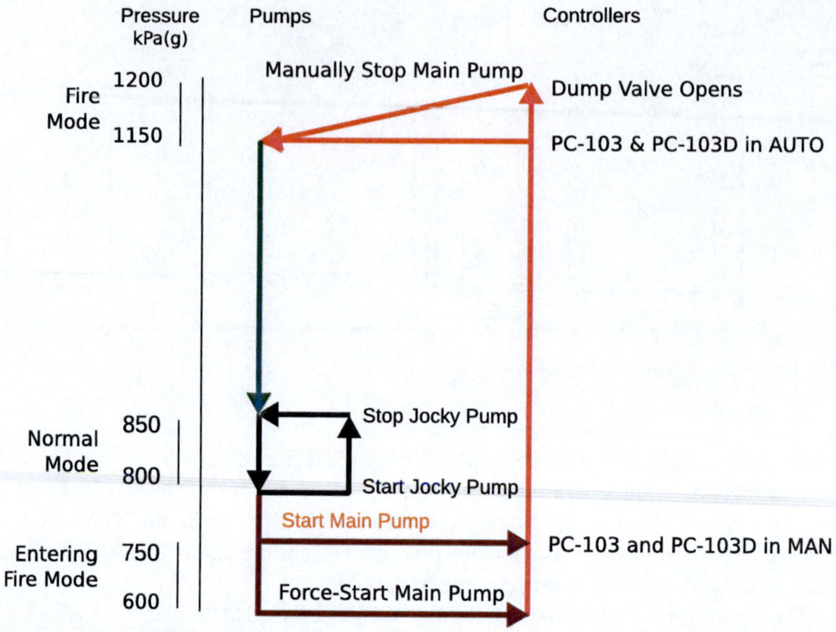

Fig. 12.21 Firewater control sequence

high pressure may be unacceptable to the piping and the pumps. Since firewater readiness is vital, a fail-close valve is preferred. The ring main piping and the pumps should be fully rated for the highest pressure to protect from high-pressure damage. Otherwise, secure instrument air (SIA) should be considered.

2. Pressure surge: When entering fire mode, the main pump automatically starts. Upon startup, the pump creates an initial pressure surge in the ring main. Depending on the pressure rating, this pressure surge may not be acceptable. Additional logic is thus required to prepare the control valve for the pressure surge.

12.5 Unit Control: Acetylene Hydrogenation Process

We continue with the acetylene hydrogenation example in Example 7.1 to demonstrate the control design of a simple unit operation. Although the process is relatively straightforward, the control design can have many variations. Again, the thinking process will follow the procedures in Chaps. 6 and 7.

12.5.1 *Process Description and Operating Requirements*

The acetylene hydrogenation process saturates the acetylene (C_2H_2) with hydrogen into ethylene and ethane. The process is commonly found in olefin plant operations:

$$\begin{cases} C_2H_2 + H_2 \rightarrow C_2H_4 & \text{reaction 1} \\ C_2H_2 + H_2 \rightarrow C_2H_6 & \text{reaction 2.} \end{cases} \tag{12.4}$$

The operating objective is to reduce the acetylene concentration to below 2 ppm through two hydrogenation converters. The reaction is facilitated by catalysts and is exothermic. See Fig. 12.22 for a process overview diagram (POD) with the key process variables.

The critical process disturbances are the variations in the feed composition and the fluctuations in the heat duty in the heaters and coolers.

12.5.2 *Supply and Demand Analysis*

The material balance is straightforward. There is only one stream and one flow path through the unit, so the operation is strictly supply-driven.

The energy balance is primarily the heating and cooling of the process flow. There is heat added from the heater to reach the reaction temperature, and there are coolers after the reactors to remove the reaction heat. The primary constraint on the supply and demand balance is the saturation of the heat exchangers; therefore, the heat duties of the heater and coolers should be adequate to prevent saturation. An

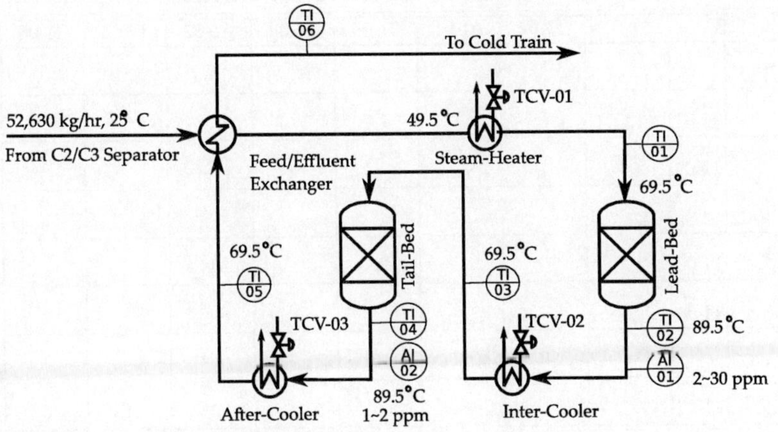

Fig. 12.22 An acetylene hydrogenation process

advanced operating requirement is the energy distribution between the heater and coolers to maximize energy efficiency.

12.5.3 *Cause and Effect Analysis*

The measurements are typically abundant in a chemical reaction process. The key process variables are the temperatures at various locations and the acetylene concentration at the outlet of the two reactors. The control handles are the temperature control valves at the heater and coolers. The cause and effect relationships among the key process variables (KPV) and the control handles are given in Table 12.6. There are more KPVs than control handles; thus, the critical decision is selecting the controlled variables.

The cause and effect table can be expanded to become Table 12.7, which is one way to show the cascade cause and effect relationship between the variables. The intermediate variable (IV) is similar in concept to the secondary variables in a PID cascade loop. For example, the heat exchange valve `TCV-01` directly affects the reactor 1 inlet temperature `TI-01`, which in turn affects the reactor outlet temperature `TI-02`, and thus impacts the outlet C_2H_2 concentration `AI-01`.

Table 12.6 Acetylene hydrogenation: preliminary cause and effect table

	TI-01	TI-02	AI-01	TI-03	TI-04	AI-02	TI-05	TI-06
TCV-01	⊕	⊕	⊖					
TCV-02				⊖	⊖	⊕		
TCV-03							⊖	⊖

Table 12.7 Acetylene hydrogenation: cause and effect table with cascade

MV	IV	TI-01	TI-02	AI-01	TI-03	TI-04	AI-02	TI-05	TI-06
TCV-01		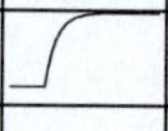							
	TI-01	1	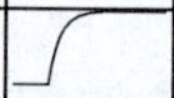						
	TI-02		1	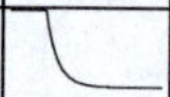					
TCV-02					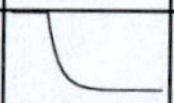				
	TI-03				1	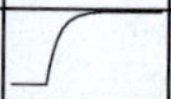			
	TI-04					1	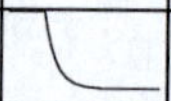		
TCV-03								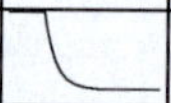	
	TI-05							1	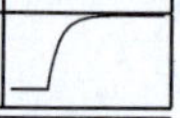

The ultimate goal of the unit operation is to achieve a 2 ppm acetylene concentration at the outlet of the second reactor, which is `AI-02`. The reactor outlet temperature `TI-04` is the preferred variable to control, considering the low availability and reliability of the online analyzers.

12.5.4 Dynamic Response Analysis

Simulation shows that the reactor outlet temperature and acetylene concentration are highly correlated (Al-Dawery and Dakhil 2012). The transit response of the outlet temperature shows a clean first-order-plus-delay-time (FOPDT) behavior.

12.5.5 Control Design

As discussed in Sect. 7.1.3, at the unit level, regulatory control is primarily concerned with quality control.

In this example, the acetylene concentration sent down to the cold train is the quality to be controlled. Based on the supply and demand analysis, capacity control is not needed for this simple supply-driven unit operation consisting of a straight flow process stream.

With the cause and effect table (Table 12.7), the control strategy has several options, from simple standalone PID controllers to complex PID schemes, to model predictive control:

1. Standalone PID controllers: The inlet temperature primarily determines the reaction, so a simple control solution is to control the reactor inlet temperature. The pairing of the variables would be `TCV-01/TI-01`, `TCV-02/TI-03`, and `TCV-03/TI-05`. The control scheme is shown in Fig. 12.23.
   However, as the concentration is also influenced by the feed flow rate and the feed composition (e.g., CO), the improved control structure would be to pair the control valves with the reactor outlet temperatures, `TCV-01/TI-02`, `TCV-02/TI-04`, and `TCV-03/TI-06`. The control scheme is shown in Fig. 12.24.
2. Cascade PID control scheme: One of the disturbances to the unit operation is the fluctuation in the heating (steam) duty. A cascade control scheme can improve the control performance, as shown in Fig. 12.25.
   Other disturbances to the unit operation include the variations in the reaction heat caused by the changes in feed flow rate, feed composition, and catalyst condition. Feedforward control can be considered if the variations are known and measurable, for example, the feed rate or the CO concentration.
   The acetylene concentration is the ultimate control target. If online analyzers are available to measure the acetylene concentration, the temperature control

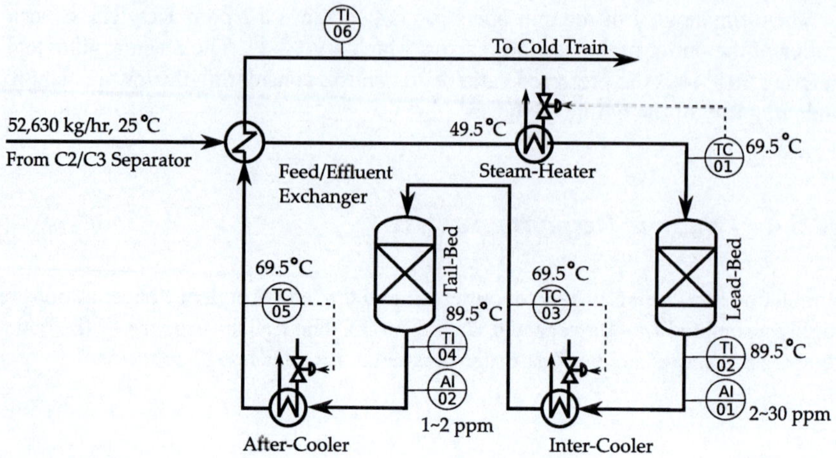

Fig. 12.23 Acetylene hydrogenation process control: design #1

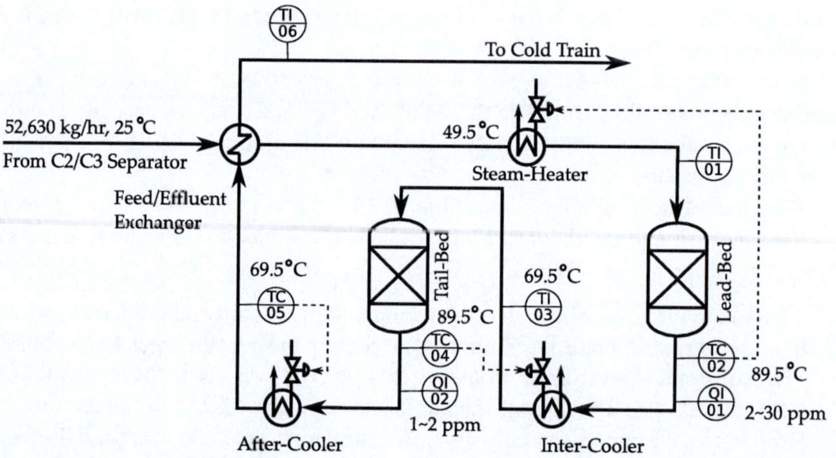

Fig. 12.24 Acetylene hydrogenation process control: design #2

can be further cascaded to the acetylene concentration for improved control. See Fig. 12.26.

3. Model predictive control (MPC): Although valve position control (see Sect. 4.1.4) can be considered to prevent the control valves from saturation, the design and configuration can be tricky. For example, Reactor 1 outlet temperature setpoint can be manipulated (within a specific range) to maintain the desired valve opening. MPC can be considered for this unit operation to achieve the best performance. MPC can optimize the heat duty balance between the heat exchangers (heaters and coolers) by including the temperature controller outputs as constraints. If one

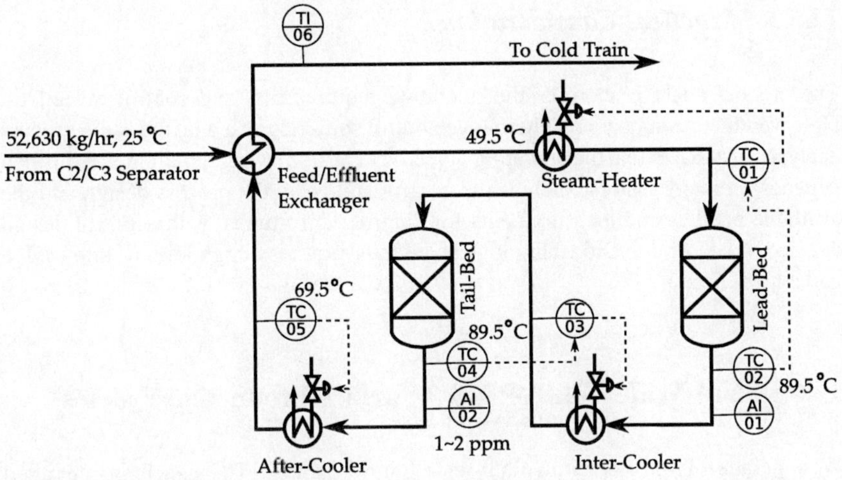

Fig. 12.25 Acetylene hydrogenation process control: design #3

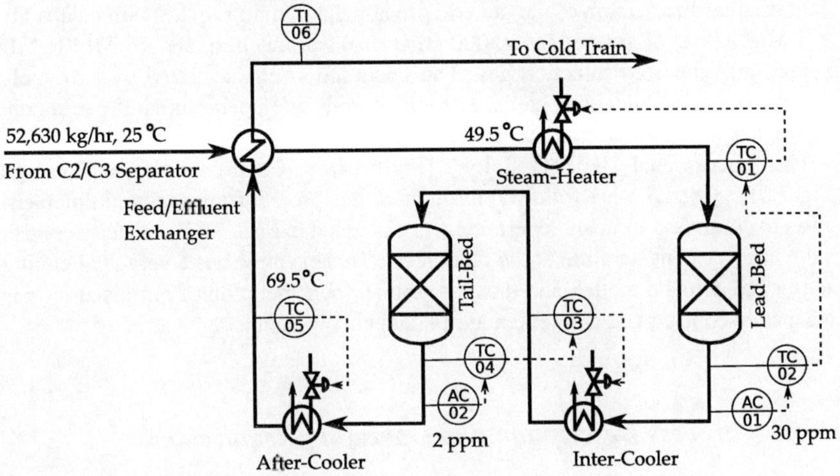

Fig. 12.26 Acetylene hydrogenation process control: design #4

of the heat exchangers saturates, MPC automatically shifts the load to other heat exchangers based on the cause and effect relationship in Table 12.7.

The two critical requirements for model predictive control are the model structure and model parameters. The model structure describes the cause and effect relationship between the manipulated variables and the controlled variables. The model parameter describes the dynamic behavior of each cause and effect relationship. Table 12.7 can serve as the model structure for the control design.

12.5.6 *Practical Consideration*

The process analysis clarifies the operating requirements and control objectives. However, there can be many alternative control solutions based on the same process analysis. To what extent the operating objectives can be met and requirements fulfilled depends on many factors, such as the controllability by the process design and the available handles and measurements for control. Unfortunately, the control design is a mixed bag of arts and technologies and thus requires a high level of knowledge and experience.

12.6 Plant-Wide Control: E&P Surface Production Process

We conclude this chapter with a plant-wide control example. The plant is an integrated oil and gas production facility built to handle oil and gas production via miscible gas injection (MGI)—an enhanced oil recovery mechanism. This multi-billion dollar plant is a real-life version of the generic process diagram in Fig. 1.3 with a capacity of 6 MMSCMD of sweet gas, 60,000 BPD of oil production, and 16 MMSCMD of high-pressure sour injection gas. The operation is characterized by extremely high-pressure and high-sour gases. For this reason, safe operation is of paramount importance in the process and control design.

The process analysis methodology (Fig. 6.14) in Chap. 6 and control design (Fig. 7.6) in Chap. 7 were followed throughout this project from design, implementation to commissioning and operation. As discussed in Sect. 7.1.2, capacity control is the primary consideration at the plant level. This example has a very challenging supply and demand model, and thus the primary consideration of our discussion is in capacity control, i.e., dynamic material and energy balancing.

12.6.1 *Process Description and Operating Requirements*

A process overview diagram (POD) is developed from hundreds of pages of PFD and P&ID and is shown in Fig. 12.27 for the plant-level overview.

Operating Objectives and Product Specifications

The operating objective is to produce oil and gas for sales and sour gas for injection. The gas processing area, which is the focus of this discussion, comprises gas production, separation, cleaning and conditioning, and gas compression. The produced gas is extraordinarily high in H_2S and CO_2. Many packaged units by various vendors are implemented to clean and condition the gas, condensate, and produced water. Each

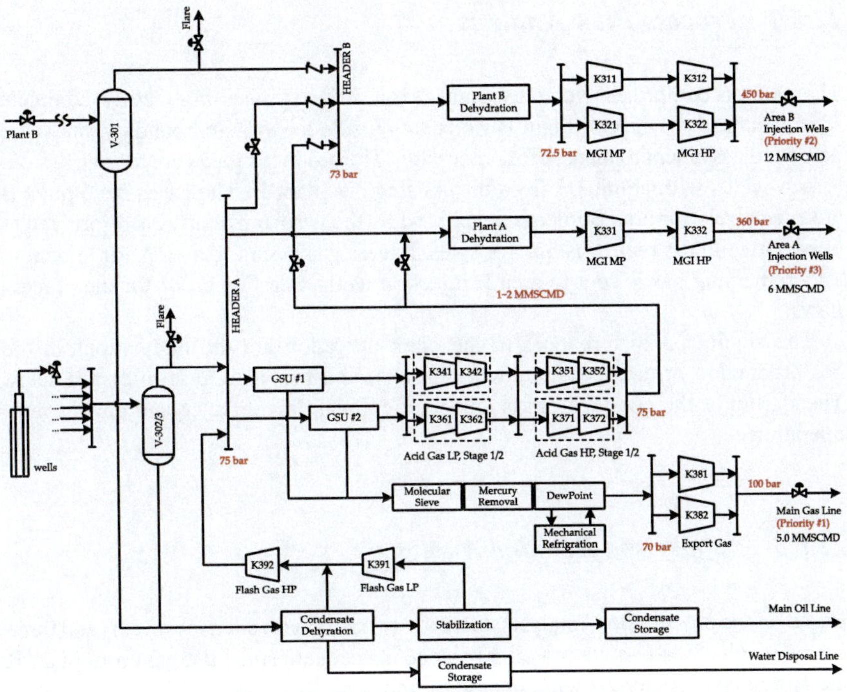

Fig. 12.27 Process overview diagram of an E&P surface facility

unit by itself is a complex unit operation, and all have their specifications to meet, such as purity, dew point, vapor pressure, and delivery pressure.

The three consumers of the processed gas include, in the order of priority: export to sales, injection wells in area B, and injection wells in area A.

Modes of Operations and Operating Constraints

In addition to the normal operating mode and startup/shutdown mode, the primary challenge is the turndown mode, where insufficient production (supply) or consumption (demand) causes the operation to run at a considerably lower capacity. Different consumers have different priorities; thus, preferential treatments need to be provided when faced with insufficient gas supply.

A commercial agreement requires the recovered off-gas from the sour gas treatment units (GSU) split 50/50 between areas A and B. Two control valves are provided to facilitate this flow split.

12.6.2 Process Flow Analysis

This process comprises two areas: plant A is a surface processing facility, complete from wellheads to export. Plant B wells are outside the system boundary, but some produced gas is sent to plant A for processing. The produced gases are primarily from area A wells, with about 1/3 from the existing gas plant B. The gases from plant B are exclusively for area B injection wells, while the gas from plant A is prioritized to be exported to the main gas line for sales. Excess gas is sent to area A for injection. Any remaining gas is sent to area B injection wells. See Fig. 12.27 for the process flows.

The significant disturbances to the plant operation are the fluctuations in the gas production in area A (shutdown of wells) and the disruption in export sales. The startup of the compressors is another significant disturbance and challenge to operation.

12.6.3 Supply and Demand Analysis

Based on the process flow analysis, there are two supplies (plants A and B) and three gas demands. Steady-state material balance is adequate since the gas from plant B has sufficient capacity to swing in case of imbalance between supply and demand.

The gas production from area A wells is usually production-driven but can be choked back in case of low demand. The gas from plant B is demand-driven, regulated by the header B pressure. Export gas is demand-driven, dictated by sales. The flows to the injection wells in both areas are demand-driven, with the flow rate decided by injection needs.

The main swing stream is the gas from plant B, which is a demand-driven operation. If necessary, the well production from area A can be reduced, but the dynamics are slow. If the supply is so low that the demands need to be reduced, the area A injection is the first stream to cut back, followed by area B injection. Export gas is the last one to be affected.

The recovered sour gas streams from the GSU units and the condensate processing unit are supply-driven, and the suction pressures regulate the capacity.

12.6.4 Cause and Effect Analysis

At the plant level, the primary concern is capacity control for balancing supply and demand. The supply and demand imbalance is reflected in the pressure–flow profile. It is thus vital to select the appropriate pressure variables to propagate the pressure–flow changes.

The two supply header pressures, Header A pressure and Header B pressure, are critical in the supply and demand propagation, and thus they are two controlled variables. The export gas is demand-driven, and thus the gas flow is a controlled variable. The gas to injection wells is also demand-driven, and thus the gas flow rates are controlled variables. The produced gas from plant B operates as a demand-driven swing stream; its flow rate is dictated by Header B pressure.

The key process variables (KPV) are shown in Fig. 12.28. The KPVs and control handles are assigned temporary tag names before they are finalized as controlled variables and manipulated variables.

The preliminary cause and effect table (CET) is given in Table 12.8. The cause and effect relationships are marked with an "⊕" or "⊖" for dominant dynamics with a positive or negative gain, respectively, and a simple "+" or "-" for the non-dominant ones.

12.6.5 Degree-of-Freedom Analysis

The pairing of the control targets and control handles are straightforward except for the two critical header pressures, PC-01 and PC-02. There are more handles

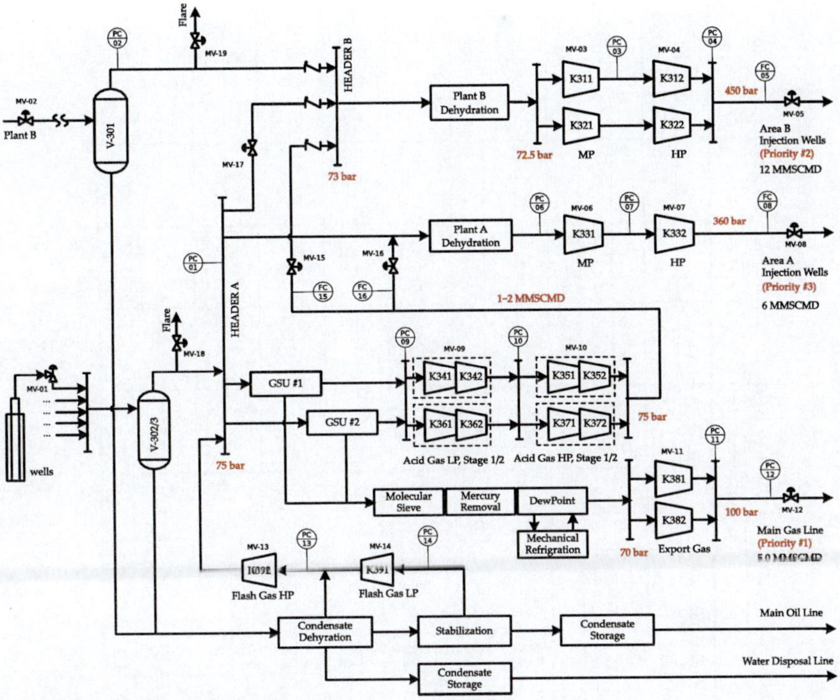

Fig. 12.28 Plant-wide control: key process variables

Table 12.8 Cause and effect table for the generic E&P process in Fig. 12.28

	PC01	PC02	PV03	PV04	FC05	PC06	PC07	FC08	PC09	PC10	FC11	PC12	PC13	PC14	FC15	PC16
MV-01	⊕															
MV-02		⊕														
MV-03			⊕													
MV-04				⊕	+											
MV-05					⊕											
MV-06						⊕										
MV-07							⊕	+								
MV-08								⊕								
MV-09									⊕							
MV-10										⊕						
MV-11											⊕	+				
MV-12												⊕				
MV-13													⊕			
MV-14														⊕		
MV-15															⊕	⊕
MV-16															⊕	⊕
MV-17	−	+														
MV-18	−															
MV-19		+														

than targets, and therefore complex PID control schemes should be considered. The analysis follows the methodology in Sect. 6.5.

12.6.6 Control Design

Our discussion here is limited to plant-level control. Each unit should have an overall control strategy of its own.

The normal regulatory control at the plant-level control is primarily for the dynamic supply and demand balance. With the cause and effect analysis above, the control overview diagram (COD) with the normal regulatory control loops is shown in Fig. 12.29.

There are 14 compressors with a total of 18 stages. Therefore, compressor control and protection are at the center of the control design. The compressor control strategy and techniques discussions in Sect. 11.3 are an integral part of the control design. What is noteworthy in this control design includes the following:

1. The complex supply and demand model: For example, the export gas compressor is demand-driven; the discharge pressure controller regulates the capacity. On the

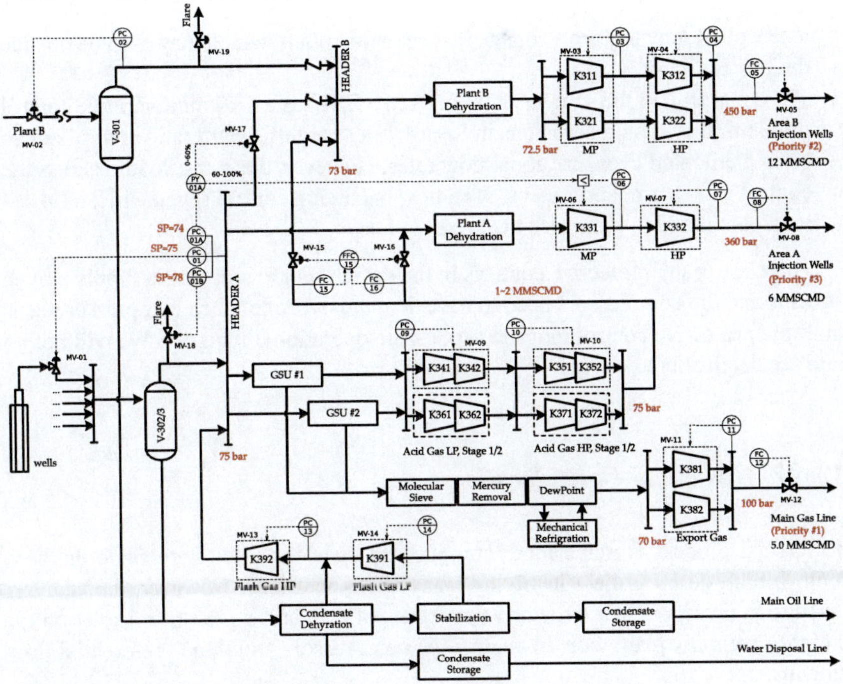

Fig. 12.29 Plant-wide control: normal regulatory control

other hand, the GSU off-gas compressor and the condensate flash gas compressors are supply-driven, and therefore, the capacity control is performed by the suction pressure controllers.

2. The two header pressures, `PC-01` and `PC-02`, are the critical nodes for overall capacity control. The fluctuations in these two pressures reflect the dynamic supply and demand swings. For Header A pressure, the normal operating pressure is maintained by adjusting the wellhead pressure. However, the dynamics from the wellhead to the pressure header are relatively slow; thus, protective controllers are provided to protect this pressure header against both high and low pressures. The low protective controller `PC-01A` is provided to cut back the gas to the injection wells if the pressure is temporarily lower than the normal header pressure by 1.0 bar. During normal operation, since the `PC-01A` setpoint of 74 bar is lower than the normal header pressure of 75 bar, the pressure controller `PC-01A` will drive the valve `MV-17` to open fully. However, if the header pressure falls below 74 bar, the protective controller `PC-01A` would start closing the flow to area A injection first, then the flow to area B, through the split-range calculation `PY-01A`.
   If a severe plant upset drives the pressure to above 78 bar during an abnormal situation, the excess gas will be sent to flare via a protective pressure controller `PC-01B`. Header B pressure is maintained by manipulating the gas flow from area B. This stream from area B is demand-driven. Sufficient swing capacity must be provided inside the area B control solution to meet the demand change.
   The supply and demand analysis in this example shows that the control design of a new plant may require a change in an existing plant, which may even be outside the project scope.
3. The equal split of the sour gas flow from two GSUs is a 2×2 multivariable control problem. Based on discussions in Sect. 4.1, a straightforward ratio controller can be implemented to ensure equal flow rates. However, the pressure loss across the control valves is not acceptable. The flow balancing control scheme in Sect. 4.1.4 is implemented to minimize the pressure loss.

There are many protective controls in the design. For example, the supply-driven streams and the compressors need to have demand override. Each compressor needs multiple protective control loops to protect the operational integrity. We will not go into the details here.

12.6.7 Practical Consideration

The actual process is much more complicated than discussed here since this is a multi-billion dollar project involving many unit operations. However, the analysis and design presented here provide an overview of the plant-wide control solution for a highly complex plant with 14 centrifugal compressors and three preferential flow streams.

With this control solution, the plant startup was smooth, and the operation has been free of significant problems. The uptime of the 800+ PID control loops in this plant has reached and sustained a remarkable 80+% several months after the plant startup.

12.7 Summary

This chapter starts with the example of tank blanketing control in Sect. 12.1. The process flow is simple, and the control requirement is clear. However, the design can still go from good, bad to ugly. Bad designs are accepted and lived with for a long time because they may not be wrong; they are just sub-optimal. Many control solutions in operation fall under this category. The incremental values in improving this type of control solution are enormous considering the large number in operation.

Section 12.2 provides a structured discussion on boiler drum level control. This example emphasizes that a good understanding of the process behavior is crucial, even for a seemingly simple process.

Section 12.3 presents the example of a well testing process and shows that process control must produce fit-for-purpose solutions to work around the practical limitations in process design. This example also demonstrates that process improvement can be the most effective way of improving control performance.

Section 12.4 presents the example of firewater process control. Firewater supply is a critical requirement for any sizable operating facility, and the control requirement is the high availability and reliability requirement. An integrated control solution requires sequential control, normal regulatory control, and abnormal protective control.

Section 12.5 shows the analysis and design of a unit-level control solution. The emphasis is on the importance of a good cause and effect relationship which is the basis for control design, either PID-based base-layer control or advanced process control.

This chapter is concluded with a complete multi-billion dollar E&P process and demonstrates the thinking process and control design at the plant level, mainly on enforcing the plant-level material balance and preferential flows.

References

Al-Dawery S, Dakhil HM (2012) Modeling and control of acetylene hydrogeneration process. Emir J Eng Res 17(1):9–16

Brannan C (2018) Rules of thumb for chemical engineers - a manual of quick, accurate solutions to everyday process engineering problems, 6th edn. Elsevier, Amsterdam

Campbell J (2004) Gas conditioning and processing - the equipment modules, vol 2, 8th edn. John M, Campbell and Company

Kinsley G (2001) Properly purge and inert storage vessels. Chem Eng Prog

Luyben WL, Tyréus BD, Luyben ML (1998) Plant-wide process control. McGraw Hill, New York
Vervalin CH (1981) Fire protection manual for hydrocarbon processing plants, 3rd edn. Gulf Publishing Co, Houston

Appendix
Special Calculations

A.1 Calculation of Vessel Volumes

For vessel level control, it is often necessary to know the vessel volume in order to calculate the PID tuning parameters (Sect. 9.5) or perform level linearization action (Sect. 10.1). This section provides the formulas for calculating the volume of some commonly encountered vessels. For a more complete list of vessel types, see Brannan (2018).

Some more complex vessels are often a combination of simpler vessels presented in this section and can be calculated by breaking them down into components. Sometimes, the vessel is so complex that it is difficult (and also unnecessary) to get the exact volume; reasonable simplification can be made to reduce the computation complexity.

The following terminologies will be used for this section:

1. Vessel height H: The vessel height from the bottom to the top in the engineering unit.
2. Liquid height h: The liquid height from the bottom to the surface of the liquid in the engineering units.
3. Liquid level LT: The liquid height from the lower tap of the level transmitter to the upper tap, in percentage (0–100%) (Fig. A.1).

A.1.1 Rectangular Tank

A rectangular tank is probably the simplest for volume calculation, although it is not common to see them in practice. The volume can be calculated as follows:

S. S. Niu and D. Xiao, *Process Control*, Advances in Industrial Control,
https://doi.org/10.1007/978-3-030-97067-3

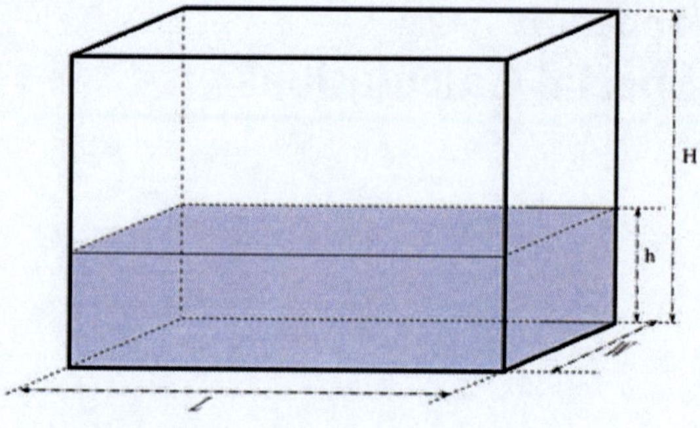

Fig. A.1 Volume of a cuboid vessel

$$\left.\begin{array}{rl}\text{horizontal cross area at height } h: & A = L \cdot W \\ \text{volume up to height } h: & V = L \cdot W \cdot h \\ \text{full volume:} & V = L \cdot W \cdot H\end{array}\right\} \quad (A.1)$$

A.1.2 Spherical Vessel

The volume of a spherical vessel can be calculated as follows. The horizontal cross area of the sphere at level h is calculated first, then the volume in the vessel up to level h is calculated (Fig. A.2):

$$\left.\begin{array}{rl}\text{horizontal cross area at level } h: & A = \pi\, h\,(D - h) \\ \text{volume up to level } h: & V = \dfrac{\pi}{6} h^2\,(3D - 2h) \\ \text{full volume:} & V = \dfrac{\pi}{6} D^3\end{array}\right\} \quad (A.2)$$

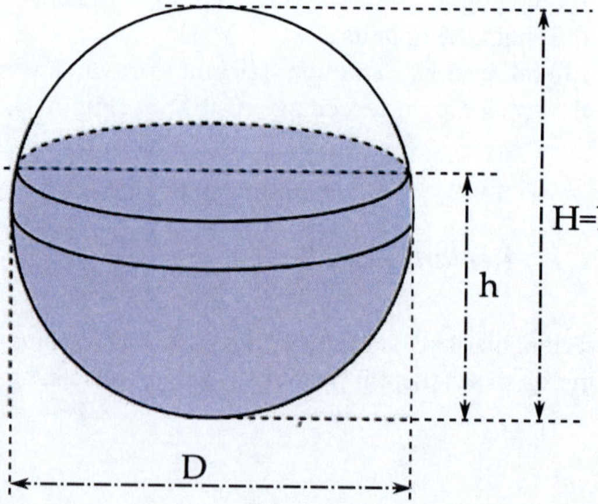

Fig. A.2 Volume of a spherical vessel

Note that the volume is a cubic function of the height. It is thus highly nonlinear. A regular PID controller with a single set of tunings would be complicated to control the level of this type of vessel.

The process gain K_p, assuming the volumetric flow rate is F_v, is given as

$$K_p = \frac{dh}{dt} = \frac{F_v}{\pi h\,(D-h)} \tag{A.3}$$

which is a function of the liquid height.

A.1.3 Vertical Cylindrical Vessel

For level control, a vertical cylindrical vessel is probably the simplest vessel found in a plant. Because of the uniform diameter along the vessel's height, the relationship between the level and volume is linear; thus, the relationship between the level and the outlet flow (Fig. A.3).

The vessel volume can be calculated as follows:

$$\left.\begin{aligned} \text{horizontal cross area at height } h: \quad & A = \frac{\pi}{4} D^2 \\ \text{volume up to height } h: \quad & V = \frac{\pi}{4} D^2\, h \\ \text{full volume:} \quad & V = \frac{\pi}{4} D^2\, H \end{aligned}\right\} \tag{A.4}$$

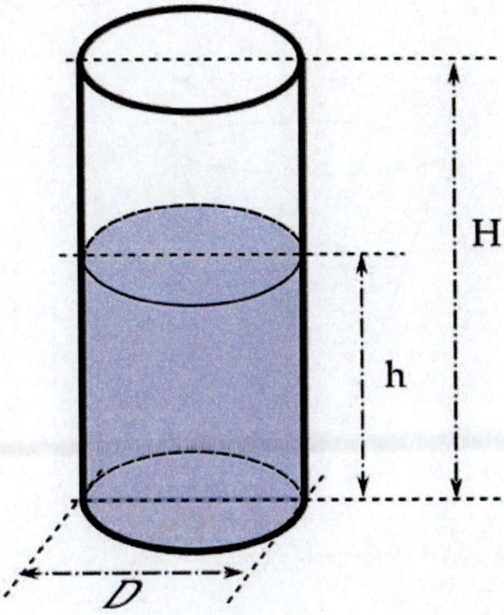

Fig. A.3 Volume of a vertical cylindrical vessel

where the area is the horizontal cross-section at level h and the volume is the bottom up to level h.

Note that the liquid volume is proportional to the liquid level $V \propto h$, so is the liquid height to flow $h \propto F$. The process gain K_p, assuming the volumetric flow rate is F_v, is given as

$$K_p = \frac{dh}{dt} = \frac{F_v}{A} = \frac{4F_v}{h\,D^2} \tag{A.5}$$

which stays constant at different heights.

A.1.4 Conical Vessel

The volume of a cone-shaped vessel can be calculated as follows. The horizontal cross area of the sphere at level h is calculated first, then the volume in the vessel up to level h is calculated (Fig. A.4):

$$\left.\begin{aligned} \text{Cross-section area at height } h\text{:}\quad A &= \frac{\pi}{4}\left(\frac{D}{H}\right)^2 h^2 \\ \text{volume up to height } h\text{:}\quad V &= \frac{\pi}{12}\left(\frac{D}{H}\right)^2 h^3 \\ \text{full volume:}\quad V &= \frac{\pi}{12} D^2\,H \end{aligned}\right\} \tag{A.6}$$

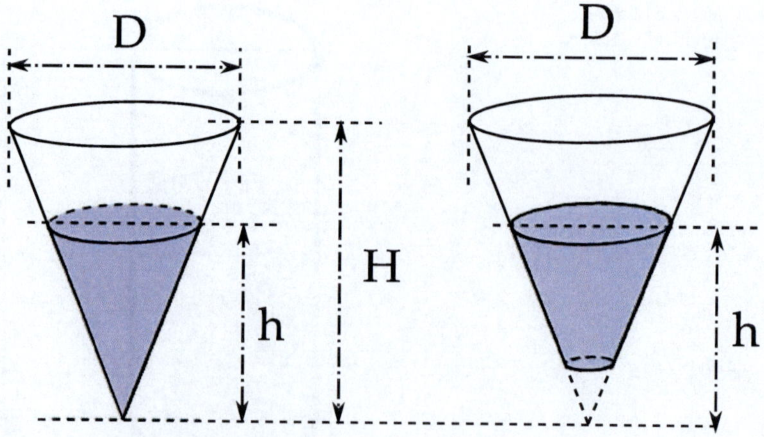

Fig. A.4 Volume of a conical vessel

Note that the volume is a cubic function of the height, and is highly nonlinear. The process gain is given by

$$K_p = \frac{dh}{dt} = \frac{F_v}{A} = \frac{4F_v H^2}{\pi D^2 h^2}. \tag{A.7}$$

A.1.5 Horizontal Cylinder

Horizontal cylindrical vessels are common in operation plants, typically with elliptical bulges. Here for simplicity, only a cylinder shown in Fig. A.5 is discussed.

The calculation of volume can be divided into two scenarios, $h < D/2$ and $h \geq D/2$. The vertical cross-section area A up to height h is calculated first. The volume is then simply the vertical cross-section area multiplied by the length of the vessel:

1. When $h < D/2$

$$\left.\begin{aligned} \text{Angle:}\quad & \theta = 2\arccos\left(1 - \frac{2h}{D}\right) \\ \text{Area up to } h\text{:}\quad & A = \frac{D^2}{8}(\theta - \sin\theta) \\ \text{volume up to } h\text{:}\quad & V = L \cdot \frac{D^2}{8}(\theta - \sin\theta) \end{aligned}\right\} \tag{A.8}$$

2. When $h \geq D/2$

Fig. A.5 Volume of a horizontal cylindrical vessel

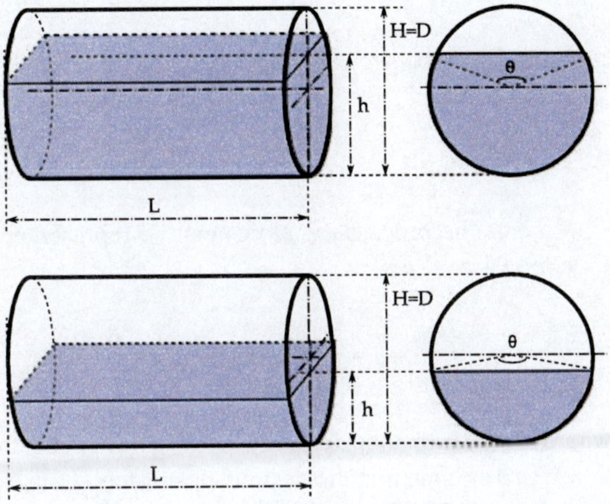

$$\left.\begin{aligned} \text{Angle:} \quad & \theta = 2\arccos\left(\frac{2h}{D} - 1\right) \\ \text{Area up to } h\text{:} \quad & A = \frac{D^2}{8}\left(2\pi - \theta + \sin(\theta)\right) \\ \text{Volume up to } h\text{:} \quad & V = \frac{L \cdot D^2}{8}\left(2\pi - \theta + \sin(\theta)\right) \end{aligned}\right\} \tag{A.9}$$

3. The full volume (at $h = D$) is given by $V = \frac{\pi}{4} D^2 L$.

Note that the volume is nonlinear to the height.

A.2 Calculation of Lambda Tuning Parameters

This section provides a simple illustration on how the λ tuning parameters are derived for first-order-plus-delay-time processes.

A.2.1 *First-Order-Plus-Delay-Time Process*

The closed-loop control is related to the process and controller as illustrated in Fig. 9.2. The close loop transfer function in Eq. 9.1 is given as follows:

$$G_{\text{cl}}(s) = \frac{G_c(s)\, G_p(s)}{1 + G_c(s)\, G_p(s)}. \tag{A.10}$$

From Eq. A.10, we can derive the controller transfer function as

$$G_c(s) = \frac{1}{G_p(s)} \cdot \frac{G_{\text{cl}}(s)}{1 - G_{\text{cl}}(s)}. \tag{A.11}$$

Assume the process dynamics can be represented with a first-order-plus-delay-time model:

$$G_p(s) = \frac{\overline{K}_p}{\tau s + 1} e^{-\theta s} \tag{A.12}$$

where $\overline{K}_p(s)$ is the normalized process gain, τ is the time constant, and θ is the delay time. Also assume that the desired closed-loop transfer function is a first-order-plus-time-delay (FOPDT) response with a desired time constant λ:

$$G_{\text{cl}}(s) = \frac{1}{\lambda s + 1} e^{-\theta s}. \tag{A.13}$$

The time delay in the closed-loop response is assumed to be the same as the process delay time since feedback control is unable to eliminate or reduce the delay time θ. From Eqs. A.11, A.12, and A.13, we have

$$\begin{aligned} G_c(s) &= \frac{\tau s + 1}{\overline{K}_p e^{-\theta s}} \cdot \frac{\frac{1}{\lambda s + 1} e^{-\theta s}}{1 - \frac{1}{\lambda s + 1} e^{-\theta s}} \\ &= \frac{\tau s + 1}{\overline{K}_p} \cdot \frac{1}{\lambda s + 1 - e^{-\theta s}}. \end{aligned} \tag{A.14}$$

If a first-order Taylor expansion is used to approximate the term $e^{-\theta s} \approx 1 - \theta s$, we have

$$\begin{aligned} G_c(s) &= \frac{\tau s + 1}{\overline{K}_p} \cdot \frac{1}{\lambda s + 1 - (1 - \theta s)} \\ &= \frac{1}{\overline{K}_p} \frac{\tau}{(\lambda + \theta)} \cdot \left(1 + \frac{1}{\tau s}\right). \end{aligned} \tag{A.15}$$

Compared with the PI controller formula in Eq. 3.8, we have

$$\begin{cases} K_c &= \dfrac{\tau}{\overline{K}_p} \dfrac{1}{(\lambda + \theta)} \\ T_i &= \tau \\ T_d &= 0. \end{cases} \tag{A.16}$$

This agrees with the tuning rule for PI controller in Table 9.6. The tuning parameter derivation for PID controller would be considerably more complicated.

A.2.2 *Pure Integrating Process*

Similarly, assume a pure integrating process with delay θ:

$$G_p(s) = \frac{\overline{K}_p}{s} e^{-\theta s}. \tag{A.17}$$

From Eqs. A.11, A.13, and A.17, we have

$$
\begin{aligned}
G_c(s) &= \frac{s}{\overline{K}_p e^{-\theta s}} \cdot \frac{\frac{1}{\lambda s+1}e^{-\theta s}}{1-\frac{1}{\lambda s+1}e^{-\theta s}} \\
&= \frac{s}{\overline{K}_p} \cdot \frac{1}{\lambda s+1-e^{-\theta s}}.
\end{aligned} \tag{A.18}
$$

With Taylor approximation $e^{-\theta s} \approx 1 - \theta s$, we have

$$
\begin{aligned}
G_c(s) &= \frac{s}{\overline{K}_p} \cdot \frac{1}{\lambda s+1-(1-\theta s)} \\
&= \frac{1}{\overline{K}_p} \cdot \frac{1}{\lambda+\theta}.
\end{aligned} \tag{A.19}
$$

This leads to a P-only controller with gain (see Table 9.7):

$$
K_c = \frac{1}{\overline{K}_p} \frac{1}{\lambda+\theta}. \tag{A.20}
$$

Reference

Brannan C (2018) Rules of thumb for chemical engineers - a manual of quick, accurate solutions to everyday process engineering problems, 6th edn. Elsevier, Amsterdam

Glossary

Actuator A component of an automatic valve used to open and close the valve. May be pneumatic or motor-driven.

Affinity Laws The affinity laws (also known as the "Fan Laws" or "Pump Laws") for pumps/fans are used to express the relationship between performance variables such as head, volumetric flow rate, shaft speed, and power. They apply to centrifugal and axial flows in pumps, fans, and hydraulic turbines.

Asset The buildings, plant, machinery, and other permanent items required to produce and supply the product.

Availability The availability of an item to be in a state to perform a required function under given conditions at a given instant of time or over a given time interval, assuming that the required external resources are provided.

Analog Controller A type of controller as distinguished from a digital controller. It operates on continuous signals such as voltages, pressures, or currents.

Automation The use of automatic control devices in a process so that human supervision is minimized.

Basic Sediment and Water (BS&W) is a technical specification of certain impurities in crude oil. When extracted from an oil reservoir, the crude oil will contain some saltwater and particulate matter from the reservoir formation. The residual content of these unwanted impurities is measured as BS&W.

Best Practice A method or technique generally accepted as superior to any alternatives because it produces results that are superior to those achieved by other means or because it has become a standard way of doing things, e.g., a standard way of complying with legal or ethical requirements. (from Wikipedia, https://wiki2.org/en/best_practice)

Constraint Variable Process output that must be maintained within an operating range.

Control Element A component of the control system that relates the error between the desired value and the manipulated variable.

Cycling Periodic changes in a process variable. Also known as oscillation.

S. S. Niu and D. Xiao, *Process Control*, Advances in Industrial Control,
https://doi.org/10.1007/978-3-030-97067-3

Deadband The largest range of values of a process variable (e.g., input variable) within which the value does not respond to changes or cause changes to other variables.

Disturbance An input signal that affects the controlled variable but is not directly manipulated. Sometimes also called load change.

Fail Safe Fail-safe allows a system to return to a safe state after a breakdown of the control. The fail-safe design allows the process to avoid hazardous conditions that may otherwise occur.

Failure The termination of the ability of an item to perform a required function.

Fault The state of an item characterized by the inability to perform a required function, excluding the inability during preventive maintenance or other planned actions, or due to lack of external resources.

Final Control Element The control element that directly changes the manipulated variable, such as a valve or motor.

HAZOP/FMEA A Hazard and Operability (HAZOP) study is a structured and systematic analysis technique for examining a system to identify potential hazards in the system and potential operability problems with the system and, in particular, the causes of operational disturbances and production deviations likely to lead to nonconforming products.

FMEA is a systematic procedure for analyzing the potential failure modes, their causes, and effects on system performance.

Highway Addressable Remote Transducer (HART) is a standard for sending and receiving digital information across analog wires between smart devices and a control or monitoring system.

Instrument Range The region in which the instrument is designed to operate.

Jockey Pump Low capacity pressure maintenance pump working on the principle of on/off control.

Life Cycle The time interval that commences with the initiation of the concept and terminates with the disposal of the asset.

Life Cycle Cost The total cost of ownership of an item considering all the costs of acquisition, personnel training, operation, maintenance, modification, and disposal.

Linear A relationship showing output proportional to input and follows the principle of superposition.

Linearize To substitute a nonlinear function with an approximate linear function.

Linear Motion Valve A valve with a sliding stem that pushes the obturator closed or pulls it open. They are capable of being adapted to bellows sealed. See also rising stem.

Load or Load Variable see disturbance.

Manipulated Variable Process input that is adjusted to maintain the controlled parameter at the setpoint.

Manned Operations Operation of a Production Plant that requires constant "human" supervision on location to manage the process and take immediate intervention action during process upsets.

Maximum Operating Pressure (MOP) The maximum pressure allowed in a device or system. The design pressure (DP) is 110% of the MOP.

Net Positive Suction Head (NPSH) To ensure that the suction conditions for a pump are such that no vaporization occurs in the pump inlet, there has to be sufficient suction pressure above the vapor pressure of the liquid. This surplus pressure is called NPSH available.

Nonlinear A relationship that does not conform to linearity;

Offset The steady-state deviation between the controlled variable and its setpoint.

Overshoot The difference between the final steady-state value and the value of the first maximum (or minimum if the response is downward) in a step response of an under-damped system, typically expressed as a fraction;

Pump Priming A pump should be primed to meet the NPSH requirements before the pump can be started. This is typically done by keeping the pump elevation below the lowest level of the liquid supply so that the pump remains primed at all times and the NPSH requirements are met.

Quarter Turn Valve A valve (such as ball, plug, butterfly) that requires a 90-degree operation of the stem to move from the fully open to the fully closed position. They will readily accept any type of actuation (pneumatic, hydraulic, electric).

Rangeability The ratio of maximum value to the minimum controllable value in a final control element.

Redundancy The existence of more than one means of performing the same intended function.

Reid Vapor Pressure Reid Vapor Pressure is the pressure exerted by a liquid to calculate the volatile nature of gasoline. RVP is defined as the vapor pressure of a substance at 100 °F (37.8 °C).

Reliability The ability of an item to perform a required function under given condition for a given time interval.

Resistance The potential required to produce change such as against a back-pressure.

Restriction Orifice (RO) A restriction orifice limits the flow through the pipe to a set flow by choking the flow at its sonic velocity.
On the other hand, a flow orifice is used to determine the flow through the pipe from the pressure drop over the plate. There a normally two pressure taps (one on each side of the plate) and a pressure transmitter that determines the pressure drop over the plate. A formula is then used to convert this pressure drop to a flow rate for the specific fluid passing through the orifice.

Risk A combination of probability, or frequency, of occurrence of a defined hazard and the magnitude of the consequences of the occurrence.

Safety Instrument System A function comprising one or more initiators, a logic solver, and one or more final control elements whose purpose is to prevent or mitigate hazardous situations

Self-Regulation The inherent characteristic of a system that produces a steady state after a disturbance.

Setpoint Value at which the controlled parameter is to be maintained by the control system.

Settling Time The time required for the absolute value of the difference between the output of the controlled variable and its final steady-state value to become and remain less than a specified amount.

Specific Heat Specific heat is the amount of energy required to increase the temperature of one unit of mass of a material by one degree.

Stable System A system whose response to a bounded input is also bounded.

Steady-State The final condition after the transient response having died out following all properties becoming constant with time.

Steady-State Error A control error at steady state.

Step Response The response of a process variable in response to a step change in another process variable.

Throttling The intentional restriction of flow by partially closing or opening a valve.

Time Constant The time required for the output of a first-order system to change 63.2% of the steady-state value.

Time to Steady State Same as Settling Time.

Transducer Any device that transmits, amplifies, or changes a signal.

Transfer Function A mathematical relationship that describes the ratio of an output of a system to the input.

Two-Position Control See on/off control.

True Vapor Pressure True vapor pressure is the equilibrium partial pressure exerted by a volatile organic liquid as a function of temperature as determined by the test method ASTM D 2879. The true vapor pressure (TVP) at 100 °F differs slightly from the Reid vapor pressure (RVP) (per definition also at 100 °F), as it excludes dissolved fixed gases such as air.

Unattended Operations Operation of a plant in which personnel are absent from the plant for short periods but operations are continuously monitored 24 h a day, e.g., during the day shift, personnel are available for the control room and the plant. During the night shift, a limited crew is available for plant intervention operations that have been subjected to a risk-based assessment.

Unmanned Operations Operation of a purpose-designed plant that allows for periods of operations in which there are no "human" supervision on location and no immediate intervention to process upset while maintaining the required design level of production availability. Periods of unmanned operations will be followed by manned periods in which the planned maintenance routines and corrective/breakdown work will be executed.

Valve Positioner A device that precisely controls the control valve stem position by adjusting the instrument air pressure to the control valve.

Valve Seat The part of the valve that the valve plug rests upon when the valve is fully closed.

Valve Stem A connecting rod between the diaphragm in a valve actuator and the valve plug that allows air pressure on the diaphragm to control the valve plug position in the valve and hence flow through the valve.

Index

S. S. Niu and D. Xiao, *Process Control*, Advances in Industrial Control,
https://doi.org/10.1007/978-3-030-97067-3

T